Springer
Tokyo
Berlin
Heidelberg
New York
Hong Kong
London
Milan
Paris

Y. Kaneda
T. Gotoh (Eds.)

Statistical Theories and Computational Approaches to Turbulence

Modern Perspectives and Applications to Global-Scale Flows

With 163 Figures

Springer

Yukio Kaneda
Professor, Department of Computational Science and Engineering
Graduate School of Engineering
Nagoya University
Furo-cho, Chikusa-ku, Nagoya 464-8603, Japan

Toshiyuki Gotoh
Professor, Department of System Engineering
Nagoya Institute of Technology
Gokiso, Showa-ku, Nagoya 466-8555, Japan

ISBN 978-4-431-67004-9 ISBN 978-4-431-67002-5 (eBook)
DOI 10.1007/978-4-431-67002-5

Library of Congress Cataloging-in-Publication Data

Statistical theories and computational approaches to turbulence : modern perspectives and applications to global-scale flows / Y. Kaneda, T. Gotoh, eds.
p. cm.
Papers presented at a workshop held Oct. 10-13, 2001 at Nagoya University, Nagoya, Japan.
Includes bibliographical references.

1. Turbulence--Mathematical models--Congresses. 2. Turbulence--Computer simulation--Congresses. 3. Geophysics--Fluid models--Congresses. I. Kaneda, Y. (Yukio), 1949- II. Gotoh, T. (Toshiyuki), 1954-

QA913 .S77 2002
532'.0527--dc21

2002030445

Printed on acid-free paper

Originally published by Springer-Verlag Tokyo in 2003.

Typesetting: Editors

SPIN: 10868573

Preface

This volume contains the papers presented at the workshop on Statistical Theories and Computational Approaches to Turbulence: Modern Perspectives and Applications to Global-Scale Flows, held October 10-13, 2001, at Nagoya University, Nagoya, Japan.

Because of recent developments in computational capabilities, the computational approach is showing the potential to resolve a much wider range of length and time scales in turbulent physical systems. Nevertheless, even with the largest supercomputers of the foreseeable future, development of adequate modeling techniques for at least some scales of motion will be necessary for practical computations of important problems such as weather forecasting and the prediction and control of global pollution. The more powerful the available machines become, the more demand there will be for precise prediction of the systems. This means that more precise and reliable knowledge of the underlying dynamics will become important, and that more efficient and precise numerical methods best adapted to the new generation of computers will be necessary. The understanding of the nature of unresolved scales then will play a key role in the modeling of turbulent motion. The challenge to turbulence theory here is to elucidate the physics or dynamics of those scales, in particular their statistical aspects, and thereby develop models on sound bases to reduce modeling ambiguity. The challenge to the computational method is to develop efficient algorithms suitable for the problems, the machines, and the developed models.

The goal of the workshop, as expressed in its title, was to identify directions for advancing the feasibility and scope of global scale flow computations, with particular emphasis on the application of statistical theories to turbulence modeling, and on the present and future roles of computational approaches. The program included 6 keynote lectures, 13 oral presentations, and 10 poster presentations. These covered a wide range of computational and statistical approaches and their application to global-scale flows and engineering flows, as seen in Part I through Part IV of this volume. In addition to the lectures and presentations, there were two panel sessions, "Advanced Computational Approaches in Turbulence Research" and "Turbulence Research for Geophysical Applications," which make up Part V of this proceedings. We hope that they will convey to readers something of the friendly and informal atmosphere of the sessions shared by the participants.

It was unfortunate that some guests from outside Japan who had planned to attend did not do so, owing to the events of September 11th which had occurred shortly before the workshop. Dr. A. Pouquet and Dr. R. Rubinstein are gratefully acknowledged for contributing their planned presentations to this volume in written form.

We are also grateful to the participants, to the authors of written contributions to this volume, and to the reviewers whose valuable work helped improve the quality of these proceedings. We wish to acknowledge, as well, the work of the members of the local organizing committee: Drs. K. Ishii, K.N. Beronov, T. Ishihara, K. Yoshida, K. Akahori, and K. Yoshimatu, in the period of many months before, during, and after the workshop, and the collaboration of Dr. R. Rubinstein in organizing the workshop. We express our gratitude to Springer-Verlag Tokyo for making possible the publication of these proceedings.

The workshop was organized under the Computational Science and Engineering for Global-Scale Flow Systems project financed by the Japanese Society for the Promotion of Science, as part of its Research for the Future program during the last year of that 5-year project (August 1998-March 2002). The activities of the local organizing committee on behalf of the workshop also were supported by the project. Other sponsors of the workshop were the Research Foundation for Electrotechnology of Chubu and the Daiko Foundation.

Nagoya, May 2002

Yukio Kaneda
Toshiyuki Gotoh

Contents

Part I

Application of the Statistical Theory to Stratified and Rotating Turbulence

Computational Challenges for Global Dynamics of Fully Developed Turbulence in the Context of Geophysical Flows

Annick Pouquet, Duane Rosenberg, and John Clyne

National Center for Atmospheric Research, Boulder, Colorado, USA

Abstract. Geophysical turbulent flows posses a very large number of degrees of freedom, and no mechanism is presently known which can reduce this number to a manageable size. In order for numerical simulations to be of use in advancing our understanding of geophysical turbulence, they must complement and make use of experiments, observations and theoretical advances. One is thus compelled to tackle numerical simulations at the highest resolution possible today, using the most powerful computers available with a heavy reliance on advances in information technology. This implies the need for developing as well powerful graphical and analysis softwares that can handle data of the order of 10 Terabytes. Such computations are to be viewed either as *gedanken* experiments, or as models for turbulence, and potentially the most promising venue is to combine both approaches. This paper reviews a few of the problems associated with these considerations, while stressing the need to maintain close contact with theoretical tools which allow for the construction of subgrid–scale models to be used in Large Eddy Simulations.

1 Introduction

Turbulent flows are ubiquitous, and as manifestations of one of the last outstanding unsolved problems of classical physics, they form today the core picture of numerous scientific and engineering inquiries and are linked to many issues in the geosciences: for example, in geology (Earth interior and dynamo problem), meteorology (cloud physics), oceanography (the role of stratification), climatology (global warming), space weather (from the Sun through the solar wind to the magnetosphere and ionosphere of the Earth), and ecology. The study of turbulence is not limited to inquiries in geophysics; turbulence plays an equally prominent–often dual–role in the understanding of nonlinear processes in physics, as well as in industrial flows, through *e.g.* the presence of seed particles or bubbles, and in studies of combustive and chemically reactive flows, and an even more pragmatic role in the area of aeronautical engineering regarding aircraft safety or in epidemiology. This interest is inter-disciplinary, and the issue of universality (or not) of physical processes and scaling laws arises, as the modeling of such complex flows becomes more realistic, but is also in more demand because of the wide range of applicability.

Although no general theory of turbulence currently exists, progress has been made recently in answering some fundamental questions (see *e.g.* Frisch, 1995).

Departure from normality in probability distribution functions (PDF) for a variety of flows is well documented, but the origin and dynamics of these "fat wings" are not understood. Such wings appear, in fact, in many nonlinear problems with a wide range of excited scales in geophysical flows (Sornette, 2000). Nonlinearities become important in turbulent flows and can overcome linear viscous dissipation when the Reynolds number, $R_e = U_o\ L_0/\nu$, is large, where U_0 and L_0 are characteristic velocity scale and length scale of the flow, and ν is the viscosity. This number is the control parameter that measures both the number of interacting modes and the ratio of active temporal or spatial scales in the problem. The number of degrees of freedom in turbulent fluids increases as $R_e^{9/4}$ for $R_e \gg 1$ and for flows in nature, R_e often exceed $\approx 10^8$. It is clear that the ability to probe large R_e, and, hence, to examine details regarding the PDFs and structures of turbulent flows, depends critically on the ability to resolve a large number of spatial and temporal scales and understand the global dynamics as well as the local interactions between modes and structures.

Theory demands that computations of turbulent flows reflect a clear scale separation between the energy-containing, inertial (self–similar) and dissipative ranges. Numerical experiments allow for fully controlled testing of a variety of models both phenomenological and basic, and they attempt to follow directly as wide a range of scales as possible in order to compute the true behavior of the flow. Detailed convergence studies show that in order to achieve the desired scale separation between the energy–containing modes and the dissipation regime, it is necessary to compute on grids with regularly–spaced points of at least 2048^3 cells (see *e.g.* Sytine *et al.*, 2000 for the compressible case). Only today can such computations be accomplished, and the subsequent data handled, although with some difficulty. For a computation of velocity, density and pressure on such a grid of 8 billion degrees of freedom and with storage using 16-bit scaled integers, 80 gigabytes (GB) of compressed data per time frame is generated. To provide good temporal resolution in a run of 3,000 time steps, a snapshot may be required every 10 time steps. Since there is no proven reduction in the degrees of freedom in a turbulent system which is guaranteed to retain all of the important characteristics of the flow, *all* of the data generated by these computations must be kept. Thus a total of 20 terabytes (TB) of storage would be required. This amount of data is sizable even for a national center, and is entirely beyond the abilities of individuals to manage at university laboratories in the near future.

In many respects, the data from numerical turbulence runs–particularly the size–is unique to turbulence studies. In the CONQUEST (CONcurrent QUErying Space and Time; Stolorz *et al.*, 1995) information system, for example, spatio-temporal features are extracted from large data sets. The features extracted may refer to cyclone tracks (or hurricanes) with distinct climatologic patterns, or they may be otherwise "indexable" (identifiable), and amenable to using learning algorithms that look for novel patterns or correlations. But in turbulence, one does not know yet *what* structures are key to our understanding the statistical properties of turbulent flows (*e.g.* vortex sheets, spiral or filaments, shocks or fronts, blobs, plumes or tetrads, knots, helices, tubes or arches) and their topology (see *e.g.* Kawahara *et al.*, 1997). Perhaps more importantly, turbulence data is 4-D because it is the spatio–temporal interactions between such structures–parallel or anti-parallel or orthogonal vortex filaments, single sheet or accumulations of such sheets embedded in a large-scale strain, kinking or intertwining of tubes; the latter is known to provide the elementary building blocks of turbulent transfer of energy to small scales in 3D and as such the source of our multi-scale problem. We note that in that light, an electronic data center for geophysical turbulent flows would be extremely useful, similar to the case of facilities for space physics observational data centers and theoretical physics experimental data centers.

Predictive models for geophysical flows require a unique synergy between computation and modeling, experiment and measurement. As an illustration, there is presently a world–wide effort to achieve an experimental dynamo with a small magnetic Prandtl number ($P_M = \nu/\eta$ where η is the magnetic diffusivity) similar to that of the Earth. Theoretical modeling (*e.g.* Pouquet, 1993; Nakayama, 2001), numerical computations (*e.g.* Nore *et al.*, 1997; Kageyama & Sato 1999) and experiments (Gailitis, 1993) must be inter–compared in order to define precisely the characteristics of the flow responsible for a viable dynamo.

Direct numerical simulations of turbulent flows can help elucidate the connection between multi–scale turbulent structures and the underlying non–Gaussian statistics. This link forms the basis of the notion of *intermittency* which plays a role in many geophysical phenomena, for example in droplet formation in clouds and in reactive flows where nonlinear interactions alter local chemical contact rates. However, intermittency is not yet included in models of these processes, nor is it clear how this should be done. A relatively new development, which may even demand new turbulence concepts, concerns the high resolution *in-situ* and remotely sensed geophysical data. These include lidar and radar measurements which can resolve turbulence structures throughout the atmospheric boundary layer to scales of a few meters, and data from such remote sensing platforms as Topex/Poseidon and QuikSCAT, which provide high resolution sea level and sea surface wind speed measurements, respectively. In principle, these data can be assimilated into global ocean and atmosphere circulation models. But the flows which these models address, and which the data represent, are non-Gaussian. An understanding of the role of intermittency will help determine how best to assimilate observational data into flow models for enhanced prediction capability, and long-term trending.

Indeed, beyond direct numerical simulations (DNS), research on parameterization of small scales for use in Large Eddy Simulations (LES) must be actively pursued as LES represents a strong link to theoretical approaches. Still, how do we proceed without being swamped by the complexity of closure schemes? Fast multi–processors enable us to compute turbulent flows at moderate Reynolds numbers in 3D and recently, a Taylor Reynolds number of $R_\lambda \sim 500$ (with $R_\lambda \sim R_e^{1/2}$) on a grid of 1024^3 points has been achieved (Kaneda, private communication). The vast amount of data associated with such a model must have its structures identified and be navigable in both space and time in order to be of use in parameterizing the small–scale behavior of the flow in LES models.

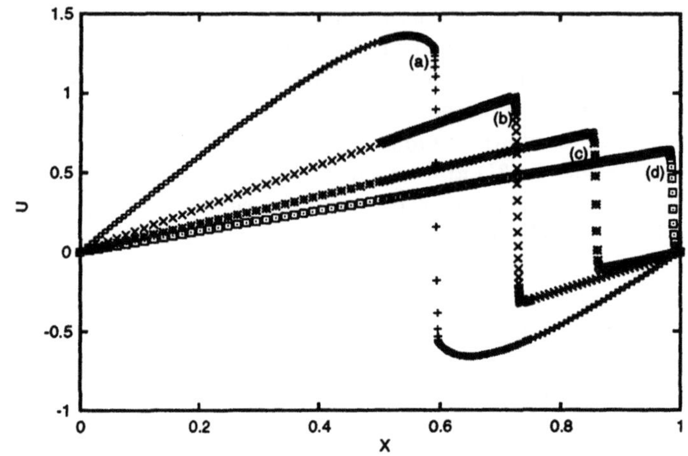

Fig. 1. 1D spectral element calculation of Burgers' equation with moving shock initial conditions (Huang, *et al.*, 1994) at $R_e = 5000$. The polynomial order is fixed at $N_p = 32$. The profiles are given at times (a) 0.2; (b) 0.6, (c) 1.0, and (c) 1.4. Note the clustering of points around the shock as the grid tracks the sharp velocity gradient there. The number of elements for each profile is 10, 8, 6, and 6, respectively. The grid is based on a binary tree, and adaptation occurs by binary splitting of an element (joining of two elements), if the velocity gradient is greater than (less than) some value.

For example, in the LANL-α model (Holm *et al.*, 1998), the small scales (which typically represent over 85% of the data) possess a dramatically reduced number of degrees of freedom when compared to conventional turbulence because it conserves the $\mathcal{H}_{1,\alpha}$ instead of the $\mathcal{L}_2$ norm: it correctly preserves the nonlinear structure of the Navier-Stokes equations for the dynamics of the large scales, but the dynamics of the small scales are limited to being swept by the larger scales. Although resolved in the computations, the information content of these scales is effectively small and we should be able to model them effectively.

2 Computational Issues in Geophysical Turbulence

2.1 A model of turbulence in one dimension, and adaptivity

The Burgers equation remains today a fertile ground for experimenting, both for numerical algorithms (Berger & Colella, 1989; Karniadakis *et al.*, 1991; Dietachmayer & Droegemeier, 1992; Gombosi *et al.*, 1994; Mavriplis, 1994) and for phenomenology for turbulence (Woyczynski, 1995; Frisch & Bec, 2000); it is also a model for many physical processes, from traffic fluctuations (Higuchi, 1978) to cosmology (Vergassola *et al.*, 1994). It reads:

$$\partial_t u + u\partial_x u = \nu\partial^2_{xx} u + f \tag{1}$$

where u is the velocity and f a forcing term which, in general, is taken to be concentrated in the large spatial scales and, for example, delta–correlated in time. When the forcing is identically zero, an exact solution is known through the Hopf–Cole transformation. Using the fact that the solution is a combination of ramps and shocks, one can show that the structure functions $\delta u(r) = u(x + r) - u(x)$ scale in the inertial range where dissipative processes can be ignored as $\langle \delta u(r)^p \rangle \sim r^{\zeta_p}$, with $\zeta_p = p$ for $p \leq 1$ and $\zeta_p = 1$ for $p \geq 1$. Furthermore, power laws for the wing of the PDFs of negative velocity gradients (the shocks) for (1) can be found analytically, but the index for such a law is in dispute (see Gotoh & Kraichnan (1998), and Frisch & Bec, *op. cit.*). The important point, though, is that such wings can now be computed analytically and thus lead to a better understanding of the role of structures in turbulent flows. Recently, it has also been shown, through an analogy with the XY spin model, that the tails of the Burgers velocity gradients can be recovered with Graner distributions (Noullez and Pinton, 2002).

Fast algorithms can be found that solve equation (1) with $L_0 \sim 1$, velocity $U_0 \sim 1$ and $\nu \sim 10^{-4}$ with only a few dozen nodal points. A criterion of choice of such points follows the concept of equidistribution according to a monitor function based in part on velocity gradients (see Figure 1; for a recent account see *e.g.* Huang & Russell (1997) and references therein). Is there a limitation on the achievable Reynolds number with grid adaptation at fixed number of points N? For a regular grid, and for Burgers equation, one can achieve $R_e \sim N^{3/2}$, a result that obtains by equating the characteristic time of nonlinear interactions based on the evaluation of the energy spectrum $E(k) \sim k^{-2}$, and of the diffusion time. This type of analysis is customary for Navier–Stokes flows using the concept of energy cascade, although for the Burgers equation the shock is a coherent structure and the time of formation of a shock is scale–independent. Let us now consider the extreme choice of taking all the available grid points in the vicinity of the shock with only a couple of points in the rarefaction wave (the ramp). Taking into account the fact that the local solution gives for the thickness of the shock a scaling $l_{min} \sim \nu^{1/2}$ because of the hyperbolic tangent profile, one now obtains $R_e \sim N^2$ (reminiscent of scaling when using irregularly spaced points with Chebychev polynomials).

2.2 Data Analysis and Manipulation

Conventional statistical analysis methods, while simple to compute, may reduce data to a relatively few numbers, filtering out information that has been computed at great expense. Still-image visualizations may reveal important instantaneous features, difficult to detect using numerical methods, but fail to capture the dynamics of time-evolving phenomena. Only temporal animations may reveal the behavior of complex evolving features. To be most effective, these animations must be realized in a highly interactive manner, using tools that permit researchers to navigate data through both time and space (Clyne *et al.*, 1998). The computational demands for interactive visual data analysis on the scales discussed are staggering. Large data visualization systems exist, such as those developed by Parker (1999) and Painter (1999), but they rely on hardware costing millions of dollars. In many ways visualization systems such as these are brute force approaches and not entirely necessary for visual data exploration, particularly if the goal is to assist in discovery of large scale features. The efficient extraction of sub-regions of data in time and space also must be possible. The extraction of data may be preceded or followed by conversion to quantities (*e.g.* temperature, vorticity, or a component of the rate of strain tensor) that are derived from the raw model data.

Multiresolution data representation methods may offer a solution to these issues. The theory of wavelets offers a natural mathematical framework for such a representation. At each successive power-of-two resolution level the wavelet transformation yields a lower frequency approximation to the previous level along with higher frequency detail coefficients representing the loss of information between levels. The accuracy of these coarser approximations is dependent on the nature of the data itself and the choice of wavelet basis function used. Exploring optimal wavelet basis functions for various forms of scientific data is an active area of research (Tao, *et al.*, 1994; Kolarov & Lynch 1997; Wilson 2000), including in turbulence (Farge *et al.*, 1999). Wavelet transforms are invertible, and lossless reconstruction of the approximation to the next higher resolution is possible (Mallat 1989). The transformation to and from wavelet space can be accomplished in linear time. Using techniques such as lifting (Sweldens, 1996), reconstruction may be accomplished using only additions and subtractions of floating point numbers, making the inverse transformation exceptionally fast. Furthermore, the transformed data occupy no more space than the original data, making multiresolution representations possible without the use of additional data storage.

Applying 3D wavelet transforms directly to data volumes in the manner first proposed by Muraki (1993) allows the efficient construction of data at different scales but it does not enable the multiresolution data sub-setting that we seek. Partitioning the data into blocks and applying the 3D transform to the individual blocks may improve the situation (Ihm 1998; Rodler 1999). Block decomposition facilitates region extraction and improves cache performance on cache-based microprocessors. However, to our knowledge these block-based efforts have all assumed static data. Furthermore, current methods, primarily aimed at large medical data sets, assume the researcher knows *a priori* which sub-region is of interest. Organization strategies that do not make these assumptions and are more appropriate for turbulence data must be explored. Similarly, analysis tools that may exploit these multiresolution encodings must be developed.

2.3 Large Eddy Simulations (LES)

There are a variety of closure schemes that have been proposed for turbulence (see *e.g.* Kraichnan, 1976; Yoshizawa, 1985; Chasnov, 1991; Kaneda *et al.*, 1999). It is clear that, as the primitive equations become more complete, more forces being taken into account, the closures become more complex since they involve evaluating a large number of triple correlations between the various variables (velocity, density, pressure, entropy, magnetic field, ...). Such equations are difficult to solve analytically and resort to numerical solutions to analyze them is unavoidable. Economical and physically-motivated methods must be developed further as they represent the only reasonable hope for modeling turbulent flows. For example, a stochastic framework of the Langevin type (Bertoglio, 1984) or a combination with inviscid Burgers noise for the unresolved scales (Majda & Kramer, 1999), enhanced or real-space eddy viscosity (Métais & Lesieur, 1992), or closures based on a Lagrangian spectral theory (Kaneda, 1981; Yoshida *et al.*, , 2002; Gotoh *et al.*, , 2002) are promising venues. In particular in the latter case, there are no adjustable parameters and the Lagrangian approach removes the problem associated with the Direct Interaction Approximation (Kraichnan, 1977) and thus leads to a classical Kolmogorov $k^{-5/3}$ spectrum (see also Nakayama (2001) for the case of weak anisotropic MHD).

On the other hand, it has been proposed that numerical methods themselves represent a closure of the primitive equations. In particular, the ability of non-oscillatory advection schemes to represent the effects of the unresolved scales of motion has already been explored (Oran and Boris, 1993, Porter *et al.*, 1994). For example, such a solver can accurately reproduce the dynamics of an atmospheric convective boundary layer. When an explicit turbulence model is implemented, the solver does not add any significant numerical diffusion and thus appears to include an effective subgrid scale model. Are the conservation/symmetry properties of the underlying equations sufficient to lead to that effect? Can such methods recover the law of the wall in channel flows? Similarly, the MPDATA methods (Smolarkiewicz and Margolin, 1998) developed for studying precipitation over mountains or forest fires can be studied to assess/quantify its LES properties exploiting a set of benchmark problems in the large R_e limit. A recent study of Burgers equation using such a code (Margolin and Rider, 2002) shows that the equivalent equation for the cell-averaged velocity $\bar{u}$ reads:

$$\partial_t\bar{u}+\bar{u}\partial_x\bar{u}-\nu_0\partial^2_{xx}\bar{u} = \delta x^2[\alpha\partial_x\bar{u}+\beta|\partial_x\bar{u}|]\partial^2_{xx}\bar{u}+\delta x^2\zeta\bar{u}\bar{u}_{xxx}+\delta x^2\eta\frac{\bar{u}^3_x}{\bar{u}}+\delta x^3\gamma\bar{u}_{xxxx} \tag{2}$$

with δx the grid spacing, and where α, β, γ, ζ and η vary according to the numerical method used. Indeed, the first term on the *r.h.s.* acts as a numerical/physical nonlinear turbulent viscosity. However, this approach remains empirical and must be backed by strong closure schemes. Furthermore, the adaptive-mesh-refinement capabilities of codes developed for turbulence studies should also be assessed in the same fashion, thereby trying to link numerically–based and physically–based parameterization schemes.

3 Can we Learn from Fully Developed Turbulence?

The hallmark of turbulence is the creation of fluid motions at ever smaller scale and faster time scales, until the energy finally has cascaded down to the microscopic scales at which it can be dissipated as heat by viscosity. This turbulent cascade process is a successive loss of stability that occurs with increasing rapidity as the Reynolds number is increased. Fully-developed turbulence at high R_e is known to be "intermittent." This means, for example, that in the decay of turbulence the correlations in the flow at different scales do not follow a simple geometric relation between scales. Instead it comes in intermittent "gusts" and the cascade of energy has fractal properties. Thus, the heart of the problem of analyzing turbulence data, *i.e.* the statistical description of many interacting degrees of freedom (infinitely so in the limit $R_e \to \infty$) is fundamentally an information technology problem. The complexity is so rich that attacking this problem will require advances in IT research that should serve many other purposes and will go hand in hand with the development of computational power for the foreseeable future.

The seminal paper of R.H. Kraichnan (1994) on intermittency for a passive scalar such as a non-reactive pollutant, has led several teams to show that there are intermittency corrections to the scaling laws stemming from dimensional analysis, even though the velocity field in that model is well behaved (Gaussian with power-law spatial correlation and delta correlated in time). Such laws can be linked to the dynamics of structures, ramps or fronts, highly concentrated in space/time (Sreenivasan *et al.*, 1997; Warhaft, 2000; Shraiman & Siggia, 2000; Falkovich *et al.*, 2001). The corrections these authors find to a linear variation of scaling exponents with order is a signature of the persistence of the dynamical influence of the large scales of the turbulent flow; it arises, as a memory of initial conditions, through the existence of statistical Lagrangian invariants (with time). An example of such an invariant involves, in the isotropic case, the one–time three–point correlation function $C_3(\boldsymbol{r}_1, \boldsymbol{r}_2, \boldsymbol{r}_3)$: even though, through Richardson law of dispersion, the average distance between the three points increases in time as $t^{3/2}$, there exists a function of the *shape* of the triangle formed by the three points at positions $(\boldsymbol{r}_1, \boldsymbol{r}_2, \boldsymbol{r}_3)$ as they evolve temporally, such that C_3 indeed remain constant (see Pumir, 1998). These invariants have physical consequences, for example concerning the problem of return to isotropy in the small scales, which is slower than thought previously and which is associated with intermittency.

The determination of the exponents ζ_p is an open problem for Navier–Stokes and MHD turbulence, and such exponents differ from the Burgers values mentioned in §2.1; however, one can also observe with both experiments and numerical simulations a saturation – Burgers being an extreme case – whereby the exponents, as p grows, tend to a limit for a passive scalar; this phenomenon is attributed to the presence of sharp fronts in the spatial distribution of the scalar density. Intermittency deals with the localization of sharp structures in space/time. Their signature is felt in the existence of fat wings in PDFs with strong events highly more probable than for the Gaussian with the same mean and variance and, as stated before, in the departure from a linear law with order of the scaling exponents of structure functions. Beyond Navier–Stokes, intermittency has been quantified in this manner in the Solar Wind (Burlaga *et al.*, 1990; Ruzmaikin *et al.*, 1995; Tu & Marsch, 1995) and in DNS for MHD (Politano *et al.*, 1998; Gomez *et al.*, 1999; Müller & Biskamp, 2000). It has been modeled successfully for Navier–Stokes flows (Castaing *et al.*, 1993; She & Leveque, 1994), as shown by numerous laboratory experiments starting with Anselmet *et al.*, (1984), and direct numerical simulations (Vincent and Meneguzzi, 1994), including in the compressible case at *r.m.s.* Mach numbers of unity (Porter *et al.*, 1999).

How much can we learn from these considerations when examining realistic flows, such as when considering anisotropy due to rotation, stratification or a uniform magnetic field, as well as in the vicinity of interfaces, or a combination of such effects, as they occur in geophysical flows? In the case of coupling to a magnetic field, in the kinematic regime where the magnetic field remains passive, anomalous scaling appears already at second order (Vergassola, 1996), and a comparison with standard two–point closures would thus be of great interest; in the case of a passive vector with pressure included, similarly corrections arise (Yoshida & Kaneda, 2001). What is the energy spectrum in anisotropic flows (see *e.g.* Ishihara *et al.*, 2002)? Is there return to isotropy at small scale for anisotropic forcing such as shear waves? Finally, the way that intermittency can be incorporated in Large-Eddy Simulations and its effects on small–scale physics is largely unknown today, although Lagrangian techniques that allow us to follow the flow locally may prove useful. The challenge is thus to retain simple modeling approaches (the only viable ones for realistic flows of industrial or geophysical interests) yet incorporating some of this knowledge recently uncovered in specific cases.

4 Conclusion

Computations should be selected to provide data sets of archetypal geophysical turbulent flows which would then enable a variety of physical conditions to be explored with as few actual runs as possible, but at the highest resolution feasible today. This kind of data would allow controlled testing of models developed in the geosciences community, and appropriate scaling laws for sub-grid scale models could then be derived from it. Such an approach could also be used to facilitate data assimilation in geophysical models from observations in order to enhance predictive capabilities. The societal impacts, with the direct effect of turbulence on geophysical flows is clear. The broader impact on education can also be mentioned, at the graduate and undergraduate levels and for the involvement of high–school teachers with exciting scientific developments at the frontier of what is possible today. An example is atmospheric turbulence and oceanic circulation, and their interaction at the air–sea interface. At intermediate scales, strong stable stratification (together with rotation) structures the flow and impedes mixing in the vertical direction (Kaneda & Ishida, 2000), destroys – at least partially – statistical isotropy, and affects variability. Questions such as whether a quasi–two dimensional approach is sufficient, or the hydrostatic approximation at the planetary scale, are wide open (see *e.g.* Danilov & Gurarie (2000) for a recent review; also, Kimura and Herring, 1996). In a broader sense, the interaction of turbulence and waves (*e.g.* here, internal gravity waves; in MHD, Alfvén and magneto-acoustic waves) remains an open problem, in some cases in part because of the non–uniformity of the weak turbulence approximation (Newell *et al.*, 2001); this approach which allows for a theoretical evaluation of scaling laws for fluxless and constant flux solutions of the equations is of great interest and has huge potential applications.

Because of the vast demands inherent to research areas in the geosciences, only a concerted effort will allow in due time for a possible breakthrough on some of the important and difficult questions remaining in the field of geophysical turbulence involving global scale dynamics. A process must be put together whereby, if successful, it would provide access to and dissemination of turbulence data with demonstrated scale separation; it would be not only immediately valuable to the geosciences community, but would also benefit industrial and commercial interests.

AP wishes to thank Aimé Fournier, Sébastien Galtier, Thomas Gomez, Steve Hammond, Jack Herring, Darryl Holm, Yoshi Kimura, Don Lenschow, Rich Loft, Sasha Madronitz, Len Margolin, Don Middleton, Hélène Politano, Yannick Ponty, David Porter, Piotr Smolarkiewicz, Paul Swarztrauber, Steve Thomas, Joe Tribbia, and Paul Woodward for useful discussions.

References

1. Anselmet, F., Gagne, Y., Hopfinger, E.J. & Antonia, R.A., *J. Fluid Mech.* **140**, 63 (1984).
2. Berger, M.J. & Colella, P. *J. Comput. Phys.* **82**, 64 (1989).
3. J. P. Bertoglio J.P., in *Macroscopic modelling of Turbulent Flows* U. Frisch & J.B. Keller Eds. pp.100 Springer (1984).
4. Bodenschatz, E. *Nature* **409**, 1017 (2001).
5. Burlaga, L.F., Mish, W.H., Roberts, D.A. *J. Geophys. Res.* **95**, 4247 (1990).
6. Chasnov J. *Phys. Fluids* **A 3** 188 (1991).
7. Clyne, J., Scheitlin, T., & Weiss, J. *Theoretical and Computational Fluid Dynamics*, **11**, 195 (1998).
8. Dietachmayer, G. & K. Droegemeier, Month. Weather Rev, 120, 1675 (1992).
9. Danilov, S.D. & Gurarie, D. *Uspekhi Fiz. Nauk* **170** 921 (2000).
10. Falkovich, G., Gawedski, K. $ M. Vergassola, arXiv:cond-mat/0105199 (2001).
11. Farge, M., Schneider, K. & Kevlahan, N. *Phys. Fluids* **11**, 2187 (1999).
12. Frisch, U., Turbulence: The Legacy of Kolmogorov, CUP (Cambridge) (1995).
13. Frisch, U. & J. Bec, Les Houches Summer School "New trends in turbulence", arXiv:nlin.CD/0012033 (2000).
14. Gailitis, A., in *Proceedings of NATO ASI "Theory of Solar and Planetary Dynamos"*, p. 91, Eds. D. Matthaeus and A. Rucklidge, CUP (1993).
15. Gombosi, T.I., K.G. Powell and D.L. De Zeeuw , J. Geophys. Res. vol. 99, 21,525 (1994).
16. Gomez, T., H. Politano & A. Pouquet: *Phys. Fluids*, **11**, 2298 (1999).
17. Gotoh, T. & R. Kraichnan, *Phys. Fluids*, **10**, 2859 (1998).
18. Gotoh T., Kakui I. & Kaneda Y., in *Statistical theories and computational approaches to turbulence*, Kaneda Y. & Gotoh T. Eds. Springer (2002).

19. Higuchi, H. *Jpn. J. Appl. Phys.* **17**, 811 (1978).
20. Holm, D. D., Marsden, J. E. & Ratiu, T. S. *Adv. in Math.* **137**, 1 (1998).
21. Huang, W., Ren, Y. & Russell, R. *J. Comput. Phys.* **113**, 279 (1994).
22. Huang, W. & Russell, R. *SIAM J. Numer. Anal.* **34**, 1106–1126 (1997).
23. Ihm, I., and Park, S., in *Proceedings of Graphics Interface "Wavelet-Based 3D Compression Scheme for Very Large Volume Data"*, p. 107 (1998).
24. Ishihara T., Yoshida, K. & Kaneda, Y. "Anisotropic velocity correlation spectrum at small scales in a homogeneous turbulent shear flow", preprint Nagoya University (2001).
25. Kageyama, A. & Sato, T. *Phys. Plasmas* **6**, 771 (1999).
26. Kaneda, Y., *J. Fluid Mech.* **107**, 131 (1981).
27. Kaneda, Y., Ishihara, T. & Gotoh, K., *Phys. Fluids* **11**, 2154 (1999).
28. Kaneda, Y. & Ishida, T. *J. Fluid Mech.* **402**, 311 (2000).
29. Karniadakis G.E., Israeli M. & Orszag S. *J. Comput. Phys.* **97**, 414 (1991).
30. Kawahara G., Kida, S., Tanaka, M., Yanase, S. *J. Fluid Mech.* **353**, 115 (1997).
31. Kimura, Y. & Herring, J.R. *J. Fluid Mech.* **328**, 523 (1996).
32. Kolarov, K., and Lynch, W., in *SPIE Conference on Applications of Digital Image Processing "Wavelet Compression for 3D and Higher-Dimensional Objects"* p. 247 (1997).
33. Kraichnan R.H., *J. Atmos. Sci.* **33**, 1521 (1976).
34. Kraichnan R.H., *J. Fluid Mech.* **83**, 349 (1977).
35. Kraichnan, R.H. *Phys. Rev. Lett.* **72**, 1016 (1994).
36. Majda, A.J. & Kramer, P.R. *Phys. Reports* **314**, 237 (1999) .
37. Mallat, S. *IEEE Trans. Pattern Anal. Machine Intell.* **11** 674 (1989).
38. Margolin, L. & Rider, W., preprint Los Alamos National Laboratory (2002).
39. Mavriplis C. *Comput. Meth. Appl. Mech. Engrg.* **116**, 77 (1994).
40. Métais O. & Lesieur M., *J. Fluid Mech.* **184** 207 (1992).
41. Müller, W.C. & Biskamp, D. *Phys. Rev. Lett.* **84**, 475 (2000).
42. Muraki, S. *IEEE Computer Graphics and Applications*, **13**, 50 (1993).
43. Nakayama, K. *Astrophys. J.* **556**, 1027 (2001).
44. Newell, A.C., Nazarenko, S. & Biven, L. *Physica D* **152–153**, 520 (2001).
45. Nore, M.E. Brachet, Politano, H. & Pouquet, A. *Phys. Plasmas Lett.*, **4**, 1 (1997).
46. Noullez, A. and Pinton, J.F., preprint Observatoire de la Côte d'Azur (2002).
47. Oran, E. and Boris, J. *Computers in Physics* **7**, 523 (1993).
48. Painter, J., Mccormick, P., and Mcpherson, A. *In NSF/DOE Workshop on Large Scale Visualization and Data Management* (1999).
49. Parker, S., Parker M,. Livnat, Y., Sloan, P., Hansen, C. & Shirley, P. *Trans. Viz. Comp. Graph.*, **5**, 238 (1999).
50. Politano, H., A. Pouquet & V. Carbone *EuroPhys. Lett.* **43**, 516 (1998).
51. Polyakov, A.M. *Phys. Rev. E.* **54**, 4896 (1996).
52. Porter, D., Pouquet, A., & Woodward, P., *Phys. Fluids* **6**, 2133 (1994).
53. Porter, D., Pouquet, A., Sytine, I. & Woodward, P., *Physica A*, **263**, 263 (1999).
54. Pouquet, A., Les Houches Summer School on Astrophysical Fluid Dynamics; Session **XLVII**, p. 139; Eds. J. P. Zahn & J. Zinn–Justin, Elsevier (1993).
55. Pumir, A., Phys. Rev. E vol. 57, 2914 (1998).
56. Rodler, R., in /sl Pacific Graphics '99 "Wavelet-based 3D Compression with Fast Random Access for Very Large Volume Data", p. 108 (1999).
57. Ruzmaikin, A., Feynman, J., Goldstein, B.E., Smith, E.J., Balogh, A. J. Geophys. Res. 100, 3395 (1995).
58. She, Z.S. & Leveque E. Phys. Rev. Lett. vol. 72, 336 (1994).

59. Shraiman B. & Siggia E. *Nature* **405** , 639 (2000).
60. Smolarkiewicz, P. & Margolin, L. *J. Comput. Phys.* **140**, 459 (1998).
61. Sornette, D., Critical Phenomena in Natural Sciences, Springer (2000).
62. Sreenivasan, K.R., Antonia, R.A. Annu. Rev. Fluid Mech. 29, 435 (1997).
63. Stolorz, P., Nakamura, H., Mesrobian, E., Muntz, R., Shek, C., Santos, J., Yi, J., Ng, K., Chien, S., Mechoso, C., Farrara, J., in *Fast Spatial–Temporal Data Mining of Large Geographical Data Sets*, AAAI Press, Menlo Park, p. 300 (1995).
64. Sytine, I., Porter, D., Woodward, P., Hodson, S. & Winkler, K-H *J. Comput. Phys.* **158**, 225 (2000).
65. Sweldens, W. *J. Appl Comp. Anal.*, **3**, 186 (1996).
66. Tao, H., and Moorhead, R., in *Proc. Visualization '94 "Progressive transmission of scientific data using biorthogonal wavelet transform"*, p. 93 (1994).
67. Tu, M. & E. Marsch, *Space Sci. Rev.* **73**, 1 (1995).
68. Vergassola, M., Dubrulle, B., Frisch, U. & Noullez, A. *Astron. Astrophys.* **289**, 325 (1994).
69. Vergassola, M. *Phys. Rev. E* **53**, R3021 (1996).
70. Vincent A. & Meneguzzi M., *J. Fluid Mech.* **258**, 245 (1994).
71. Warhaft Z. *Ann. Rev. Fluid Mech.* **32**, 203 (2000).
72. Wilson, J., in *IEEE Data Compression Conference Poster Session " Wavelet-Based Lossy Compression of Turbulence Data* (2000)
73. Woyczynski, W.A. in "Nonlinear Waves and Weak Turbulence with applications in oceanography and condensed matter physics" pp. 279-311 (Birkhauser, Berlin) (1995).
74. Yoshida, K. & Kaneda, Y. *Phys. Rev. E* **63**, 1 (2001).
75. Yoshida K., Ishihara T. & Kaneda Y., in *Statistical theories and computational approaches to turbulence.* Kaneda Y. & Gotoh T. Eds. Springer (2002).
76. Yoshizawa, A. *Phys. Fluids* **28**, 3313 (1985).

Structural and Statistical Aspects of Stably Stratified Turbulence

Jackson R. Herring[1] and Yoshifumi Kimura[2]

[1] N.C.A.R., Boulder CO 80307, USA
[2] Graduate School of Mathematics, Nagoya University, Furo-cho, Chikusa-ku, Nagoya 464-8602, Japan

Abstract. Stably stratified turbulence is examined *via* direct numerical simulation (DNS) and elementary statistical theory. Such flows decay more slowly than isotropic turbulence with the same initial conditions. We offer an explanation in terms of the diminution of energy transfer to small scales because of phase mixing nature of gravity waves. Structures in stratified flows (pancakes) are distinctly different from those of isotropic turbulence (vortex tubes). Thus waves of smaller frequencies survive. Finally we compare probability distribution functions (PDFs) for the Eulerian acceleration for stratified flow to isotropic PDF's. The PDF's of stable stratification are significantly more Gaussian than those of isotropic turbulence, suggesting that the underlying assumptions of two–point closures may be more applicable to stratified turbulence than to isotropic flows.

1 Stably Stratified Turbulence

The striking feature of stably stratified turbulence is that it decays slower than isotropic three dimensional turbulence with the same initial conditions. Fig. 1 illustrates this. It shows the decay of energy for the standard initial energy spectrum $E(k) \sim k^4 \exp(-k^2)$. Here, $R_\lambda(0) = 200$, and the resolution is 512^3. Stratification (with N=10.) is turned on when the skewness reaches its maximum (about 1 eddy circulation time).The subsequent decay rapidly evolves into self–similarity at a near constant Taylor microscale Reynolds number $R_\lambda = 40$.

The non–dimensional equations of motion under discussion are:

$$(\partial_t - \nabla^2)\mathbf{u} = -\nabla p - \mathbf{u}\cdot\nabla\mathbf{u} - \hat{\mathbf{g}}\theta - 2\mathbf{\Omega}\times\mathbf{u} \tag{1}$$

$$(\partial_t - \sigma\nabla^2)\theta = N^2 w - \mathbf{u}\cdot\nabla\theta \tag{2}$$

$$\nabla\cdot\mathbf{u} = 0 \tag{3}$$

Here $\mathbf{u} = (u, v, w)$ and θ the deviation of the temperature field from its mean, whose constant vertical gradient is nondimensionalized to -1. We focus on the case $\mathbf{\Omega} = 0$ in this paper, except to note later how this study has possible implications for geophysical flows. Fig. 1 shows that the decay of kinetic energy, $E(t)$, is $\sim t^{-1}$ at late times.

We recall that for unstratified turbulence $E(t) \sim t^{-1.4}$. The first order question is, then, how to explain the slow–down in the decay of $E(t)$ from that found

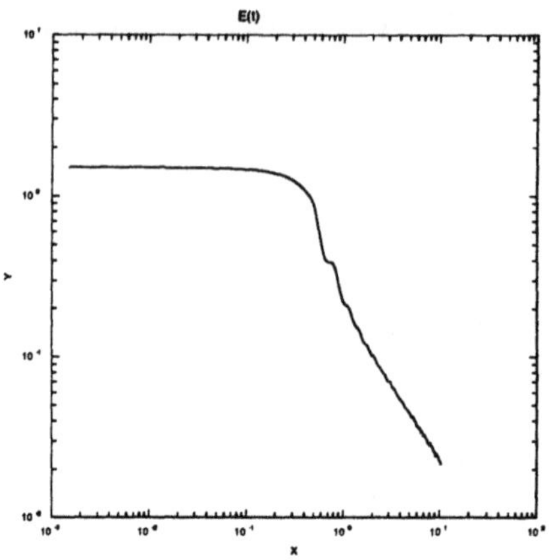

Fig. 1. Decay of kinetic energy, $E(t)$ as a function of time, t for $N^2 = 100$. Resolution is 512^3 using Kimura's code "Ranryu" as extended to C++ by Rodney James (NCAR, SCD).

for isotropic DNS. A significant feature of stratified flows is its strong anisotropy, with the vorticity organized into "scattered pancakes" as shown in Fig. 2.

Here we show the late–time organization of the spacial distribution of enstrophy, $\mathcal{E}(\mathbf{r}, t) \equiv |\nabla \times \mathbf{u}(\mathbf{r}, t)|^2$. The flow is strongly anisotropic, but its degree of anisotropy tends to saturate as $N \to \infty$. This is indicted in Fig. 3, which shows the angular distribution of the vorticity vector.

It is remarkable that although stratified flows contain waves, the evolution of the enstrophy patches shows little wave–like fluctuations. Perhaps this is because pancake organization signifies strong vertical variability, for which the wave frequency,

$$\omega = \sqrt{N^2 \sin^2(\vartheta) + 4\Omega^2 \cos^2(\vartheta)} \tag{4}$$

is near zero. Here $\cos\vartheta = k_z/k$, and we include in ω a possible rotation rate, Ω. Indeed, if we formulate a perturbation theory in which the zeroth–order state is Gaussian turbulence, then we see that the resonant triadic interactions are dominant, and these are just those waves for which $N \sin\vartheta = 0$. Thus, for stratified turbulence there is no clean separation of the flow into "slow" and "fast" components, as there would be for $\Omega \neq 0$.

We now comment on the reason for the slow–down in the decay of stratified turbulence beyond that of isotropic turbulence. We have already noted the almost self–similar nature of the decay. Of course for self–similar decay of isotropic turbulence, the decay law is $E(t) \sim t^{-1}$, but such requires $E(k, t) \sim k$, $k \to 0$, which is never observed in DNS. The present resolution of our DNS (512^3) is insufficient to test the idea that as $k \to 0$, $E(k) \sim k^4$: the k^4 dissolves rapidly into a progressively more shallow spectrum as time advances.

The following is a rough estimate of the decay of kinetic energy for stratified turbulence, based on closure. Here we assume a preservation of the k^4, $k \to 0$ spectrum, for simplicity. We recall first a simple wave–number diffusion equation

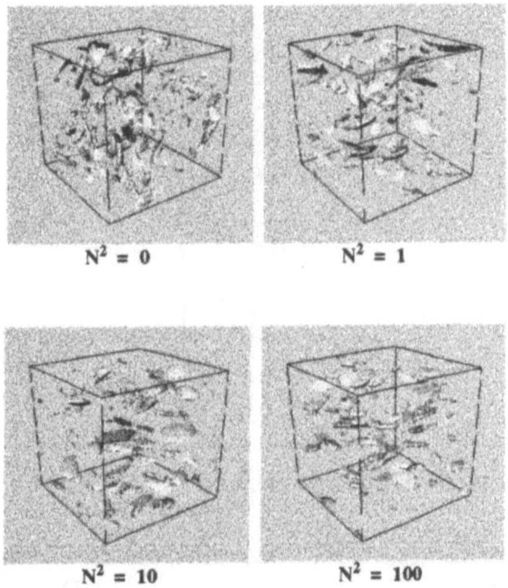

Fig. 2. Iso surfaces of enstrophy for various values of N. From [4]

for the evolution of the energy spectrum, $E(k,t)$ as proposed by Leith (1968)[1]:

$$(\partial_t + \nu k^2)E(k,t) = \partial_k \left\{ k^4 \partial_k \{ \frac{\int_0^k p^2 dp E(p)}{\eta(k)} \frac{E(k)}{k^2} \} \right\}. \quad (5)$$

Here, $\eta(k)$ is the eddy relaxation rate, which for isotropic turbulence is $\sqrt{\int_0^k p^2 dp E(p)}$. (Actually, Leith proposed $k^{3/2}\sqrt{E(k,t)}$). For strongly stratified turbulence, we expect $\eta \sim N$. Why, though should waves act to dampen the energy? An essential point here is that gravity waves "phase mix", and thus produce an attenuation of correlations in time. Kaneda & Ishida (2000)[2] have stressed this point in their application of rapid distortion theory to stratified turbulence. See also Britter *et al.* (1993)[3]. Then by integrating (5) over [0,k], with k in the inertial range there follows $E(k) \sim \epsilon^{2/3}k^{-5/3}$, if we take $\eta(k) \sim \sqrt{\int_0^k p^2 dp E(p)}$. If, on the other hand $\eta(k) \sim N$, as for strongly stratified flow, there follows,

$$E(k) \sim (N\epsilon)^{1/2}/k^2. \quad (6)$$

Note that we have ignored anisotropic effects here, so our argument is rough. But, according to Fig. 3 anisotropy is not overwhelming, even for large N. We may now estimate the decay of $E(t)$ extending the analysis of Kolmogorov (1942)[5] to the stratified domain: $E(k) \sim k^4, (k < k_0)$, $E(k) \sim \sqrt{N\epsilon}/k^2$, $k > k_0$. There results,

$$E(t) \sim t^{-5/7} \quad (7)$$

The exponent is the average of 3D decay (10/7) and that of 2D (0). The exponent in Fig. 1 is -1 instead of -5/7. This could be attributable to finite $R_\lambda \sim 35$ in the DNS, just as our estimates of the decay of isotropic turbulence is somewhat

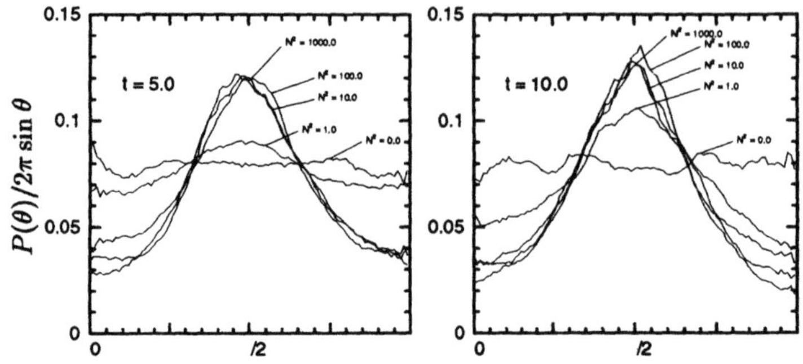

Fig. 3. Polar angle distribution of vorticity vector. (a) t=5, (b)t=10. From [4]

faster than indicated by the EDQNM calculations (-1.5 instead of -1.37). Clearly, such estimates should be replaced by more secure EDQNM or DIA calculations, such as those proposed by Godeferd & Cambon (1994)[6] and Sanderson *et Al.* (1991)[7]. Finally, we remark that the spectral form (6) has been compared (favorably ?) to DNS for rotating turbulence by Yeung and Zhou (1998)[8].

2 Relation of Stratified Turbulence to Atmospheric Flows

On the more practical level, stably stratified turbulence was thought to serve as a model for important scales of atmospheric and oceanographic motion. In this connection, consider the observations of Nastrom *et al.* (1984)[9] that the horizontal kinetic energy spectrum of atmosphere motions in the range (100km–1 km) conform nicely to a $k^{-5/3}$ power spectrum. Such observations were used by Gage (1979)[10] to argue that the underlying dynamics of the energy spectrum in this range was the inverse–cascading of essentially two dimensional turbulence. Such an interpretation seems natural since the flow is both random and near two–dimensional. We know that if strictly two–dimensional flows are randomly forced at small scales (k_f, say), then at scales $k > k_f$ the energy spectrum becomes $k^{-5/3}$. But what do the dynamics of stratified turbulence have in common with that of two–dimensional turbulence? First, the small scale forcing could be attributable to thunderstorms and their anvil outflow at great heights. With respect to the dynamics of the flow, the analysis of Riley *et al.* (1981)[11], Lilly (1983)[12], and Majda and Grote (1997)[13] showed that as ($N \to \infty$) stratified turbulence becomes two–dimensional in vertical layers. In each of these layers the flow is very nearly two—dimensional, but the layers are roughly statistically independent. Thus it may be possible to invoke two–dimensional turbulence arguments to explain the observed atmospheric spectrum in the 100km–1km range.

We should note, however, that such dynamics are essentially different from the near two–dimensional dynamics of quasi geostrophic turbulence (QGT), in which both rotation and stratification play an essential role. In the latter case, layering does not occur. Rather the various vertical layers are strongly coupled.

QGT is valid if several conditions are met, among which is the dominance of rotation over the advection term in (1): $R_O \equiv| \mathbf{u} \cdot \nabla \mathbf{u} \mid / \mid \boldsymbol{\Omega} \times \mathbf{u} |<< 1$. This condition is met at large, planetary scales of motion but as the scale–size approaches 100km, $R_O \sim 1$, and arguments based on quasi geostrophic dynamics cease to be valid. Earlier attempts to examine this issue *via* DNS (Herring and Métais (1989))[14] showed that the layering did indeed happen, but that the inverse cascade was absent. Its absence was interpreted as the result of frictional effects associated with the irregular shape of the edges of the layers. More recently, Lindborg (1999)[15] examined aircraft data for evidence of inverse cascade dynamics in this range. His analysis focused on the third order structure function, rather on energy spectra. Thus, his question of the data was: is the energy cascade in the atmospheres closely similar to that of two dimensional turbulence? He found little evidence for inverse cascade dynamics. Thus, the reason for a $k^{-5/3}$ range is still open. It could be that VanZandt's (1982)[16] proposal that this spectrum are simply buoyancy waves is more satisfactory.

3 Linearity and Nonlinearity in turbulence

Stratified and rotating turbulence furnishes the simplest example in which both linear and nonlinear terms play important dynamical roles. Recently several authors have noted that many aspects of stratified flow may be understood by focusing nearly exclusively on the linear terms. In brief, such methods discard the fluctuating self interactions ($\mathbf{u} \cdot \nabla \mathbf{u}$ and ($\mathbf{u} \cdot \nabla \theta$)) in the equations of motion (1)–(3). Such is called *Rapid Distortion* theory (RDT). In this category is the study of Britter *et al.* (1993)[3], and Hanazaki & Hunt (2001)[17]. Considerable insight into dispersion of particles in stratified flows was obtained by Kaneda and Ishida (2000)[3], who used RDT in conjunction with Corrsin's independence principle (Corrsin, (1959))[18]. The latter allowed estimates of Lagrangian covariances from the Eulerian fields as computed by RDT.

Here we begin with a simple question: given a turbulent signal, how do we know that it is produced by a linear or nonlinear process? One avenue to this question formulated by Hasselman *et al.* (1963)[19] is to examine the bispectrum, $B(x_1, x_2)$ of the fluctuation, $u(x)$. Here

$$B(x_1, x_2) =< u(x_0)u(x_0 + x_1)u(x_0 + x_2) > \tag{8}$$

We may expand B about $(x_1 - x_0, x_2 - x_0) \to 0$. To quintic order:

$$B(x_1, x_2) = \langle (\frac{\partial u}{\partial x_0})^3 \rangle \left\{ \frac{1}{6}(x_1^3 + x_2^3) - \frac{1}{4}(x_1 x_2^2 + x_1^2 x_2) \right\} +$$
$$\langle (\frac{\partial^2 u}{\partial x_0^2})^2 (\frac{\partial u}{\partial x_0}) \rangle \left\{ \frac{1}{24}(x_1^5 + x_2^5) - \frac{5}{48}(x_1 x_2^4 + x_1^4 x_2) + \frac{1}{24}(x_1^2 x_2^3 + x_1^3 x_2^2) \right\} + \cdots \tag{9}$$

In this derivation, we have eliminated many terms by imposing the statistical symmetries that derive from the Navier–Stokes symmetry of (1)–(3):

$$(\mathbf{u}(\mathbf{x}, t), \theta(\mathbf{x}, t)) \to -[\mathbf{u}(-\mathbf{x}, t), \theta(-\mathbf{x}, t))] \tag{10}$$

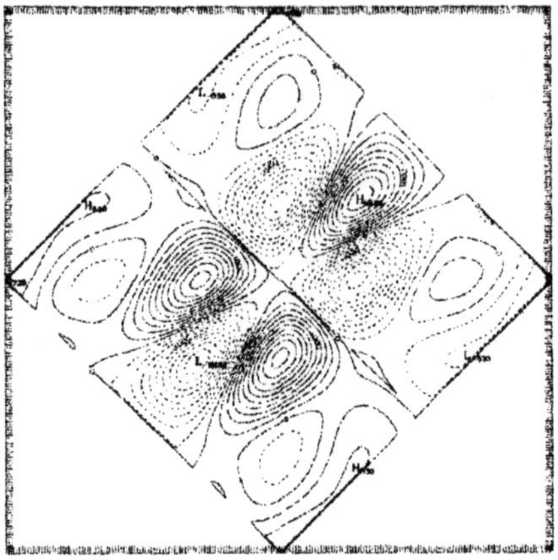

Fig. 4. $B(x,y)$ for simple model of nonlinear time series as described in the text.

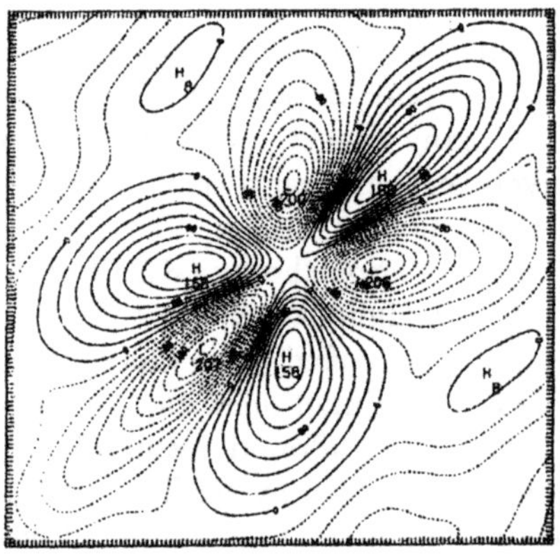

Fig. 5. $B(x,y)$ for isotropic DNS of Herring and Métais (1992)[14]. Here, $R_\lambda \sim 30$.

We also used homogeneity: $< Q(\mathbf{x}) >=< Q(\mathbf{x}+\mathbf{x}') >$ for any $\mathbf{x}$, $\mathbf{x}'$. From (9), we see immediately that B at small displacements depends on the skewness, ($S =< (\partial u/\partial x)^3 > / < (\partial u/\partial x)^2 >^{3/2}$) which–in turn–measures the overall rate of energy transfer to small scales in the sense that if we write the evolution equation for kinetic energy as $(\partial_t + 2\nu k^2)E(k,t) = T(k,t)$ then $S \sim \int_0^\infty k^2 T(k,t)$. In case $S = 0$ (as for 2–D turbulence), the quintic term takes over at small displacements, and its coefficient is closely similar to the two–dimensional skewness, $S_{2D} \sim< (\partial u/\partial x)(\nabla^2 u)^2 >$.

We next propose a simple model of $B(x_1, x_2)$ based on an elementary "turbulent" pulse, $\mathcal{V}(\zeta, \zeta_1, \zeta_2)$:

$$\begin{aligned} &= \zeta, \ (0 < \zeta < \zeta_1), \\ &= \zeta_1 - b(\zeta - \zeta_1) \ (\zeta_1 < \zeta < \zeta_3), \\ &= A + c(1 - \zeta), \ (\zeta_3 < \zeta < 1) \end{aligned} \tag{11}$$

with constants $(A, b, c) u \in \mathcal{C}^0$, and $\int_0^1 u(\zeta) d\zeta = 0$. Note that $S = 0$ for $\mathcal{V}(\zeta, 1/4, 3/4)$. Consider now what happens to $\mathcal{V}(\zeta, 1/4, 3/4)$ if it is self-advected–as is Navier–Stokes: $(\partial_t + \mathcal{V}\partial/\partial\zeta)\mathcal{V} = 0$. We know from studies of Burgers' equation that the effect is to move ζ_1 towards ζ_3. In fact, Burgers' dynamics moves ζ_1 toward 1/2 from below, and ζ_3 towards 1/2 from above by equal amounts. In the latter case, $S = (1 - 4x_1)/\sqrt{2x_1(1 - 2x_1)}$. The model consists in a series of pulses, $\mathcal{V}(\zeta, 1/4 + \epsilon, 3/4 - \epsilon)$, together with its spacial reflection on the interval $0 \leq \zeta \leq 1)$, $\mathcal{V}^r(\zeta)$. ϵ is then adjusted to give the desired skewness. The model is thus defined on contiguous closed intervals $x \in n\zeta \leq (n+1)\zeta$, $(0 < \zeta < 1)$.

$$u(x) = A_n \mathcal{V}(\frac{x - x_n}{n}) \tag{12}$$

where A_n is a Gaussianly distributed random amplitude.

Fig. 4 displays $B(x_1, x_2)$ for our simple one dimensional model described above, while Fig. 5 shows $B(x_1, x_2)$ for isotropic turbulence ($R_\lambda \sim 30$) from Herring and Métais (1992)[14] . The comparison of these figures suggests that the simple Burgers' dynamics is sufficient to explain the B–pattern. Stratified turbulence–on the other hand evolves toward $S \sim< (\partial u/\partial x)^3 >\sim 0$, so that its numerical estimate is more noisy. Thus, higher resolution ($> 512^3$) is needed to discern its shape. Such calculations are now in progress.

We next compare single point probability distribution function (PDF) information for unstratified and stratified turbulence. Here we examine the PDF for the Eulerian acceleration,

$$\mathcal{P}(\mathbf{a}), \ \mathbf{a} \equiv \partial \mathbf{u}/\partial t \tag{13}$$

$\mathcal{P}$ is presented in Fig. 6 for unstratified turbulence, while Fig. 7 gives $\mathcal{P}$ for stratified turbulence. Fig. 6 is quite similar to experimental measurements beginning with that of Wyngaard and Tennekes(1970)[19]. Fig. 7 for stratified turbulence shows a distinct anisotropy, but notice that for all components (u, v, w) $\partial\mathbf{u}/\partial t$ are quite close to Gaussian. Such information suggests that closure approximations based on close–to–Gaussian PDF's may have more validity for stratified turbulence than for unstratified.

4 Concluding Comments

In conclusion, we call attention to several aspects of stable stratified turbulence for which further research could be useful. First, we have noted the approximate rate of decay of stratified turbulence $E(t) \sim t^{-1}$, and have suggested that

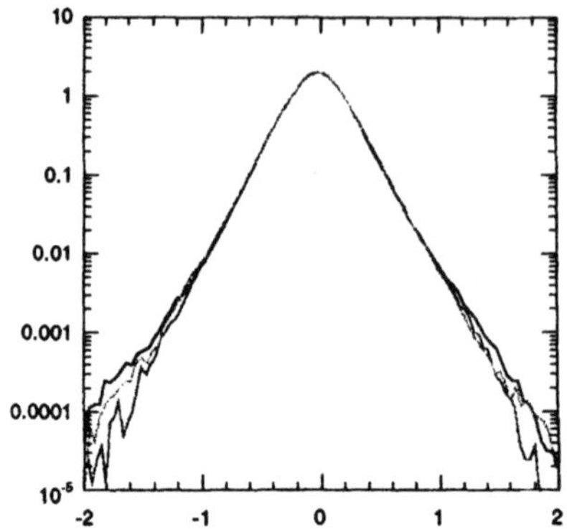

Fig. 6. PDF of Eulerian acceleration for unstratified turbulence. Here, $R_\lambda \sim 30$.

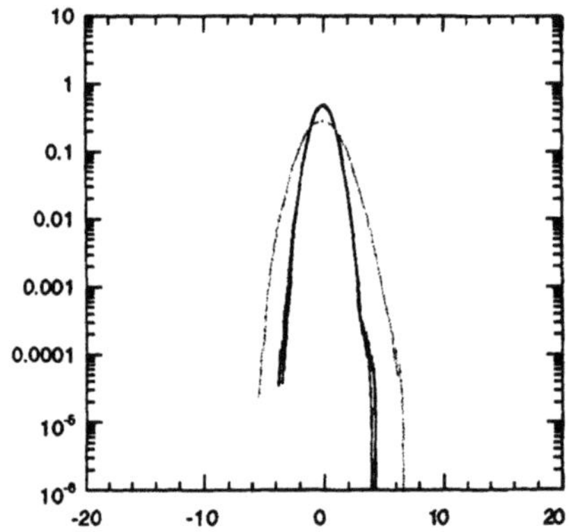

Fig. 7. PDF for Stratified Turbulence, with N=10. Again, $R_\lambda \sim 30$.

simple closure estimates may be able to comprehend this decay law. The closure estimates presume a power law for the energy spectrum, which is so far unsubstantiated, largely because the DNS resolution is so far too small. Perhaps studies currently underway at resolution $512^3 \rightarrow 1024^3$ will resolve this issue. An important point here is that the DNS must resolve accurately both the k^4 spectrum at small k, and the inertial range at large k.

Associated with the weakened decay rate is a weakening of the energy transfer rate to small scales. We proposed a simple measure of this rate based on the Bispectra, which is simply the correlation of the velocity (or temperature field) at three points. For isotropic turbulence, earlier results showed agreement between theory and DNS. Good DNS statistics for stratified flows are more difficult because of the presence of very low frequency waves ($\omega = N\sin\vartheta$, $\vartheta \sim 0$).

Finally, we examined probability distribution functions (PDFs) for both isotropic turbulence and stratified turbulence. Here we noted that for the Eulerian acceleration, the PDF of stratified flows is anisotropic, but closely Gaussian, while that for isotropic flows is near exponential. This result may be consistent with the dominance of the linear term in stratified flow. Of perhaps more dynamical importance would be Lagrangian PDFs. In any case we should note that the degree of anisotropy in stratified flows is determined by the presence of the nonlinear terms in (1)–(2): rapid distortion theory would predict less anisotropy than the full dynamics.

Acknowledgment. This work was supported by the U. S. NSF–ATM grant 9840693 and the Japanese grant JSPS MPCR375.

References

1. Leith, C. E. 1968: Diffusion approximation for turbulent scalar fields. *Phys. Fluids*, **11**, 1612–1617.
2. Kaneda, Y. and Ishida, T. 2000: Suppression of vertical diffusion in strongly stratified turbulence. *J. Fluid Mech.*, **402**, 311–327.
3. Britter, R. E., Hunt, J.C.R., G.,L. Marsh, and W.,H. Snyder 1993: The effects of stable stratification on the turbulent diffusion and the decay of grid turbulence. *J. Fluid Mech.*, **127**, 27–44.
4. Kimura, Y. and Herring, J. R. 1996: Diffusion in stably stratified turbulence. *J. Fluid Mech.*, **328**, 253–263.
5. Kolmogorov, A. N. 1942: On the degeration of isotropic turbulence in an incompressible viscous liquid. *Dok. Akad. Nauk. SSSR*, **31**, 538–541.
6. Godeferd, F.S. and C. Cambon 1994: Detailed investigation of energy transfer in homogeneous stratified turbulence *Phys. Fluids*, **6** (6), 2084–2100.
7. Sanderson, R. C., A. D. Leonard, J. R. Herring, and J. C. Hill, 1991: 'Fossil and active turbulence'. In: *Turbulence and coherent structures: 89: Organized structures and turbulence in fluid mechanics, Grenoble, 18–21 September 1989*, O. Métais and M. Lesieur, Eds., Kluwer Academic Press, pp 429–448.
8. Yeung, P. K. and Zhou, Y. 1998: Numerical study of rotating turbulence with external forcing. *Phys. Fluids*, **10**, 2895–3244.
9. Nastrom, G.D., Gage, K.S., and Jasperson, W.H. 1984: Atmospheric kinetic energy spectrum, $10^0 - -10^4$km. *Nature*, **310**, 36–38.
10. Gage, K.S. 1979: Evidence for a $k^{-5/3}$ law inertial range in mesoscale two- dimensional turbulence. *J. Atmos. Sci.*, **36**, 1950-1954.
11. Riley, J.J., Metcalfe, R.W., and Weissman, M.A. 1982: Direct numerical simulations of homogeneous turbulence in density stratified fluids. In: *Proc. AIP Conf. on Nonlinear Properties of Internal Waves* Bruce J. West Ed, 79–112.
12. Lilly, D.K. 1983: Stratified turbulence and the Mesoscale variability of the atmosphere. *J. Atmos. Sci.*, **40**, 749-761.
13. Majda, A. J., Grote, M. J. 1997: Model dynamics and collapse in decaying strongly stratified flows. *Phys. Fluids*, **9** (10), 2932–2940.
14. Herring, J. R., and O. Métais 1992: Spectral Transfer and Bispectra for Turbulence with Passive Scalars. *J. Fluid Mech.*, **235**, 103-121.
15. Lindborg, E. 1999: Can the atmospheric energy spectrum be explained by two-dimensional turbulence. *J. Fluid Mech.*, **388**, 259–288.
16. VanZandt, T.E. 1982: A universal spectrum of buoyancy waves in the atmosphere. *Geophys. Res. Lett.* **9**, 575-578.
17. Hanazaki, H. and Hunt, J. C. R. 2001: Linear processes in unsteady stably stratified turbulence. *J. Fluid Mech.*, **318**, 303–337.

18. Corrsin, S. 1959: Progress report on some turbulent diffusion research. In: *Atmospheric Diffusion and Air Pollution* ed. F. N. Frenkiel & P. A. Shepard. Advances in Geophysics, **6**, 161–164. Academic Press.
19. Hasselman, K., Munk, W. and Mcdonald, G. 1963: Bispectra of ocean waves. In: *Time Series Analysis* M. Rosenblatt, ed., 125–139. Wiley & Sons, New York.
20. Wyngaard, J. C. and Tennekes, H. 1970: Measurements of the small–scale structure of turbulence at moderate Reynolds numbers. *Phys. Fluids*, **13**, 1962–1969.

Dynamics of Rotating Stably Stratified Flows

Claude Cambon[1]

Laboratoire de Mécanique des Fluides et d'Acoustique
UMR 5501, BP 163, Ecole Centrale de Lyon
69131 Ecully Cedex, France.

Abstract. In the absence of preexisting organised eddies and without geometric constraints, a turbulent motion in a vertically stably stratified and rotating fluid is analysed as a combination of steady and oscillating normal modes. These modes correspond to Quasi-Geostrophic (QG) and Ageostrophic (AG) motions, the latter reflecting inertia-gravity dispersive waves. Linear and nonlinear dynamics are then discussed; the amplitudes of the eigenmodes, which are kept constant in the inviscid linear, or RDT, limit, are treated as 'slowly' evolving variables, whose long-time evolution is governed by nonlinear triadic interactions. Statistical *multimodal and anisotropic* models of EDQNM-type ([10], [17]) are revisited, in close connection with recent Eulerian Wave-Turbulence theories for weak interaction. Emphasis is placed on the case of pure rotation, in which the inertial wave modes form a complete basis for the velocity field (no QG contribution, actually pure wave-turbulence at small Rossby number). Pure inviscid linear dynamics, even if irrelevant for predicting the rise of columnar structures, yield interesting new results for diffusivity of rapidly rotating turbulence, if two-time velocity correlations are calculated and time-integrated [8], according to [27]. Partial two-dimensionalisation, however, is only triggered by nonlinear interactions, with a dominant role of triadic resonances. In the general case with stable stratification, the role of the QG mode appears to be pivotal; it is responsible for the dominantly horizontal diffusivity, that can be analysed in the linear limit, and it drives the energy and anisotropy transfers mediated by nonlinear interactions, at least if its energy is initially significant at large scale. At dominant stratification, the nonlinear tendency to create the layering, or pancake structuring, is explained by a wave-released spectral transfer, which tends to concentrate the kinetic energy towards vertical wave-vectors (this limit correspond to the Vertically Sheared Horizontal Flow, very different from the two-dimensional, or barotropic, limit, which is concerned in pure rotation). Comparing different studies, including recent DNS [46] and stability analyses, a new insight is given to nonlinear dynamics, including both *angular energy drain* and *inverse cascades* in Fourier space, with typical eigenmodes called into play.

1 Introduction

Atmospheric and oceanic flows are affected by density stratification and, provided the lengthscale under consideration is large enough, by Coriolis forces. Relevant elements of their dynamics can be studied in the absence of geometric constraints, as in homogeneous anisotropic turbulence subjected to an uniform vertical density gradient in a rotating frame. For the sake of simplicity, the angular velocity of the rotating frame is chosen aligned with the gravitational acceleration, and with the mean density gradient (all are vertical). The two important 'external' frequencies are N, the Brunt-Väisälä frequency, and 2Ω, the system vorticity.

In the linear, or 'Rapid Distortion' limit, the turbulent flow can be seen as a superimposition of one steady and two oscillating modes of motion. The first one (steady) corresponds to the 'vortex' or 'Quasi-geostrophic' (QG hereinafter) mode, and involves only the toroidal component of the velocity field, whereas the second (oscillating) correspond to inertia-gravity dispersive waves, and incorporate also the poloidal component. Pure linear dynamics is somewhat poor since it cannot explain the emergence of cigar (dominant rotation) or pancake (dominant stratification) structure from initially isotropic turbulence. Nevertheless, the decomposition in terms of eigenmodes of the linear regime remains of interest for analysing nonlinear interactions. For instance, it is possible to consider the amplitudes of the eigenmodes, which are kept constant in the linear limit, as slowly evolving new variables, and to construct equations for their statistical correlations. A common background follows for wave-turbulence weak interaction theory, and for multimodal anisotropic two-point closure theory (essentially of EDQNM-type in the present paper). Results of statistical theories are compared to those of available DNS, in order to evaluate which nonlinear interactions, in terms of which eigenmodes, are dominantly involved for orienting the evolution of initially isotropic turbulence towards quasi two-component or two-dimensional structure. This yields supporting a dynamical, not phenomenological, interpretation of classical observations, such as collapse and layering in stratified flows, and partial transition towards two-dimensional structure in rotating flows. The definition of eigenmodes can be derived from a poloidal-toroidal decomposition of the velocity field, and this is dramatically facilitated using Fourier space (in the absence of geometric constraints, of course). In addition to the QG-AG decomposition, and the analysis of relevant nonlinear interactions in term of them, is shown the crucial importance of the angular dependence in Fourier space, as an essential characteristic of detailed anisotropy. For instance, two-dimensionalization in rotating turbulence is directly reflected by an angular drain of energy which tends to concentrate this energy towards the horizontal wave-plane. Conversely, horizontal layering in stably stratified turbulence corresponds to concentration towards vertical wave-vectors. The horizontal wave-plane is both the slow (vanishing frequency for dispersive inertial waves) manifold for pure rotation and the two-dimensional one (sometime denoted barotropic). Vertical wave-vectors corresponds to the slow (zero frequency) manifold for gravity waves; they characterize the Vertically Sheared Horizontal Flow (VSHF hereinafter, after [46]), which has nothing to do with a two-dimensional state (it is two-component instead).

Going back to pure linear inviscid, or RDT, evolution, all can be predicted from the conservation of both total energy and slow amplitude of the steady mode, but an apparently nontrivial evolution derives from angular integration of the wavy modes. The subsequent 'phase-mixing' (after [28]) is not very important looking at single-time second-order velocity correlations. For pure rotation, no evolution is found if the initial data are isotropic, and RDT yields nontrivial results only for initially anisotropic turbulence (see e.g. [9]). In the presence of both stratification and rotation, and for isotropic initial data, no evolution is found for initial equipartition of 'wave' (half the kinetic energy in pure stratification, depends on $N/(2\Omega)$ in the general case) and potential energy, and the RDT solution asymptotically reaches this equipartition, if not stated initially. Accordingly, the asymptotic anisotropy which results from RDT for single-time second order correlations is a byproduct of linear phase-mixing and is easily predicted from the initial distribution in terms of QG, wave and potential energy. This anisotropy has nothing to do with the emergence of pancake or cigar structures. More interesting, however, is the anisotropy resulting from RDT for two-time correlations, which can give relevant information on turbulent diffusion, as shown by [28], even if disconnected again from cigar-pancake dynamics.

Note that pure rotation is somewhat singular, in the sense that the steady (QG) mode disappears for $N = 0$, so that the modes of inertial waves form a complete basis for expanding the velocity field. In addition, it is no longer possible to account for a potential energy, which is directly connected to the density variance in the presence of stratification. Even if the general case with stratification is more complex, the presence of QG and potential energy allow to present simple explanations based on energetics and/or phenomenology. In our survey, with emphasis on dynamical aspects, the case of pure rotation is then considered as pivotal, and linear-nonlinear approaches are presented in detail only for $N = 0$. Accordingly, This paper is organised as follows. Section 2 is devoted to linear theory (not very new), with some recent results for turbulence diffusion in a rotating flow. Linear and nonlinear analysis are presented in section 3 for pure rotation. Stratified turbulence with and without rotation is discussed in section 4. Concluding comments are given in section 5, and details for algebra (poloidal-toroidal decomposition, related use of Craya-Herring frame of reference, application to anisotropic, or semi-axisymmetric, second order spectral tensor) are gathered in Appendix.

2 Linear theory

2.1 Background for Rapid Distortion Theory and Eigenmodes

Navier-stokes equations, with buoyancy force b within the Bousinesq assumption, are given below in the presence of a uniform mean density gradient and in a rotating frame. [1] For the sake of simplicity, the mean flow is restricted to a

[1] The problem of rotating turbulence can be directly related to the context of turbulence in the presence of a mean flow with space-uniform gradients A_{ij}, provided a pure antisymmetric form be chosen, or $A_{ij} = \epsilon_{ikj}\Omega_k$, with $\boldsymbol{\Omega}$ the angular velocity, but it is simpler to work with coordinates system and velocity vectors seen in the rotating frame. In this *non-Galilean* frame, rotation of the frame only introduces inertial forces, centrifugal and Coriolis. Since the former can be incorporated in the pressure term, only the latter has to be taken into account.

uniform vertical gradient of density and to a solid body rotation in the horizontal direction, with typical parameters N (the Brunt-Väisälä frequency) and Ω (the angular velocity).

$$(\partial_t + \boldsymbol{u}\cdot\boldsymbol{\nabla})\,\boldsymbol{u} + 2\Omega\boldsymbol{n}\times\boldsymbol{u} + \boldsymbol{\nabla}p - \nu\nabla^2\boldsymbol{u} = \boldsymbol{b} \tag{1}$$

$$(\partial_t + \boldsymbol{u}\cdot\boldsymbol{\nabla})b - \kappa\nabla^2 b = -N^2\boldsymbol{n}\cdot\boldsymbol{u} \tag{2}$$

$$\boldsymbol{\nabla}\cdot\boldsymbol{u} = 0 \tag{3}$$

Basic variables are fluctuating velocity $\boldsymbol{u}$ and intensity of the buoyancy force b; p denotes the pressure fluctuation divided by a mean reference density, together with a contribution from centrifugal acceleration.

The vector $\boldsymbol{n}$ denotes the vertical unit upward direction aligned with both the gravitational acceleration $\boldsymbol{g} = -g\boldsymbol{n}$ and the angular velocity of the rotating frame $\boldsymbol{\Omega} = \Omega\boldsymbol{n}$. The buoyancy force is related to the fluctuating temperature field τ by $\boldsymbol{b} = -\boldsymbol{g}\beta\tau = b\boldsymbol{n}$, through the coefficient of thermal expansivity β. With temperature stratification characterized by the vertical gradient γ, the Brunt-Väisälä frequency $N = \sqrt{\beta g\gamma}$ appears as the characteristic frequency of buoyancy-stratification. Hence the linear operators in equations (1) and (2) display the two frequencies N and 2Ω. Without loss of generality the fixed frame of reference is chosen such that $n_i = \delta_{i3}$.

Analysis of the linear limit, mathematical treatment of the full nonlinear equations above in terms of eigenmodes, and closure methods for statistics in homogeneous anisotropic turbulence, were extensively carried out in the particular case of pure rotation ($N = 0$), with more details in section 3.

By discarding nonlinear advection terms for $\boldsymbol{u}$ and b, solutions of linearized equations are found as a combination of steady and wavy modes.

If the pressure-related term is ignored in (1), the Coriolis force only affect the horizontal motion through 2Ω-periodic motion, whereas buoyancy and stratification terms yield oscillations for vertical velocity and buoyancy term, with frequency N. These simple motions reflects that the buoyancy force acts as a restoring force in the case of stable stratification. The pressure term in (1), however, is needed to satisfy (3), and its role in the complete linear solution consists of coupling vertical and horizontal motion and of generating dispersive inertia-gravity waves. In the case of pure rotation, all the motion in the linear limit consists of superimposed inertial waves. Only in the general case with stable statification, a part of the horizontal motion remains steady in the linear limit, and therefore decoupled from 3D wave motion: this is related to the quasi-geostrophic mode, which is pivotal in dynamical meteorology.

In the unbounded case, or for periodic boundary conditions, the different modes, wavy and steady, are easily found in Fourier space, and a tractable linear solution is found in terms of them. For instance

$$\boldsymbol{u}(\boldsymbol{x},t) = \int \hat{\boldsymbol{u}}(\boldsymbol{k},t)\exp[\boldsymbol{k}\cdot\boldsymbol{x}]d^3\boldsymbol{k} \tag{4}$$

for an infinite unbounded domain, and a discrete summation replaces the volumic integral if the problem is seen in a finite box with periodic boundary conditions.

Solution of the general initial value problem for $(\hat{\boldsymbol{u}}(\boldsymbol{k},t),\ \hat{b}(\boldsymbol{k},t))$ is the starting point for both 'Rapid Distortion Theory' (RDT hereinafter) or stability analysis. In the absence of mean shear, there is no distortion, so that RDT is perhaps misunderstanding, but it remains a useful reference in the turbulence community. Linear solutions can be expressed for $\hat{u}_1, \hat{u}_2, \hat{u}_3, \hat{b}$ in the fixed frame of reference without introducing the basis of eigenmodes ([22], [23]), but it is simpler to develop these solutions and to interpret them in term of eigenmodes, which have some physical meaning (QG and inertia-gravity modes of motion) and yield more tractable algebra. In addition to classical RDT solutions for second order single-time statistics, eigenmode decomposition is invaluable for exploiting two-time RDT solutions (see [28] and subsection 2.2), and especially to address nonlinear analysis and computation (see [17], [2], [7], [46]).

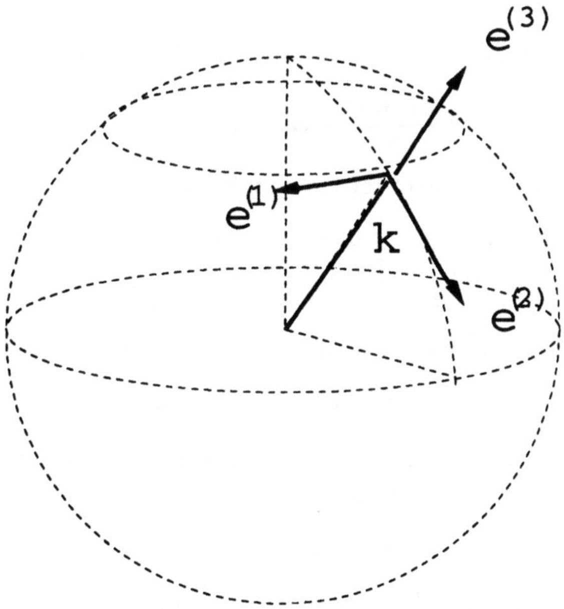

Fig. 1. Polar-spherical system of coordinates for $\boldsymbol{k}$ and related 'Craya-Herring' frame of reference.

As a simple prerequisite to use the eigenmodes decomposition, the pressure fluctuation term is removed from consideration in the Fourier-transformed equations by using the local frame in the plane normal to the wave vector (often denoted Craya-Herring, after Craya (1958) and Herring (1974), see Figure 1 and Appendix)

$$\hat{\boldsymbol{u}}(\boldsymbol{k},t) = u^{(1)}\boldsymbol{e}^{(1)} + u^{(2)}\boldsymbol{e}^{(2)} \tag{5}$$

taking advantage of (3), which reduces to $\boldsymbol{k}\cdot\hat{\boldsymbol{u}}$ in Fourier space. In so doing, the problem in five components (u_1, u_2, u_3, p, b) in physical space is reduced to a problem in three components, two solenoidal velocity components ($u^{(1)}$, $u^{(2)}$) and a component for $\hat{b}$, in Fourier space. The three-component set ($u^{(1)}$, $u^{(2)}$, $\hat{b}$) is not a true vector, and this can complicate further mathematical developments in terms of its eigenmodes and statistical correlations. So it is more convenient to gather these three components into a new vector $\hat{v}$, whose inverse 3D Fourier transform, v, is real. $\hat{v}$, can be written as

$$\hat{v} = \hat{u} + \imath \frac{1}{N}\hat{b}\frac{k}{k} \tag{6}$$

so that its three components are $u^{(1)}$, $u^{(2)}$, and $\imath\frac{1}{N}\hat{b}$ in the Craya-Herring frame of reference (Appendix, (44)). The scaling of the contribution from the buoyancy force allows one to define twice the total energy spectral density as

$$\widehat{v}_i^*\widehat{v}_i = \widehat{u}_i^*\widehat{u}_i + N^{-2}\widehat{b}^*\widehat{b} \tag{7}$$

The linear equation for $\hat{\mathbf{v}}$, derived from (1)- (3), using (4) is of the form

$$\frac{\partial \hat{v}(k,t)}{\partial t} + \mathbf{M}(k)\hat{v}(k,t) = 0$$

In the Craya-Herring frame of reference, this writes

$$\partial_t \begin{pmatrix} \hat{v}_1 \\ \hat{v}_2 \\ \imath\hat{v}_3 \end{pmatrix} + \begin{pmatrix} 0 & -\sigma_r & 0 \\ \sigma_r & 0 & -\sigma_s \\ 0 & \sigma_s & 0 \end{pmatrix} \begin{pmatrix} \hat{v}_1 \\ \hat{v}_2 \\ \imath\hat{v}_3 \end{pmatrix} = \begin{pmatrix} 0 \\ 0 \\ 0 \end{pmatrix} \tag{8}$$

with a simple antisymmetric real matrix, which displays the two basic frequencies

$$\sigma_r = 2\Omega\cos\theta_k, \quad \sigma_s = N\sin\theta_k, \tag{9}$$

where θ_k is the polar angle of k in Figure 1. Accordingly the RDT solution can be found in terms of the three eigenmodes of $\mathbf{M}$, as

$$\hat{v} = \xi^{(0)}N^{(0)} + \xi^{(1)}N^{(1)} + \xi^{(-1)}N^{(-1)} \tag{10}$$

in which the eigenmodes $N^{(s)}, s = 0, \pm 1$ are simple linear combinations of the vectors in (44) (their definitions are given in [7]). The general inviscid linear solution is given by

$$\hat{v}_i(k,t) = G_{ij}(k,t,t_0)\hat{v}_j(k,t_0) \tag{11}$$

in terms of any initial data, in which the Green's function can be expressed in terms of eigenmodes and eigenvalues:

$$G_{ij}(k,t,t_0) = \sum_{s=0,\pm1} N_i^s(k)N_j^{-s}(k)\exp[\imath s\sigma_k(t-t_0)], \tag{12}$$

The eigenvalues are 0, σ_k and $-\sigma_k$, with $\sigma_k^2 = \sigma_r^2 + \sigma_s^2$, or

$$\sigma_k = \sqrt{N^2\left(\frac{k_\perp}{k}\right)^2 + 4\Omega^2\left(\frac{k_\parallel}{k}\right)^2} \tag{13}$$

for the unsigned value of the frequency given by the dispersion law of inertia-gravity internal waves. Because of the form of the eigenvectors and of the dispersion law, the structure of **G** in (12) is consistent with axisymmetry around the axis of reference (chosen vertical here), without mirror symmetry, and $k_{\parallel}$ and $k_{\perp}$ hold for axial (along the axis) and transverse (normal to the axis) components of $\boldsymbol{k}$. As a consequence, the inviscid RDT solution for the velocity only is simply expressed as

$$\hat{\boldsymbol{u}}(\boldsymbol{k},t) = \mathbf{A}^{(0)}(\boldsymbol{k}) + \mathbf{A}^{(1)}(\boldsymbol{k})\exp(\imath\sigma_k t) + \mathbf{A}^{(-1)}(\boldsymbol{k})\exp(-\imath\sigma_k t) \qquad (14)$$

in which the $(\mathbf{A}^{(s)}, s = 0, \pm 1)$ have only two components in the Craya-Herring frame, in agreement with (5), and involve the projection of the initial velocity field onto the three eigenmodes.

Relevance of homogeneous RDT Isotropy can be significantly broken through axisymmetrical response function for triple correlations only, or possibly for two-time second order statistics (see subsection 2.2), but the linear limit exhibits no interesting creation of structural anisotropy in classic RDT for predicting second order single-point statistics. However in practice there is two-dimensionalisation in rotating turbulence and a horizontal layering tendency in the stably stratified case. In other words, RDT only alters phase dynamics, and conserves exactly the spectral density of typical modes (full kinetic energy for the rotating case, total energy and 'vortex', or potential vorticity, energy for the stably stratified case), so that two-dimensionalization or 'two-componentalization' (horizontal layering), which affect the distribution of this energy, are typically nonlinear effects. Related nonlinear structuring in terms of cigar-pancake is illustrated by Figures 2 and 3 from DNS.

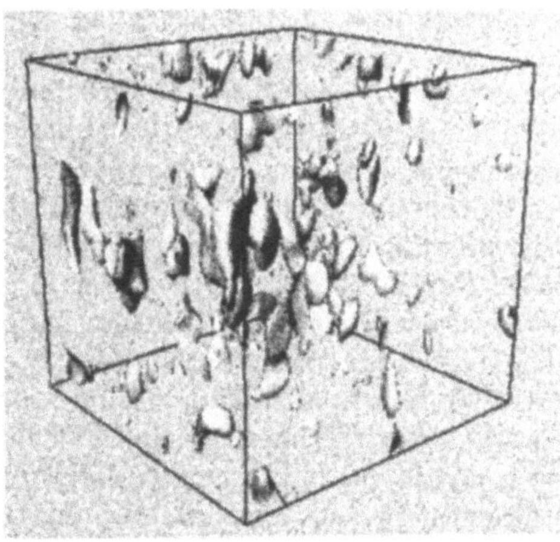

Fig. 2. Isovalues of vorticity for dominant rotation. Courtesy of Kimura and Herring (private communication)

The situation is completely different in the presence of a mean shear. In this case, RDT can predict qualitative 'structural' trends, and even quantitative ones for single and two-point statistical correlations, which are often dimensionless and characterize anisotropy. Most usual quantities are Reynolds Stress components $\overline{u_i'u_j'}$, with nondimensional deviatoric tensor b_{ij}, and integral lengthscales $L_{ij}^{(n)}$ for different velocity components (subscripts i and j) and different directions of two-point separation (superscript n) for them. The anisotropy reflected in the latter lengthscales can be very different from the Reynolds Stress anisotropy, and therefore cannot be derived from the knowledge of b_{ij}. Qualitative relevance of RDT solutions can appear even for particular realizations (snapshots) of the fluctuating velocity field, when compared to full DNS. This is illustrated by Figure 4, which is borrowed from [35], in the case of pure plane shear and plane channel flow near the wall. Accordingly, the tendency to create elongated 'streaklike' structures by a strong mean shear is inherent to the 'homogeneous RDT' operator, independently of the presence of a wall.

In contrast, the tendency to create columnar/pancake structures in a rotating stratified flow cannot be understood by means of RDT alone. For instance, the typical snapshots of isovorticity shown in Figures 2 and 3 cannot be found if one cancels the nonlinear terms in the DNS.

2.2 Linear dynamics for turbulent diffusion

The understanding of transport in anisotropic turbulence, dominated by body forces, is a topic of great importance for industrial and geophysical flow phenomena. Given the large scale disparity of the turbulent eddies and the filaments of tracer material, the single-particle diffusion in rotating stratified turbulence involves complex Lagrangian and Eulerian statistics.

Lagrangian stochastic models are often used to model turbulent diffusion, but it is not possible to incorporate in them the 'true' dynamics from vortex-dispersive waves motions, and all consequences of incompressibility, for a flow filling all space, are not satisfied. By contrast, an exact description of the linear dynamics, as reminded in the previous section, yields an interesting new dynamical insight. From previous analyses, it is clear that the anisotropic diffusivity in rotating stratified turbulence involves two different effects. The first one is a direct effect of the body forces, and can be tackled by only considering the linear dynamics (or so-called Rapid Distortion Theory, RDT) of the velocity field. Non trivial results were obtained, either by connecting Lagrangian two-time correlations to their RDT Eulerian counterparts ([28]; [27]), or by computing Lagrangian statistics in a KS (Kinematic Simulation) in which the unsteadiness of the velocity field incorporates RDT linear dynamics ([43]).

An additional effect is due to the increasingly anisotropic structuring of the velocity field itself, with the emergence of pancake or cigar structures. The latter effect is essentially nonlinear in quasi-homogeneous rotating stratified flows, and thus can be only accounted for using full DNS or nonlinear two-point closure theories (as rediscussed in sections 3 and 4).

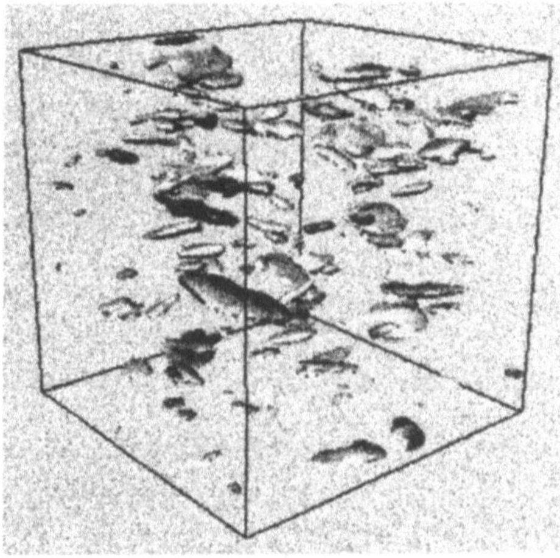

Fig. 3. Isovalues of vorticity for dominant stratification. Courtesy of Kimura and Herring (private communication)

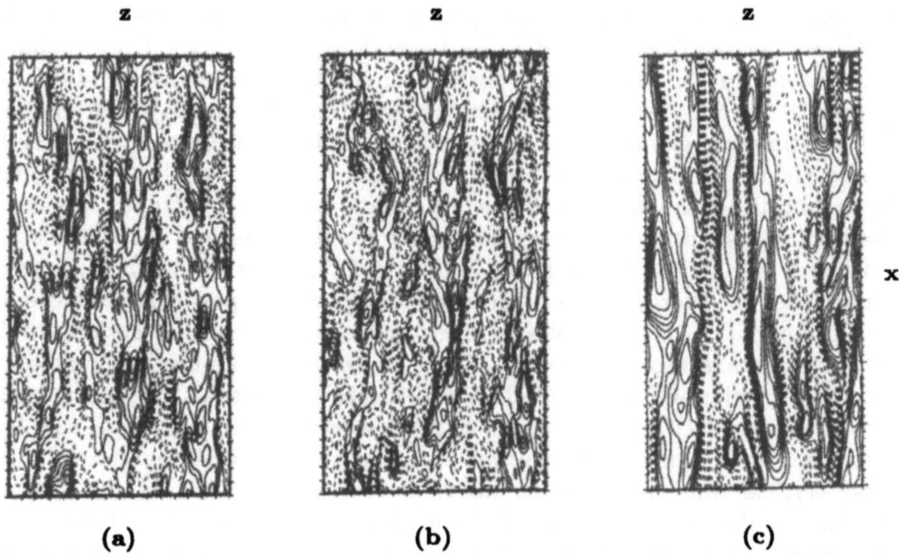

Fig. 4. Contours of streamwise fluctuating velocity from (a) direct numerical simulation (DNS), and (b) rapid distortion theory (RDT) calculations for uniformly sheared homogeneous turbulence, and (c) direct numerical simulation of plane channel flow near a wall (horizontal plane $y^+ \sim 10$. The streamwise elongation of turbulent structures resulting from shear appears clearly, as does the strong similarity between RDT and DNS results. From Lee, Kim and Moin (1990).

In this subsection, we revisit the models and theories which make use of RDT Eulerian dynamics, with emphasis on the case of rapid rotation without stratification which has been neglected in the previous studies and which is the one case where conservation of energy arguments are not conclusive with regard to turbulent diffusion. In [8], new DNS results are compared with those of RDT and KS, in order to help validating some assumptions, and to have a quantitative insight of the role of nonlinear dynamics.

The starting point is the equation related to (14)

$$u(x,t) = \sum \exp[\imath(k \cdot x + \omega t)] \\ \times \left[\mathbf{A}^{(0)}(k) + \mathbf{A}^{(1)}(k)\exp(\imath\sigma_k t) + \mathbf{A}^{(-1)}(k)\exp(-\imath\sigma_k t)\right] \quad (15)$$

for the velocity field, and similar equations for p and b. Either Fourier integral or discrete summation can be used, depending on the model used (analytical RDT, KS). Random frequencies, denoted ω in (15), were possibly introduced in KS realizations, in order to mimic nonlinear time decorrelation.

Results from RDT are firstly briefly recalled. In (15) with $\omega = 0$, the three contributions from the quasi-geostrophic mode $\mathbf{A}^{(0)}$, and from the wavy modes $\mathbf{A}^{(\pm 1)}$ are derived from initial data for $\mathbf{u}$ and b in Fourier space. Corresponding initial-value, linear, solutions, are derived for spectral (two-point) two-time Eulerian velocity and buoyancy correlations, and finally for single-point, one-time and two-time correlations. The last step involves integrating over the angles of $\mathbf{k}$, and yields damped oscillations for certain correlations, due to the angular-dependent dispersion law (13). If the initial data are chosen isotropic, this *linear phase-mixing* yields a moderate anisotropy when single-time correlations are considered (see detailed asymptotic results in [23], in terms of $2\Omega/N$, for zero initial potential energy). This anisotropy disappears for particular initial data with equipartition of potential and kinetic wave energy, and has nothing to do with the pancake/cigar structuring. More interesting is the result obtained for two-time correlations: when stratification is present, vertical Eulerian velocity correlations $< u_3(t')u_3(t) >$ tends to zero, whereas their horizontal counterpart tends to a nonzero plateau value, for large $t - t'$ time-separations. The latter vertical time-decorrelation due to the linear phase mixing was used by [28] for calculating an estimate of the vertical rms excursion length for a single particle, by equating RDT Eulerian and Lagrangian two-time correlations, and time-integrating the latter according to Taylor procedure [48]. The (strong) assumption which yielded replacing Lagrangian correlations by Eulerian ones is referred to as 'simplified Corrsin hypothesis' (SCH).

The KS procedure reported here is also based on dynamical equations of type (15), and the velocity field remains very simple (with respect to the 'true' one expected from a DNS run), but with incompressibility constraint strictly respected (because $\mathbf{k} \cdot \mathbf{A}^{(s)} = 0$) and a precise description of polar angular dependence. Linear (RDT) dynamics is accounted for because of σ_k-terms in (15), and related linear phase-mixing. The essential interest of KS is to give access to Lagrangian statistics by computing full trajectories ($x_i = \tilde{x}_i(\mathbf{X}, t)$

from various initial position, labelled by $\mathbf{X}$, from the simplified velocity field (15). The RDT cannot give direct access to fluctuating trajectories, since their equation is nonlinear, and the SCH can be relaxed in KS. Accordingly the KS can offer databases for checking the SCH, prior to the use of (more cumbersome and time-consuming) DNS. [2]

As soon as the stratification is present ($N \neq 0$), a plateau can be found for the square of the vertical rms excursion length, or $\Delta_v(t) =< \tilde{x}_3^2(t) >$, with similar scaling obtained from analytical (RDT+SCH) and KS procedures. More generally, the presence of the plateau, irrespectively of its refined scaling, can be expected from the link of Δ_v to a —necessarily bounded— potential energy. In this case, SCH is not questionned by KS or DNS. For the horizontal direction, the results are quite different: with $N \neq 0$, the analytical (RDT+SCH) procedure yields an eventual ballistic behaviour $\Delta_h \propto t^2$. This ballistic behaviour is simply explained because the steady part of the motion, connected with $\mathbf{A}^{(0)}$ in (15), or quasi-geostrophic part, contributes only to the horizontal part of the flow, and is eventually responsible for free ballistic dispersion. The vertical motion, in contrast, is affected by linear phase mixing through $\mathbf{A}^{(\pm 1)}$ terms. Without additional decorrelation terms using random frequencies ω in (15), KS results give also the eventual ballistic behaviour, whereas a more realistic (as in DNS) Brownian behaviour is recovered thanks to these random frequencies. Almost all the analytical results of RDT+SCH can be found in [27] for rotating stratified flow, except the case of pure rotation, discussed below.

The most interesting case is rotation without stratification, since the steady velocity part $\mathbf{A}^{(0)}$ vanishes identically in (15) and Δ_v is no longer connected to a potential energy. Accordingly, all the velocity field is affected by inertial waves, but, surprisingly, the RDT calculation using SCH does not yield a plateau for $\Delta_{h,v}$, as for the stratified case, but an eventual (large-time) Brownian behaviour which follows the (short-time) ballistic one. This illustates the subtelties of the linear phase-mixing, and its strong sensitivity to the dispersion law. By time-integrating the RDT solutions for two-time velocity correlations, [8] found:

$$< \tilde{x}_3^2(s) >= \Delta_{33}(s,0) = \frac{q_0^2}{8\Omega^2}\left[4S(2\Omega s)\Omega s + 2\cos 2\Omega s - 4 + \frac{\sin 2\Omega s}{\Omega s}\right] \qquad (16)$$

and

$$\begin{aligned} < \tilde{x}_1^2(s) + \tilde{x}_2^2(s) > &= \Delta_{11}(s,0) + \Delta_{22}(s,0) \\ &= \frac{q_0^2}{8\Omega^2}\left[4S(2\Omega s)\Omega s + 2\cos 2\Omega s - \frac{\sin 2\Omega s}{\Omega s}\right] \end{aligned} \qquad (17)$$

[2] Kimura and Herring, for instance, have shown that the plateau for vertical excursion, obtained by DNS, is not dramatically altered by cancelling nonlinear terms. This procedure is close to KS incorporating RDT, but the result cannot be attributed to RDT only: DNS without nonlinearity ought to reproduce RDT results (up to discretisation errors) for velocity snapshots and Eulerian statistics, but fluctuating Lagrangian trajectories cannot be derived from RDT, since the Fourier mode $\exp(\imath k_i \tilde{x}_i)$ in (15) depends on an unknow $\tilde{x}_i$.

where q_0^2 is twice the (constant) kinetic energy, and $S(t) = \int_0^t u^{-1} \sin u \, du$. These equations describe an initial ballistic behaviour, followed by a Brownian one at larger Ωt, or

$$\Delta_h / L^2 = \frac{\pi}{4} Ro \frac{q_0 t}{L} \tag{18}$$

$$\Delta_v / L^2 = \frac{\pi}{2} Ro \frac{q_0 t}{L} \tag{19}$$

where $Ro = q_0/(2\Omega L)$ is the Rossby number based on the large lengthscale L. The latter two equations show that the vertical square displacement becomes twice the horizontal one at large time. This contradicts some simplified phenomenological arguments, in which the horizontal displacement should be blocked at a typical 'Rossby radius' scale.

3 Pure rotation: approach to nonlinear dynamics

Rotation of the reference frame is an important factor in certain mechanisms of flow instability, and the study of rotating flows is interesting from the point of view of turbulence modelling in fields as diverse as engineering (*e.g.* turbomachinery and reciprocating engines with swirl and tumble), geophysics and astrophysics. Effects of mean curvature or of advection by a large eddy can be tackled using similar approaches. The governing equation reduces to

$$\frac{\partial \boldsymbol{u}}{\partial t} + 2\boldsymbol{\Omega} \times \boldsymbol{u} - \nu \nabla^2 \boldsymbol{u} + \boldsymbol{\nabla} p = -\mathbf{u} \cdot \boldsymbol{\nabla} \boldsymbol{u} \tag{20}$$

From several experimental, theoretical and numerical studies, in which rotation is suddenly applied to homogeneous turbulence, some agreed statements are summarised as follows ([1]; [25]; [9]; [10]; [7]; [40]).

- Rotation inhibits the energy cascade, so that the dissipation rate is reduced.
- The initial three-dimensional (3D) isotropy is broken through nonlinear interactions modified by rotation, so that a moderate anisotropy, consistent with a transition from a 3D to a 2D state, can develop.
- Both previous effects involve non-linear or 'slow' dynamics, and the second is relevant only in an intermediate range of Rossby numbers as found by Jacquin et al. (1990). This intermediate range is delineated by $Ro_L = u_{\rm rms}/(2\Omega L) < 1$ and $Ro_\lambda = u_{\rm rms}/(2\Omega\lambda) > 1$, in which $u_{\rm rms}$ is an axial rms velocity fluctuation, whereas L and λ denotes a typical integral lengthscale (macroscale) and a typical Taylor microscale respectively.
- If the turbulence is initially anisotropic, the 'rapid' effects of rotation (linear dynamics tackled in a RDT fashion) conserve a part of the anisotropy (called directional, $b_{ij}^{(e)}$) and damp the other part (called polarization anisotropy $b_{ij}^{(z)}$), through phase-mixing, resulting in a spectacular change of the anisotropy b_{ij} of the Reynolds Stress tensor (see equations (48) and (49) in Appendix). Note that the asymptotic anisotropy obtained by RDT corresponds to an equipartition of toroidal and poloidal kinetic energy.

These effects, which are not at all taken into account by current one-point second order closure models (from $k-\varepsilon$ to $\overline{u'_i u'_j}-\varepsilon$ models), have motivated new modelling approaches by [9], [10], and to a lesser extent by [29] for linear (or 'rapid') effects only. It is worth noticing that the modification of the dynamics by the rotation ultimately comes from the presence of inertial waves ([21]), having an anisotropic dispersion law, which are capable of changing the initial anisotropy of the turbulent flow and also can affect the non-linear dynamics. Contrary to a well-known interpretation, the Proudman theorem shows only that the 'slow manifold' (the stationary modes unaffected by the inertial waves) is the two-dimensional manifold at small Rossby number, but cannot predict the transition from 3D to 2D turbulence, which is a non-linear mechanism of transfer from all possible modes towards the 2D ones. This can be discussed from the vorticity equation, which is derived from (20)

$$\frac{\partial \omega_i}{\partial t} - 2\Omega_l \frac{\partial u_i}{\partial x_l} = \omega_l \frac{\partial u_i}{\partial x_l} + \nu \nabla^2 \omega_i \tag{21}$$

Nonlinear and viscous terms are gathered on the right-hand-side. The linear inviscid limit is found by discarding the right-hand-side, assuming very low Rossby number and very high Ekman number. The two-dimensional limit $\partial u_i/\partial x_\parallel = 0$, however, is only found *if the additional assumption of slow motion is done.* In Fourier space, the slow — and 2D — manifold corresponds to the wave plane normal to the rotation axis, or $k_\parallel = 0$.

3.1 The linear wave-regime

Linearized inviscid equation writes

$$\frac{\partial \boldsymbol{u}}{\partial t} + 2\boldsymbol{\Omega} \times \boldsymbol{u} + \boldsymbol{\nabla} p = 0, \qquad \boldsymbol{\nabla}\cdot\boldsymbol{u} = 0 \tag{22}$$

Since the Coriolis force is not divergence-free, the pressure term has a nontrivial contribution to maintain the incompressibility constraint. The velocity can be eliminated between the latter equation and the Poisson equation for the pressure, for which a closed equation is found

$$\partial_t^2 \left(\nabla^2 p\right) + 4\Omega^2 \nabla_\parallel^2 p = 0 \tag{23}$$

Even if the primitive Poisson equation $\nabla^2 p = f$ is of the parabolical type, the equation (23)admit solutions in terms of propagating waves. Very surprising properties of these inertial waves are illustrated by the St Andrew-cross shaped structures from the experiment by [39] (see Figure 5 and 6). If a local harmonic forcing takes place, with frequency σ_0, in the tank rotating at angular velocity Ω, simplified solutions can be sought under normal modes, or $p = e^{\imath\sigma_0 t}\mathcal{P}$, so that the spatial part is governed by

$$[\sigma_0^2 \nabla_\perp^2 + (\sigma_0^2 - 4\Omega^2)\nabla_\parallel^2]\mathcal{P} = 0$$

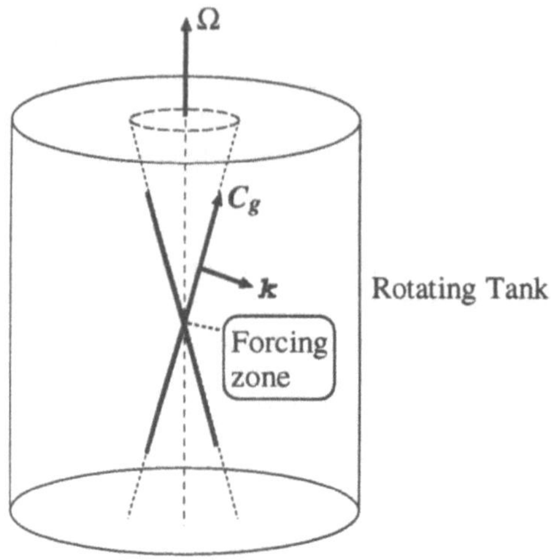

Fig. 5. Sketch of the experiment by Mc Ewan (1970).

which shows the possible transition from elliptic to hyperbolic nature, when σ_0 crosses the threshold 2Ω by decreasing values. Hence the sudden appearance of the cross-shaped structure for $\sigma_0 < 2\Omega$. In spite of the rather complex geometry, one can assume, in addition, that the disturbances are plane waves, or $p \sim e^{\imath(\boldsymbol{k}\cdot\boldsymbol{x}-\sigma t)}$. Injecting in (23), the classical dispersion law of inertial waves is recovered as

$$\sigma_k = \pm 2\Omega\frac{k_\parallel}{k} = \pm 2\Omega\cos\theta. \tag{24}$$

If one interprets the rays emanating from the small forcing zone in the figure as trace of isophase surfaces, so that the wave vector is normal to them, equation (24) with $\sigma_k = \sigma_0$ gives the angle θ (angle of $\boldsymbol{k}$ with respect to the vertical axis) in excellent agreement with the directions of the rays. Note that, without pressure, only the horizontal part of the flow is affected by circular periodic (constant frequency 2Ω) motion, but propagating waves cannot occur. Hence, fluctuating pressure (through its linkage to divergence-free condition) is responsible both for anisotropic dispersivity and for horizontal-vertical coupling.

Going back to velocity, an equation similar to (23) can be found for both poloidal and toroidal potentials in (42) (Appendix). Without forcing and boundary conditions, the related 'RDT' problem writes

$$\frac{\partial \hat{u}_i}{\partial t} + 2\Omega P_{in}\epsilon_{n3j}\hat{u}_j = 0 \tag{25}$$

in which $P_{in} = \delta_{in} - \frac{k_i k_n}{k^2}$ is the projection operator.

Given the incompressibility constraint $\hat{\boldsymbol{u}}\cdot\boldsymbol{k} = 0$, it is easier to project the equation onto the local frame $(\boldsymbol{e}^{(1)}, \boldsymbol{e}^{(2)})$ normal to $\boldsymbol{k}$ in Appendix (44), as in (5). The solution expresses that the initial Fourier component $\hat{\boldsymbol{u}}(\boldsymbol{k}, 0)$ is rotated about the axis $\boldsymbol{k}$ of an angle $2\Omega t k_3/k$ (also equal to $\sigma_k t$). A tractable diagonal form

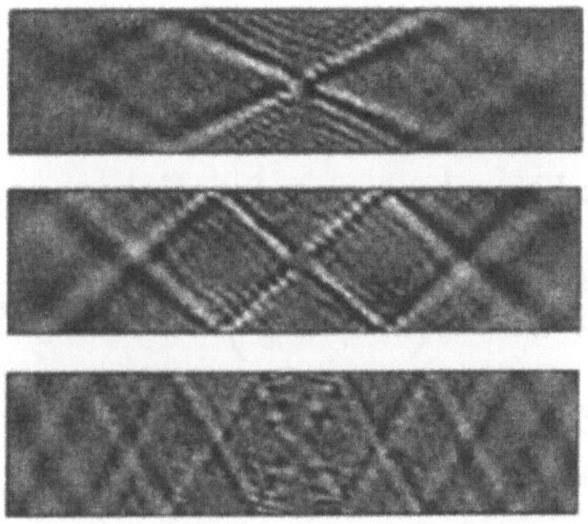

Fig. 6. Saint Andrew's Cross shaped structures in a rotating flow, from DNS in a plane channel, rotating around the vertical axis. From top to bottom $2\Omega/\sigma_0 = 1.10, 1.33, 2.$. See Godeferd and Lollini (1998)

of the RDT Green's function is found in terms of the two complex eigenvectors $\boldsymbol{N} = \boldsymbol{e}^{(2)} - \imath \boldsymbol{e}^{(1)}$ and $\boldsymbol{N}^* = \boldsymbol{N}(-\boldsymbol{k}) = \boldsymbol{e}^{(2)} + \imath \boldsymbol{e}^{(1)}$ in the plane normal to $\boldsymbol{k}$, or

$$G_{ij}^{RDT}(\boldsymbol{k}, t, t') = \sum_{s=\pm 1} N_i(s\boldsymbol{k}) N_j(-s\boldsymbol{k}) e^{\imath s \sigma_k (t-t')} \tag{26}$$

Diagonal decomposition is particularly useful in the context of pure rotation, since $\boldsymbol{N}$ and $\boldsymbol{N}^*$ more generally generate the eigenmodes of the Curl operator, and directly appear in the (e, Z, h) decomposition in (47) (Appendix). Accordingly, the RDT solution for the second-order spectral tensor writes

$$e(\boldsymbol{k}, t) = e(\boldsymbol{k}, t_0), \quad h(\boldsymbol{k}, t) = h(\boldsymbol{k}, t_0), \quad Z(\boldsymbol{k}, t) = e^{4\imath \sigma_k (t-t_0)} Z(\boldsymbol{k}, 0) \tag{27}$$

3.2 Linear-nonlinear or rapid-slow ?

Simplified equations projected on the basis of eigenmodes $\boldsymbol{N}$ and $\boldsymbol{N}^*$ can be used for discussing both linear and nonlinear operators, as well as to develop closure theories for rotating turbulence. In terms of the amplitudes $\xi_s, s = \pm 1$, which are defined by

$$\hat{\boldsymbol{u}}(\boldsymbol{k}, t) = \xi_+(\boldsymbol{k}, t) \boldsymbol{N}(\boldsymbol{k}) + \xi_-(\boldsymbol{k}, t) \boldsymbol{N}(-\boldsymbol{k}) \tag{28}$$

the complete equation writes

$$\begin{aligned} &[\frac{\partial}{\partial t} + \nu k^2 - \imath s \underbrace{\left(2\Omega \frac{k_\parallel}{k}\right)}_{\sigma_k}] \xi_s \\ &= \sum_{s', s'' = \pm 1} \int_{k+p+q=0} m_{ss's''}(\mathbf{k}, \mathbf{p}) \xi^*_{s'}(\mathbf{p}, t) \xi^*_{s''}(\mathbf{q}, t) d^3\mathbf{p} \end{aligned} \tag{29}$$

with a diagonal form for the linear operator, and a modified quadratic form for the nonlinear term. The linear inviscid solution writes

$$\xi_s(\mathbf{k},t) = \exp\left(2\imath s\Omega t\frac{k_\parallel}{k}\right)\xi_s(\mathbf{k},0), \quad s = \pm 1$$

Replacing the initial condition by a new function $\boldsymbol{a}_s$, or

$$\xi_s(\mathbf{k},t) = \exp\left(2\imath s\Omega t\frac{k_\parallel}{k}\right)a_s(\mathbf{k},t), \quad s = \pm 1 \tag{30}$$

An equation for $\boldsymbol{a}_s$ is derived, in which linear operators are absorbed in the nonlinear one, as integrating factors.

$$\dot{a}_s = \sum_{s',s''=\pm 1}\int_{k+p+q=0}\exp\left[2\imath\Omega\left(s\frac{k_\parallel}{k}+s'\frac{p_\parallel}{p}+s''\frac{q_\parallel}{q}\right)t\right]$$

$$m_{ss's''}(\mathbf{k},\mathbf{p})a^*_{s'}(\mathbf{p},t)a^*_{s''}(\mathbf{q},t)d^3\mathbf{p} \tag{31}$$

Note that the $\boldsymbol{a}_s$ play the role of slow amplitudes, and the problem could be analyzed by a technique of multiple (two) time-scale, setting $\boldsymbol{a}_s = \boldsymbol{a}_s(\boldsymbol{k},\epsilon t)$, with ϵ a small parameter for asymptotic expansion, that can be directly connected to a Rossby number. Such a refined analysis is not needed here, and we will just retain from equation (31) the importance of resonant triads, $\sigma_k \pm \sigma_p \pm \sigma_q = 0$, which correspond to zero value of the phase term in the right-hand-side of (31), or

$$\frac{k_\parallel}{k} \pm \frac{p_\parallel}{p} \pm \frac{q_\parallel}{q} = 0 \quad \text{with} \quad \boldsymbol{k}+\mathbf{p}+\mathbf{q}=0. \tag{32}$$

These resonant, or quasi-resonant, triads are found to dominate nonlinear slow motion, since the effect of the phase term in the left-hand-side of (31), if far from zero, is to severely damp the nonlinearity by a ‘scrambling effect’. The complexity of the resonant surfaces is illustrated by Figure 7.

Going back to second order statistics, equation for the second order spectral tensor $\hat{R}_{ij}$, which is directly connected to $< \hat{\boldsymbol{u}}^*_i\hat{\boldsymbol{u}}_j >$ and involves four real scalars (Appendix, (47)), reduces to the simple system

$$\left(\frac{\partial}{\partial t} + 2\nu k^2\right)e = T^{(e)} \tag{33}$$

$$\left(\frac{\partial}{\partial t} + 2\nu k^2 + 2\imath\sigma_k\right)Z = T^{(Z)} \tag{34}$$

$$\left(\frac{\partial}{\partial t} + 2\nu k^2\right)\mathcal{H} = T^{(h)} \tag{35}$$

in terms of the set energy-polarization-helicity $(e, Z, \mathcal{H})$. Contribution from triple velocity correlations are gathered into the generalized transfer terms $T^{(e,Z,h)}$, which derive from equations for any anisotropic two-point closure. If the above system of equation is started with 3D isotropic initial data, or $e(k,0) = E(k)/(4\pi k^2)$,

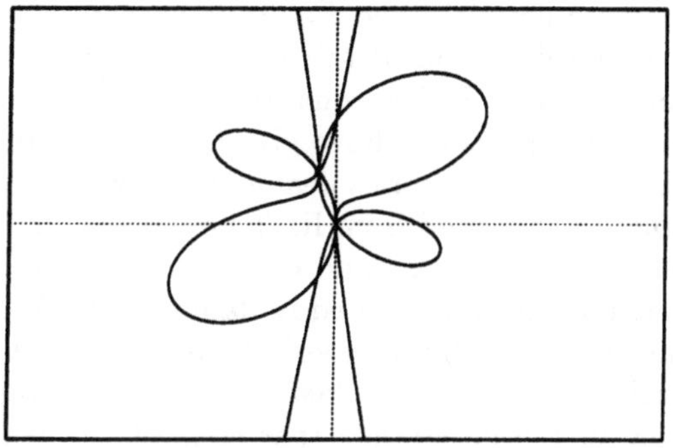

Fig. 7. Visualization of resonant surfaces of inertial waves, given by (32), for a given orientation of $\boldsymbol{k}$. The locus of $\boldsymbol{p}$ is seen in the plane (p_1, p_3) at $p_2 = 0$, for $\theta_k = 1.1$. Complex loops appear for $\pi/3 < \theta_k < \pi/2$. Courtesy of F.S. Godeferd.

$Z = \mathcal{H} = 0$, the anisotropy which should reflect the transition towards 2D structure, can be created by the nonlinear transfer terms only. This anisotropy consists of axisymmetry without mirror symmetry, leading to $e = e(k, \cos\theta = \frac{k_\parallel}{k}, t)$, $Z = Z(k, \cos\theta = \frac{k_\parallel}{k}, t)$, with $Z = 0$ if $\boldsymbol{k}$ is parallel to the vertical axis, in agreement with the symmetries of basic (rotating Navier-Stokes) equations , which ought to be satisfied by the closure theory.

It is possible to derive an optimal, multimodal and anisotropic, version of the EDQNM theory (see [44] for background and isotropic case) if the correlations tensors are directly built in terms of the 'slow' amplitudes $a_s(\boldsymbol{k}, t)$ in (30), with corresponding equations for double and triple correlations. More classically, in the general case of homogeneous anisotropic turbulence in the presence of uniform mean velocity gradients, a generic form for the transfer tensor T_{ij}, which gather the contributions from triple velocity correlations in the equation governing the second order spectral tensor $\hat{R}_{ij}$, is derived as follows:

$$T_{ij}(\boldsymbol{k}, t) = \tau_{ij}(\boldsymbol{k}, t) + \tau^*_{ji}(\boldsymbol{k}, t) \tag{36}$$

where

$$\tau_{ij}(\boldsymbol{k}, t) = P_{jkl}(\boldsymbol{k}) \int_{-\infty}^{t} \int_{\mathbf{k}+\mathbf{p}+\mathbf{q}=\mathbf{0}} G_{im}(\boldsymbol{k}, t, t') G_{kp}(\mathbf{p}, t, t') G_{lq}(\mathbf{q}, t, t')$$
$$\times \hat{R}_{qn}(\mathbf{q}, t')[\frac{1}{2} P_{mnr}(\boldsymbol{k}) \hat{R}_{pr}(\mathbf{p}, t') + P_{pnr}(\mathbf{p}) \hat{R}_{mr}(\boldsymbol{k}', t')]\, d^3\mathbf{p}\, dt' \tag{37}$$

in which the triple product of Green's functions arises from the Green's function solution for the third-order moments and the notation $\mathbf{k}+\mathbf{p}+\mathbf{q}=\mathbf{0}$ on the integral sign means that $\mathbf{q}$ should be replaced by $-\boldsymbol{k}-\mathbf{p}$ throughout the integrand, representing interacting triads of wavenumbers $\boldsymbol{k}$, $\mathbf{p}$, $\mathbf{q}$ which form triangles (see [11] for more details). The P_{jkl}-type operators combine projections onto and normal to the direction of the related wavevector, as in the primitive nonlinear term of Navier-Stokes equations. Only quasi-normal assumption (QN) was used, but the equations above give the generic anisotropic structure of most of generalised classical theories dealing with two-point closure, provided the 'bare' Green function inherited from RDT be replaced by a modified version, for instance including viscous terms and eddy damping as in EDQNM theory. As we will see later, the choice of the eddy damping (ED) is not crucial if the rotation is dominant (low Rossby number). More important is a correct introduction of the 'Markovianisation' (M, last letter of the acronym EDQNM), or the time-dependency in the integrands that connect the transfer term to second order correlations. Closed EDQN equations for the spectral transfer display three kinds of time-dependent terms:

1. Viscous, or viscous + damping, terms
$$\exp\left(\int_{t'}^{t} \mu dt''\right) \rightarrow V(t,t')$$
(not displayed in (37), simple exponential factor multiplying $\mathbf{G}$).
2. Terms from the RDT Green's function
$$\mathbf{G}(t,t') \rightarrow \exp[\pm \imath\sigma(t-t')],$$
under diagonal form.
3. Terms from the second-order spectral tensor (though quasi-normal assumption)
$$\hat{\mathbf{R}}(t') \rightarrow (e,Z,h)(t'),$$
using (47) in Appendix.

According to the markovianization procedure in classical EDQNM, we can state that $V(t,t')$ is so rapidly decreasing in term of time-separation $\tau=t-t'$ that it is only concerned by the time integral in the closure equations (37), whereas the other terms take their instantaneous value, at $t'=t$, so that they are replaced by $\mathbf{G}(t,t)$ and $\hat{\mathbf{R}}(t)$, respectively. In other words, one considers only $V(t,t')$ as rapid, and the other terms as slow. This procedure, say EDQNM1, is not convenient for rotating turbulence, since the presence of $\mathbf{G}(t-t')$ in the closure relationship is responsible for breaking the initial isotropy. Using EDQNM1 started with isotropic initial data, isotropy is maintained, and no effect of system rotation can appear. A second step, say EDQNM2, consists of only 'freezing' (setting $t'=t$ in them) the (e,Z,h) terms, whereas the complete 'readjusted' response function, with both $V(t,t')$ and $\mathbf{G}(t,t')$ terms, is conserved in the time-integrand with its

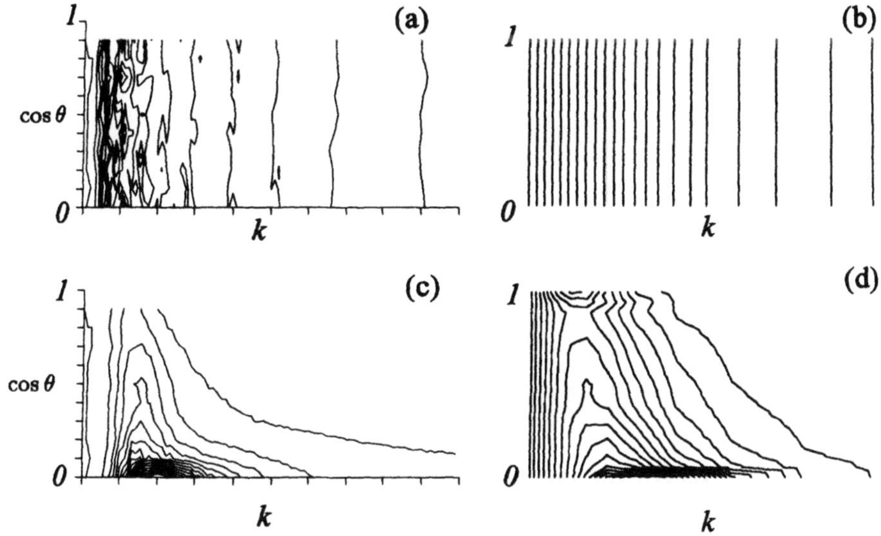

Fig. 8. Isolines of kinetic energy $e(k, \cos\theta, t)$ for 512 × 128 × 128 LES computations (a) at $\Omega = 0$ at time $t/\tau = 427$, (b) EDQNM2 with $\Omega = 0$; (c) LES with $\Omega = 1$ at $t/\tau = 575$; and (d) EDQNM2 calculation with $\Omega = 1$ at time $t/\tau = 148$. The vertical axis bears $\cos\theta_k$ (from 0 to 1 upwards) and the horizontal one the wave number k. (see Cambon *et al.*, 1997, section 5.)

detailed time-dependency. An interesting result is that the time integral of the three-fold product of response functions yields a generic closure relationship as

$$T^{(e,Z,h)} = \sum_{s=\pm 1, s'=\pm 1, s''=\pm 1} \int \frac{S^{ss's''}(e,Z,h)}{\mu_k + \mu_p + \mu_q + \imath(s\sigma_k + s'\sigma_p + s''\sigma_q)} d^3\mathbf{p} \quad (38)$$

in which μ_k is a simple scalar (viscous+ eddy damping)-term, and σ_k is the 'bare' dispersion frequency. (see [10], Appendix, for detailed closure relationship). Typical results from EDQNM2, regarding the creation of directional anisotropy, or description of the transition from 3D isotropy to 2D structure, are shown in Figure 8, and compared with high-resolution 528 × 128 × 128 LES. Recall that the development of angular dependency in $e(k, \cos\theta = \frac{k_\parallel}{k}, t)$, which amounts to a concentration towards the 2D slow manifold (sketched in Figure 11), results from nonlinear interactions mediated by $T^{(e)}$ in (33) and (38).

As pointed out by Julian Scott (private communication), the latter procedure can be questioned, in spite of its excellent numerical results, since it is not completely consistent with the basic rapid-slow decomposition suggested by (30). All the terms in the set (e, Z, h), which generates $\hat{\mathbf{R}}$, have not to be considered as 'slow', regarding RDT solution (27). Therefore, it is necessary to write

$$Z(t') = \exp(2\imath\sigma t') Z0(t'), \quad (39)$$

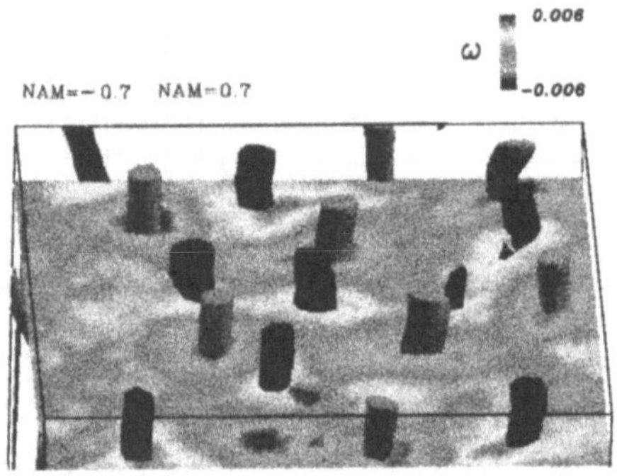

Fig. 9. NAM isovalues. See Godeferd *et al.* (1999).

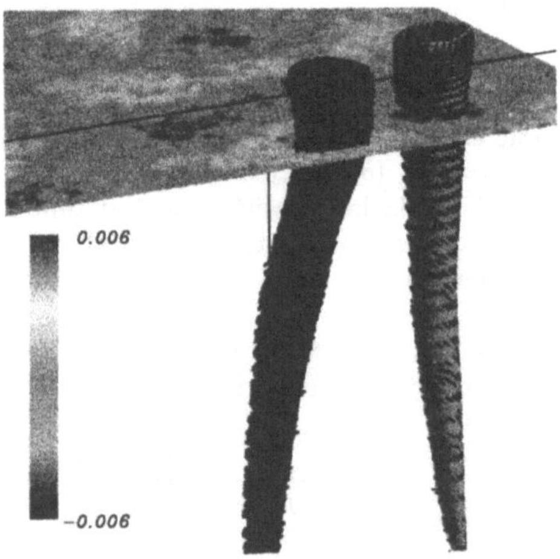

Fig. 10. Visualisation of helicoidal circulation structure. See Godeferd *et al.* (1999).

so that $Z0$ only appears as a slow variable, in complete agreement with (30). In so doing, this optimal procedure, say EDQNM3, yields freezing $e(t') = e(t)$, $h(t') = h(t)$, $Z0(t') = Z0(t)$ with keeping the (t') dependency under the integral for $Z(t') = \exp(2\imath\sigma t')Z0(t)$, and other terms $V(t,t')$, $\mathbf{G}(t,t')$, as before. This EDQNM3 version only slightly differs from EDQNM2, but presents decisive advantages. It is completely consistent with EDQNM in terms of slow amplitudes using (30). As a bonus, an asymptotic development can be derived in the limit $\mu \ll 2\Omega$, which exactly coincides with Eulerian wave-turbulence theory (see [6] in the similar case of stratified turbulence, and concluding remarks). Realizability can be demonstrated in this limit, and is not ensured in the EDQNM2 version.

3.3 Conclusion

Turbulence in a rotating frame illustrates the subtle interplay between linear and nonlinear processes and the significance of spectral anisotropy, especially the angular dependence of spectral energy which reflects the dimensionality. As evidenced by equations (33) to (35), the 3D isotropy, if initially stated, can only be broken by nonlinear interactions. In contast with isotropic free turbulence, Renormalized Perturbation Theories can be based on an actual small parameter, the Rossby number, for exploring the 'slow' nonlinear tendencies. EDQNM2-3 models illustrate the role of resonant triads of inertial waves, as should do an Eulerian 'wave-turbulence' theory. In the limit of small Rossby number, all these approches give essentially the same equations, but a sophisticated version of the eddy damping would be desirable to describe a broader range of Rossby and Reynolds numbers. RPT's, such as LRA [26], could be used for renormalising the eddy damping μ_k and the 'bare' frequency $\sigma_k = 2\Omega\frac{k_\parallel}{k}$ in (38). At the leading order, however, a modification of the frequency is expected in the quasi-2D case for interacting Rossby waves [24], but likely not in the 3D case considered here.

In spite of the evidence of the transition towards 2D structure in homogeneous decaying turbulence, the creation of large strong vortices from initially unstructured turbulence by the Coriolis force, is difficult to observe in numerical simulations (free of numerical bias !) in the absence of some forcing effects (see [20]; [7]; [46] and references therein). For instance, some structures of circulation are shown in figures 9 and 10, from DNS in a rotating domain bounded by two horizontal planes, in the presence of a forcing located around an intermediate horizontal plane. It is shown in both figures that isovorticity surfaces (noisy spots) are not very informative, whereas much more coherent structures are visualized by Normalised Angular Momentum. Two structures, cyclonic and anticyclonic, are isolated in figure 10, and typical helicoidal streamlines are drawn on them.

4 Stably stratified turbulence with and without rotation

4.1 Pure stratification

The case of stably stratified turbulence is different from the one of pure rotation, even if the gravity waves present strong analogies with inertial waves. An additional element is the presence of the 'vortex', or potential vorticity (PV) mode, which is a particular case of the quasi-geostrophic mode $\mathbf{N}^0$, which is related to $s = 0$ steady motion in (12). Hence the motion in the linear limit is not completely dominated by wavy modes, in contrast with the case of pure rotation. Regarding dispersive gravity waves, the general relationship (13) reduces to

$$\sigma_k = N\frac{k_\perp}{k} \tag{40}$$

and cross-shaped structures can be found in a stratified tank ([41]), from a local harmonic forcing, as discussed in the last section for pure rotation.

Focusing on nonlinear effects, interactions which do not involve the wavy mode, i.e. the part of the flow which is steady in the linear limit (PV mode + VSHF mode) have been found to be dominant in triggering the loss of isotropy, as a prerequisite to orient the evolution of the initially isotropic velocity field towards a two-component state. EDQNM2 ([17]) and DNS results have shown that the spectral energy concentrates towards vertical wavenumbers $k_\perp \sim 0$. Because $\boldsymbol{k}$ and $\boldsymbol{u}$ are perpendicular, these wavenumbers correspond to predominantly horizontal, low-frequency motions (called VSHF from [46]). As for the partial transition towards 2D structure shown in pure rotation, a new dynamical insight is given to the collapse of vertical motion expected in stably stratified turbulence, but the long-time behaviour essentially differs from a two-dimensionalisation.

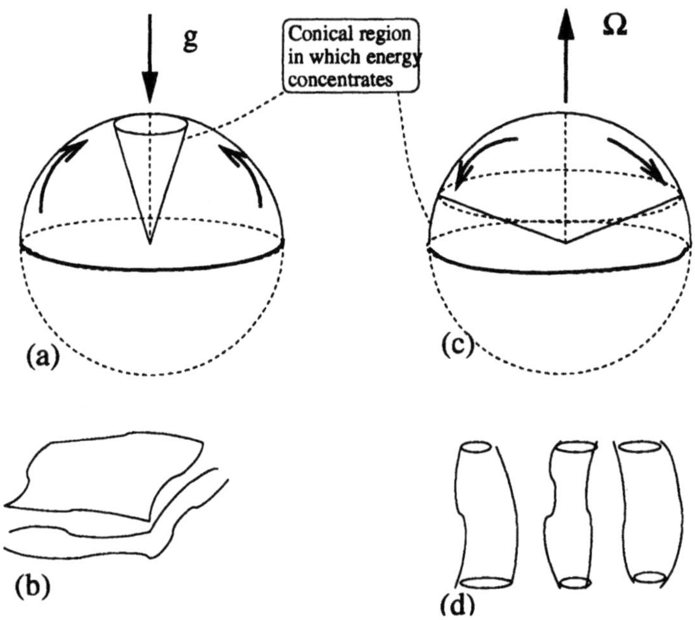

Fig. 11. Sketch of the angular energy drain in spectral space (top) and related structure in physical space (bottom). Pure stratification (left) and pure rotation (right).

A sketch of the different nonlinear effects of pure rotation and pure stratification is shown in Figure 11. Previous EDQNM studies ([24]) focused on triple correlation characteristic times modified by wave frequencies, whereas wave-turbulence theories proposed scaling laws for wave-part spectra. None of them, however, was capable of connecting wave-vortex dynamics to the vertical collapse and layering. Only recently, by re-introducing a small but significant vortex part in their wave turbulence analysis, [6] found that: *'The vortex part obeys a limiting slow dynamics equation exhibiting vertical collapse and layering which may contaminate the wave-part spectra'.* This is in complete agreement with the main

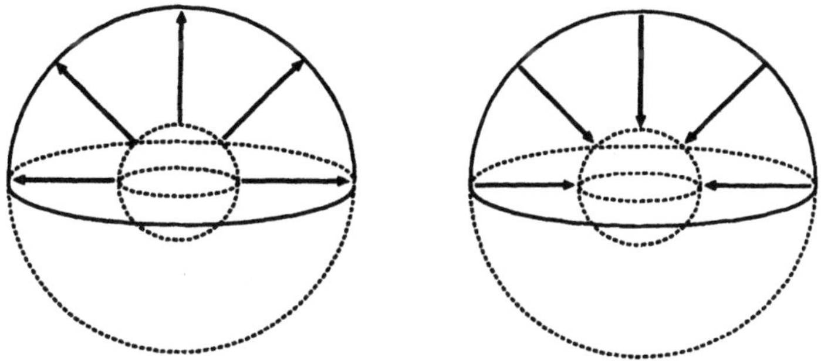

Fig. 12. Isotropic energy drain in spectral space. Classical direct cascade is sketched on the left.

finding of [17], where this result reflects a scrambling of any triadic interactions, including at least one wave mode, so that the pure PV (+VSHF) interaction becomes dominant. The corresponding 'wave-released' triadic energy transfer is strongly anisotropic. It does not yield a classic direct cascade, as sketched in Figure 12, (which would contribute to dissipate the energy) but instead yields the angular drain of energy which condenses the energy towards vertical wave-vectors, in agreement with vertical collapse and layering (Figure 11). The latter effect is reflected in physical space by a pancake structure, as illustrated in Figure 13, in which isovalues of velocity gradients are obtained from a snapshot of 256^3-DNS. This layering can be statistically quantified by the development of two different integral length scales, as shown in Figure 14 (from EDQNM2) and 15 (from DNS), with excellent agreement. The integral lengthscale related to horizontal velocity components and horizontal separation $L_{11}^{(1)}$ is shown to develop similarly to isotropic unstratified turbulence, whereas the one related to vertical separation $L_{11}^{(3)}$ is blocked. In the same conditions, with initial equipartition of potential and wave energy, linear calculation (RDT) exhibits no anisotropy, or $L_{11}^{(1)} = 2L_{11}^{(3)}$.

The mode related to vertical wave-vectors, which is linked to the VSHF mode in physical space, appears to be very important, since the concentration of spectral energy on it gives the most sensible identification of the development of vertical collapse and layering. It corresponds to the limit of the wavy mode, when the dispersion frequency tends to zero. Strictly speaking, this mode is a slow mode, which cannot be strictly referred to 'vortex' or 'wave'. It is absorbed in any decomposition based on the Craya-Herring frame (see equation (44) in Appendix and Figure 1), provided that some care is taken to extend by continuity the definition of the unit vectors $(\mathbf{e}^{(1)}, \mathbf{e}^{(2)})$ towards $\boldsymbol{k}$ aligned with the polar (vertical here) axis of the frame of reference. In so doing, the mode related to $\mathbf{e}^{(1)}$ coincides with a toroidal, or 'horizontal vortex', mode, but for vertical wave vectors, where it includes half the energy of the vertical slow (VSHF) mode. In the same way, the mode related to $\mathbf{e}^{(2)}$ coincides with a poloidal mode, affected

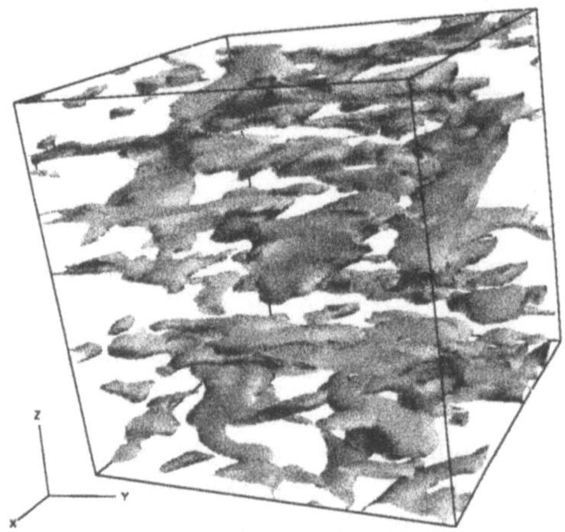

Fig. 13. Isovalues of vertical gradient of horizontal velocity fluctuation. Pure stratification. 256^3 DNS with isotropic initial data. Courtesy of F. S. Godeferd and C. Staquet

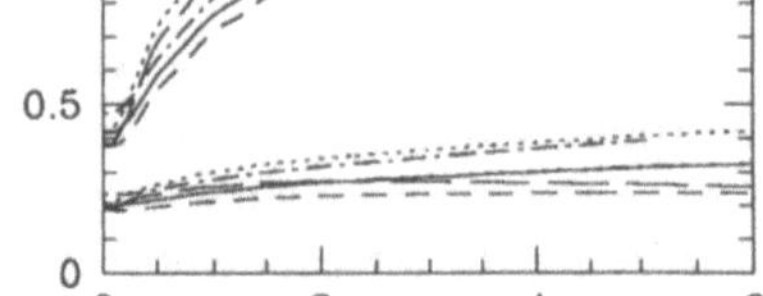

Fig. 14. Development of typical integral lengthscales from EDQNM2. $L_{11}^{(1)}$ (top) and $L_{11}^{(3)}$ (bottom), with 1 and 3 horizontal and vertical directions, respectively. (Initial) isotropy implies $L_{11}^{(1)} = 2L_{11}^{(3)}$. Courtesy of F. S. Godeferd and C. Staquet.

by the wavy motion, but for vertical wave vectors, where it includes the other half of the energy of the VSHF mode.

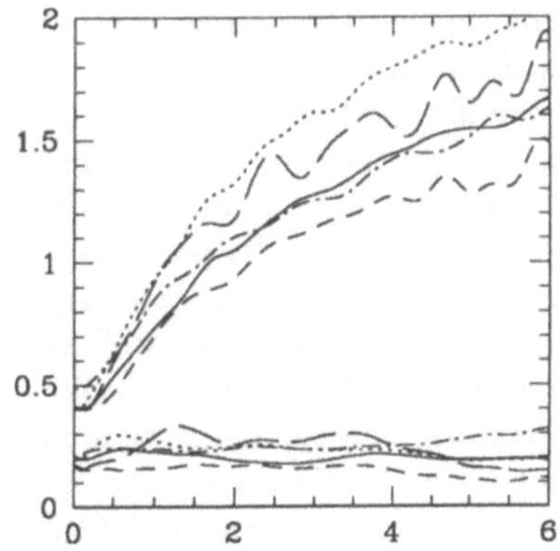

Fig. 15. The same as Figure 14, from 256^3 DNS.

4.2 Rotating stratified turbulence

The basis of eigenmodes, as discussed in section 2, is used in different nonlinear approaches, statistical theories ([7] and references therein) as well as full DNS (e.g. [46]). As seen before, dynamics from pure RDT is of poor interest in this case, at least if Eulerian single-time correlations are concerned, since strict conservation of the amplitudes $\boldsymbol{A}^{(s)}$ in (15) prevents interesting structuring to occur. These effects, two-dimensionalisation in pure rotation and two-componentalization in pure stratification, are thus ultimately controlled by nonlinear interactions, even weak. Typical 'cigar' and 'pancake' structure are shown in Figures 2 and 3, respectively, and refined calculations are in progress in different teams. As for the case of pure rotation, the $A^{(s)}$ in (15) have to be considered as *slowly* evolving amplitudes, in order to account for nonlinearity. For instance, the nonlinear models in section 3 may be based on $\boldsymbol{A}^{(s)} = \boldsymbol{A}^{(s)}(\boldsymbol{k}, \epsilon t), s = \pm 1$, with $\epsilon \sim (Ro)^2$. A particular problem, however, comes from the fact that the QG contribution in the general case, is not necessarily slow at vanishing Froude or Rossby number. Likely, the leading order (in terms of small Froude and Rossby numbers) for second order correlations of 'slow' amplitudes is not RDT (small time), or nonlinearity dominated by resonant wave interactions (larger time), but RDT for all second order correlations, except for the one displaying only the QG (+VSHF) contribution, in which the transfer mediated by triple QG-QG-QG interactions does not scale with Rossby and Froude numbers and is relevant at short time. Anyway, it is worthwhile to derive exact nonlinear equations for all the $\boldsymbol{A}^{(s)}(\boldsymbol{k}, t), s = 0, \pm 1$ and their statistical correlations, from the background Navier Stokes equations for u_i and b, using (14). In other words, even if RDT is not relevant in itself, it suggest to substitute to primitive velocity-temperature variables a set of projections onto a convenient basis of eigenmodes. Analysis of long-time effects (or short time for the QG part) of nonlinear interactions is facilitated in terms of these eigenmodes. In addition to DNS, closure models can be constructed for predicting detailed energy distribution in terms of different modes and different wavevectors, along the line of what was done in pure rotation and pure stratification.

In order to give more physical meaning to the mathematical procedures introduced here, consistent results from (anisotropic multimodal) EDQNM and DNS are summarized as follows. Important parameters are the system vorticity of the rotating frame 2Ω, or the Coriolis parameter f in a geophysical context, the Brunt-Väisälä frequency N which characterizes the gravity wave frequency and the strength of density-stratification, as well as nondimensional Rossby (or Froude) and Reynolds numbers. In all cases, nonlinear dynamics amounts to a concentration of energy towards a typical slow mode, and is dominated by resonant or quasi-resonant triads in 3D wavespace. But the relevant slow mode is not always the same, depending on the ratio $N/(2\Omega)$. The relevant slow mode is either the quasi-geostrophic one, corresponding to $\boldsymbol{A}^{(0)}$ in (14), or a wavy one, found in the limit of vanishing dispersion frequency (corresponding to $\boldsymbol{A}^{(\pm 1)}$ if σ_k can reach a zero value). Of course, wavy slow modes fill a very small part of the configuration space, so as do resonant wave interactions ($\sigma_k \pm \sigma_p \pm \sigma_q = 0$ for a triad of wavevectors $\boldsymbol{k} + \boldsymbol{p} + \boldsymbol{q} = 0$). If the QG part is initially relevant, with all related nonlinear interactions being resonant for any triad, then this shows the relevance of a nonlinear quasi-geostrophic model, with other contributions (not purely QG) governed by RDT. The latter situation is likely true as far as the energy of the QG mode is initially important (e.g. for isotropic turbulence with energy at large scale). Since the slow 'wave' mode is the 2D mode in pure rotating turbulence, and the VSHF mode plays the same role in pure stratified turbulence, concentration of energy towards this mode is consistent with two-dimensionalisation in pure rotating turbulence and two-componentalisation in pure stratified turbulence, or related tendency to create cigar or pancake elongated structures in physical space, respectively. This concentration involves a highly anisotropic angular energy drain in wavespace. Another aspect is the presence of an inverse cascade for these typical slow modes, as evidenced by [46] using DNS forced isotropically at small scale.

4.3 Forced versus unforced case

Finally, it is important to point out that the building of large scale with an obvious, but very slow, concentration towards the VSHF mode in the latter DNS at dominant stratification, was attributed by the authors to resonant triads of gravity waves, and no significant creation of large scale PV (or QG) mode was found. The mechanisms of concentration towards the VSHF mode are thus radically different in ([17], [6]), on the one hand, and in [46], on the other hand. Only in the case in which a significant large-scale PV part of the flow exists initially, the process of layering is relatively rapid and is mediated by those interactions which exclude wavy contributions.

Another interesting case was found for $N/(2\Omega)$ not far from 1. Because The QG mode has initially no energy at large scale, its possible building would involve long time inverse cascade. Accordingly, the relevance of a nonlinear model only based on QG, as expected in turbulence decaying with a classical isotropic energy spectrum, is not obvious. This situation (nonlinear QG model), however, was shown by [46] to prevail when $1/2 < N/(2\Omega < 2$, the triadic 'wave' resonances being forbidden in this range of parameter. The fact that the essential feature is an inverse cascade for the QG mode confirms preliminary DNS results by [2].

As a conclusion of this section, a complete understanding of dominant nonlinear interactions in all these cases requires to have a combined description of the cascade (energy drain between different wave numbers) and of the angular energy drain (which is exclusively emphasized in [17] and related Figure 11).

5 Achievements and open problems

5.1 Recap of dominant modes of motion and their nonlinear interactions

The basis of eigenmodes of the linearized equations is very useful for analysing the full dynamics. Even RDT solutions for single-time and two-time Eulerian correlations are easier to calculate and interprete. Pure RDT results reflect conservation of the steady, QG, mode, and phase-mixing due to dispersive inertia-gravity waves. The more informative decomposition of the velocity field is not in terms of horizontal and vertical flow, but in terms of poloidal and toroidal velocity components, since the steady QG mode only involves the toroidal component. Fourier space, and related use of the Craya-Herring frame of reference, are only mathematical conveniences to render this decomposition more tractable. Only for pure rotation, the QG mode has no contribution to the velocity field, so that rotating turbulence at low Rossby number is purely inertial-wave-turbulence. Nonlinear interactions can trigger a slow transition from 3D to 2D structure through triadic resonances. At dominant stratification, the nonlinear evolution reflects the concentration towards the VSHF (Vertically Sheared Horizontal Flow), which is the zero-frequency mode of gravity waves in pure stratification, in agreement with vertical collapse and layering. This tendency is mediated by interactions that excludes waves (only QG and VSHF), if the QG mode has initially a significant energy at large scale, and possibly resonant wave interactions if this condition is not fulfilled. If stratification is significant, in the general case with or without rotation, a good prediction can be expected by taking into account the relevant nonlinearity in term of the QG mode only. The latter situation was shown to prevail, even without initial QG energy at large scale, in the range $1/2 < N/(2\Omega) < 2$ where no triadic resonance of inertia-gravity waves can occur ([46]).

5.2 Wave-Turbulence theory versus two-point closure thories

Mathematical developments in the area of wave-turbulence theory (WT), have recently renewed interest in flows that consist of superimposed dispersive waves, in which nonlinear interactions drive the long time behaviour. Individual modes are of the kind

$$u_i(\boldsymbol{x},t) = a_i(t)\exp[\imath(\boldsymbol{k}\cdot\boldsymbol{x} - \sigma_k t)] \tag{41}$$

with a known analytical dispersion law for $\sigma_k = \sigma(\boldsymbol{k})$. Similar averaged nonlinear amplitude equations can be found using either WT or TPC (Two-point closures), the advantages and drawbacks of which are briefly discussed below.

In the case of wave-turbulence, statistical homogeneity and quasi-normal assumption have equivalent counterparts, obtained by assuming *a priori* Gaussian random phases for the wave fields. As a consequence, an isotropic version of the quasi-normal assumption (used in TPC) is derived as

$$\overline{a_i a_j a_k^* a_l^*} = \overline{a_i a_i^*}.\overline{a_j a_j^*}\,(\delta_{ik}\delta_{jl} + \delta_{il}\delta_{jk}),$$

as discussed in [47]. In addition, *isotropic* dispersion laws such as $\sigma_k = |\boldsymbol{k}|^\alpha$ in (41) are almost exclusively treated in WT for deriving Kolmogorov spectra, with the key hypothesis of constant and isotropic energy fluxes across different scales associated with a wavenumber $|\boldsymbol{k}|$ ([50]). By contrast, in geophysical flows, dispersion laws are anisotropic, with for instance $\sigma = \pm\beta k_x/k^2$ in the case of Rossby waves, $\sigma = \pm 2\Omega k_\parallel/k$ for inertial waves and $\sigma = \pm N k_\perp/k$ for gravity waves (k_x, $k_\parallel$ and $k_\perp$ are the components of the associated wavevector respectively in the zonal direction, and the directions parallel or perpendicular to the rotation/gravity vectors). In the latter two 3D cases, this anisotropy is reflected by the strange conical — 'St Andrew cross' — shape of iso-phase surfaces in typical experiments with a localized point forcing (see views of this type in Figure 6, and in [39]; [41]; [20]), and by angular-dependent energy drains when looking at nonlinear interactions, as illustrated throughout this paper.

At least if Eulerian correlations are considered, The TPC and WT theories share in general an important background. Kinetic equations for mean spectral energy densities of waves are found in WT, similar to homogeneous TPC. Their slow evolution is governed by similar energy transfer terms, which are cubic in terms of wave amplitudes (triads). There is also a possibility that these transfers involve fourth-order interactions (quartets) in WT when triple resonances are forbidden by the dispersion laws and/or by geometric constraints (*e.g.* shallow waters). Resonant quartets seem to be particularly relevant when resonances are seen in a Lagrangian description. When triple resonances are allowed, for instance in cases of rotating, stably stratified and MHD turbulence ([6]; [16]), WT kinetic equations have exactly the same structure as their counterpart in elaborated (anisotropic multimodal) EDQNM. Hence, WT and TPC have a common limit at very small interaction parameter (*e.g.* Rossby number, Froude number, magnetic number in MHD). The shape of the typical eddy damping parameter, which remains the heuristic correction to quasi-normal transfer in EDQNM, is unimportant in this limit. Even if weak nonlinear interactions are addressed by WT in the inviscid case, a similar damping term is introduced as a mathematical convenience. Accordingly, even a vanishing damping term is needed in both EDQNM and Eulerian WT, at least for regularising the resonance operators. Significant eddy damping, or more generally the nonlinear contribution to

Kraichnan's response tensor, regains some importance for moderate and large interaction parameters, allowing extrapolation from WT via TPC towards the case of strong interactions (e.g. isotropic turbulence without wave effects, for which TPC are known to work satisfactorily).

In the three-dimensional case of stably stratified turbulence, with or without rotation, a part of the motion is decoupled from dispersive waves, at least in the linear limit. The presence of this QG mode, including potential vorticity, is not consistent with WT asymptotic expansions of nonlinearities in terms of small, Froude or Rossby, parameters. In [6], for instance, the 'vortex' mode is marginally treated with only qualitative results. A (multimodal anisotropic) EDQNM model, close to the ones developed for pure rotation and pure stratification, would present particular interest to complement WT in the general rotating stratified case. As a first useful simplified version, it is possible to restrict nonlinear transfer terms to pure QG (or QG+VSHF with zero rotation) triadic interactions, with wavy modes only governed by viscous linear dynamics.

5.3 Interest of stability analysis around organised eddies

Pure rotation If the flow includes preexisting large-scale quasi two-dimensional vortices, the stability analysis shows the importance of the Coriolis force, which alters, but even can create or inhibit classical elliptical or centrifugal three-dimensional instabilities. This can be analysed by using a WKB short-wave expansion for the disturbance, which amounts to a zonal RDT analysis at the leading order ([18]). Regarding a pair of counterrotating vortices in a rotating frame, the anticyclonic vortex is more affected by 3D instabilities modified by system rotation than the cyclonic one. In contrast with nonlinear interactions of inertial waves, the most energetic instabilities tends to amplify spanwise modes, which correspond to vertical (along the rotation axis) wavevectors [34].

Pure stratification The zig-zag instability ([5]) affects a vortex pair, initially two-dimensional in the plane normal to the density gradient. If the stable stratification is strong enough, the cooperative instabilities (elliptical, Crow) are inhibited, and the vertical coherence of the eddies is broken via a typical zig-zag motion. It seems that the tall-column instability shown in a quasi-geostrophic flow ([14]) is of the same nature. If one consider the whole flow, without splitting into base flow and disturbance, the mechanism of zig-zag instability is perhaps not disconnected from the growth of the energy of the VSHF mode, though nonlinear interactions excluding gravity waves, as discussed in section 4. Indeed, dominant 'vortex' contribution is given by the vortex pair, whereas some analogy can be found between the zig-zag motion and the VSHF. Qualitatively, the zig-zag and tall-column instabilities give an explanation for the layering in stratified flow, as does the growth of the VSHF energy through nonlinear interactions. The typical thickness of the slices predicted by the inviscid stability analysis, however, is likely much larger than the one seen in a typical turbulent stratified flow.

5.4 Towards (statistically) inhomogeneous flows

In addition to the use of zonal RDT for stability analysis ([36], [18]), it is interesting to derive weakly inhomogeneous RDT using a Gabor transform and related WKB development [42]. The small parameter of the WKB expansion is the ratio of the wave-length of the Fourier mode to the length of its Gaussian envelope. The interest of the method is not to derive the equations for the wave vector and the amplitude of the fluctuating velocity field (the method in [18] does the job in a simplest and more general way), but to calculate a space-dependent Reynolds stress tensor by integrating $\hat{R}_{ij}(\boldsymbol{x}, \boldsymbol{k}, t)$. Consequently, the nonlinear term which expresses the feedback from the Reynolds stress tensor in the mean flow equation can be evaluated (it is zero in pure homogeneous RDT). An extension of this procedure is in progress with application to subgrid-scale modelling (Bérangère Dubrulle, private communication). Related calculations, using both weakly inhomogeneous RDT and DNS, were performed by [15] in stratified flows. Their results are consistent with increasing the energy of the VSHF (related to vertical wavevectors), even if their splitting in terms of mean/fluctuating and homogeneous/inhomogeneous is different of the approach in section 4.

Another classical version of geometric optics [37] can be used, in which the small parameter of the WKB expansion affects not only the phase of the wavepacket disturbances, as in [36], but also 'slow' time and space variables. An hamiltonian formalism can be developed, which supports a mixed physical/spectral description. Space variables denote slow variation, the wavevector holds for the gradient of the phase, and the Hamiltonian H includes both the dispersion frequency of waves and the doppler frequency due to convection by the mean flow, or $\boldsymbol{k}\cdot\bar{\boldsymbol{u}}$, and the related characteristic curve is the ray along with energy propagates. In the context of gravity waves, promising perspectives, with transport of statistical spectra with nonlinear effects and diffusion, are offered by [12]. In the latter work, the Hamiltonian function is affected by nonlinear dynamics in connection with a simplified version of DIA, and the role of resonant triad interactions is displayed.

6 Appendix

6.1 Poloidal-toroidal decomposition, and Craya-Herring frame of reference

The poloidal-toroidal decomposition is classically used to represent a three-component divergence-free velocity field in terms of two independent scalar terms, taking advantage of the presence of a privileged direction $\boldsymbol{n}$.

$$\boldsymbol{u}' = \boldsymbol{\nabla} \times (s_{pol}\boldsymbol{n}) + \boldsymbol{\nabla} \times [\boldsymbol{\nabla}(s_{tor}\boldsymbol{n})] \tag{42}$$

the axial vector $\boldsymbol{n}$ being chosen along the vertical direction, without loss of generality. As a caveat, some care is needed to represent a Vertically Sheared Horizontal Flow (VSHF hereinafter) or $\boldsymbol{u}'(\boldsymbol{x}\cdot\boldsymbol{n}, t)$, with $\boldsymbol{u}'\cdot\boldsymbol{n} = 0$, with this decomposition.

In Fourier space, the above decomposition yields a pure geometrical one, or

$$\hat{\boldsymbol{u}} = \boldsymbol{k} \times \boldsymbol{n}(\imath \hat{s}_{pol}) - \boldsymbol{k} \times (\boldsymbol{k} \times \boldsymbol{n})(\hat{s}_{tor}) \tag{43}$$

and it appears immediately that the Fourier mode related to vertical wavevector direction, or $\boldsymbol{k} \parallel \boldsymbol{n}$, has zero contribution; this 'hole' in the spectral description yields missing the VSHF mode in physical space. In order to complete the decomposition, one can define an orthonormal frame of reference, which is nothing else than the local reference frame of a polar-spherical system of coordinates for $\boldsymbol{k}$ (see Figure 1).

$$\boldsymbol{e}^{(1)} = \frac{\boldsymbol{k} \times \boldsymbol{n}}{|\boldsymbol{k} \times \boldsymbol{n}|} \quad \boldsymbol{e}^{(2)} = \boldsymbol{e}^{(3)} \times \boldsymbol{e}^{(1)} \quad \boldsymbol{e}^{(3)} = \frac{\boldsymbol{k}}{k} \tag{44}$$

for $\boldsymbol{k} \times \boldsymbol{n} \neq 0$, and $\boldsymbol{e}^{(1)}, \boldsymbol{e}^{(2)}, \boldsymbol{e}^{(3)}$ may coincide with the fixed frame of reference, with $\boldsymbol{e}^{(3)} = \boldsymbol{n}$ for $\boldsymbol{k} \parallel \boldsymbol{n}$. In the turbulence community, the local frame $(\boldsymbol{e}^{(1)}, \boldsymbol{e}^{(2)})$ of the plane normal to the wavevector is often referred to as Craya-Herring frame. Accordingly, the divergence-free velocity field in wave-space has only two components in the Craya-Herring frame, or

$$\hat{\boldsymbol{u}}(\boldsymbol{k}, t) = u^{(1)}\boldsymbol{e}^{(1)} + u^{(2)}\boldsymbol{e}^{(2)} \tag{45}$$

For $\boldsymbol{k} \times \boldsymbol{n} \neq 0$, $u^{(1)}$ and $u^{(2)}$ are directly linked to the toroidal mode and the poloidal mode, respectively. For $\boldsymbol{k} \times \boldsymbol{n} = 0$, they correspond to the VSHF mode. RDT equations can be written in the Craya-Herring frame, resulting in a reduced Green's function with only four independent components (Cambon's thesis, 1981). A similar decomposition is used in [3]. Finally, the 'wave-vortex' decomposition introduced by [45] in the particular context of stably stratified turbulence, is also a particular case of (42).

6.2 Anisotropic second order spectral tensor

Independently of closure, the second order spectral tensor $\hat{R}_{ij}(\boldsymbol{k}, t)$, which is directly connected to $< \hat{\boldsymbol{u}}_i^* \hat{\boldsymbol{u}}_j >$, is not a general complex matrix, but has a number of special properties, including the fact that it is Hermitian, positive-definite, and satisfies $\hat{R}_{ij} k_j = 0$, obtained from the incompressibility condition $k_j \widehat{u}_j = 0$. Taken together, these properties mean that, instead of the 18 real degrees of freedom of a general complex tensor, $\hat{R}_{ij}$ has only four. Indeed, using a spherical polar coordinate system in $\boldsymbol{k}$-space, or (44) and (5), the tensor takes the form (see [10] for details).

$$\hat{R} = \begin{pmatrix} e + Z_r & Z_i - \imath\mathcal{H}/k & 0 \\ Z_i + \imath\mathcal{H}/k & e - Z_r & 0 \\ 0 & 0 & 0 \end{pmatrix} \tag{46}$$

where the scalars $e(\boldsymbol{k}, t)$ and $\mathcal{H}(\boldsymbol{k}, t)$ are real, and $Z(\boldsymbol{k}, t) = Z_r + \imath Z_i$ is complex. The quantity $e(\boldsymbol{k}, t) = \frac{1}{2}\hat{R}_{ii}$ is the energy density in $\boldsymbol{k}$-space, whereas $\mathcal{H}(\boldsymbol{k}, t) =$

$\imath k_l \epsilon_{lij} \hat{R}_{ij}$ is the helicity spectrum and, along with Z, is zero in the isotropic case. Recall that the local frame of reference $\boldsymbol{e}^{(1)}, \boldsymbol{e}^{(2)}, \boldsymbol{e}^{(3)}$, in (44) is chosen direct and orthonormal, with $\boldsymbol{e}^{(3)} = k_i/k$.

Anisotropy is expressed through variation of these scalars with the direction of $\boldsymbol{k}$, as well as departures of $\mathcal{H}$ and Z from zero at a given wavenumber. Whatever spectral closure is used, the number of real unknowns may be reduced to the above four when carrying out numerical calculations, and presentation of the results can be simplified using these variables, particularly when the turbulence is axisymmetric.

In order to display the different contributions from $e, Z, \mathcal{H}$ with their related tensorial operators, (46) is rewritten as follows

$$\hat{R}_{ij} = e(\boldsymbol{k},t)P_{ij}(\boldsymbol{k}) + Re[Z(\boldsymbol{k},t)N_i(\boldsymbol{k})N_j(\boldsymbol{k})] + \imath\mathcal{H}(\boldsymbol{k},t)\epsilon_{ijn}\frac{k_n}{k} \tag{47}$$

in which P_{ij} denotes the projection operator, as in (25), and $\boldsymbol{N} = \boldsymbol{e}^{(2)} - \imath\boldsymbol{e}^{(1)}$. The anisotropic structure is then analyzed by isolating in (47) the pure isotropic contribution, so that

$$Re\left(\hat{R}_{ij}\right) = \underbrace{\frac{E(k)}{4\pi k^2}P_{ij}}_{\text{Isotropic part}} + \underbrace{\left(e(\mathbf{k}) - \frac{E(k)}{4\pi k^2}\right)P_{ij}}_{\text{Directional anisotropy}} + \underbrace{Re\left(Z(\mathbf{k},t)N_iN_j\right)}_{\text{Polarization anisotropy}} \tag{48}$$

A three-fold splitting follows for any single-point correlation, for instance

$$\overline{u_iu_j} = \int \hat{R}_{ij}(\boldsymbol{k},t)d^3\boldsymbol{k} = q^2\left(\frac{\delta_{ij}}{3} + \underbrace{b_{ij}^{(e)} + b_{ij}^{(z)}}_{b_{ij}}\right) \tag{49}$$

for the Reynolds Stress tensor,

$$D_{ij} = \int \frac{k_ik_j}{k^2}2e(\boldsymbol{k},t)d^3\boldsymbol{k} = q^2\left(\frac{\delta_{ij}}{3} + 2b_{ij}^{(e)} + 0\right)$$

for the 'Dimensionality structure tensor' ([29]), and

$$\overline{\omega_i\omega_j} = \omega^2\left(\frac{\delta_{ij}}{3} + b_{ij}^{(k^2e)} - b_{ij}^{(k^2z)}\right)$$

for the vorticity correlations tensor.[3]

The above relationship shows that the deviatoric tensor b_{ij} is the sum of two very different contributions, $b_{ij}^{(e)}$ from the directional anisotropy (or dimensionality) and $b_{ij}^{(z)}$ from polarization anisotropy. Surprising RDT results in rotating

[3] Rather than the vorticity correlations tensor, [29] introduced a 'Circulicity tensor' C_{ij}, which involves larger scales. This tensor corresponds to $C_{ij} = q^2\left(\frac{\delta_{ij}}{3} + b_{ij}^{(e)} - b_{ij}^{(z)}\right)$, with our notations

flows are explained by this decomposition (see, e.g. [9]), and the formalism of [29] appears as a byproduct of (47) in homogeneous turbulence, with the decomposition in terms of directional and polarization anisotropy lending support to componental/ dimensional anisotropy.

In conclusion, it is worthwhile to point out that a fully anisotropic spectral (or two-point) description carries a very large amount of information, even if restricted to second-order statistics. In the inhomogeneous case, the POD (proper orthogonal decomposition, [38]) has renewed interest in second-order two-point statistics, but this technique is applied to strongly inhomogeneous quasi-deterministic flows. It is only said that POD spatial modes are Fourier modes in the homogeneous turbulence, without considering that a spectral tensor such as $\hat{\mathbf{R}}$ ought to be diagonalized in order to exhibit its eigenmodes as POD modes in the *anisotropic* case. In fact, diagonalising the real part of the tensor $\hat{\mathbf{R}}$, is an easy task using the above $e - Z$ decomposition (46): the principal components (nonzero eigenvalues) are $e + |Z|$ and $e - |Z|$, and the angular position of the principal axes (eigenvectors) in the Craya-Herring frame of reference ($e^{(1)}, e^{(2)}$, see (44)), is given by the phase of Z, at each $\boldsymbol{k}$.

References

1. J. Bardina, J.M. Ferziger, R.S. Rogallo: Effect of rotation on isotropic turbulence: computation and modelling. J. Fluid Mech. **154** (1985)
2. P. Bartello: Geostrophic adjustment and inverse cascades in rotating stratified turbulence. J. Atmos. Sci. **52** (1995)
3. B.J. Bayly, D.D. Holm, A. Lifschitz: Three-dimensional instability of elliptic vortex columns in external strain flows. Phil. Trans. R. Soc. London A **354** (1996)
4. D.J. Benney, P.G. Saffman: 'Nonlinear interactions of random waves in a dispersive medium'. In: *Proc. R. Soc. London, Ser. A* **289** (1966), pp. 301–320.
5. P. Billand, J.M. Chomaz: Experimental evidence for a new instability of a columnar vortex pair in a strongly stratified fluid. J. Fluid Mech. **418** (2000)
6. P. Caillol, Zeitlin W: Kinetic equations and stationary energy spectra of weakly nonlinear internal gravity waves. Dyn. Atm. Oceans **32** (2000)
7. C. Cambon: Turbulence and vortex structures in rotating and stratified flows. Eur. J. Mech. B (fluids) **20** (2001)
8. C. Cambon, F.S. Godeferd, F. Nicolleau, J.C. Vassilicos: Turbulent diffusion in rapidly rotating turbulence with or without stable stratification. J. Fluid Mech. submitted in (2002)
9. C. Cambon, L. Jacquin, J.L. Lubrano: Towards a new Reynolds stress model for rotating turbulent flows. Phys. Fluids A **4** (1992)
10. C. Cambon, N.N. Mansour, F.S. Godeferd: Energy transfer in rotating turbulence. J. Fluid Mech. **337** (1997)
11. C. Cambon, J. F. Scott: Linear and nonlinear models of anisotropic turbulence. Ann. Rev. Fluid Mech. **31** (1999)
12. G. Carnevale, J.S. Frederiksen: A statistical dynamical theory of strongly nonlinear internal gravity waves. Geophys. Astrophys. Fluid Dyn. **20**, 8 (1983)
13. S. Corrsin: In *Atmospheric diffusion and air pollution*. Advances in geophysics, ed. by F.N. Frenkiel and P.A. Sheppard, Academic Press, New York **6** (1959)

14. D.G. Dritschel, M. de la Torre Juarez, M.H.P. Ambaum: The three-dimensional vortical nature of atmospheric and oceanic turbulent flows. Phys. Fluids **11** (16) (1999)
15. M. Galmiche, J.C.R. Hunt, O. Thual, P. Bonneton: Turbulence-mean field interactions and layer formation in a stratified fluid. Eur. J. Mech. B (fluids) **20** (2001)
16. S. Galtier, S. Nazarenko, A.C. Newell, A. Pouquet: A weak turbulence theory for incompressible MHD. J. Plasma Phys. **63** (2000)
17. F.S. Godeferd, C. Cambon: Detailed investigation of energy transfers in homogeneous stratified turbulence. Phys. Fluid **6** (1994)
18. F.S. Godeferd, C. Cambon, S. Leblanc: Zonal approach to centrifugal, elliptic and hyperbolic instabilities in Suart vortices with external rotation. J. Fluid Mech. **449** (2001)
19. F.S. Godeferd, C. Cambon, J.F. Scott: Report on the workshop: Two-point closures and their application. J. Fluid Mech. **436** (2001)
20. F.S. Godeferd, L. Lollini: Direct numerical simulations of turbulence with confinment and rotation. J. Fluid Mech. **393** (1999)
21. H.P. Greenspan: *The theory of rotating fluids.* (Cambridge University Press, 1968)
22. H. Hanazaki: Linear processes in stably and unstably stratified rotating turbulence. J. Fluid Mech. To appear in (2002)
23. H. Hanazaki: present volume.
24. G. Holloway, M.C. Hendershot: Statistical closure for nonlinear Rossby waves. J. Fluid Mech. **82** (1977)
25. L. Jacquin, O. Leuchter, C. Cambon, J. Mathieu: Homogeneous turbulence in the presence of rotation. J. Fluid Mech. **220** (1990)
26. Y. Kaneda: Renormalized expansions in the theory of turbulence with the use of the Lagrangian position function. J. Fluid Mech. **107** (1981)
27. Y. Kaneda: Single-particle diffusion in strongly stratified and/or rotating turbulence. J. Phys. Soc. Japan. **69** (2000)
28. Y. Kaneda, T. Ishida: Suppression of vertical diffusion in strongly stratified turbulence. J. Fluid Mech. **402** (2000)
29. S. Kassinos, W.C. Reynolds, M.M. Rogers: One-point turbulence structure tensors. J. Fluid Mech. **428** (2000)
30. Y. Kimura, J.R. Herring: Diffusion in stably stratified turbulence. J. Fluid Mech. **328** (1996)
31. Y. Kimura, J.R. Herring: Particle dispersion in rotating stratified turbulence. In: *Proceedings of FEDSM99*, FEDSM99-7753 (1999)
32. Y. Kimura, J.R. Herring: Present volume.
33. R.H. Kraichnan: The structure of turbulence at very high Reynolds numbers. J. Fluid Mech. **5** (1959)
34. S. Leblanc, C. Cambon: On the three-dimensional instabilities of plane flows subjected to Coriolis force. Phys. Fluids **9** 5 (1997)
35. J.M. Lee, J. Kim, P. Moin: Structure of turbulence at high shear rate. J. Fluid Mech. **216** (1990)
36. A. Lifschitz, E. Hameiri: Local stability conditions in fluid dynamics. Phys. Fluids A **3** (1991)
37. M.J. Lighthill: *Waves in Fluids.* (Cambridge University Press, 1978)
38. J.L. Lumley: The structure of inhomogeneous turbulent flows. In: *Atmospheric turbulence and radio wave propagation*, ed. by A.M. Yaglom and V.I. Tatarsky, NAUCA, Moscow (1967)
39. A.D. Mc Ewan: Inertial oscillations in a rotating fluid cylinder. J. Fluid Mech. **40** (1970)

40. Y. Morinishi, K. Nakabayashi, S.Q. Ren: Dynamics of anisotropy on decaying homogeneous turbulence subjected to system rotation. Phys. Fluids **13**, 10 (2001)
41. D.E. Mowbray, B.S.H. Rarity: A theoretical and experimental investigation of the phase configuration of internal waves of small amplitude in a density stratified liquid. J. Fluid Mech. **28** (1967)
42. S. Nazarenko, N.N. Kevlahan, B. Dubrulle: A WKB theory for rapid distortion of inhomogeneous turbulence. J. Fluid Mech. **390** (1999)
43. F.C.G.A. Nicolleau, J.C. Vassilicos: Turbulent diffusion in stably stratified non-decaying turbulence. J. Fluid Mech. **410** (2000)
44. S.A. Orszag: Analytical theories of turbulence. J.Fluid Mech. **41**: (1970)
45. J. Riley, R.W. Metcalfe, M.A. Weisman: DNS of homogeneous turbulence in density stratified fluids. In: *Proc. AIP conf. on nonlinear properties of internal waves*, ed. by B.J. West, (AIP, New York, 1981)
46. L.M. Smith, F. Waleffe: Generation of slow, large scales in forced rotating, stratified turbulence. J. Fluid Mech. **451** (2002)
47. C. Staquet, J. Sommeria: Internal gravity waves: from instabilities to turbulence. Annu. Rev. Fluid Mech. **34** (2002)
48. G.I. Taylor: Diffusion by continuous movements. Proc. London Math. Soc. A **20** (1921)
49. A.A. Townsend: *The structure of turbulent shear flow.* (Cambridge University Press, 1956, 1976)
50. V.E. Zakharov, V.S. L'vov, G. Falkowich: *Kolmogoroff spectra of turbulence I. Wave turbulence.* (Springer series in nonlinear dynamics. Springer Verlag, 1992)

An Introduction to Mixing in a Stably Stratified Fluid

Chantal Staquet

Laboratoire des Ecoulements Géophysiques et Industriels, BP 53, 38041 Grenoble cdx 9, France

Abstract. We provide an account of mixing in stably stratified fluids: we present the basic concepts of mixing, recall how mixing is quantified and analyze mixing in the stably stratified shear layer along these lines. We also discuss the importance of mixing in the ocean and describe how it is modeled in this context, focusing upon the commonest approaches.

1 Introduction

General circulation models of the ocean, which provide a numerical description of the mean currents at oceanic basin scales, use resolution of a few km along the horizontal and of tens-to-hundreds of meters along the vertical. It is therefore necessary to model the motions below the grid scale, the so-called subgrid-scale motions, at which kinetic energy is dissipated and mixing occurs. This modeling is crucial because the dissipating processes contribute fundamentally to the large scale dynamics of the ocean. Let us illustrate this important point through two arguments, related to the thermohaline circulation.

The thermohaline circulation is a large scale convective motion formed by the sinking at high latitude of cold water masses; these cold water masses spread throughout the bottom of ocean basins and slowly upwell. This upwelling is due to mixing processes: without deep mixing, the ocean would have turned, within a few thousand years, into a stagnant pool of cold salty water extending to within one meter from the surface. The thermocline, which separate the bottom waters from the surface waters mixed by the wind, would have reduced to a very thin layer. This yields the fundamental idea that the thermohaline circulation is not driven by buoyancy effects but by the mechanical energy that mix the water masses, namely wind and tidal energy (see [15] and [24] and references therein).

A global estimate of the turbulent diffusivity associated with mixing in the ocean was proposed by Munk [23], by assuming a balance between the upwelling water masses and the downgradient mixing of heat. Using the density profile measured in the ocean below the thermocline and an estimate of the vertical velocity of upwelling from the distribution of chemical tracers, Munk [23] obtained a value for the turbulent diffusivity equal to $\kappa_t \simeq 10^{-4} m^2/s$ (that is, one thousand times the molecular diffusivity). The method used by Munk was subjected to many discussions (and some controversy) but κ_t is now considered as the canonical value for the global diffusivity in the ocean.

As shown by Bryan [3] already in 1987, a constant diffusivity is certainly too crude an approximation for the parameterization of mixing in general circulation models. A refined parameterization is needed, namely a simple relation between a turbulent diffusivity and a dynamical parameter characterizing the large resolved scales of the model (the same need arises for the turbulent viscosity). This is still an open problem: for instance, Beckmann [2] provides results of three numerical simulations of the thermohaline circulation using general circulation models differing by the representation of the bottom topography and the modeling of the subgrid scales. Currents for depths greater than 500 m are plotted in Fig. 1: strong discrepancies between these large scale currents are visible from one model to the other, which illustrates their strong sensitivity to small scale processes.

It is necessary at this stage to provide a definition of mixing. Mixing is the destruction by molecular diffusion of density gradients generated by turbulence. Some authors (such as Toole [37]) also include in this definition of mixing the destruction of velocity gradients by viscosity. We shall rather keep the usual terminology of kinetic energy dissipation for the latter process. In the ocean, mixing takes place at scales of 1 mm to 1 m.

Mixing may also be referred to as vertical transport: because of the existence of the restoring force of buoyancy along the vertical direction, no net vertical transport can occur in a stably stratified fluid without mixing. By contrast, no restoring force exists along the horizontal direction (ignoring the Coriolis force), so that horizontal motions are not constrained. Horizontal transport thus leads to characteristic length scales that are much larger than those associated with vertical transport, because the former process is controlled by advective effects while the latter results from dissipating processes. Finally, we point out that mixing should be made distinct from stirring. Stirring results from the straining by large scale motions, which increases density gradients without mixing. An example of stirring is provided in Fig. 2.

This introduction mainly focused upon oceanic processes so that the active scalar field, which is mixed, is the density. The concept of mixing also applies in the atmosphere, by considering the potential temperature field in place of the density field. However, all concepts presented in the present paper rely upon the Boussinesq approximation so that they must be modified when compressible effects become important (see [1]).

The outline of the paper is the following. In the next section we present the basic concepts of mixing along with their modeling in the ocean. We provide an example of mixing analysis in section 3, by considering three-dimensional numerical simulations of the stably stratified shear layer (most results presented in this section stem from [32] and [33]. We also briefly address the question of parameterization of mixing resulting from shear instability, both in the numerical simulations and in large scale oceanic models. Conclusions are drawn in the final section.

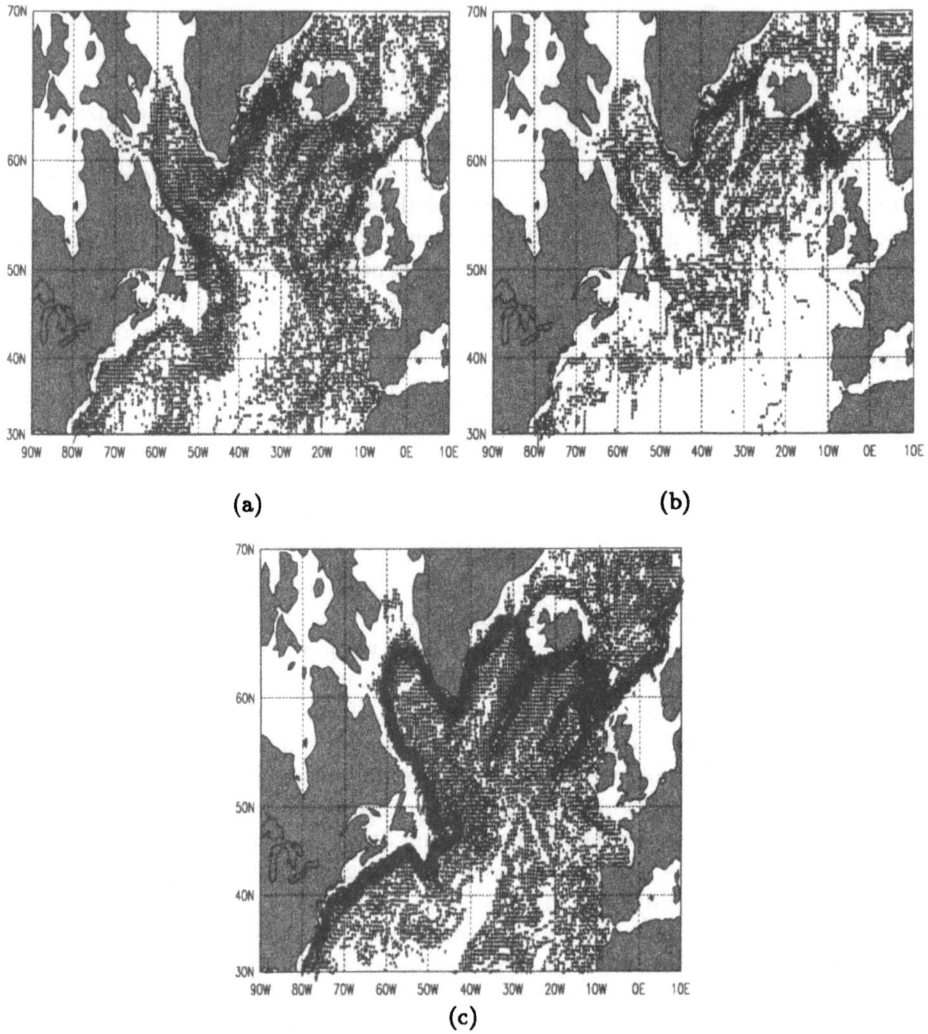

Fig. 1. (a) Time-mean bottom layer circulation in the subpolar North Atlantic from the 1/3 degree DYNAMO-SPEM general circulation model. (b) Same as (a), for the DYNAMO-MICOM general circulation model. (c) Same as (a), for the DYNAMO-MOM general circulation model (from [2]).

2 Basic concepts of mixing

2.1 Background and available potential energy

Energy is required to mix a stably stratified fluid. Even if we consider a fluid at rest, made of two layers of different density, the denser one being at the bottom, the interface mixes through molecular effects : the density gradient smooths, the source of energy being the internal energy of the fluid. The diffusion of

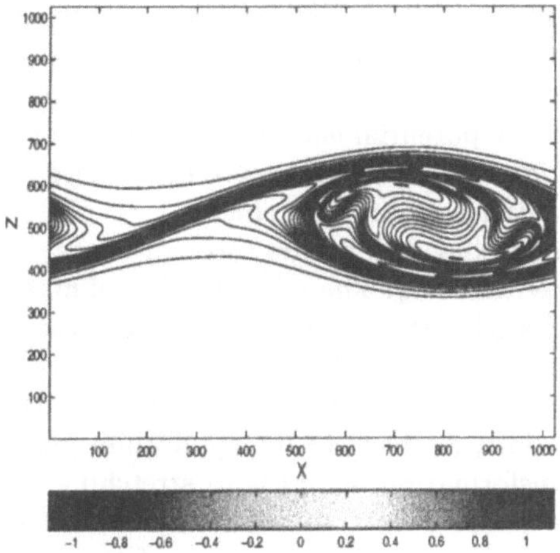

Fig. 2. The large scale vortex that develops from the Kelvin-Helmholtz instability induces a strain field which results in a thin layer of strong density gradient outside the vortex. The wrapping of dense and light fluid as the vortex forms creates intertwined thin layers of different density, thus yielding strong density gradients within the vortex. The large scale motion associated with the vortex mainly stirs the fluid and hardly mixes it (from [33]).

density at the interface occurs through a diapycnal flux of mass (i.e. across the constant density surfaces, or isopycnals). This is expressed by the standard diffusion equation:

$$\frac{\partial \rho_b}{\partial t} + \frac{\partial}{\partial z}\phi_{lam} = 0. \tag{1}$$

$\rho_b(z)$ is the density profile of the fluid and

$$\phi_{lam}(z) = -\kappa \, d\rho_b/dz \tag{2}$$

is the local diapycnal mass flux. This flux results in an increase of the center of mass of the fluid so that the potential energy of the fluid E_b (usually defined per unit volume) increases as well. The expression for E_b is

$$E_b = \frac{1}{V}\int_V \rho_b(z) g z dV, \tag{3}$$

where V is the volume of the fluid, assumed in the following to be contained in a fixed parallelepipedic rigid box, bounded by a surface S. The evolution equation for E_b is easily inferred from Eq. (1):

$$\frac{d}{dt}E_b = \Phi_{lam}, \tag{4}$$

with

$$\Phi_{lam} = \frac{\kappa g}{H}\Delta\rho = g\frac{1}{V}\int_V \phi_{lam} dV \quad > 0. \tag{5}$$

κ is the molecular diffusivity for density changes, g is the acceleration of gravity, H is the height of the fluid volume and $\Delta\rho = \rho_b(0) - \rho_b(H)$ is the (positive) density difference between the bottom and the top of the container. Φ_{lam} is the rate of production of the potential energy of the fluid at rest. Equivalently, Φ_{lam} is the conversion of internal energy into potential energy. We shall refer to this rate as the laminar mixing rate for simplicity.

In a moving fluid, mixing still occurs through a local diapycnal *diffusive* flux of mass, which we refer to as ϕ_d. The *advective* flux of mass $g\rho\mathbf{u}$ only distorts the isopycnals without changing their value. In order to access the diffusive mass flux ϕ_d, let us do a thought experiment: we assume that the fluid relaxes instantaneously toward a rest state through an adiabatic redistribution of density. Each fluid element is thus brought back to its equilibrium position without changing its density. This transformation amounts to stretching out each isopycnal horizontally; it yields a stable density profile (denoted ρ_b again), whose potential energy (E_b) is the minimum potential energy of the fluid at that time. These concepts have been introduced by Lorenz [22]. Hence, the above situation of a fluid at rest is recovered instantaneously through the adiabatic transformation so that the same equations should hold formally: the changes in ρ_b are only due to the diapycnal mass flux ϕ_d

$$\frac{\partial \rho_b}{\partial t} + \frac{\partial}{\partial z}\phi_d = 0. \tag{6}$$

This result has been demonstrated rigorously by Winters & D'Asaro [39]. The expression for ϕ_d is:

$$\phi_d(z) = \kappa \frac{< |\nabla\rho|^2 >_I}{-d\rho_b/dz} > 0, \tag{7}$$

where $<>_I$ denotes an average along an isopycnal and ρ refers to the total density field. The equation for the background potential energy is analogous to Eq. (4)

$$\frac{dE_b}{dt} = \Phi_d + Fb^{diff}_{surf} + Fb^{adv}_{surf}, \tag{8}$$

up to diffusive (Fb^{diff}_{surf}) and advective (Fb^{adv}_{surf}) mass fluxes across the bounding surface S (exact expressions of the latter fluxes are provided in [38]). As before, Φ_d is the volume averaged value of ϕ_d times g:

$$\Phi_d = g\frac{1}{V}\int_V \phi_d \, dV \tag{9}$$

and is the rate of production of the background potential energy. Hereafter, we simply refer to Φ_d as the mixing rate. When the fluid is at rest, $|\nabla\rho| = -d\rho_b/dz$ and Φ_d reduces to Φ_{lam}. In general $\Phi_d \geq \Phi_{lam}$ (≥ 0), expressing the well-known fact that turbulent motions (more precisely: density fluctuations) increase mixing. Note that density fluctuations refer here to deviations about the background density field ρ_b.

Lorenz' concepts were first applied by Thorpe [35] for the case of one-dimensional density profiles measured in a lake: the idea was to sort out the elementary fluid volumes contributing to the fluid column by order of increasing density, so as to get an estimate of the vertical length of turbulence. Indeed, the sorting method only operates when the fluid has overturned, fluid volumes being left at their position when the fluid is locally stable. Thorpe recorded the distance each particle has to be displaced to reach its position in the stable profile and defined a length scale from the root mean square value of the distances. This length scale, now referred to as the Thorpe scale, is thus a measure of the vertical extent of overturning events in the original density profile. It also provides an estimate of the size of the turbulent eddies.

In a three-dimensional fluid volume, the use of an adiabatic transformation to analyse mixing has been performed only recently, starting with the work [38]; references are provided in the next section. The practical method to compute the stable density profile ρ_b is described in this section as well; we also show that $\rho_b(z)$ is distinct in general from a mean (horizontally averaged) density profile.

The adiabatic transformation that leads to ρ_b releases the energy associated with the density fluctuations. This energy is the available potential energy, introduced by Lorenz [22] also:

$$E_a = E_p - E_b, \tag{10}$$

where E_p is the potential energy per unit volume of the fluid before the transformation is applied. (Closed expressions for E_a are provided by Holliday & McIntyre [14]). The minimum potential energy state can also be characterized by $E_a = 0$.

The evolution equation for E_a is simply derived from that for E_b and E_p. We recall the evolution equation for the total potential energy per unit volume (see f.i. [38])

$$\frac{d}{dt}E_p = g\frac{1}{V}\int_V \rho w dV + Fp^{adv}_{surf} + \Phi_{lam} + Fp^{diff}_{surf}; \tag{11}$$

the first term on the right-hand-side is the volume averaged buoyancy flux, the second term is the advective mass flux through the boundaries, Φ_{lam} is the laminar mixing rate and the last term is the mass flux across the boundaries due to molecular effects. Using (8), (10) and (11), one gets the evolution equation for E_a:

$$\frac{d}{dt}E_a = \frac{1}{V}\int_V g\rho w dV - (\Phi_d - \Phi_{lam}) + (Fp^{diff}_{surf} - Fb^{diff}_{surf}) + (Fp^{adv}_{surf} - Fb^{adv}_{surf}). \tag{12}$$

This equation shows that the dissipation rate of E_a within the fluid volume, $\Phi_d - \Phi_{lam}$, is the source for turbulent fluid mixing. The study of mixing thus amounts to decomposing the total potential energy E_p into a reversible part E_a, available for mixing, and an irreversible part E_b, which increases through mixing.

2.2 Quantification of mixing: diapycnal diffusivity and mixing efficiency

The rate of change of the background potential energy E_b *per se* does not provide any usable information on mixing. dE_b/dt needs to be related to another quantity to provide some quantitative information. Equation (12) for E_a shows that the mixing rate due to density fluctuations, $\Phi_d - \Phi_{lam}$, should rather be used in place of Φ_d when mixing resulting from the flow dynamics is to be studied. $\Phi_d - \Phi_{lam}$ is referred to as the turbulent mixing rate hereafter.

One common measure of mixing is provided by a flux Richardson number, Rf_b, which relates the rate of production of potential energy due to turbulent mixing within the fluid volume $\Phi_d - \Phi_{lam}$ to the rate of energy input to the fluid $\dot{M}$:

$$Rf_b|_{forced} = \frac{\Phi_d - \Phi_{lam}}{\dot{M}}. \tag{13}$$

When the fluid is not forced, the total sink of energy within the fluid volume is used in place of $\dot{M}$. The energy is lost through mixing and through kinetic energy dissipation ϵ so that, in this case, the flux Richardson number within the fluid volume is expressed as:

$$Rf_b = \frac{\Phi_d - \Phi_{lam}}{(\Phi_d - \Phi_{lam}) + \epsilon}. \tag{14}$$

The changes of energy from the beginning of the flow evolution may also be used in place of the instantaneous rates of change, leading to the definition of a global flux Richardson number :

$$Rf_b|_{global} = \frac{\Delta E_b}{\Delta E_b + \Delta E_k}. \tag{15}$$

The interest in the latter definition is that it leads to a smoother evolution of the flux Richardson number and accounts for the boundary mass fluxes. E_k is the volume averaged kinetic energy. Note that either definition of the flux Richardson yields a value comprised between 0 and 1. A very small value ($\Delta E_b \ll \Delta E_k$) implies that very weak mixing occurs while a value close to 1 ($\Delta E_k \ll \Delta E_b$) implies that mixing is very efficient. In the latter case indeed, much more energy is lost into mixing than into heat. In the literature, the important concept of mixing efficiency is rather defined by the quantity $\gamma_b = 1/(Rf_b^{-1} - 1)$, instead of Rf_b itself.

Mixing is also most commonly characterized by a turbulent diffusivity K_ρ. When the scale of the background density gradient is much larger that the overturning scale (as measured by the Thorpe scale for instance), the turbulent flux is assumed to depend linearly on the gradient

$$\phi_d(z) = -K_b(z)\frac{\partial \rho_b}{\partial z}, \tag{16}$$

the coefficient defining the turbulent diffusivity. The definition ensures that $K_b(z)$ is positive at any time and for any value of z. When the background density gradient is linear, a volume averaged turbulent diffusivity can be defined

$$\frac{1}{V}\int_V K_b(z)dV = \kappa \,\frac{\Phi_d}{\Phi_{lam}}. \tag{17}$$

One important issue is the dependency of the turbulent diffusivity K_b upon the molecular diffusivity κ. When fluid motions are strictly horizontal, $K_b = \kappa$. When fluid motions displace the isopycnals but remain laminar, as for internal gravity waves of low amplitude, K_b is proportional to κ. When fluid motions are turbulent, a prediction may be conjectured when $\kappa \to 0$, the Prandtl number being set to a constant value, from analogy with ordinary turbulence. In the latter case, when $\nu \to 0$, the dissipation rate of kinetic energy remains finite and ν-independant. The analogy suggests that, as $\kappa \to 0$, the mixing rate Φ_d remains finite and κ-independent, and so is the volume averaged value of K_b, from Eq. (17). When the flow is not laminar and κ is not small, no theory exists, to our knowledge, and one should rely upon (numerical or laboratory) experiments to compute K_b.

2.3 Characterization of mixing in geophysical fluids: traditional approach

Mixing has been investigated for several decades in the ocean. As we discuss it in the present subsection, the major difficulty in such studies is the computation of the turbulent mixing rate ($\Phi_d - \Phi_{lam}$) so that models have always been used to estimate it. Hence, the exact computation of the mixing rate from the background density profile described above is a novel approach of mixing.

In oceanic models, the turbulent mixing rate is usually estimated from the average value of the advective flux of density:

$$\Phi_d - \Phi_{lam} \simeq g\overline{\rho' w'}, \tag{18}$$

the average being a temporal, a spatial or an ensemble one; it is denoted by an overbar in the following (thus $\rho' = \rho - \overline{\rho}$, $w' = w - \overline{w}$, etc). The idea of taking the averaged advective density flux to estimate the turbulent mixing rate is that the average filters out the oscillations of the advective flux and yields the residual contribution of diffusive processes. Unlike in the atmosphere, the direct measurement of $\overline{\rho' w'}$ has been attempted only rarely in oceanography. The main reason is that, in addition to the smallness of this residual contribution, most instruments profile vertically in the ocean and therefore cannot measure vertical velocities (see, e.g., [11] and [37] for a review). Therefore the oceanic turbulent mixing rate is usually inferred from microstructure measurements (at scales of about 1 m) through statistical models. In this approach, the turbulent diffusivity is defined by

$$K = \frac{\overline{\rho' w'}}{-d\overline{\rho}/dz} = \frac{\frac{g}{\rho_0}\overline{\rho' w'}}{N^2}, \tag{19}$$

where $N^2 = -(g/\rho_0)d\overline{\rho}/dz$ is the Brunt-Väisälä frequency.

A now classical statistical model was proposed by Osborn & Cox [26]. It relates the turbulent mixing rate to the dissipation rate of the variance of the density fluctuations. Assuming that the mean density gradient is about constant over the integral scale of the turbulence and that a statistically stationary regime has been reached, the equation for the variance of the density fluctuations reduces to:

$$\overline{\rho' w'}\ \frac{d\overline{\rho}}{dz} = -\kappa\ \overline{|\nabla \rho'|^2}. \tag{20}$$

This equation states that the production of fluctuating density variance by the action of the turbulent vertical flux of density $\overline{\rho' w'}$ on the mean gradient $d\overline{\rho}/dz$ is balanced by the dissipation rate of this variance $\kappa\ \overline{|\nabla \rho'|^2}$. Using definition (19) for the turbulent diffusivity, Eq. (20) yields the expression:

$$K' = \kappa \frac{\overline{|\nabla \rho'|^2}}{(d\overline{\rho}/dz)^2}. \tag{21}$$

This relation was originally established by Osborn & Cox [26] for the variance of the temperature fluctuations. It can be expressed in terms of the density field, as done here, when the Boussinesq approximation is valid.

The same approach was used by Osborn [25] when the turbulence is sustained by a mean shear flow $\overline{U}(z)$. Under the same assumption of a statistically steady state, the equation for the velocity variance is (written per unit mass):

$$\overline{u' w'}\ \frac{d\overline{U}}{dz} = -\epsilon - \frac{g}{\rho_0}\ \overline{\rho' w'}. \tag{22}$$

As above, this equation states that the production of turbulent kinetic energy per unit mass by the action of the Reynolds stress $\overline{u'w'}$ onto the mean velocity gradient $d\overline{U}/dz$ is balanced in the mean by the sinks of turbulent kinetic energy; these are the dissipation rate of kinetic energy ϵ and and the turbulent mixing rate per unit mass $g/\rho_0 \overline{\rho' w'}$ (which increases the potential energy resulting from mixing). Introducing a flux Richardson number:

$$Rf = \frac{\frac{g}{\rho_0}\overline{\rho' w'}}{-\overline{u' w'}\frac{d\overline{U}}{dz}}, \tag{23}$$

Eq. (22) can be rewritten in a simplified way:

$$\frac{g}{\rho_0}\overline{\rho' w'} = \frac{Rf}{1 - Rf}\epsilon. \tag{24}$$

Using the definition of the mixing efficiency $\gamma = \frac{Rf}{1-Rf}$, the turbulent mixing rate (per unit mass) becomes related to the dissipation rate of kinetic energy (instead of the dissipation rate of fluctuating density variance):

$$\frac{g}{\rho_0}\overline{\rho' w'} = \gamma\epsilon. \tag{25}$$

The turbulent diffusivity within this model is thus defined by, using (19) and (25)

$$K_\rho^{''} = \gamma \frac{\epsilon}{N^2} = \frac{R_f}{1 - R_f} \frac{\epsilon}{N^2}. \tag{26}$$

From a practical point of view, ϵ is measured from a single component, assuming isotropy (the validity of this assumption has been addressed in [40] and [36]). A constant value of 0.2 for γ is generally adopted. This value has been inferred from a conjecture by Osborn [25] (stemming from observations) that $Rf \leq 0.15$. Because the maximum value for Rf is not firmly established, an empirical value for γ equal to 0.2 is used. As pointed out by Caldwell & Moum [4], γ is not constant: values comprised between 0.05 and 0.48 have indeed been computed from microstructure measurements in the ocean (assuming $K_\rho^{''} = K_\rho^{'}$), which seem to depend upon the instability that creates the turbulence. In the atmosphere, γ displays the same variability about the 0.2 value.

3 Mixing in a stably stratified shear layer

In section 2, we introduced basic concepts of mixing along with two estimators of mixing (the flux Richardson number and the turbulent diapycnal diffusivity). We now illustrate these theoretical points and consider for this purpose the stably stratified shear layer. Most of the results presented in this section stem from [32] and [33].

3.1 The stably stratified shear layer: a brief introduction

The shear layer is the simplest case of a inhomogeneous shear flow. It is formed by two layers of fluids having different densities and horizontal velocities. The interface between the layers, which thickens by molecular diffusion, enlarges most efficiently if the interface is unstable. A necessary condition for instability is the Richardson number

$$J = \min_z \frac{g}{\rho_0} \frac{-d\bar{\rho}/dz}{\left(d\overline{U}/dz\right)^2} \tag{27}$$

to have a value smaller than 0.25 somewhere in the fluid (see e.g. [10]). The overbar in Eq. (27) refers to a horizontal average. If this condition is met, the Kelvin-Helmholtz instability develops in the vertical plane of the shear flow. It is therefore a two-dimensional instability. As the amplitude of the perturbation grows, nonlinear effects gain importance and makes the instability saturate while the flow vorticity organizes into quasi-horizontal vortices (with axis parallel to the mean shear). The vortices are most unstable to a (two-dimensional again) subharmonic instability which makes them pair (see e.g. [8] and, for a review, [13]). Secondary three-dimensional instabilities may also occur, at smaller scales than the two-dimensional instabilities. Further detail about the three-dimensional instabilities in a stably stratified shear layer may be found in, e.g., [28], [16] or [5].

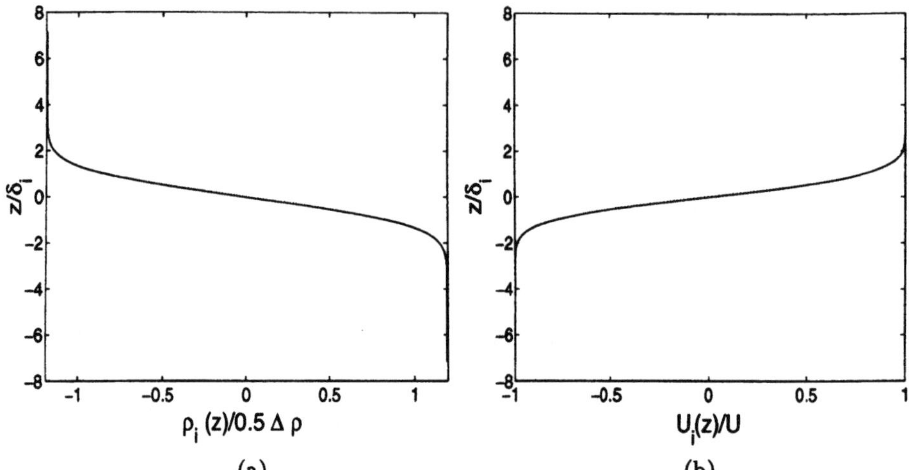

Fig. 3. Initial density profile (a) and velocity profile (b) of the runs. δ_i is half the (so-called) vorticity thickness, U and $\Delta\rho/\sqrt{Pr}$ respectively are half the velocity and density difference across the horizontal boundaries; Pr is the Prandtl number, equal here to 0.7 (from [32]).

3.2 Mathematical model and numerical method

To investigate the shear layer dynamics and resulting mixing properties, we solve the three-dimensional Boussinesq equations:

$$\frac{\partial \boldsymbol{u}}{\partial t} + (\boldsymbol{u}.\nabla)\boldsymbol{u} = -\nabla P + J\,\rho + \frac{1}{Re}\nabla^2 \boldsymbol{u} \tag{28}$$

$$\frac{\partial \rho}{\partial t} + (\boldsymbol{u}.\nabla)\rho = \frac{1}{Pr Re}\nabla^2 \rho \tag{29}$$

$$\nabla.\boldsymbol{u} = 0. \tag{30}$$

$\boldsymbol{u} = (u, v, w)$ denotes the velocity field with u, v and w being respectively the streamwise, spanwise and vertical components; P and ρ are the dynamical pressure deviation and the density deviation about a hydrostatic balanced state of reference density ρ_0; ν is the kinematic viscosity.

The initial velocity and density fields, $U_i(z)$ and $\rho_i(z)$ respectively, are of the error function type, the thickness of the density profile being $\sqrt{Pr}$ larger than that of the velocity profile (Fig. 3). A two-dimensional perturbation is superimposed upon U_i to destabilize the flow (the Kelvin-Helmholtz instability is thus able to grow). Three-dimensional perturbations are also added to create three-dimensional motions. The boundary conditions are periodic along the streamwise (x) and spanwise (y) directions and of free slip type along the vertical (z).

The Boussinesq equations (28–30) are written in non dimensional form. We take half the thickness of the initial velocity profile as a length scale, which we denote as δ_i, half the velocity difference across the flow as a velocity scale (U) and half the density difference across the flow times $\sqrt{Pr}$, as a density scale

$(\Delta\rho\sqrt{Pr}/2)$. The time unit is δ_i/U. This scaling sets three parameters J, Re and Pr, which control the flow dynamics. They are defined by:

$$J = \frac{g}{\rho_0}\frac{\Delta\rho\sqrt{Pr}}{2}\frac{\delta_i}{U^2} \qquad Re = \frac{U\delta_i}{\nu} \qquad Pr = \nu/\kappa. \tag{31}$$

A pseudo-spectral method is used to solve the Boussinesq equations: the equations are solved in Fourier space except for the nonlinear terms, which are computed in physical space; aliasing errors are removed by a truncation scheme; symmetry conditions are imposed along the z-direction so that the fields can be expanded in cosine or sine series along this direction. A third-order Adams-Bashforth scheme is used to advance the equations in time.

We have performed several runs in which J ranges from 0.04 to 0.16 and Pr has a value of 0.7 or 1.4. The resolution is $(N_x, N_y, N_z) = (128, 128, 257)$ and the domain size is $(\lambda, \lambda, 2\lambda)$, where λ is the most amplified wavelength predicted by linear theory (λ is nearly independent upon J, as long as $J < 1/4$ [12]). Hence, only one Kelvin-Helmholtz vortex can develop in the numerical domain. The Reynolds number $Re \simeq 300$. The approximate value for Re stems from the fact that a uniform but time varying viscosity is used. Indeed, in order not to have the flow dominated by both stratification and viscous effects at the end of the runs, the viscosity is allowed to vary in time according to the constraint that the Kolmogoroff scale $\eta = (\nu^3/\epsilon)^{1/4}$ remains equal to the grid size at any time. Computing ϵ (the dissipation rate of kinetic energy) at each time step provides the value of ν at that time. In practice, ν smoothly decreases by a factor 3 from its initial value (equal to 1/300) up to the time ϵ has reached a maximum value (the flow has become strongly turbulent) and eventually reaches a value lower than $\nu(0)$ at the end of the run (equal to 1/500).

An example of the flow development is displayed in Fig. 4. At $t = 34$ (Fig. 4a), the Kelvin-Helmholtz instability has grown and saturated: it is manifested as one Kelvin-Helmholtz vortex, already distorted by the growth of the three-dimensional perturbation. The vortex is located in between sheets, which are the usual braids reinforced by the baroclinic torque (see [7], [30]). The saturation of the Kelvin-Helmholtz instability is followed by the growth of a three-dimensional secondary instability. Fig. 4b displays the total vorticity field at the time the dissipation rate of kinetic energy ϵ is maximum: the three-dimensional instability has led to small scale structures, whose dominant vorticity component lies along the streamwise direction. This instability is thus of the Rayleigh-Taylor type, having occurred when heavy fluid lies over light fluid ([16]). The end of the run is visualized in Fig. 4c: the central part of the shear layer is relaminarizing and the turbulent activity, while decaying, is expulsed further outward of the central region.

In the following, the Kelvin-Helmholtz vortex will be referred to as a *large* scale structure, as opposed to the *small* scale three-dimensional structures. These adjectives may be quantified by the Reynolds number associated with each structure: $Re \simeq 800$ in the former case and $Re \simeq 25$ in the latter.

(a)

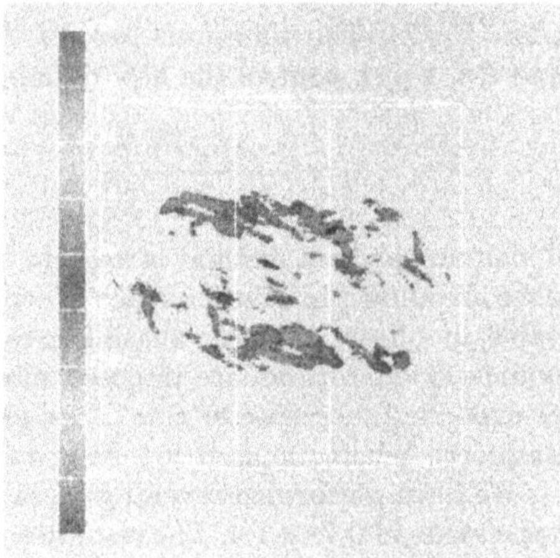

(b)

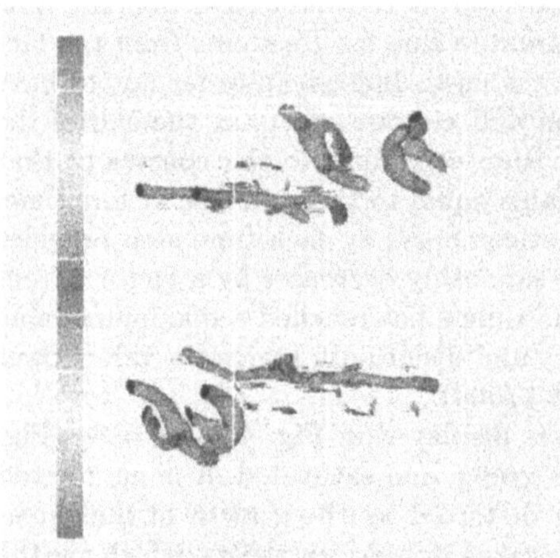

(c)

Fig. 4. Temporal evolution of a three-dimensional shear layer simulation with $J = 0.125$, $Re \simeq 1/300$ and $Pr = 1.4$. A surface equal to $0.5 \max_D(|\omega_x| + |\omega_y| + |\omega_z|)$, where $|\omega_i|$ is the i-component of the vorticity and D the fluid domain, is shown at successive times: (a) $t = 34$. (b) $t = 108$. (c) $t = 227$, at the end of simulation; only 3/4 of the numerical domain along the vertical direction is shown for clarity. The time unit is δ_i/U (δ_i and U are defined in section 3.2) (from [32]).

3.3 Analysis of mixing

The mixing properties of a stably stratified shear layer have been studied by a few authors. To our knowledge, the only laboratory experiments are those of Thorpe [34] and Koop & Browand [17]. All the other works are numerical ([29], [9], [31], [32], [6]). The method derived by Winters et al. [38] is used in the two latter papers only (we still note that background density profiles are computed in [29]).

• *Computational method of the background density profile*

The method to compute the background density profile ρ_b from the three-dimensional density field is as follows. The fluid domain is partitioned into elementary volumes of side equal to the grid size. Each elementary volume represents one fluid particle. These fluid particles are next sorted according to the value of their density. Two equivalent methods can be used to construct the stable density profile. In the first method ([38], [32]), the profile is constructed by filling the successive horizontal planes with the ordered particles, the bottom plane being filled with the heaviest particles. The value of the density slightly varies from particle to particle in a given plane (by a few percents at most in our computations) so that the value of the background density at each plane is computed by a horizontal average. In the second method ([6]), each particle is stretched horizontally so as to fill successively each horizontal plane; no horizontal average is therefore required. In numerical computations however, the density field needs to be interpolated if combined with other quantities, to match the vertical resolution.

• *Horizontally averaged and background density profiles*

The horizontally averaged and background density profiles, $\overline{\rho}(z)$ and $\rho_b(z)$ respectively, are plotted in Fig. 5 at the same key times as in Fig. 4, but for a different run (see caption). Figure 5a shows that $\rho_b(z)$ is stable, consistently with its definition, and remains close to the initial density profile. By contrast, $\overline{\rho}(z)$ is not monotonic and strongly differs from $\rho_b(z)$. This indicates that (i) the large scale Kelvin-Helmholtz vortex hardly mixes the fluid (because $\rho_b \simeq \rho_i$); (ii) a horizontal average does not remove reversible effects (because $\overline{\rho} \neq \rho_b$). Hence, the use of the horizontally averaged density profile is not a good approximation to the background density profile. In Fig. 5b, the two profiles are displayed when ubiquitous small scale three-dimensional motions have developed. The striking feature of the figure is the broadening of the sorted density profile: this reveals the dominant contribution to fluid mixing by the three-dimensional small scale structures. The horizontally averaged density profile displays the same trend but it fluctuates about the background profile and its gradient changes sign. Only at the end of the run (Fig. 5c), when the flow starts to relaminarize in its central part and turbulence decays in its outer part, do both profiles coincide. They have nearly become linear and strongly depart from the initial profile.

• *Turbulent mixing rate*

The temporal evolution of the turbulent mixing rate $\Phi_d - \Phi_{lam}$ offers a synthetic view of the respective contribution to mixing by the Kelvin-Helmholtz vortex and the three-dimensional structures. Fig. 6 thus shows that $\Phi_d - \Phi_{lam}$ displays a relative maximum at $t = 35$, when the Kelvin-Helmholtz instability saturates; $\Phi_d - \Phi_{lam}$ further increases when three-dimensional instabilities develop and, at the time turbulence is the strongest ($t \simeq 85$), reaches an absolute

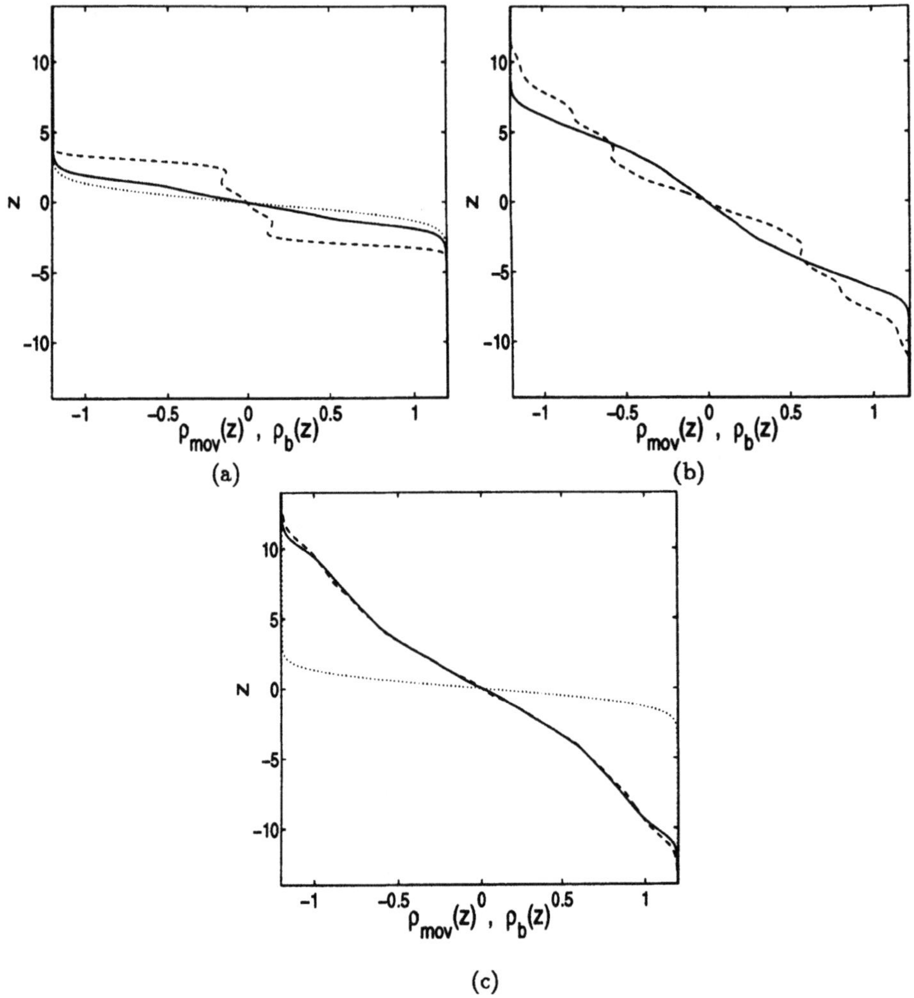

Fig. 5. Horizontally averaged $\overline{\rho}(z)$ (dashed line) and background $\rho_b(z)$ (full line) density profiles. In (a) and (c), the initial density profile is plotted for comparison (dotted line). (a) t=28 (when the Kelvin-Helmholtz vortex has developed); (b) t=108 (when three-dimensional small scale structures have formed); (c) t=227 (end of simulation). The parameters of this run are $J = 0.04$, $Pr = 0.7$, $Re \simeq 300$ (from Staquet [32]).

maximum which is nearly three times higher than the relative maximum (note that $\Phi_d/\Phi_{lam} \simeq 20$ when these maxima are reached).

• *Mixing efficiency*

The previous results demonstrate that mixing is promoted by small scale three-dimensional structures, and not by a large scale Kelvin-Helmholtz vortex. But what about the efficiency of mixing? To address this important point, and in order to interpret it properly, we first present in Fig. 7 the dissipation rate of

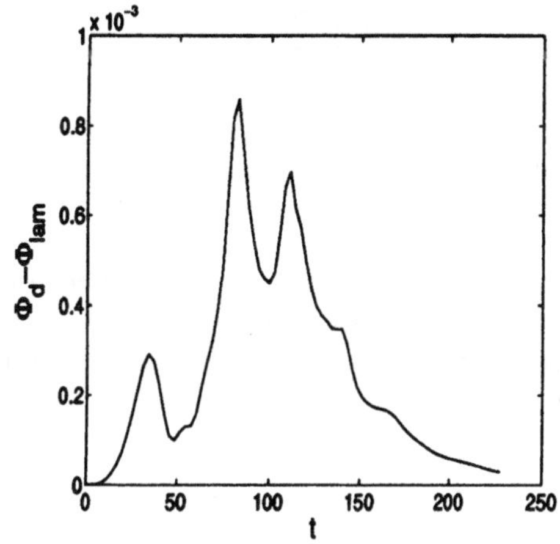

Fig. 6. Turbulent diapycnal flux $\Phi_d - \Phi_{lam}$ versus time for the same run as in Fig. 5.

kinetic energy ϵ for the same run as before. The behavior of ϵ is not surprising: it remains close to its initial value when the Kelvin-Helmholtz instability grows and saturates, because this instability, and the associated Kelvin-Helmholtz vortex, are two-dimensional. As is well-known (e.g. [20]), the material conservation of vorticity in a two-dimensional flow prevents energy from being transferred toward small scales, where dissipation is acting. By contrast, Fig. 7 shows that the growth of a three-dimensional instability, and its nonlinear manifestation as small scale structures, makes ϵ strongly increase, up to a value equal to nearly 20 times its initial value.

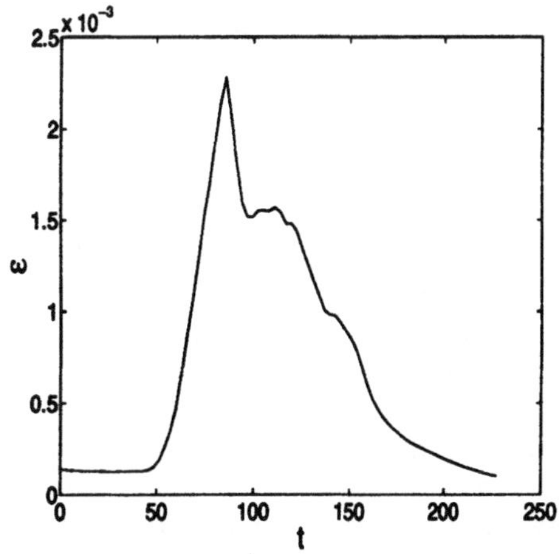

Fig. 7. Dissipation rate of the turbulent kinetic energy, ϵ, versus time for the same run as in Fig. 5.

The behavior of the mixing efficiency or, equivalently, of the flux Richardson number Rf_b can be immediately inferred from Figs. 6 and 7. Figure 8, where Rf_b is displayed versus time, thus shows that the efficiency of mixing is high (close to 0.7) during the initial two-dimensional stage of the flow and moderate (close to 0.3) when three-dimensional secondary structures have developed. In other words, the large scale Kelvin-Helmholtz vortex contributes weakly to mixing, but in an efficient manner (because nearly no energy is dissipated) while the three-dimensional structures are responsible for the mixing of the fluid, but much less efficiently (because they occur at scales close to dissipative scales). Hence, the Kelvin-Helmholtz vortex mainly stirs the fluid while the three-dimensional structures effectively mix it.

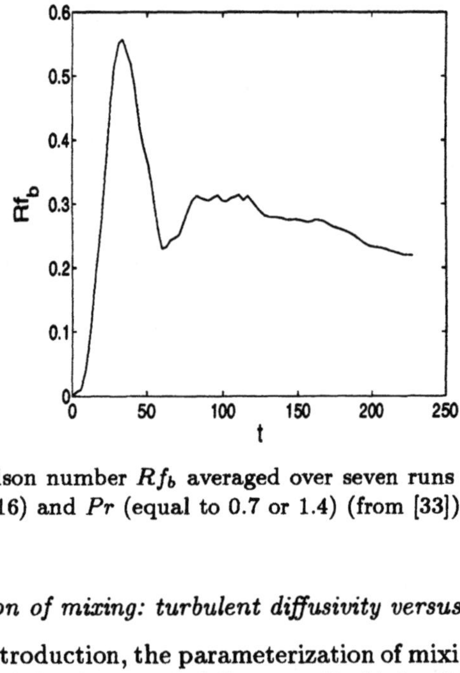

Fig. 8. Flux Richardson number Rf_b averaged over seven runs differing by the value of J (from 0.04 to 0.16) and Pr (equal to 0.7 or 1.4) (from [33]).

• *A parameterization of mixing: turbulent diffusivity versus Richardson number*

As said in the Introduction, the parameterization of mixing consists of a functional dependency of the diapycnal diffusivity K_b (defined by Eq. (16)) upon a dynamical parameter characterizing the largest scales; in a general circulation model, the latter scales belong to the resolved scales. Such a functional dependency may be seeked for in the numerical simulations of a stably stratified shear layer we presented because the largest scale is a shear flow (that is, with no horizontal scale) and is well separated from the scale of the three-dimensional motions that mix the fluid. One may thus assume the former motion to be associated with a current that would lie among the resolved scales in an oceanic circulation model while the latter motions are at subgrid scale.

The presence of a mean shear implies that a Richardson number should be taken for the dynamical parameter. We make use of the monotonicity of the background density profile ρ_b to introduce a Richardson number that remains one-signed at any time:

$$Ri_b(z) = -J \frac{d\rho_b/dz}{\left(d\overline{U}/dz\right)^2}. \tag{32}$$

The normalized diapycnal diffusivity $K_b(z)/\kappa$ is plotted as a function of $Ri_b(z)$ in Fig. 9a, when the turbulence has started to decay (for $t \geq 196$; no clear dependency could be found before this time). Two parts may be distinguished in Fig. 9a, where a log-log scale is used: for $Ri_b < 0.2$, $K_b(z)/\kappa$ decays quasi-linearly with $Ri_b(z)$. This decaying behavior is expected: when Ri_b decreases, stratification effects weaken relative to nonlinear effects so that the flow becomes more turbulent and mixing increases. This part of the curve is contributed by the outer region of the shear layer, where turbulence remains. Figure 9a shows that values of the turbulent diffusivity K_b up to 200 times the molecular diffusivity are found in this region. For $Ri_b(z) > 0.2$ by contrast, that is, in the central relaminarizing part of the shear layer, no clear dependency is observed.

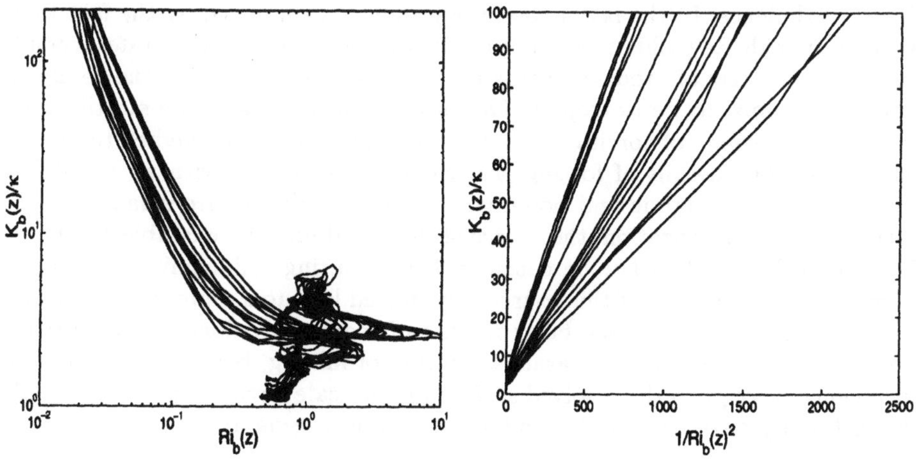

(a) (b)

Fig. 9. Diapycnal diffusivity $K_b(z)$ (defined by Eq. (16)) normalized by the molecular diffusivity κ as a function of a gradient Richardson number $Ri_b(z)$ (defined by Eq. (32)), for $t \geq 196$ (each curve corresponds to a different time). (a) Log-log scale; (b) Lin-lin scale. For this run, $J = 0.125$, $Pr = 1.4$.

The quasi-linear dependency of $K_b(z)/\kappa$ with $Ri_b(z)$ displays a -2 slope. This is confirmed in Fig. 9b, where $K_b(z)/\kappa$ is plotted versus $1/Ri_b^2(z)$: the linear dependency is recovered but the slope varies in time. In the ocean, below the upper mixed layer, several parameterizations are used depending upon the process that creates the turbulence: shear instability, internal gravity wave breaking and double diffusion (due to the simultaneous diffusion of heat and salt). For the former process, a parameterization of the form $K/K_{max} = [1 - (Ri/Ri_0)^2]^3$

is commonly used in large scale circulation models for $0 < Ri < Ri_0$, with $K_{max} = 5\ 10^{-3} m^2/s$ and $Ri_0 = 0.7$. The high value of Ri_0 stems from the fact that the measurements of Ri rarely fall as low as the theoretical value of 1/4. According to Large et al. [18] , this parameterization is supported by limited observations ([27]). Our numerical results suggest that this is a too sharp dependency.

4 Conclusion

As discussed in this article, basic concepts of mixing have been known for a long time (dating from Lorenz [22]) and are now properly formalized. The parameterization of mixing still remains a difficult problem. Parameterizing implies a fundamental understanding between mixing and flow dynamics. The existence of a link between mixing, which occurs at dissipative scales and the flow dynamics, whose energy reservoir is at large scale, is not surprising: mixing is controlled by these large scales, as is kinetic energy dissipation in (three-dimensional) ordinary turbulence. In the latter case, the problem is simplified by the fact that, at high enough Reynolds number, the kinetic energy dissipation rate depends only upon the rate of input of energy at large scale, and not upon the details of forcing (assumed to be isotropic). This is not true when a stable stratification exists: the memory of forcing is all the more important the higher the stratification is. The details of forcing is most often associated with an instability that triggers the (kinetic and potential) energy transfer toward small scale. To derive the parameterization, the source of the instability (an unstable shear flow for instance) should rather be considered. Computing a dynamical parameter from the characteristics of this source may indeed lead to a more general behavior than considering the instability itself, as is commonly done. This approach would also provide a usable parameterization of mixing, because the source of turbulence belongs to the resolved scales in large scale circulation models, while the instability that leads to fluid mixing does not in general.

Acknowledgments

I thank A. Beckman for having kindly provided Fig. 1 and A.-M. Treguier for useful discussions about parameterization of mixing in the ocean.

References

1. D.G. Andrews: J. Fluid Mech. **107**, 227 (1981)
2. A. Beckmann: 'Turbulent mixing in the ocean: intensity, causes and consequences'. In: *Proccedings of Les Houches Winter School, France, January 1998*, ed. by E. Chassignet, J. Verron (Kluwer Academic Publishers, 1998) pp. 135–154
3. F. Bryan: J. Phys. Oceanogr. **17**, 970 (1987)
4. D.R. Caldwell, J.N. Moum: Rev. Geophys. **Suppl.**, 1385 (1995)

5. C.P. Caulfield, W.R. Peltier: Phys. Fluids **6**(12), 3803 (1994)
6. C.P. Caulfield, W.R. Peltier: J. Fluid Mech. **413**, 1 (2000)
7. G.M. Corcos, F.S. Sherman : J. Fluid Mech, **73**, 241 (1976)
8. G.M. Corcos, F.S. Sherman : J. Fluid Mech, **139**, 29 (1984)
9. A.B. Cortesi, B.L. Smith, G. Yadigaroglu, S. Banerjee: Phys. Fluids **11**-1, 162 (1999)
10. P.G. Drazin, W.H. Reid: *Hydrodynamic stability* (Cambridge University Press, Cambridge 1981)
11. M.C. Gregg: J. Geophys. Res. **92-C5**, 5249 (1987)
12. P. Hazel: J. Fluid Mech. **51**, 39 (1972)
13. H. Ho, P. Huerre: Ann. Rev. Fluid Mech. **16**, 365 (1984)
14. D. Holliday, M.E. McIntyre: J. Fluid Mech. **107**, 221 (1981)
15. R.X. Huang: J. Phys. Oceanogr. **29**, 727 (1999)
16. G.P. Klaassen, W.R. Peltier: J. Fluid Mech. **227**, 71 (1991)
17. C.G. Koop, F.K. Browand: J. Fluid Mech. **93**, 135 (1979)
18. W.G. Large, J.C. McWilliams, S.C. Doney: Reviews of Geophysics **32-4**, 363 (1994)
19. J.R. Ledwell, A.J. Watson, C.S. Law: Nature **364**, 701 (1993)
20. M. Lesieur: 'Turbulence in fluids'. Kluwer Academic Pub. (1990)
21. P.N. Lombard, D.D. Stretch, J.J. Riley: 'Energetics of a stably stratified mixing layer'. In: *Ninth Symposium on Turbulence and Diffusion, Roskilde, Danemark, April 30–May 3, 1990*, (A.M.S. Publ. 1990), pp. 202–206
22. E.N. Lorenz: Tellus **7**, 157 (1955)
23. W. Munk: Deep-Sea Research **13**, 707 (1966)
24. W. Munk, C. Wunsch: Deep-Sea Research I **45**, 1977 (1998)
25. T.R. Osborn: J. Phys. Oceanography **4**, 83 (1980)
26. T.R. Osborn & C.S. Cox: Geophys. Fluid Dyn. **3**, 321 (1972)
27. H. Peters, M.C. Gregg, J.M. Toole: J. Geophys. Res. **93**, 1199 (1988)
28. D.G. Schowalter, J.C. Lasheras, C.W. van Atta: J. Fluid Mech. **281**, 247 (1994)
29. J.F. Scinocca: J. Atmos. Sciences **52**-14, 2509 (1995)
30. C. Staquet: J. Fluid Mech **314** (1995)
31. C. Staquet, K.B. Winters: 'Mixing in a stably-stratified shear layer'. In: *Proc. of the XIth Symposium on Turbulent Shear Flows, Grenoble, France, September 1997*, ed. by G. Binder, vol. 2, session 20, p. 25–30
32. C. Staquet: Fluid Dyn. Research **27**, 367 (2000)
33. C. Staquet, P. Bouruet-Aubertot: Dyn. Atmos. Oceans **34**, 81 (2001)
34. S.A. Thorpe: J. Fluid Mech. **61**, 731 (1973)
35. S.A. Thorpe: Phil. Trans. R. Soc. Lond. **A 286**, 125 (1977)
36. S. Thorroddsen, C.W. van Atta: J. Geophys. Res. **97-C3**, 3647 (1992)
37. J. Toole: 'Turbulent mixing in the ocean: intensity, causes and consequences'. In: *Proccedings of Les Houches Winter School, France, January 1998*, ed. by E. Chassignet, J. Verron (Kluwer Academic Publishers, 1998) pp. 171–190
38. K.B. Winters, P.N. Lombard, J.J. Riley, E.A. D'Asaro: J. Fluid Mech., **289**, 115 (1995)
39. K.B. Winters K., E.A. D'Asaro: J. Fluid Mech. **317**, 179 (1996)
40. H. Yamazaki, T. Osborn: J. Geophys. Res. **95**–C6, 9739 (1990)

Linear Processes in Stratified Turbulence with Rotation or Mean Shear

Hideshi Hanazaki[1]

Institute of Fluid Science, Tohoku University, 2-1-1 Katahira, Aoba-ku, Sendai, 980-8577, Japan

Abstract. Linear processes of the unsteady turbulence in stratified flow with system rotation or mean shear are analysed by the rapid distortion theory (RDT). In stratified *rotating* turbulence, the ratio of Coriolis parameter f to Brunt-Väisälä frequency N, i.e. f/N, determines the steady components of energy and the phase of energy/flux oscillation. On the other hand, unsteady aspects are dominated by stratification, and the energy/flux oscillates at frequency $\sim 2N$ as in the flow with only stratification. For stratified *shear* flow, the energy or the flux again oscillates at frequency $2N$, although the shear distorts the energy spectra so that they are more localized to smaller streamwise wavenumbers ($k_1 \rightarrow 0$). The results show that neither rotation nor mean shear affects the wave number components which dominate the buoyancy oscillation of the energy and the flux.

1 Introduction

The clarification of the transport and mixing mechanisms in stratified turbulence is important not only in understanding the fundamental dynamics of the flows in the atmosphere and ocean, but also in modeling and parameterising the subgrid-scale phenomena necessary for the numerical weather prediction. Simple stratification often significantly modifies the fundamental aspects of the isotropic turbulence, in particular at large scale components where values of the practically important turbulent energies and fluxes are determined. Stratification causes, for example, the wave motion due to the buoyancy force, anisotropy in the structure of turbulence, and inhibits the energy cascade to smaller scales.

In the first direct numerical simulation (DNS) of stratified turbulence, Riley, Metcalfe & Weissman (1981) found the periodic exchange of energy between the kinetic energy and the potential energy. The corresponding periodic oscillation of vertical density flux with zero mean has also been found. For strongly stratified turbulence they proposed the notion of wave/vortex decomposition, which decomposes the velocity into the *vortex* components with vertical vorticity and the *wave* components without the vertical vorticity. The decomposition is similar to the Helmholz decomposition which decomposes an arbitrary vector into non-divergent and irrotational components. While the vortex components comprises the unoscillatory components as found in the usual turbulence, the wave components show oscillations due to the restoring bouyancy force. So far there have been many DNS studies including those by Métais & Herring(1989), Gerz & Yamazaki(1993), Kimura & Herring(1996) and Staquet & Godeferd (1998).

These studies have clarified the effects of initial conditions on the energetics of turbulence, unsteady spectral behaviours, and the eddy structures of turbulence such as the pancake vortex.

In laboratory experiments, Komori, Ueda, Ogino & Mizushina (1983) found a counter-gradient density flux, a vertical turbulent flux in opposite direction to the molecular diffusion, which makes the turbulent diffusion coefficient negative. In succeeding experiments, Itsweire, Helland & Van Atta (1986), Lienhard & Van Atta (1990), Yoon & Warhaft (1990) and Komori & Nagata (1996) investigated the time development of the kinetic/potential energies, vertical density flux and their spectra.

In theoretical attempts to predict the special characters of stratified turbulence, Hunt, Stretch & Britter (1988) used the rapid distortion theory (RDT), which is a linear theory, and succeeded in reproducing the periodic oscillation of the energies and the fluxes. Hanazaki & Hunt (1996) solved the RDT equations analytically and showed explicitly that the peoriodic oscillation is the result of linear buoyant oscillation. They also showed, by applying the method of steepest descents or stationary phase, that the decaying oscillation observed in DNS and experiments occurs even in the inviscid fluid. This is because only the horizontal wavenumber components ($k_3 = 0$) with vertical fluid motion, which oscillates most slowly (at frequency N), contribute to the energy and the flux. The phenomena, in particular those in the context of diffusion, have been called later as 'phase mixing' (Kaneda & Ishida 2000). The studies based on RDT have shown the importance of initial conditions, such as the initial ratio of kinetic energy to the potential energy in determining the subsequent time development of turbulence. They explained the difference in the direction and amplitude of the vertical density flux observed in the DNS (e.g., Gerz & Yamazaki 1993; Métais & Herring 1989) where the different initial energy ratio gave different subsequent time developments. The RDT succeeded in predicting also the characteristics of the unsteady spectral behaviour which depend on the Prandtl number.

The effect of rotation on stratified turbulence is important in predicting the rotation effects of the earth in the atmosphere and the ocean. Since the Coriolis parameter f determined by the period of earth's rotation ($\sim O(10^{-5})$ sec^{-1}) is usually much smaller than the Brunt-Väisälä frequency N ($\sim O(10^{-2})$ sec^{-1}), effect of even a small f/N needs to be clarified to consider the realistic effects of rotation in the geophysical context. In the analysis of DNS data, Bartello (1995) and Métais, Bartello, Garnier, Riley & Lesieur (1996) utilized the wave/vortex decomposition and investigated the nonlinear energy-transfer in the spectral space. Recently Iida & Nagano (1999) and Tsujimura, Iida & Nagano (1998) investigated the rapidly rotating ($f/N \geq 1$) stratified turbulence by DNS and found periodic oscillation of the vertical density flux as in the stratified non-rotating turbulence. Some important basic mechanisms of the stratified and/or rotating turbulence are reviewed in Cambon (2001).

In this paper we show the analytical solutions of the RDT equations whcih clarify a number of important mechanisms of stratified rotating turbulence (cf. Hanazaki, 2000, 2002). In particular the effects of parameter ratio f/N, which would become important in the geophysical context, are investigated. While the linear theory has limitations in its applicability conditions, the analytical solutions are useful to clarify some essential mechanisms of turbulence with an external force, and will become a useful guide for the future DNS and experiments.

The effects of *mean shear* on stratified turbulence are also of vital importance in the realistic flows which appear in the atmosphere, ocean or engineering since the mean shear usually feeds energy to the turbulence and is necessary to maintain the fluctuations. One of the important parameters here is the Richardson number Ri $(= N^2/\alpha^2)$, which is the square of the ratio of Brunt-Väisälä frequency to mean vertical shear α $(= dU_1/dx_3)$. For Richardson numbers smaller than the critical value $Ri_{cri}(= 0.25)$ the mean flow has been said to be unstable to small fluctuations. Then the change of structure of turbulence with the variation of Ri is of much interest.

There have been many laboratory, theoretical and numerical studies (e.g., water tank experiments by Komori, Ueda, Ogino & Mizushina (1983) and Rohr, Itsweire, Helland & Van Atta (1988), wind-tunnel experiments by Piccirillo & Van Atta (1997), Direct numerical simulations (DNS) by Gerz, Schumann & Elghobashi (1989) and Holt, Koseff & Ferziger (1992), Jacobitz, Sarkar & Van Atta (1997) and large eddy simulation (LES) by Kaltenbach, Gerz & Schumann (1994)) on turbulence in stratified shear flows which investigated the unsteady aspects of the kinetic/potential energies, the vertical density flux and their spectra.

In this paper, the analytical solutions of RDT equations for stratified shear flow at arbitrary times $(0 < \alpha t < \infty)$ are discussed (cf. Hanazaki & Hunt 2001, 2002). The results are consistent, in many aspects, with the previous DNS and the experiments, clarifying the combined effects of shear and stratification on turbulent flows.

2 Stratified Rotating Flows

2.1 Governing Equations and the Applicability Conditions of RDT

First we consider a stratified rotating turbulence. The Boussinesq equations for the homogeneous turbulent flow with vertical density stratification $(d\overline{\rho}/dx_3)$ and system rotation around the vertical axis are

$$\frac{\partial \boldsymbol{u}}{\partial t} + (\boldsymbol{u} \cdot \nabla)\boldsymbol{u} + 2\boldsymbol{\Omega} \times \boldsymbol{u} = -\frac{1}{\rho_0}\nabla p - g\hat{\boldsymbol{x}}_3\frac{\rho}{\rho_0} + \nu\nabla^2\boldsymbol{u}, \tag{1}$$

$$\frac{\partial \rho}{\partial t} + (\boldsymbol{u} \cdot \nabla)\rho + u_3\frac{d\overline{\rho}}{dx_3} = \kappa\nabla^2\rho, \tag{2}$$

and

$$\mathrm{div}\boldsymbol{u} = 0, \tag{3}$$

where ρ is the density perturbation from $\overline{\rho}(x_3)$, $\boldsymbol{u}$ is the velocity fluctuations, $\boldsymbol{\Omega} = (0, 0, \Omega)$ is the angular velocity of the system rotation, g is the acceleration due to gravity, $\hat{\boldsymbol{x}}_3$ is the unit vector in the vertical upward direction, ρ_0 is the representative density, and ν and κ are the viscosity and diffusion coefficient respectively.

We then substitute the following spectral decompositions (Batchelor & Proudman,1954; Townsend,1976)

$$u_i = \sum_{\boldsymbol{k}} \hat{u}_i(\boldsymbol{k}, t) \exp(i\boldsymbol{k} \cdot \boldsymbol{x}) \quad (i = 1, 2, 3), \tag{4}$$

and

$$\frac{g}{\rho_0}\rho = \sum_{\boldsymbol{k}} \hat{\rho}(\boldsymbol{k}, t) \exp(i\boldsymbol{k} \cdot \boldsymbol{x}), \tag{5}$$

into the governig equations (1)–(3) and neglect the nonlinear terms so that we obtain a set of linear ordinary differential equations (RDT equations) given by

$$\left(\frac{d}{dt} + \nu k^2\right) \hat{u}_i + \left(\delta_{ij} - \frac{k_i k_j}{k^2}\right) \epsilon_{j3l} f \hat{u}_l = \left(\frac{k_i k_3}{k^2} - \delta_{i3}\right) \hat{\rho}, \tag{6}$$

and

$$\left(\frac{d}{dt} + \kappa k^2\right) \hat{\rho} = N^2 \hat{u}_3. \tag{7}$$

Here $f = 2\Omega$ is the Coriolis parameter and N is the Brunt-Väisälä frequency defined by $N^2 = -(g/\rho_0)(d\overline{\rho}/dx_3)$. With no mean shear, the wavenumber does not change with time, i.e.

$$\boldsymbol{k}(t) = \boldsymbol{k}(0) = (k_1, k_2, k_3). \tag{8}$$

Some discussions are necessary here on the conditions for which the linearisation (RDT) is valid. Formally, they are given by the conditions that the nonlinear term $(\boldsymbol{u} \cdot \nabla)\boldsymbol{u}$ ($\boldsymbol{u} = (u_1, u_2, u_3)$, $|\, \boldsymbol{u} \,| = O(u)$) in the Navier-Stokes equations is small compared to either the buoyancy term $g\rho/\rho_0$ (Derbyshire & Hunt, 1993) or the Coriolis term $2\boldsymbol{\Omega} \times \boldsymbol{u}$. At the same time the term $(\boldsymbol{u} \cdot \nabla)\rho$ must be small compared to $u_3 d\overline{\rho}/dx_3$ in the equation for the density. The above conditions are reduced to (Hanazaki 2002)

$$Fr_l \ll 1 \text{ (if } f/N < 1), \tag{9a}$$

$$Ro_l \left(\frac{f}{N}\right)^{1/2} = Fr_l \left(\frac{N}{f}\right)^{1/2} \ll 1 \text{ (if } f/N > 1). \tag{9b}$$

where eddy Froude number $Fr_l = u/Nl$ and eddy Rossby number $Ro = u/fl$ are used.

If we define also the turbulent Froude number $Fr (\equiv u_0/Nl_0)$ and the turbulent Rossby number $Ro (\equiv u_0/fl_0)$ based on the length (l_0) and velocity (u_0) scale of the energy-containing eddies, the applicability conditions can be rewritten using Fr and Ro. At low and moderate Re, $Fr \sim Fr_l$ while at high Re, $Fr_l \sim Fr(l_0/l)^{2/3}$ assuming that the energy dissipation rate is a constant at all scales ($\epsilon = u^3/l = u_0^3/l_0$), we find that at low and moderate Re, the applicability conditions become

$$Fr \ll 1 \text{ (if } f/N < 1), \tag{10a}$$

$$\text{or} \quad Ro\left(\frac{f}{N}\right)^{1/2} = Fr\left(\frac{N}{f}\right)^{1/2} \ll 1 \text{ (if } f/N > 1), \tag{10b}$$

while at high Re they become

$$Fr\left(\frac{l_0}{l}\right)^{2/3} \ll 1 \text{ (if } \frac{f}{N} < 1), \tag{11a}$$

$$\text{or} \quad Ro\left(\frac{l_0}{l}\right)^{2/3}\left(\frac{f}{N}\right)^{1/2} = Fr\left(\frac{l_0}{l}\right)^{2/3}\left(\frac{N}{f}\right)^{1/2} \ll 1 \text{ (if } f/N > 1), \tag{11b}$$

where l is the length scale of the eddy of interest.

Therefore, at high Re, the applicability condition has a restriction on the eddy size. In laboratory experiments and DNS for low- or moderate- Re flows, RDT is valid for low values of Fr (or $Ro(f/N)^{1/2}$). On the other hand, at high Re, for the smaller scales of turbulence with $l/l_0 < Fr^{3/2}$ $(f/N < 1)$ or $l/l_0 < Fr^{3/2}(N/f)^{3/4}$ $(f/N > 1)$, RDT is not valid at small scales even if $Fr \ll 1$ $(f/N < 1)$ or $Fr(N/f)^{1/2} \ll 1$ $(f/N > 1)$ is satisfied.

We note here that if the rotation is weak $(f/N < 1)$, the condition (9a), i.e. $Fr_l \ll 1$, is equivalent to the *length* scale relation $l \gg (\epsilon/N^3)^{1/2} (\equiv L_o$: Ozmidov scale) if the relation $\epsilon(l) = u(l)^3/l$ (=const) still holds at large scales for which the turbulence is affected by stratification. Then, if the energy containing scale is larger than the Ozmidov scale ($l_0 \gg L_o$), the energy containing eddies are dominantly governed by the linear processes. The equivalent condition $Fr \ll 1$, which can be derived by assuming $\epsilon = u_0^3/l_0$, is the condition that linear theory gives good approximation in the evaluation of the energies and the fluxes even when the small scales ($l < L_o$, i.e. $Fr_l > 1$) are affected by the nonlinear effects.

2.2 Inviscid fluid

When the fluid inviscid ($\nu = \kappa = 0$), the solutions of the RDT equations (6) and (7) become, e.g.,

$$\hat{\rho} = \hat{\rho}_0\left[\cos at + \frac{f^2k_3^2}{a^2k^2}(1-\cos at)\right] + \frac{N^2}{a}\hat{u}_{30}\sin at + \frac{fN^2k_3}{a^2k^2}(1-\cos at)(k_2\hat{u}_{10} - k_1\hat{u}_{20}), \tag{12}$$

and

$$\hat{u}_3(t) = \hat{u}_{30}\cos at + \frac{fk_3}{ak^2}(k_2\hat{u}_{10} - k_1\hat{u}_{20})\sin at - \frac{k_1^2+k_2^2}{ak^2}\hat{\rho}_0\sin at, \tag{13}$$

with

$$a = \frac{\left(N^2(k_1^2 + k_2^2) + f^2 k_3^2\right)^{1/2}}{k}, \tag{14}$$

where k denotes $|\boldsymbol{k}|$, the subscript 0 denotes the initial values, and we have also used the incompressibility condition $k_i \hat{u}_{i0} = 0$ (Hanazaki 2002; see Iida & Nagano 1999 for the case of $\hat{\rho}_0 = 0$).

We assume that the initial density fluxes are zero, i.e.,

$$\Phi_{\rho i}(\boldsymbol{k}, 0) = \frac{1}{2}\overline{\hat{\rho}_0^* \hat{u}_{i0} + \hat{\rho}_0 \hat{u}_{i0}^*} = 0 \; (i = 1, 2, 3), \tag{15}$$

where overline denotes the ensemble average. Then, the three-dimensional spectra become, e.g.,

$$\begin{aligned}
\Phi_{\rho 3}(\boldsymbol{k}, t) &= \frac{1}{2}\overline{\hat{\rho}^* \hat{u}_3 + \hat{\rho}\hat{u}_3^*} \\
&= -\frac{k_1^2 + k_2^2}{k^2 a}\left[\cos at + \frac{f^2 k_3^2}{k^2 a^2}(1 - \cos at)\right] \sin at \Phi_{\rho\rho}(\boldsymbol{k}, 0) \\
&+ \frac{N^2}{2a}\Phi_{33}(\boldsymbol{k}, 0) \sin 2at + \frac{f^2 N^2 k_3^2}{a^3 k^4} \\
&\times \left(k_2^2 \Phi_{11}(\boldsymbol{k}, 0) + k_1^2 \Phi_{22}(\boldsymbol{k}, 0) - 2k_1 k_2 \Phi_{12}(\boldsymbol{k}, 0)\right)(1 - \cos at) \sin at \\
&+ \frac{f N^2 k_3}{a^2 k^2}\left(k_2 \Phi_{13}(\boldsymbol{k}, 0) - k_1 \Phi_{23}(\boldsymbol{k}, 0)\right)(\cos at - \cos 2at),
\end{aligned} \tag{16}$$

where we have used $k_i \Phi_{ij} = k_j \Phi_{ij} = 0$ derived from the incompressibility condition $k_i \hat{u}_{i0} = 0$.

2.3 Initially isotropic turbulence

Initial conditions Since the turbulence is initially isotropic in usual laboratory experiments for grid turbulence and in most of the DNS, we consider here an initially isotropic turbulence. Then, the initial conditions for the three-dimensional spectra are given by

$$\Phi_{ij}(\boldsymbol{k}, 0) = \frac{E(k)}{4\pi k^2}\left(\delta_{ij} - \frac{k_i k_j}{k^2}\right), \tag{17}$$

and

$$\Phi_{\rho\rho}(\boldsymbol{k}, 0) = \frac{S(k)}{4\pi k^2} 2N^2, \tag{18}$$

where

$$KE_0 = \int_0^\infty E(k) dk, \tag{19}$$

and

$$PE_0 = \frac{1}{2N^2}\int \Phi_{\rho\rho}(\boldsymbol{k}, 0) d\boldsymbol{k} = \int_0^\infty S(k) dk, \tag{20}$$

are the initial turbulent kinetic and potential energy.

We write the wavenumber vector in spherical coordinates as

$$\begin{aligned} k_1 &= k \sin\theta \cos\phi, \\ k_2 &= k \sin\theta \sin\phi, \\ k_3 &= k \cos\theta, \end{aligned} \tag{21}$$

with

$$k^2 = k_1^2 + k_2^2 + k_3^2, \tag{22}$$

so that the frequency a defined by (14) can be rewritten as

$$a^2 = f^2 \cos^2\theta + N^2 \sin^2\theta. \tag{23}$$

Variances and covariances Using (21), we can calculate the vertical density flux and the energy components as

$$\begin{aligned} \overline{\rho u_3}(t) &= \int \Phi_{\rho 3}(\boldsymbol{k}, t) 2\pi k^2 dk \sin\theta d\theta \\ &= \frac{N^2}{2} E_0^{(C)} \int_0^\pi d\theta \frac{\sin^3\theta}{a^3} \sin at (N^2 \sin^2\theta \cos at + f^2 \cos^2\theta), \end{aligned} \tag{24}$$

$$\begin{aligned} \overline{u_1^2}(t) = \overline{u_2^2}(t) = \frac{2}{3} KE_0 &+ \frac{N^2}{8} E_0^{(C)} \int_0^\pi d\theta \frac{\sin^3\theta \cos^2\theta}{a^4} \\ &\times \left(4f^2 \cos^2\theta (\cos at - 1) + (f^2 - N^2) \sin^2\theta (1 - \cos 2at)\right), \end{aligned} \tag{25}$$

$$\overline{u_3^2}(t) = \frac{2}{3} KE_0 - \frac{N^2}{2} E_0^{(C)} \int_0^\pi d\theta \frac{\sin^5\theta}{a^2} \sin^2 at, \tag{26}$$

and

$$\begin{aligned} \overline{\rho^2}(t) = 2N^2 PE_0 &+ \frac{N^4}{4} E_0^{(C)} \int_0^\pi d\theta \frac{\sin^3\theta}{a^4} \\ &\times \left(N^2 \sin^2\theta (1 - \cos 2at) + 4f^2 \cos^2\theta (1 - \cos at)\right), \end{aligned} \tag{27}$$

where $E_0^{(C)} = KE_0 - 2PE_0$ is the deviation from the equilibrium state of turbulence.

We note that the sign of $E_0^{(C)}$ determines the direction of the oscillation. In other words, initial partition of energy between the kinetic and potential energy $KE_0/PE_0 < 2$ or > 2, determines the subsequent direction of the vertical density (heat) transport at a specific time.This is identical to the results for pure stratification ($f = 0, N \neq 0$) by Hanazaki & Hunt (1996).

In the case of pure rotation ($N = 0, f \neq 0$), energy partition does not change with time and $\overline{u_1^2}(t) = \overline{u_2^2}(t) = \overline{u_3^2}(t) = \frac{2}{3} KE_0$ holds at all times. This agrees with the DNS results for initially isotropic rotating turbulence which retained initial isotropy (Bardina, Ferziger & Rogallo 1985).

2.4 Long-time asymptotics

General case of $N \neq f$ At large times ($Nt \gg 1$ or $ft \gg 1$), the most slowly oscillating components in the integrand, the component which satisfies $\partial a/\partial\theta = 0$, most contribute to the integral. This is the essence of the method of stationary phase and it can be applied as has been done in the pure stratification (Hanazaki & Hunt 1996). Since the stationary phase $\theta_0 = \pi/2$ satisfies $\cos\theta_0 = 0$, the rotation effects do not last for a long time, since the Coriolis parameter f appears only in the form of $f\cos\theta$, and vanishes when $\cos\theta = 0$. Then, the vertical density transport is maintained essentially by stratification even with the rotation effects. Specifically, the long-time behaviours become, e.g.,

$$\overline{\rho u_3}(t) = \frac{N}{4}E_0^{(C)}\left(\frac{\pi N}{t|N^2-f^2|}\right)^{1/2}\sin\left(2Nt \pm \frac{\pi}{4}\right), \tag{28}$$

$$\overline{u_1^2}(t) = \overline{u_2^2}(t) = \frac{2}{3}KE_0 + E_0^{(C)}\left[\frac{N^2(23f^2-2N^2)}{24(f^2-N^2)^2} - \frac{N^2f^2(3f^2+4N^2)}{8(f^2-N^2)^3}I_A\right],$$

$$\overline{u_3^2}(t) = \frac{2}{3}KE_0 + E_0^{(C)}\left[\frac{N^2(5f^2-2N^2)}{6(f^2-N^2)^2} - \frac{N^2f^4}{2(f^2-N^2)^3}I_A\right]$$
$$+ \frac{1}{4}E_0^{(C)}\left(\frac{\pi N}{t|N^2-f^2|}\right)^{\frac{1}{2}}\cos\left(2Nt \pm \frac{\pi}{4}\right), \tag{30}$$

$$\overline{\rho^2}(t) = 2N^2PE_0 + E_0^{(C)}\left[-\frac{N^4(11f^2-2N^2)}{4(f^2-N^2)^2} + \frac{N^4f^2(5f^2+4N^2)}{4(f^2-N^2)^3}I_A\right]$$
$$- \frac{N^2}{4}E_0^{(C)}\left(\frac{\pi N}{t|N^2-f^2|}\right)^{\frac{1}{2}}\cos\left(2Nt \pm \frac{\pi}{4}\right), \tag{31}$$

where

$$I_A = \int_0^1 \frac{dx}{x^2 + \frac{N^2}{f^2-N^2}}$$
$$= \frac{(f^2-N^2)^{1/2}}{N}\tan^{-1}\frac{(f^2-N^2)^{1/2}}{N} \quad (f > N)$$
or
$$= \frac{(N^2-f^2)^{\frac{1}{2}}}{N}\log\frac{N-(N^2-f^2)^{1/2}}{f} \quad (f < N), \tag{32}$$

and the sign $\pm$ represents $+$ when $f > N$, and $-$ when $f < N$.

In the real atmosphere and ocean, typical values of f and N are $f \sim 7.3 \times 10^{-5}\text{sec}^{-1}$ and $N \sim 10^{-2}\text{sec}^{-1}$. Then, if $f \ll N$ is satisfied, the rotation effects would become negligible in a rather short time, since the method of stationary phase gives good approximations for $Nt \geq O(1)$.

It is important to note that there is a *phase shift* ($\pm\pi/4$) depending on the relative values of f and N, i.e. whether $f > N$ or $f < N$. This explains the apparent difference in the oscillation 'period' between the non-rotating results ($N > f$) and other rapidly rotating results ($f > N$) observed in DNS.

We note that, if $f \ll N$ is satisfied as in the real atmosphere, the energy ratio $ER(= PE/VKE)$ of the potential energy PE to the vertical kinetic energy VKE approximately becomes

$$ER = \frac{\frac{1}{2N^2}\overline{\rho^2}}{\frac{1}{2}\overline{u_3^2}} \sim \frac{3}{2} \quad (Nt \gg 1), \tag{33}$$

as in the case of pure stratification (Hanazaki & Hunt 1996), although weak dependence on f might be observed when $f \neq 0$.

We clearly note that the final 'steady' values of the energy components depend on the ratio of the Coriolis parameter f to the Brunt-Väisälä frequency N, i.e. f/N, in addition to the initial kinetic energy KE_0 and the initial potential energy PE_0.

Bartello (1995) did DNS for decaying stratified rotating turbulence under the conditions of $N = 5.3, f = 8.7, PE_0 = 0$ and $Pr = 1$. Substituting $f/N = 1.64$ and $PE_0 = 0$ into (29)-(31), we obtain the long-time limit values as $\overline{u_1^2} = \overline{u_2^2} = 0.54KE_0$, $\overline{u_3^2} = 0.44KE_0$ and $PE = (1/2N^2)\overline{\rho^2} = 0.24KE_0$. Then we obtain $KE/PE(t \to \infty) = 3.2$ (and $ER(t \to \infty) = 1.1$), in good agreement with the DNS results which gave $KE/PE(t \to \infty) \sim 3.0$ (figure 5 of Bartello 1995).

Special case of $N = f$ In the special case of $N = f$, the exact solution becomes simple and the integration can be done exactly to give the variances and the covariances at arbitrary times. The results become

$$\overline{\rho u_3}(t) = \frac{2}{15} N E_0^{(C)} (\sin Nt + 2\sin 2Nt), \tag{34}$$

$$\overline{u_1^2}(t) = \overline{u_2^2}(t) = \frac{2}{3}KE_0 + \frac{2}{15}E_0^{(C)}(\cos Nt - 1), \tag{35}$$

$$\overline{u_3^2}(t) = \frac{2}{3}KE_0 + \frac{4}{15}E_0^{(C)}(\cos 2Nt - 1), \tag{36}$$

and

$$\overline{\rho^2}(t) = 2N^2 PE_0 + \frac{4}{15}N^2 E_0^{(C)}(2 - \cos Nt - \cos 2Nt). \tag{37}$$

We note that, when $N = f$, oscillations in the variances and the covariances do not decay with time in contrast to the case of $N \neq f$ (see also Kaneda 2000). In this case, $a(= N = f)$ is independent of θ so that all the spectral components oscillate at the same period $2\pi/N(= 2\pi/f)$ independent of the direction of the wave number vector. Then there is no phase difference which leads to the inviscid decay of oscillation (Hanazaki & Hunt 1996; Kaneda & Ishida 2000), and the energies never reach constant values.

It is also of interest to note that $\overline{u_1^2}(= \overline{u_2^2})$ contains only the $\cos Nt$ component and not $\sin 2Nt$ or $\cos 2Nt$, showing that the horizontal kinetic energy oscillates slower than the vertical kinetic energy and the potential energy. As will be discussed later, the slow oscillation with frequency N has been observed in DNS (Iida & Nagano 1999) when the initial turbulence is anisotropic and $N = f$ is satisfied. The present results identify that it is not the result of initial anisotropy but because of $N = f$.

2.5 Examples

Figure 1 shows the time development of the vertical density flux in the stablly stratified rotating turbulence obtained by RDT. There is a phase difference between the case of $N = 2, f = 0$ (bold solid line) and all the other cases. When there is no rotation, $N > f$ is satisfied and the flux is proportional to $\sin(2Nt - \pi/4)$, showing the phase delay. On the other hand, when rapid rotation is imposed as in all the other cases, $N < f$ is satisfied and the flux is proportional to $\sin(2Nt + \pi/4)$, showing the phase advance.

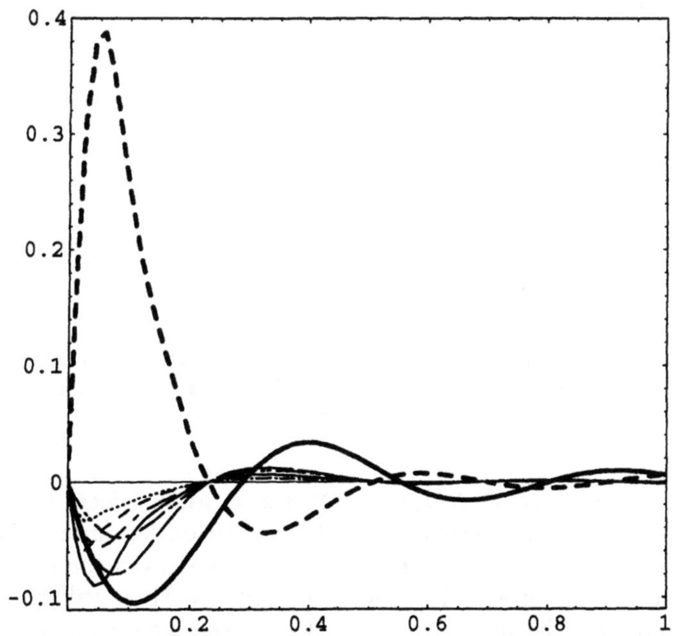

Fig. 1. Time development of the verticel density flux $-\overline{\rho u_3}/N^2$ for initially isotropic stably stratified rotating turbulence ($PE_0 = 0$, $Pr = 1$) obtained by RDT. The abscissa is $Nt/2\pi$ and the ordinate is $-\overline{\rho u_3}/N^2$. ————(bold line), $N = 2, f = 0$; ———, $N = 1, f = 10$; - - - -, $N = 1, f = 20$; — — —, $N = 2, f = 10$; —·—·—, $N = 2, f = 20$; ...., $N = 2, f = 40$; —··—··—, $N = 4, f = 20$; - - - -(bold line), $PE_0 = 4KE_0$, $N = 2, f = 20$.

When $E_0^{(C)} > 0$ or $KE_0 > 2PE_0$ is satisfied, e.g. when there is no initial potential energy, $-\overline{\rho u_3}$ initially goes negative, while if large $PE_0(> KE_0/2)$ is applied, the flux goes positive. In one case ($N = 2, f = 20$: bold dashed line), initial potential energy PE_0 is not zero and $PE_0 = 4KE_0(> KE_0/2)$. In that case the flux reverses its sign.

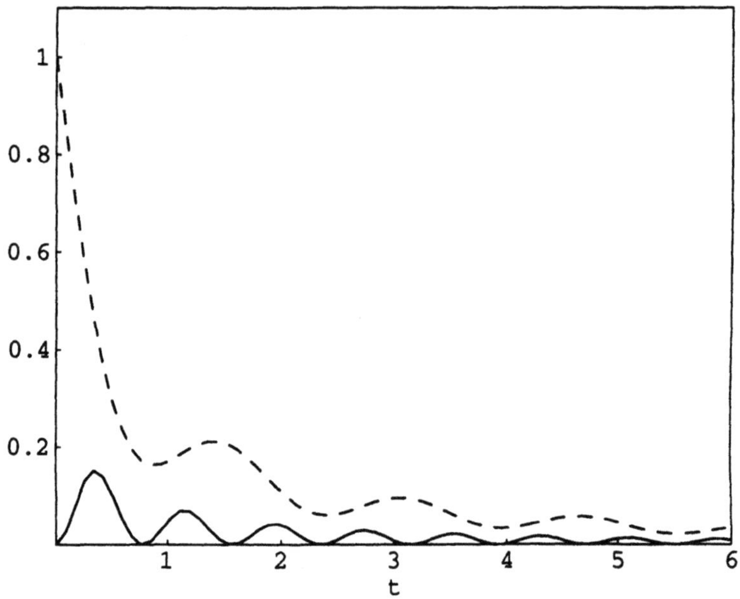

Fig. 2. Time development of the horizontal and vertival kinetic energy for initially axisymmetric and purely horizontal stably stratified rotating turbulence calculated by RDT with $N = f(= 4)$ $(Pr = 1)$. – – – –, $\overline{u_1^2}(= \overline{u_2^2})$; ———, $\overline{u_3^2}$.

At larger times ($0.05 < Nt/2\pi < 0.2$) the rotation effects begin to work, but as (28) suggests, the amplitude of the oscillation becomes approximately proportional to f^{-1}, provided that $f \gg N$ is satisfied as in the cases described in Fig. 1. This is also observed in Fig. 1, noting that three values are used for f (=10,20 and 40). The results for $f = 10$ show the largest amplitude and the results for $f = 40$ show the smallest amplitude. However, the rotation effect diminishes rapidly with time except for its effect on the amplitude and the time oscillation period ($= 2N$) is equal to twice the stratification parameter N as shown by (28). The amplitude difference for different values of f (and N) in a long time is smaller in DNS compared to RDT. One possible explanation is the effect of nonlinearity.

Figure 2 shows the RDT results for the initially axisymmetric and purely horizontal turbulence. Although the results are not for the initially isotropic turbulence, it is important to note that the results given here are for a special case of $N = f$ and the exact RDT solutions have forms similar to (34)-(37).

Therefore, the horizontal kinetic energy $\overline{u_1^2}(=\overline{u_2^2})$ oscillates with frequency N in contrast to the other variances and covariances which oscilltate at frequency $2N$. We should also note here that the decay of the energy with time when $N = f$ is purely the viscosity/diffusion effect and not due to the inviscid mechanism. When $a = N = f$ holds, all the wave number components with different directions (θ) oscillate in phase. Therefore, if the fluid is inviscid, the amplitude does not decay with time, in contrast to the general case of $N \neq f$.

3 Stratified Shear Flows

3.1 Governing Equations and the Applicability Conditions of RDT

We next consider a homogeneous turbulent flow both stratified ($d\overline{\rho}/dx_3$) and sheared ($\alpha = dU_1/dx_3$) in the vertical (x_3) direction (Hanazaki & Hunt 2002). The governing equations of rapid distortion theory (RDT) (Batchelor & Proudman 1954; Townsend 1976; Stretch 1986; Hunt, Stretch & Britter 1988) which describe the turbulence in the frame of reference moving with the mean shear flow are

$$\left(\frac{d}{dt} + \nu k^2\right)\hat{u}_i = \alpha\hat{u}_3\left(\frac{2k_ik_1}{k^2} - \delta_{i1}\right) + \left(\frac{k_ik_3}{k^2} - \delta_{i3}\right)\hat{\rho}, \tag{38}$$

$$\left(\frac{d}{dt} + \kappa k^2\right)\hat{\rho} = N^2\hat{u}_3. \tag{39}$$

When there is vertical shear, the wave number vector develops with time as

$$\frac{dk_i}{dt} = -\alpha k_1\delta_{i3}, \tag{40}$$

giving

$$\boldsymbol{k}(t) = (k_1, k_2, k_3) = (k_{10}, k_{20}, k_{30} - \alpha t k_{10}), \tag{41}$$

where we have denoted the initial wave number by $\boldsymbol{k}_0 = (k_{10}, k_{20}, k_{30})$.

We again use a spherical coordinate in the spectral space defined by

$$k_1 = k\sin\theta\cos\phi, \quad k_2 = k\sin\theta\sin\phi, \quad k_3 = k\cos\theta. \tag{42}$$

where $0 \leq \theta \leq \pi$ and $0 \leq \phi \leq 2\pi$. Then, initial wave number vector satisfies

$$k_{10} = k_0\sin\theta_0\cos\phi_0, \quad k_{20} = k_0\sin\theta_0\sin\phi_0, \quad k_{30} = k_0\cos\theta_0, \tag{43}$$

with $k_0 = (k_{10}^2 + k_{20}^2 + k_{30}^2)^{1/2}$, and the distorted wave number at time t becomes

$$k_1(t) = k(t)\sin\theta(t)\cos\phi_0, \quad k_2(t) = k(t)\sin\theta(t)\sin\phi_0, \quad k_3(t) = k(t)\cos\theta(t). \tag{44}$$

Since only the k_3 component changes with time, θ changes with time, while $\phi(=\phi_0)$ and the horizontal wavenumber $k_H \equiv (k_1^2 + k_2^2)^{1/2} = (k_{10}^2 + k_{20}^2)^{1/2} = k_0\sin\theta_0$ are invariant with time.

The applicability conditions of RDT for stratified shear flow are given by the conditions that the nonlinear term $(\boldsymbol{u}\cdot\nabla)\boldsymbol{u}$ ($\boldsymbol{u}=(u_1,u_2,u_3)$, $|\boldsymbol{u}|=O(u)$) in the Navier-Stokes equations is small compared to either the buoyancy term $g\rho/\rho_0$ (Derbyshire & Hunt 1993) or the mean shear term $u_3 dU/dx_3 (=\alpha u_3)$. At the same time the term $(\boldsymbol{u}\cdot\nabla)\rho$ must be small compared to $u_3 d\overline{\rho}/dx_3$ in the equation for the density.

For not small $t(Nt \geq 1)$, $Fr_l = u/Nl \ll 1$ is the necessary condition for $O((\boldsymbol{u}\cdot\nabla)\rho) < O(u_3 \mathrm{d}\overline{\rho}/\mathrm{d}x_3)$, and this also assures $O((\boldsymbol{u}\cdot\nabla)\boldsymbol{u}) < O((g/\rho_0)\rho)$.

The ratio of the nonlinear term to the mean shear term is given by $1/Sh_l$, where the eddy shear number Sh_l is defined by $Sh_l = \alpha l/u$. Since Sh_l and Fr_l has a simple relation given by $1/Sh_l = Fr_l Ri^{1/2}$, for a fixed Richardson number ($Ri \leq O(1)$), small $1/Sh_l$ is assured automatically when $Fr_l \ll 1$. Then, the condition $1/Sh_l \ll 1$ is of only secondary importance for the validity of RDT when the condition $Fr_l \ll 1$ is already satisfied. Therefore, the applicability conditions for RDT are essentially the same as the stratified flow with no shear which have been given by Hanazaki & Hunt (1996).

We should note that, at low and moderate Reynolds numbers as in laboratory experiments and DNS, the eddy Froude number $Fr_l = u/Nl \ll 1$ is of the same order as the Froude number defined by $Fr_0 = u_0/Nl_0$, where u_0 is the rms velocity and l_0 is the scale of the energy containing eddy as in §2. Then the RDT equations (38) and (39) are valid for $Fr_0 \ll 1$. On the other hand, at high Re, RDT is not valid for the small scales ($l < l_0$) of turbulence even if Fr_0 is small. We finally mention that for small Richardson numbers ($Ri < 0.25$) when linear instability occurs (Miles 1961), the comparison of RDT with DNS generally gives poorer agreement. This is because the condition of $Fr_l \ll 1$ is violated over a wide wavenumber range.

3.2 Inviscid Fluid

As is clear in (38) and (39), the equations for $\hat{u_3}$ and $\hat{\rho}$ do not contain $\hat{u_1}$ and $\hat{u_2}$. Therefore these two equations can be solved independently from the other equations in (38) and (39). Eliminating $\hat{u_3}$, we obtain the single equation for $\hat{\rho}$ as

$$\left(k_{10}^2 + k_{20}^2 + (k_{30} - \alpha t k_{10})^2\right)\frac{d^2\hat{\rho}}{dt^2} - 2\alpha k_{10}(k_{30} - \alpha t k_{10})\frac{d\hat{\rho}}{dt} + (k_{10}^2 + k_{20}^2)N^2\hat{\rho} = 0. \tag{45}$$

This equation becomes a Legendre equation

$$(1-z^2)\frac{d^2\hat{\rho}}{dz^2} - 2z\frac{d\hat{\rho}}{dz} + \nu(\nu+1)\hat{\rho} = 0, \tag{46}$$

if we define a new variable z by

$$\begin{aligned} z(t) &= \mathrm{i}\frac{k_{30} - \alpha t k_{10}}{(k_{10}^2 + k_{20}^2)^{1/2}} = \mathrm{i}(\cot\theta_0 - \alpha t \cos\phi_0) \\ &= \mathrm{i}\frac{k_3(t)}{k_H} = \mathrm{i}\cot\theta(t), \end{aligned} \tag{47}$$

and select ν as

$$\nu(\nu+1) = -\frac{(k_{10}^2 + k_{20}^2)N^2}{\alpha^2 k_{10}^2}, \tag{48}$$

i.e.,

$$\nu = \frac{1}{2}\left[-1 \pm \left(1 - 4\frac{(k_{10}^2 + k_{20}^2)N^2}{\alpha^2 k_{10}^2}\right)^{1/2}\right] = \frac{1}{2}\left[-1 \pm \left(1 - 4\frac{Ri}{\cos^2\phi_0}\right)^{1/2}\right]. \tag{49}$$

The solution of (46) consists of two Legendre functions of the first and second kind, i.e., P_{ν_1} and Q_{ν_1}. Since the two possible values of ν, i.e. ν_1 and ν_2 which takes the sign of + and − respectively in (49), satisfy the relation $\nu_2 = -\nu_1 - 1$, the general relations between the Legendre functions $Q_{\nu_1} = P_{-\nu_1-1} = P_{\nu_2}$ and $Q_{\nu_2} = P_{-\nu_2-1} = P_{\nu_1}$ show that the solution is independent of the choice of ν. Then we hereafter use ν to represent ν_1, and write the solution as

$$\hat{\rho}(t) = C_1\, P_\nu(z) + C_2\, Q_\nu(z), \tag{50}$$

and substitute (50) into (39) to obtain $\hat{u}_3$ as

$$\hat{u}_3(t) = -\frac{\mathrm{i}\alpha\cos\phi_0}{N^2}\left(C_1 P'_\nu(z) + C_2 Q'_\nu(z)\right). \tag{51}$$

Here, C_1 and C_2 are the constants determined from the initial conditions $\hat{\rho}_0$ and $\hat{u}_{30}$, i.e.,

$$C_1 = \frac{1}{\sin^2\theta_0} Q'_\nu(z_0)\hat{\rho}_0 + \frac{N^2}{\mathrm{i}\alpha\cos\phi_0\sin^2\theta_0} Q_\nu(z_0)\hat{u}_{30}, \tag{52}$$

$$C_2 = -\frac{1}{\sin^2\theta_0} P'_\nu(z_0)\hat{\rho}_0 - \frac{N^2}{\mathrm{i}\alpha\cos\phi_0\sin^2\theta_0} P_\nu(z_0)\hat{u}_{30}, \tag{53}$$

where $z_0 = z(0) = \mathrm{i}\cot\theta_0$ denotes the initial value of z and the primes denote differentiation by z, i.e.,

$$P'_\nu(z_0) = \left.\frac{dP_\nu}{dz}\right|_{z=z_0}. \tag{54}$$

Next we calculate the three-dimensional spectra. In this study we assume that the initial density fluxes are zero, i.e.,

$$\Phi_{\rho i}(\boldsymbol{k}_0, 0) = \frac{1}{2}\overline{\hat{\rho}_0^*\hat{u}_{i0} + \hat{\rho}_0\hat{u}_{i0}^*} = 0 \ (i = 1, 2, 3). \tag{55}$$

Then, the results become, e.g.,

$$\begin{aligned}\Phi_{\rho 3}(\boldsymbol{k}(t), t) &= \frac{1}{2}\overline{\hat{\rho}^*\hat{u}_3 + \hat{\rho}\hat{u}_3^*}\\ &= \mathrm{Re}\left[\frac{\mathrm{i}\alpha\cos\phi_0}{N^2\sin^4\theta_0}\left(P_\nu(z)Q'_\nu(z_0) - P'_\nu(z_0)Q_\nu(z)\right)\right.\\ &\times \left(P'_\nu(z)Q'_\nu(z_0) - P'_\nu(z_0)Q'_\nu(z)\right)^* \Phi_{\rho\rho}(\boldsymbol{k}_0, 0)\\ &+ \frac{\mathrm{i}N^2}{\alpha\sin^4\theta_0\cos\phi_0}\left(P_\nu(z)Q_\nu(z_0) - P_\nu(z_0)Q_\nu(z)\right)\\ &\times \left.\left(P'_\nu(z)Q_\nu(z_0) - P_\nu(z_0)Q'_\nu(z)\right)^* \Phi_{33}(\boldsymbol{k}_0, 0)\right],\end{aligned} \tag{56}$$

where the asterisks denote the complex conjugates. This shows that the three-dimensional spectra are determined by Ri and αt, which are contained in ν and z respectively.

It is important to note that for the components of $\cos\phi = \cos\phi_0 = 0$ ($k_1 = k_{10} = 0$), the spectra are identical with the no shear results since α is always coupled with $k_1 (= k_{10})$ in (41). For example, the spectrum for the vertical density flux becomes

$$\Phi_{\rho 3}(\boldsymbol{k}(t), t) = -\frac{\sin\theta}{2N}\Phi_{\rho\rho}(\boldsymbol{k}_0, 0)\sin(2Nt\sin\theta) + \frac{N}{2\sin\theta}\Phi_{33}(\boldsymbol{k}_0, 0)\sin(2Nt\sin\theta), \tag{57}$$

in agreement with the solution for no-shear flows (Hanazaki & Hunt 1996). This shows that the oscillations near $\phi = \pi/2, 3\pi/2$ ($\cos\phi = 0$) are not affected by the shear so that the energies and the fluxes oscillate at period π/N even in a shear flow.

3.3 Initially Isotropic Turbulence

We consider the initially isotropic turbulence here. Using (17)-(20) with k being replaced by k_0 and noting that the integration by k_0 can be done separately, we obtain the vertical flux of density as

$$\begin{aligned}
\overline{\rho u_3}(t) &= \int \Phi_{\rho 3}(\boldsymbol{k}(t), t) d\boldsymbol{k}_0 \\
&= \int_0^\infty k_0^2 dk_0 \int_0^\pi \sin\theta_0 d\theta_0 \int_0^{2\pi} d\phi_0 \Phi_{\rho 3}(\boldsymbol{k}(t), t) \\
&= \frac{\alpha}{2\pi} PE_0 \int d\theta_0 d\phi_0 \frac{\cos\phi_0}{\sin^3\theta_0} \\
&\times \mathrm{Re}\left[\mathrm{i}\left(P_\nu(z)Q'_\nu(z_0) - P'_\nu(z_0)Q_\nu(z)\right)\left(P'_\nu(z)Q'_\nu(z_0) - P'_\nu(z_0)Q'_\nu(z)\right)^*\right] \\
&+ \frac{\alpha}{4\pi} RiKE_0 \int d\theta_0 d\phi_0 \frac{1}{\sin\theta_0\cos\phi_0} \mathrm{Re}[\mathrm{i}\left(P_\nu(z)Q_\nu(z_0) - P_\nu(z_0)Q_\nu(z)\right) \\
&\times \left(P'_\nu(z)Q_\nu(z_0) - P_\nu(z_0)Q'_\nu(z)\right)^*].
\end{aligned} \tag{58}$$

Other variances and covariances can be calculated similarly, i.e.,

$$\begin{aligned}
\overline{\rho^2}(t) &= \frac{N^2}{2\pi} PE_0 \int d\theta_0 d\phi_0 \frac{1}{\sin^3\theta_0} \left|P_\nu(z)Q'_\nu(z_0) - P'_\nu(z_0)Q_\nu(z)\right|^2 \\
&+ \frac{N^2}{4\pi} RiKE_0 \int d\theta_0 d\phi_0 \frac{1}{\sin\theta_0\cos^2\phi_0} \left|P_\nu(z)Q_\nu(z_0) - P_\nu(z_0)Q_\nu(z)\right|^2
\end{aligned} \tag{59}$$

$$\begin{aligned}
\overline{u_3^2}(t) &= \frac{1}{2\pi Ri} PE_0 \int d\theta_0 d\phi_0 \frac{\cos^2\phi_0}{\sin^3\theta_0} \left|P'_\nu(z)Q'_\nu(z_0) - P'_\nu(z_0)Q'_\nu(z)\right|^2 \\
&+ \frac{1}{4\pi} KE_0 \int d\theta_0 d\phi_0 \frac{1}{\sin\theta_0} \left|P'_\nu(z)Q_\nu(z_0) - P_\nu(z_0)Q'_\nu(z)\right|^2.
\end{aligned} \tag{60}$$

We note here that the time development of the variances and the covariances, such as the kinetic and potential energies and the vertical density flux, are determined by the initial total kinetic (KE_0) and potential (PE_0) energies and not by their specific spectral forms as far as the fluid is inviscid. This is the same as the stratified turbulence with no shear (Hanazaki & Hunt 1996). Since the expressions contain ν and z, the variances and the covariances are determined by Ri and αt, similarly to the three-dimensional spectra.

Note that the integrand is the same as the no shear flow at $\cos\phi = 0$ ($k_1 = 0$). Therefore, if the energies and the fluxes have dominant contributions near $\cos\phi = 0$ ($\phi = \pi/2, 3\pi/2$), the energies and the fluxes oscillate with period π/N just like the no shear flow.

From (38) and (39) we can directly derive the following useful general relations among the three-dimensional spectra:

$$\left(\frac{d}{dt} + 2\nu k^2\right) \sum_i \Phi_{ii} = -2\alpha\Phi_{13} - 2\Phi_{\rho 3}, \tag{61}$$

$$\left(\frac{d}{dt} + 2\kappa k^2\right) \Phi_{\rho\rho} = 2N^2\Phi_{\rho 3}. \tag{62}$$

These equations show that $\overline{u_1 u_3}$ extracts kinetic energy from the mean shear, while the potential energy is supplied only through the exchange between the kinetic energy by $\overline{\rho u_3}$. Adding these two equations and integrating, we obtain the total energy budget equation for the inviscid fluid as

$$KE(t) + PE(t) = -\alpha \int_0^t \overline{u_1 u_3} dt + KE_0 + PE_0. \tag{63}$$

3.4 Effects of Viscosity and Diffusion

When there is viscosity or diffusion, the solutions of (38) and (39) generally become much more complicated, but if the Prandtl number $Pr (= \nu/\kappa)$ is equal to 1, the solutions take a much simpler form. For example, the viscous and diffusive three-dimensional spectra $\Phi_{\rho 3}(\boldsymbol{k}, t)_{\mathrm{inv}}$ becomes

$$\Phi_{\rho 3}(\boldsymbol{k}, t) = \Phi_{\rho 3}(\boldsymbol{k}, t)_{\mathrm{inv}} D^2(t), \tag{64}$$

where subscript inv denotes the inviscid solutions and the damping function $D^2(t)$ is given by

$$\begin{aligned} D^2(t) &= e^{-2\nu t\left(k_0^2 - \alpha t k_1 k_{30} + \frac{1}{3}(\alpha t)^2 k_1^2\right)} \\ &= e^{-2\nu t k_0^2\left(1 - \alpha t \sin\theta_0 \cos\theta_0 \cos\phi_0 + \frac{1}{3}(\alpha t)^2 \sin^2\theta_0 \cos^2\phi_0\right)}. \end{aligned} \tag{65}$$

The calculation of the energy and the fluxes (58)-(60) requires integrations like

$$\overline{\rho u_3}(t) = \int_0^\infty k_0^2 dk_0 \int_0^\pi \sin\theta_0 d\theta_0 \int_0^{2\pi} d\phi_0 \Phi_{\rho 3}(\boldsymbol{k}, t)_{\mathrm{inv}} D^2(t). \tag{66}$$

Then, if the initial energy spectrum is given by

$$E(k_0) = KE_0 \left(\frac{2}{9\pi}\right)^{1/2} \left(\frac{2}{k_p}\right)^5 k_0^4 e^{-\frac{2k_0^2}{k_p^2}}, \tag{67}$$

which gives $E(k_0) \propto k_0^4$ for $k_0/k_p \ll 1$ (Saffman, 1967) as has been often used in DNS to model the final decay of turbulence, integration of $E(k_0)D^2(t)$ by k_0 gives

$$KE_0 \frac{1}{\left[1 + \nu t k_p^2 \left(1 - \alpha t \sin\theta_0 \cos\theta_0 \cos\phi_0 + \frac{1}{3}(\alpha t)^2 \sin^2\theta_0 \cos^2\phi_0\right)\right]^{5/2}}. \tag{68}$$

Then, with viscosity and diffusion but for $Pr = 1$, KE_0 and PE_0 in (58)-(60) are replaced by the function of the form (68), which should be integrated by θ_0 and ϕ_0. The density flux for viscous and diffusive fluid ($Pr = 1$) finally becomes

$$\begin{aligned}
\overline{\rho u_3}(t) = &\frac{\alpha}{2\pi} PE_0 \int d\theta_0 d\phi_0 f_{\mathrm{vis}}(\theta_0, \phi_0) \frac{\cos\phi_0}{\sin^3\theta_0} \\
&\times \mathrm{Re}\left[\mathrm{i}\left(P_\nu(z)Q'_\nu(z_0) - P'_\nu(z_0)Q_\nu(z)\right)\left(P'_\nu(z)Q'_\nu(z_0) - P'_\nu(z_0)Q'_\nu(z)\right)^*\right] \\
&+ \frac{\alpha}{4\pi} RiKE_0 \int d\theta_0 d\phi_0 f_{\mathrm{vis}}(\theta_0, \phi_0) \frac{1}{\sin\theta_0 \cos\phi_0} \\
&\times \mathrm{Re}[\mathrm{i}\left(P_\nu(z)Q_\nu(z_0) - P_\nu(z_0)Q_\nu(z)\right) \\
&\times \left(P'_\nu(z)Q_\nu(z_0) - P_\nu(z_0)Q'_\nu(z)\right)^*],
\end{aligned} \tag{69}$$

where

$$f_{\mathrm{vis}}(\theta_0, \phi_0) = \frac{1}{\left[1 + \nu t k_p^2 \left(1 - \alpha t \sin\theta_0 \cos\theta_0 \cos\phi_0 + \frac{1}{3}(\alpha t)^2 \sin^2\theta_0 \cos^2\phi_0\right)\right]^{5/2}}. \tag{70}$$

The results show that the decay rate of the energies and the fluxes indicated by the powers of $-5/2$ depends on the initial spectral forms of $E(k_0)$ and $S(k_0)$ for $k_0/k_p \ll 1$ as in the no shear flow (Hanazaki & Hunt 1996). However, the decay also depends on the peak wave number k_p and the shear α, where the effects of the latter have the directional ($\theta - \phi$) preference. For example, when $k_1 = k_{10} = 0$ or equivalently $\cos\phi = 0$, there is no additional viscous decay due to the shear. Since ν is coupled only with α and not with N, the viscous decay is controlled only by the shear even in the stratified fluid. This verifies the conjecture by Hunt, Stretch & Britter (1988) that decay would be mainly controlled by the shear.

3.5 Examples

Figure 3 shows the RDT results for the time development of the normalized vertical density flux $\overline{\rho u_3}/(\overline{\rho^2}\,\overline{u_3^2})^{1/2}$ as a function of Nt for finite $Re \sim 43$ (cf. Hanazaki & Hunt 2002). The oscillation period is equal to π/N irrespective of α

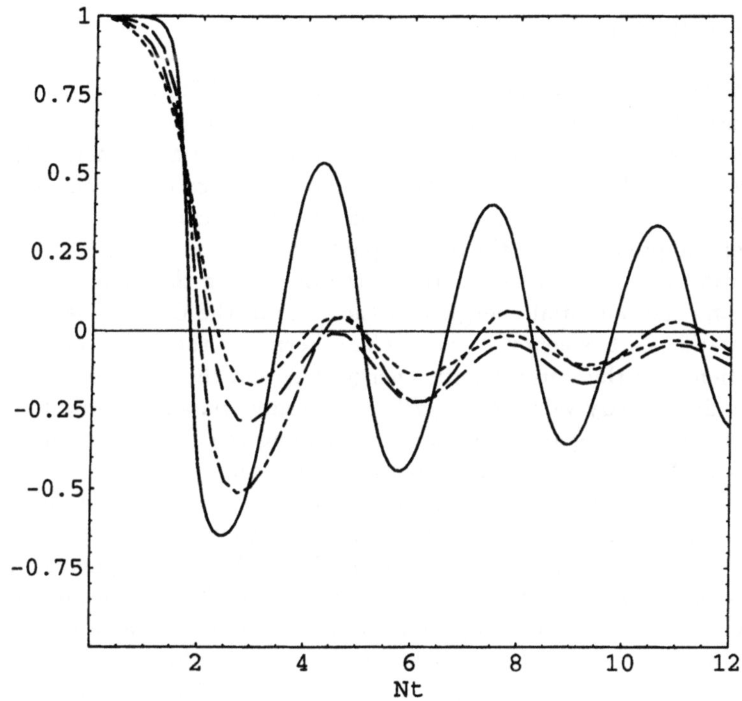

Fig. 3. Time development of the normalised vertical density flux $\overline{\rho u_3}/(\overline{\rho^2 u_3^2})^{1/2}$ for fixed $N(= 2.729)$ and different α calculated by RDT. The shear rate α varies according to $Ri = N^2/\alpha^2$ while N is fixed. ———, $Ri = \infty$; —·—·—, $Ri = 1$; - - - -, $Ri = 0.4$; ----------, $Ri = 0.25$.(Hanazaki & Hunt 2002)

(or Ri) because the spectral components unaffected by the shear, i.e. components with $k_1 = 0$, become dominant in a rather short time ($Nt \sim O(1)$) and those components oscillate at an angular frequency $2N$.

The results for $Ri = 0.25$ and 0.4 show a *persistent* counter-gradient flux. We note that having smaller Ri, i.e. larger α, makes the counter-gradient flux weaker, showing that the shear reduces the counter-gradient flux. This is in agreement with the experiments by Komori & Nagata (1996) for $Ri = 0.4$, 1 and ∞, where the stronger shear with the same vertical temperature difference gave a weaker counter-gradient flux. We should note that, although the flux for no shear ($Ri = \infty$) gives large amplitude oscillations, it oscillates in proportion to $\sim \sin(2Nt)$ (Hanazaki & Hunt 1996) so that the time averaged value becomes zero. Then, as regards the time averaged value, the larger Ri does not necessarily give a stronger counter-gradient flux. In Fig. 3 ($0 \leq Nt \leq 12$), $Ri \sim 0.4$ gives the strongest persistent counter-gradient flux.

4 Conclusions

For stratified rotating turbulence, the solutions of the RDT equations showed the importance of initial conditions as in the stratified non-rotating turbulence. For example, the ratio of the initial potential energy to the kinetic energy determines the direction of the vertical density flux. We found good agreement with the previous DNS by Iida & Nagano (1999) in many aspects and also with Bartello (1995) for the prediction of the energy partition.

The rotation modifies the energy partition among the kinetic energy components and the potential energy, and the ratio of the Coriolis parameter f to the Brunt-Väisälä frequency N, i.e. f/N, determines the final steady values, in combination with the initial kinetic energy KE_0 and the initial potential energy PE_0. The final steady value of KE/PE for the same parameters used in DNS by Bartello (1995) gave good agreement, confirming that the linear processes are dominant in decaying stratified rotating turbulence.

However, the effects of rotation on the *unsteady* aspects of the stratified turbulence are not large. The long-time asymptotics show that energy components and the fluxes oscillate at frequency $2N$ (N: Brunt-Väisälä frequency), the same frequency as the non-rotating stratified turbulence. This is because the Coriolis parameter f is always coupled with $\cos\theta$ or the vertical wavemumber k_3 in the form of $f\cos\theta$ or fk_3, which vanishes for the dominantly contributing angle $\theta = \pi/2$ to the variances and the covariances. The most contributing direction of the wavenumber direction, i.e. the horizontal direction ($k_3 = 0$) or the barotropic mode, is the same as the non-rotating stratified turbulence. Therefore, the stratification dominates the unsteady aspects of turbulence even with the system rotation.

Long-time limit steady values of the energies and the fluxes, of course depend on the ratio of N and f. However, if $f \ll N$ holds as in the atmosphere and the ocean, those asymptotic values agree approximately with the pure stratification results ($N \neq 0, f = 0$) given by Hanazaki & Hunt (1996). For example, energy ratio *ER*, the ratio of the potential energy to the vertical kinetic energy, approaches to 3/2 in a long time, irrespective of the initial conditions. One interesting effect of the ratio f/N is that it determines the *phase* of oscillation in the energy and the fluxes. If $f/N > 1$, there is a phase advance ($+\pi/4$), while if $f/N < 1$, there is a phase delay ($-\pi/4$).

In the special case of $N = f$, the time oscillations of the energies and the fluxes do not show inviscid decay like $\propto t^{-1/2}$ as observed in the general case of $N \neq f$, which includes the case of pure stratification ($f = 0$) (Hanazaki & Hunt 1996). Then the energy components never reach constant values. This is because all the wavenumber components oscillate in phase, irrespective of their directions so that the contributing components are not restricted to the horizontal wavenumbers. We note that, in this case, the exchange between the horizontal kinetic energy and the potential energy occur at low frequency N, and the horizontal energies $\overline{u_1^2}$ and $\overline{u_2^2}$ oscillate at that frequency. On the other hand, the energy exchange at the normal high frequency $2N$ occurs exclusively between the vertical kinetic energy and the potential energy.

We should note that the DNS data mainly used for comparison with RDT in this study have been done for low Reynolds numbers, which will make the comparison better. As shown in §2.1, the RDT is not applicable to the small scales at high *Re*. Then, in interpreting the results given in this study in the context of real atmosphere or ocean, we should take care about the applicability conditions, since the *Re* is usually very high there.

For stratified sheard turbulence, we have shown some of the main mechanisms using again the linear rapid distortion theory (RDT). The analytical solutions of RDT equations are found to be described by the Legendre functions of complex degree with pure-imaginary argument. The numerical evaluation of the analytical results showed good agreement with DNS for stable mean flow ($Ri \geq 0.25$) particularly at low turbulent Froude numbers or at high shear numbers when the nonlinear effects are not significant.

The counter-gradient density and momentum fluxes also occur with shear ($Ri \leq 1.0$), and are described by the linear theory. However, at large Ri (≥ 1) when the time oscillation becomes larger, the mean value of this flux approaches zero in the limit of no shear ($Ri \to \infty$). This is likely to be the main mechanism for the suppression of turbulence in stratified shear flows.

The effect of shear on viscous decay is not affected by stratification and is identical to that in pure shear flow, i.e. the components of larger streamwise wavenumber k_1 decays faster, leaving the components of smaller k_1. In combination with the effect of spectrum deformation, the energy and the flux are increasingly dominated by the small k_1 components as time elapses. Since the components of $k_1 = 0$ are unaffected by the shear and oscillate in time with period π/N, this explains why the energies and the fluxes oscillate at period π/N even in a shear flow.

Another unresolved problem is the Prandtl number (Pr) dependence of the energy ratio *ER* and the counter-gradient flux, since only the case of $Pr = 1$ is considered here. The energy ratio *ER* will also depend on Pr, as the viscosity affects the vertical kinetic energy *VKE* and the potential energy *PE* differently.

References

1. Bartello, P. Geostrophic adjustment and inverse cascades in rotating stratified turbulence. J. Atmos. Sci. **52**, 4410–4428 (1995)
2. Batchelor, G. K., Proudman, I. The effect of rapid distortion on a fluid in turbulent motion. Q. J. Mech. Appl. Maths. **7**, 83–103 (1954)
3. Bardina, J., Ferziger, J. H. & Rogallo, R. S. Effect of rotation on isotropic turbulence: computation and modelling. J. Fluid Mech. **154**, 321–336 (1985)
4. Cambon, C. Turbulence and vortex structures in rotating and stratified flows. Eur. J. Mech. B-Fluids **20**, 489–510 (2001)
5. Derbyshire, S. H. & Hunt, J. C. R. 'Structure of turbulence in stably stratified atmospheric boundary layers; Comparison of large eddy simulations and theoretical results'. In: ***Waves and Turbulence in Stably Stratified Flows***, ed. by S. D. Mobbs and J. C. King, (Clarendon Press 1993) pp. 23–59
6. Gerz, T. Schumann, U. & Elghobashi, S. E. Direct numerical simulation of stratified homogeneous turbulent shear flows. J. Fluid Mech. **200**, 563–594 (1989)

7. Gerz, T. & Yamazaki, H. Direct numerical simulation of buoyancy-driven turbulence in stably stratified fluid. J. Fluid Mech. **249**, 415–440 (1993)
8. Hanazaki, H. 'On the transport mechanisms in stably stratified rotating turbulence'. In: *Turbulence, Heat and Mass Transfer 3*, ed. by Y. Nagano, K. Hanjalic & T. Tsuji, (Aichi Shuppan 2000) pp. 639-644
9. Hanazaki, H. Linear processes in stably and unstably stratified rotating turbulence. J. Fluid Mech. in press, (2002)
10. Hanazaki, H. & Hunt, J. C. R. Linear processes in unsteady stably stratified turbulence. J. Fluid Mech. **318**, 303–337 (1996)
11. Hanazaki, H. & Hunt, J. C. R. 'Linear processes in unsteady stratified sheared turbulence'. In: *IUTAM Symp. on Geometry and Statistics of Turbulence*, ed. by T.Kambe, T.Nakano & T.Miyauchi, (Kluwer 2001) pp. 291–296
12. Hanazaki, H. & Hunt, J. C. R. Structures of unsteady stably stratified turbulence with mean shear. submitted to J. Fluid Mech. (2002)
13. Holt, S. E., Koseff, J. R. & Ferziger, J. H. A numerical study of the evolution and structure of homogeneous stably stratified turbulence. J. Fluid Mech. **237**, 499–539 (1992)
14. Hunt, J. C. R., Stretch, D. D. & Britter, R. E. 'Length scales in stably stratified turbulent flows and their use in turbulence models'. In: *Stably Stratified Flow and Dense Gas Dispersion*, ed. by J. S. Puttock, (Clarendon Press 1988), pp. 285–321
15. Iida, O. & Nagano, Y. Coherent structure and heat transfer in geostrophic flow under density stratification. Phys. Fluids **11**, 368–377 (1999)
16. Itsweire, E. C., Helland, K. N. & Van Atta, C. W. The evolution of grid-generated turbulence in a stably stratified fluid. J. Fluid Mech. **162**, 299–338 (1986)
17. Jacobitz, F. G., Sarker, S. & Van Atta, C. W. Direct numerical simulation of the turbulence evolution in a uniformly sheared and stably stratified flow. J. Fluid Mech. **342**, 231–261 (1997)
18. Kaltenbach, H. -J., Gerz, T. & Schumann, U. Large eddy simulation of homogeneous turbulence and diffusion in stably stratified shear flow. J. Fluid Mech. **280**, 1–40 (1994)
19. Kaneda, Y. Single-particle diffusion in strongly stratified and/or rapidly rotating turbulence. J. Phys. Soc. Japan **69**, 3847–3852 (2000)
20. Kaneda, Y. & Ishida, T. Suppression of vertical diffusion in strongly stratified turbulence. J. Fluid Mech. **402**, 311–327 (2000)
21. Kimura, Y. & Herring, J. R. Diffusion in stably stratified turbulence. J. Fluid Mech. **328**, 253–269 (1996)
22. Komori, S. & Nagata, K. Effects of molecular diffusivities on counter-gradient scalar and momentum transfer in strongly stable stratification. J. Fluid Mech. **326**, 205–237 (1996)
23. Komori, S., Ueda, H., Ogino, F. & Mizushina, T. Turbulence structure in stably stratified open-channel flow. J. Fluid Mech. **130**, 13–26 (1983)
24. Lienhard, J. H. & Van Atta, C. W. The decay of turbulence in thermally stratified flow. J. Fluid Mech. **210**, 57–112 (1990)
25. Métais, O., Bartello, P., Garnier, E., Riley, J. J. & Lesieur, M. Inverse cascade in stably stratified rotating turbulence. Dyn. Atmos. Oceans **23**, 193–203 (1996)
26. Métais, O. & Herring, J. Numerical simulations of freely evolving turbulence in stably stratified fluids. J. Fluid Mech. **202**, 117–148 (1989).
27. Miles, J. W. On the stability of heterogeneous shear flow, J. Fluid Mech. **10**, 496–508 (1961)
28. Piccirillo, P. & Van Atta, C. W. The evolution of a uniformly sheared thermally stratified turbulent flow. J. Fluid Mech. **334**, 61–86 (1997)

29. Riley, J. J., Metcalfe, R. W. & Weissman, M. A. 'Direct numerical simulations of homogeneous turbulence in density stratified fluids'. In: *Nonlinear Properties of Internal Waves, AIP Conference Proc.* vol.76, (American Institute of Physics 1981), pp. 79–112
30. Rohr, J. J., Itsweire, E. C., Helland, K. N. & Van Atta, C. W. Growth and decay of turbulence in a stably stratified shear flow. J. Fluid Mech. **195**, 77–111 (1988)
31. Saffman, P. G. The large-scale structure of homogeneous turbulence. J. Fluid Mech. **27**, 581–593 (1967)
32. Staquet, C. & Godeferd, F. S. Statistical modelling and direct numerical simulations of decaying stably stratified turbulence. Part 1. Flow energetics. J. Fluid Mech. **360**, 295–340 (1998)
33. Stretch, D. D. The dispersion of slightly dense contaminations in a turbulent boundary layer. Ph.D. Thesis, Dept. of Engineering, University of Cambridge (1986)
34. Townsend, A. A. *The Structure of Turbulent Shear Flow.* (Cambridge University Press 1976), pp. 429
35. Tsujimura, S. Iida, O. & Nagano, Y. 'Effects of rotation on unstably stratified turbulence'. In: *Proc. Int. Conf. on Turbulent Heat Transfer 2, Manchester, U. K.* vol. 1, 5-58 – 5-71 (1998)
36. Yoon, K. & Warhaft, Z. The evolution of grid generated turbulence under conditions of stable thermal stratification. J. Fluid Mech. **215**, 601–638 (1990).

Part II

Wall-Bounded Flows

Very Large Anisotropic Scales in Turbulent Wall-Bounded Flows

Javier Jiménez[1,2] and Juan C. del Álamo[1]

[1] School of Aeronautics, Universidad Politécnica
28040 Madrid, Spain
[2] Centre for Turbulence Research, Stanford University, Stanford CA 94305, USA

1 Introduction

It has long been realized that turbulent flows contain a wide range of scales, from small viscous ones responsible for the viscous dissipation, to large ones which carry the turbulent energy and the Reynolds stresses. The former are believed to be roughly universal among different flows, while the latter vary with the geometry and with the flow conditions. Evidence has accumulated for some time that in an intermediate layer of wall-bounded shear flows, including the logarithmic region and part of the outer layer, these large scales are very anisotropic and very large, with streamwise lengths that may be of the order of 100 times their distance to the wall [2,4,8]. At their longest, somewhat above the top of the logarithmic layer, this amounts to 20-30 times the boundary layer thickness. The earliest detailed study of these structures was done by Perry [9,10], who identified them as an $E_{uu} \sim k^{-1}$, long-wavelength, spectral range. Since the turbulent energy is proportional to $\int kE_{uu}\mathrm{d}(\log k)$, a k^{-1} spectral range essentially contains most of the fluctuating energy in the flow (see figure 1). Moreover, because the size of these structures requires either very large computational boxes or very long experiments, relatively little was known about them until recently. There is for example very little information on their spanwise dimensions, or on the relation of the different velocity components. The goal of the simulations discussed here is to study the large anisotropic scales in turbulent channels, their origin and structure, and their possible influence on other flow properties.

2 Procedure

Recent investigations of the small scales of the turbulent near-wall region, using direct numerical simulations of autonomous minimal channels [7], have resulted in the identification of self-sustaining structures which survive in the absence of interactions with the external flow. The basic technique for computing such autonomous wall flows was introduced in [6], and consists of a numerical mask that damps vorticity fluctuations above a given distance from the wall. Irrotational fluctuations are not masked, and extend into the outer flow. That the near-wall region could generate turbulence by itself had already been established in [6], but it was still unclear whether large-scale interactions among different structures through the outer flow, whether potential or not, was needed to maintain

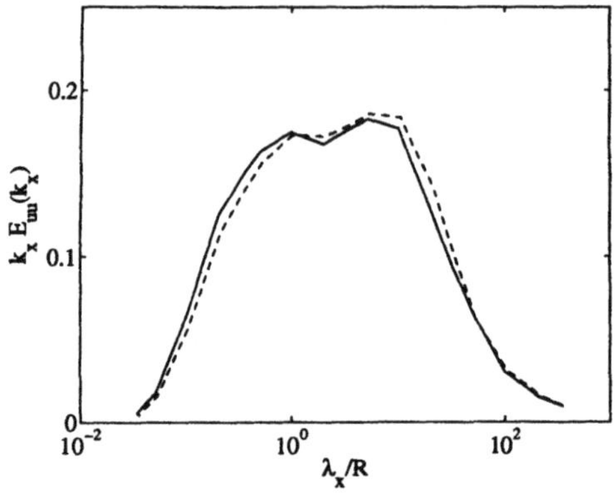

Fig. 1. Premultiplied longitudinal energy spectra in the logarithmic region of a smooth pipe. Note that the k_x^{-1} part of the spectrum, which in this representation appears as approximately constant, is at wavelengths which are considerably longer than the radius R of the pipe, and that most of the energy, which is the area underneath the spectra, is contained in that region. —, $y/R = 0.065$; – – –, $y/R = 0.09$. $Re_\tau = 2300$. Data from [9].

it. The experiments in [7] showed that this was not so, and that self-sustaining structures, essentially similar to those educed by other investigators from fully turbulent flows, can exist without either inner or outer interactions. Related structures have been identified using different methods by other investigators [11,12]. These results suggest that the near-wall region acts in the wall-bounded flows as a turbulence-generating layer, which may be responsible, at least in part, for the properties of the turbulence of the outer flow. A similar conclusion was reached in [3] from the consideration of global energy fluxes. While the existence of these structures can be considered as relatively well established, little is known about the reason for their large-scale organization.

To inquire into this question we discuss two numerical experiments:

1. To establish a complete experimental data base of the characteristics of the large scales in a full turbulent flow, we undertook a large direct numerical simulation of a turbulent channel at moderate Reynolds number, $Re_\tau = 550$. Although computations at somewhat higher Reynolds numbers were available in the literature, we believe that this is the first simulation in which the computational box is large enough not to interfere with the behaviour of the largest flow structures. After preliminary simulations at lower Reynolds numbers ($Re_\tau = 180$ and 360), it was decided to use a box of length of $8\pi h \times 2h \times 4\pi h$ in the streamwise, cross-channel and spanwise directions [1]. The associated grid is $1536 \times 257 \times 1536$ (6×10^8) collocation points, using a fully pseudospectral code.
2. To investigate the possible large-scale organization of the autonomous wall structures described above, we undertook autonomous wall simulations in large computational boxes. Except for the presence of the numerical mask, the numerical method and resolutions used in these simulations were the

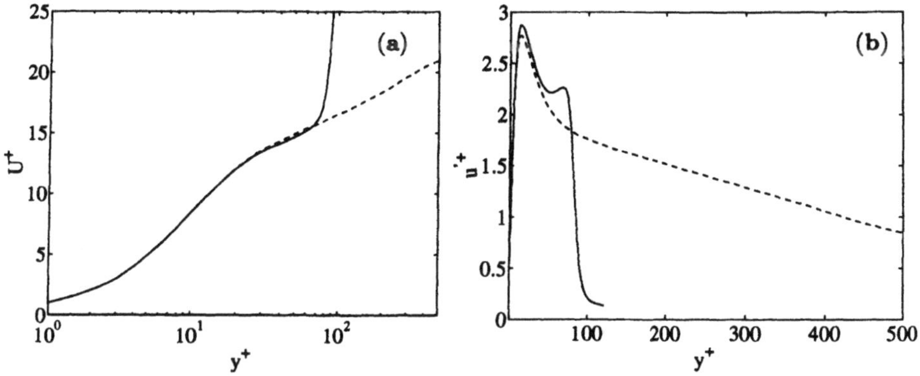

Fig. 2. Mean profiles of: (**a**) Mean velocity. (**b**) Streamwise velocity fluctuations. Solid lines are for an autonomous wall simulation masked above $\delta_1^+ = 72$, on a box of size $L_x^+ = 1.2 \times 10^4$, $L_z^+ = 10^3$ [5]. Broken lines are for a full turbulent channel, $Re_\tau = 550$, $L_x^+ = 1.4 \times 10^4$, $L_z^+ = 7 \times 10^3$ [1].

same as in the previous experiment but, instead of a full flow extending across a channel with $L_y^+ = 1100$, turbulence was restricted to a thin layer whose height is $\delta^+ \approx 80$.

3 Results

As an example of the results obtained from the simulations of autonomous wall turbulence, we will discuss a experiments in which the flow is masked above $y^+ \approx 72$, but the computational box is made long and wide, $x^+ = 12,000$, $z^+ = 1,000$, to allow for any possible self-organization of the wall. The one-point statistics are in excellent agreement with those of full channel simulations below roughly half the mask height, as shown in figure 2. The two-dimensional premultiplied spectra for the three velocity components, $k_x k_z E^{2D}(k_x, k_z)$ are shown in figure 3 as functions of the corresponding wavelengths $\lambda = 2\pi/k$. They have substantial energy in structures which are several thousand wall units long, suggesting the large-scale organization of several streaks. It is fascinating that the spectra of the wall-normal velocity are almost identical, below $y^+ \approx 50$, to those of the full direct simulation of the channel at $Re_\tau = 550$, even if the outer flow is completely missing in the autonomous case. This result strongly suggests that the organization of the wall is essentially independent of the outer flow, even at these large scales. The same is true of the streamwise velocity, although in this case the structures of the autonomous wall appear to be longer than those of the full channel. This is puzzling, and suggests that the effect of the outer flow is to limit the length of the near-wall structures, rather than to impose on them the large-scale organization.

A visualization of the computational channel is shown in figure 4, where the large-scales appear in the form of clusters of vortical structures. They are predominantly long, and can be approximately classified into longitudinal low-velocity active 'streaks', and high-speed relatively quiescent 'sweeps'. Note that

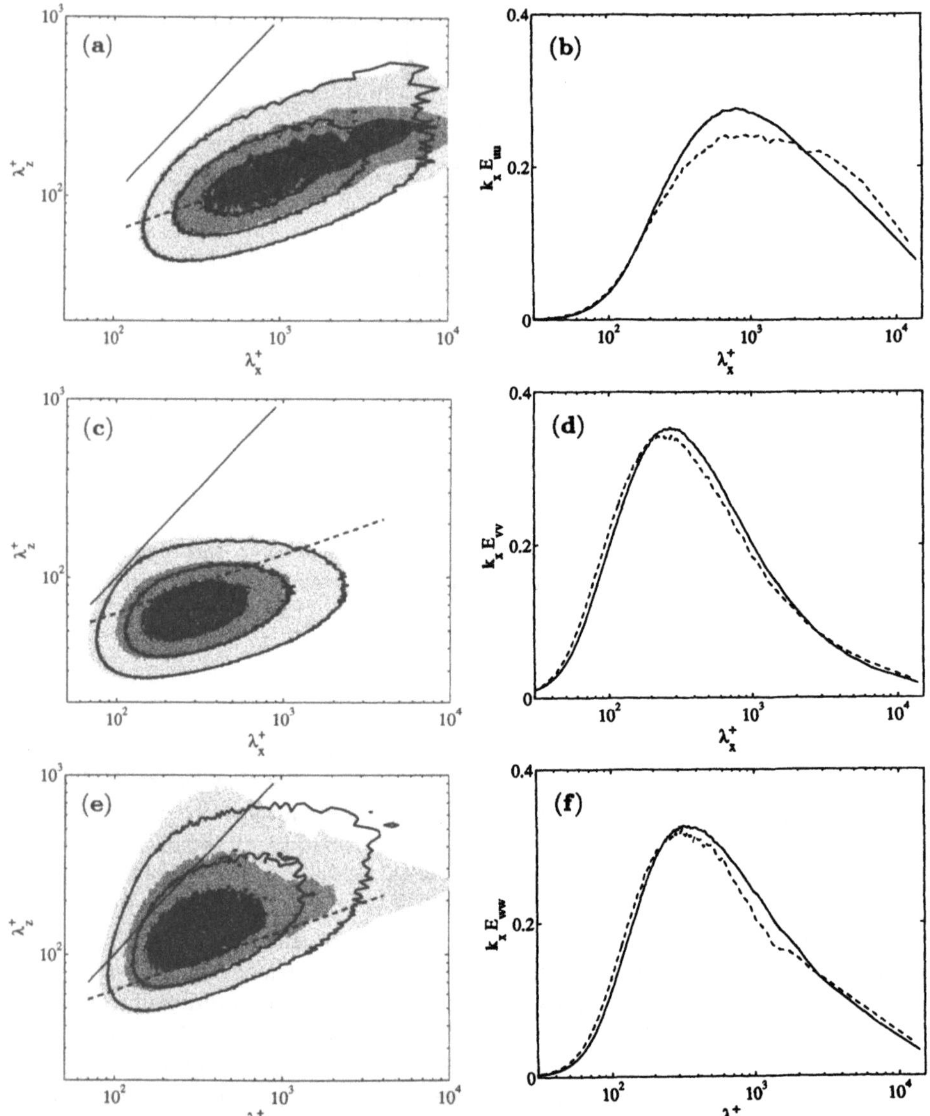

Fig. 3. (**a**), (**c**), (**e**) Premultiplied two-dimensional velocity spectra, $k_x k_z E^{2D}(\lambda_x, \lambda_z)$, as functions of the streamwise and spanwise wavelengths. $y^+ = 16$. Shaded contours are the autonomous wall simulation in figure 2. Lines are the full turbulent channel also used in that figure. The solid lines are $\lambda_x = \lambda_z$, and correspond to horizontally isotropic structures. The broken lines have logarithmic slopes 1/3, and pass through $\lambda_x^+ = \lambda_z^+ = 50$. The contours are (0.25, 0.5, 0.75) times the maximum value of each spectrum. (**b**), (**d**), (**f**) One-dimensional streamwise spectra, $k_x E(\lambda_x)$, for the same data. The solid line is the full channel; the dashed one is the autonomous wall. (**a**), (**b**) Streamwise velocity. (**c**), (**d**) Wall-normal velocity. (**e**), (**f**) Spanwise velocity.

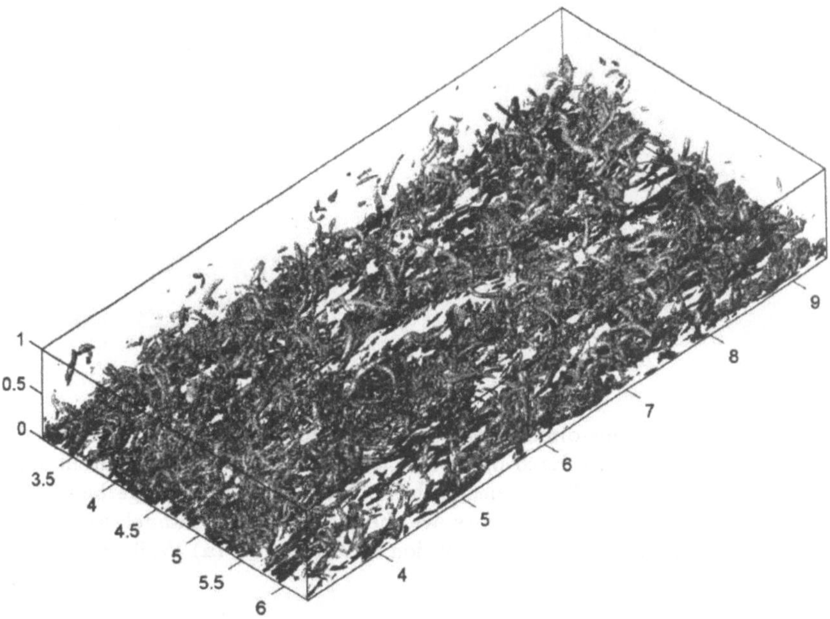

Fig. 4. Vortical structures in a computational channel from [1]. $Re_\tau = 550$. The horizontal size of the displayed box is $L_x \times L_z = 3400^+ \times 1700^+ = 2\pi h \times \pi h$. The computational box is four times larger in both directions. Only the lower half-channel is shown for clarity. The large-scale clustering seen in the figure corresponds to the size of the structures described in the text. The vortex-populated regions are low-speed zones, while the gaps contain higher-speed fluid coming from above.

the width of the clusters in the figure is approximately 500 wall units, and that they span the length of the visualization, being therefore much larger than the classical buffer-layer streaks. They are also fully turbulent structures, as opposed to the relatively smooth classical ones.

The long structures near the wall are also wide, following an apparent power relation $\lambda_z \sim \lambda_x^\alpha$, with an exponent $\alpha = 1/3$. Similar power laws are also found at other distances from the wall, although the exponent becomes $\alpha = 1/2$ as we move away from it (figure 6). This suggests that the size of those structures is controlled by dissipation, probably due to some eddy viscosity. In fact, it is possible to recover these power laws by considering the spreading of an initially localized perturbation under the effect of advection by the mean flow and of a constant viscosity [1,5]. The exponent 1/2 is found when the advection velocity is approximately constant, which is a reasonable approximation away from the wall, while the exponent 1/3 corresponds to advection by a linear shear, which would be more appropriate very close of the wall.

Note that the structures described by the long-wavelength end of these spectra are not the classical buffer-layer streaks. They are both at least an order of magnitude longer and wider than the latter, and should be interpreted as a next larger stage in the organization of turbulence.

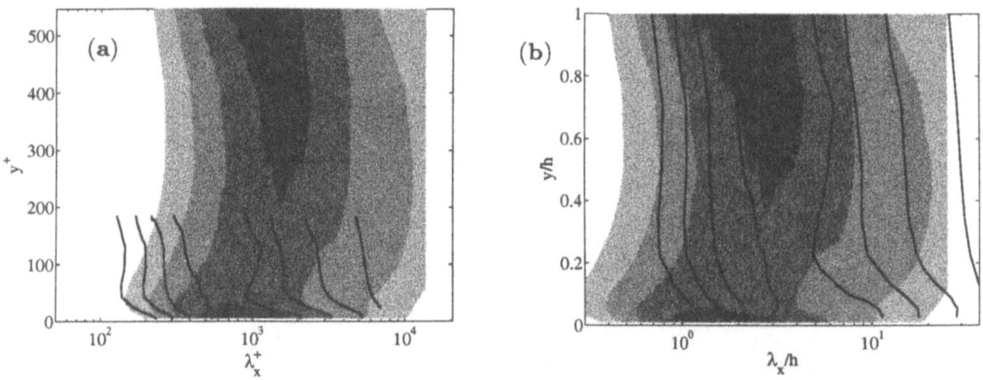

Fig. 5. Premultiplied one-dimensional velocity spectra, $k_x E_{uu}^{1D}(\lambda_x)$, as functions of the streamwise wavelength. Each horizontal level corresponds to a different spectrum at a fixed wall distance. All the spectra have then been stacked together as a single function of λ_x and y. Shaded contours are the full turbulent channel used in figure 2 at $Re_\tau = 550$. Lines are a similar simulation at $Re_\tau = 180$. **(a)** Wall units. **(b)** Outer units.

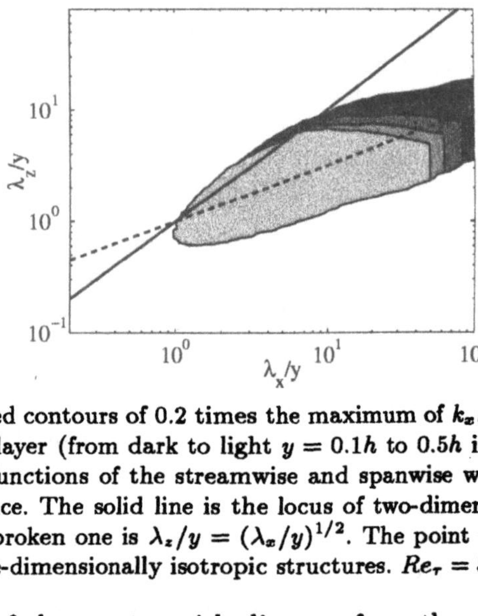

Fig. 6. Superimposed contours of 0.2 times the maximum of $k_x k_z E_{uu}^{2D}$ at five wall distances in the outer layer (from dark to light $y = 0.1h$ to $0.5h$ in steps of $0.1h$). They are represented as functions of the streamwise and spanwise wavelengths normalized with the wall distance. The solid line is the locus of two-dimensional isotropic structures $\lambda_z = \lambda_x$; the broken one is $\lambda_z/y = (\lambda_x/y)^{1/2}$. The point where both lines cross corresponds to three-dimensionally isotropic structures. $Re_\tau = 550$.

The variation of the spectra with distance from the wall in full turbulent channels is investigated using direct simulations at $Re_\tau = 180$ and $Re_\tau = 550$. In figure 5 we show longitudinal one-dimensional spectra $k_x E^{1D}(k_x)$ of the streamwise velocity. There is an inner spectral peak, which scales in wall units, and an outer one which scales in outer units. The longest velocity component in the streamwise direction is the streamwise velocity u, and reaches a maximum length of about $\lambda_x = 4h$ somewhat above the upper edge of the logarithmic layer.

As noted above, the power law relating the streamwise with the spanwise wavelengths of the two-dimensional spectra also holds away from the wall, with an exponent 1/2. As shown in figure 6 the prefactor of this law scales well with y, in agreement with the classical self-similar 'logarithmic' scaling, although the significance of this fact is not immediately clear.

Of the three premultiplied velocity spectra, that of the spanwise velocity is the widest, with a peak which widens steadily from about $\lambda_z^+ = 100$ near the wall, to about $\lambda_z = 1.5h$ at the centre of the channel, but it remains relatively short. The overall structure of the flow away from the wall is therefore also one of streamwise velocity 'streaks', elongated for the streamwise velocity component, but relatively isotropic for the other two.

Acknowledgments

The $Re_\tau = 550$ simulation was run at the CEPBA/IBM research centre at Barcelona, with time that has been graciously donated by IBM and by the U. Politécnica de Catalunya. Their contribution and the help of J.M. Cela at CEPBA in optimizing the parallel performance of the code are gratefully acknowledged.

References

1. J. C. del Álamo: Direct numerical simulation of the largest scales in a turbulent channel. Ph. D. Thesis. U. Politécnica de Madrid (in progress), also J.C. del Álamo, J. Jiménez: Direct numerical simulation of the very-large anisotropic scales in a turbulent channel, CTR Ann. Res. Briefs, 329–342 (2001)
2. M.H. Hites: Scaling of high-Reynolds number turbulent boundary layers in the National Diagnostic Facility. Ph. D. Thesis, Illinois Inst. of Technology (1997)
3. J. Jiménez: The physics of wall turbulence, Physica A **263**, 252–262 (1998)
4. J. Jiménez: The largest structures in turbulent wall flows. CTR Ann. Res. Briefs, 137–154 (1998)
5. J. Jiménez, O. Flores, M. García-Villalba: The large scale organization of autonomous turbulent walls, CTR Ann. Res. Briefs, 317–329 (2001)
6. J.Jiménez, A. Pinelli: The autonomous cycle of near-wall turbulence, J. Fluid Mech. **389**, 335–359 (1999)
7. J. Jiménez, M. P. Simens: Low-dimensional dynamics in a turbulent wall, J. Fluid Mech. **435**, 81–91 (2001)
8. K.C. Kim, R.J. Adrian: Very large-scale motion in the outer layer. Phys. Fluids A. **11**, 417-422 (1999)
9. A. E. Perry, C. J. Abell, Asymptotic similarity of turbulence structures in smooth- and rough-walled pipes, J. Fluid Mech. **79**, 785-799 (1977)
10. A.E. Perry, S. Henbest, M.S. Chong: A theoretical and experimental study of wall turbulence. J. Fluid Mech. **165**, 163–199 (1986)
11. S. Toh, T. Itano: 'On the regeneration mechanism of turbulence in the channel flow - role of the traveling-wave solution'. In: *Proc. IUTAM Symp. on Geometry and Statistics of Turbulence at Hayama, Japan, November 1998*, ed. by T. Kambe, T. Nakano, T. Miyauchi (Kluwer, 2001) pp. 305–310.

12. F. Waleffe: Exact coherent structure in channel flow. J. Fluid Mech. **435**, 93–102 (2001)

Turbulent Plume Diffusion in a Pipe Flow by the PDF Method

Yasuhiko Sakai[1], Haruki Suzuki[1], and
Hiroyuki Tsunoda[2]

[1] Department of Mechano-Informatics and Systems, Graduate School of Engineering, Nagoya University, Furo-cho, Chikusa-ku, Nagoya, Aichi 464-8603, Japan
[2] Department of Mechanical System Engineering, Yamanashi University, Takeda 4-3-11, Kofu, Yamanashi 400-8501, Japan

Abstract. This paper gives the new simulation results of the axisymmetric point source plume in a fully developed turbulent pipe flow by the Lagrangian Probability Density Function (PDF) method. The Lagrangian velocities of stochastic particles are modeled by a generalized Langevin equation expressed in the cylindrical coordinates. The generalized Langevin equation was originally suggested by Haworth and Pope [Phys. Fluids, **29**-2, 387(1986)] and its cylindrical expression was derived by Sakai et al. [JSME Int. J., B**39**-4, 667(1996)]. For the molecular mixing of the scalar, two different models (i.e. the Dopazo's deterministic model and the modified Curl's model) are adopted. The simulated radial profiles of the mean concentration and concentration variance by two models show good agreements with the experimental data by Becker et al. [AIChE J., **15**-5, 964(1966)]. It is also shown that the mixing process can be characterized well by the downstream variation of the scalar PDF profile.

1 Introduction

Recently, the prediction of turbulent transport processes of pollutant in the atmosphere and in the ocean has become an important research subject because of great interest in environmental problems. Also, for improvement of various environmental equipments (e.g., combustion apparatus, a chemical reactor) it is necessary to understand turbulent diffusion phenomena. If the experiments with conditions adapted to several turbulent diffusion phenomena would be planned, it apparently needs very hard labor and a lot of cost. On the other hand, the numerical simulations by computers have become effective to predict turbulent diffusion phenomena because of the improvement of computers and the development of numerical methods.

The purpose of the present study is to predict numerically the turbulent diffusion field of material in a fully developed turbulent pipe flow, which was chosen because of its basic and industrial importance.

Generally, there are Eulerian and Lagrangian methods for the simulation of the turbulent diffusion field. In the present study, we adopted the Lagrangian PDF (Probability Density Function) method[1]-[3] which is one of Lagrangian methods and a kind of Monte-Carlo method. In the Lagrangian PDF method, the exact velocity-scalar joint PDF is approximated by the discrete PDF for a set of many stochastic particles, and the variations of velocity and scalar are modeled by the Lagrangian equation (See [1]-[3] in detail of the method). In this study, as the velocity model of the stochastic particles, a generalized Langevin model expressed in the cylindrical coordinate system[4] is adopted. The generalized Langevin model was first proposed by Haworth and Pope[5] and expressed in the cylindrical coordinate system by Sakai et al.[4]. This model is constructed to satisfy the consistency condition of velocity field[6] and thermodynamic constraint[7]. On the other hand, as the molecular mixing model which determine the change of scalar, two models (i.e. the Dopazo's deterministic model[8] and the modified Curl's model[9]-[11]) are attempted.

In this paper, the scalar diffusion field from the point source at the center of the pipe is simulated and the effectiveness of the models used in the present study is evaluated by comparing the simulation results with experimental data.

Nomenclature

r, θ, x : radial, azimuthal and main streamwise coordinate, respectively (See Fig.1)
R : inner radius of the pipe
y : distance from the wall ($= R - r$)
ν : kinematic viscosity
U_{av} : cross sectional mean velocity
Re : Reynolds number($= 2RU_{av}/\nu$)
U_i : instantaneous velocity components ($i = r, \theta, x$)
u_i : velocity fluctuation components ($i = r, \theta, x$)
u_τ : friction velocity
ϱ : density
$\bar{\ }, \langle\rangle$: mean value or ensemble averaged value
Γ : instantaneous concentration
γ : concentration fluctuation
$\hat{\gamma}$: concentration fluctuation r.m.s value ($= \sqrt{\overline{\gamma^2}}$)

2 Coordinate system

In the present research, coordinate system (r, θ, x) is used as shown in Fig.1, where r, θ, x are the radial, azimuthal and main streamwise coordinate, respectively.

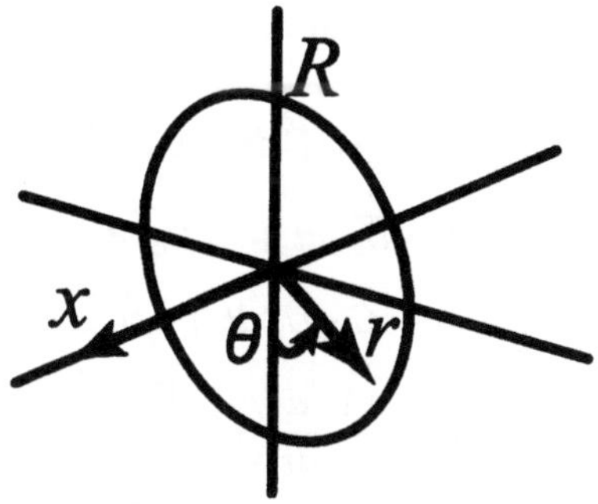

Fig. 1. Coordinate system

3 Probability Density Function (PDF) method

The Lagrangian PDF method is a kind of Monte-Carlo method in which the modeled equations for the velocity and scalar of stochastic particle are constructed and they are solved numerically by integrating the state of stochastic particles under the given initial conditions and boundary conditions for the problems. In the following sections, the models of the velocity and scalar of the stochastic particles are explained.

3.1 Conditions of the Lagrangian velocity model

In this study, the semiempirical Lagrangian stochastic model is adopted for the simulation of the velocity field to save the computing time and improve the efficiency of the algorithm. At first, we assume that the evolutions of the moments of Euler velocity pdf in a pipe flow (the first and second moment etc.) are known from the experimental data, then the generalized Langevin equation is constructed so that the evolutions of the moments calculated by the model may be equal to those (the consistency condition[6]). Another important condition for the stochastic model is the so-called "thermodynamic constraint" that an initially uniform distribution of material be maintained[7]. In other words, if, initially, the number density of marked stochastic particles is uniform, then it remains uniform. According to Pope[6], the above condition is equivalent to the condition that the calculated mean velocity satisfies the continuity equation.

3.2 A generalized Langevin model in the cylindrical coordinates

In the present study, the motion of fluid particle is modeled and tracked by the generalized Langevin model. Here we show only the equations used (See [4] in detail).

Assuming axisymmetricity and the condition of full development for the turbulent pipe flow, we can derive the following final form of the generalized

Langevin model for each velocity component,

$$dU_r^* = -\frac{1}{\varrho}\frac{\partial \overline{P}}{\partial r}dt + \frac{1}{r^*}U_\theta^{*2}dt + G_{rr}U_r^*dt$$
$$+G_{rx}(U_x^* - \overline{U_x})dt + \sqrt{C_0\varepsilon}dW_r, \tag{1a}$$
$$dU_\theta^* = -\frac{U_r^*U_\theta^*}{r^*}dt + G_{\theta\theta}U_\theta^*dt + \sqrt{C_0\varepsilon}dW_\theta, \tag{1b}$$
$$dU_x^* = -\frac{1}{\varrho}\frac{\partial \overline{P}}{\partial x}dt + \nu\frac{1}{r^*}\frac{d}{dr}\left(r^*\frac{d\overline{U_x}}{dr}\right)dt$$
$$+G_{rr}U_r^*dt + G_{xx}(U_x^* - \overline{U_x})dt + \sqrt{C_0\varepsilon}dW_x, \tag{1c}$$

where * denotes the random variable attendant on the particle, dt: the time increment, dU_i: the increment of velocity component ($i = r, \theta, x$), ε: mean dissipation rate per unit mass, C_0: Kolmogorov constant, dW_i: increment of an isotropic Wiener process in the i direction with the mean 0 and variance 1. The tensor G_{ij} is determined on the basis of the consistency condition[6] up to the second-order moments of the velocity field (See [4] in detail on G_{ij}). And the gradient of the mean pressure can be expressed as

$$\frac{1}{\varrho}\frac{\partial \overline{P}}{\partial r} = -\frac{1}{r}\frac{\partial}{\partial r}\left(r\overline{u_r^2}\right) + \frac{\overline{u_\theta^2}}{r}, \tag{2a}$$
$$\frac{1}{\varrho}\frac{\partial \overline{P}}{\partial x} = -2\frac{u_\tau^2}{R}. \tag{2b}$$

Here, to satisfy consistency condition up to the second-order moments of the velocity, it is necessary to specify the radial distributions of the following mean quantities up to the third-order.

First-order moment	: $\overline{U_x}$
Second-order moments	: $\overline{u_r^2}$, $\overline{u_\theta^2}$, $\overline{u_x^2}$, $\overline{u_r u_x}$
Third-order moments	: $\overline{u_r^3}$, $\overline{u_r^2 u_x}$, $\overline{u_r u_\theta^2}$, $\overline{u_r u_x^2}$, $\overline{u_\theta^2 u_x}$
mean dissipation rate per unit mass	: ε

In the following, we explain the way of giving these parameters. Firstly, the radial distribution of the mean velocity is given by the following equation in which the wake function[12] is added to the equation of mean velocity by Reichardt[13],

$$\frac{\overline{U_x}}{u_\tau} = \frac{1}{\kappa}\ln(1+\kappa y^+) + c\left[1 - \exp\left(-\frac{y^+}{\delta_l{}^+}\right) - \frac{y^+}{\delta_l{}^+}\exp(-0.33y^+)\right]$$
$$+d\left[1 - \cos\left(\pi\frac{y}{\delta}\right)\right],$$
$$\kappa = 0.4,\ c = 6.0,\ \delta_l{}^+ = 11.0,\ d = 0.5,\ \delta = 0.77R, \tag{3}$$

where $y^+ = u_\tau y/\nu$ and the third term of the right-hand side is the wake function. The parameters c and δ are chosen for the mean velocity at the center of the pipe $\overline{U_c}$ and cross-sectional mean velocity U_{av} to agree with those given by Becker et al.[14]. The experimental study by Becker et al. is the subject of the simulation in the present research.

With regard to the second-order moments, i.e. $\overline{u_r^2}, \overline{u_\theta^2}$ and $\overline{u_x^2}$ and the mean dissipation rate ε, Laufer's data[15] (Reynolds number $Re_c=2R\overline{U_c}/\nu$=500,000) are used. Here it is noted that the Reynolds number of Laufer's experiment is different from that of the experiment by Becker et al.[14] ($Re_c=2R\overline{U_c}/\nu$=796,000; This value corresponds to $Re=2RU_{av}/\nu$=684,000). However, both Re_c are in the same order, so distributions of the second moments and the mean dissipation rate are not so different with each other. From this reason, the Laufer's experimental results are used as the specified data.

The Reynolds stress $\overline{u_r u_x}$ is readily determined by integrating the x-direction mean velocity equation, which is given by

$$\overline{u_r u_x} = \nu \frac{\partial \overline{U_x}}{\partial r} + \frac{{u_\tau}^2}{R} r. \tag{4}$$

With regard to the third-order moments of the velocity, although there exist Laufer's experimental data, the scattering of the data is quite large. Thus, we judged that those data are not suitable for inputting to the present model, and decided to omit the input of the third-order moments to the model. This means that the present model does not satisfy the consistency condition exactly. Thus we need to check the consistency condition in the practical simulation in order to judge whether the present model gives us reliable results or not. This check is made by comparing the calculated statistics of the velocity field with the experimental (prescribed) data.

3.3 Molecular mixing model

When one particle mixes with other particles, the molecular mixing model gives the rule about the change of the scalar attendant on the particle. In the present study, the Dopazo's deterministic model[8] and the modified Curl's model[9]-[11] are adopted as the molecular mixing model. In the following, we explain these molecular mixing models.

Dopazo's deterministic model. In the Dopazo's deterministic model[8], when the scalar (concentration) attendant on the particle is represented by Γ, the increment of the scalar $d\Gamma$ is given by

$$d\Gamma = -\frac{1}{2} C_\phi \Big(\Gamma - \langle \Gamma \rangle\Big) \frac{dt}{\tau}, \tag{5}$$

where $\langle \Gamma \rangle$ is the ensemble average of the scalar over particles within each cell which is fixed on the spatial area. C_ϕ is the parameter which determines the decay rate of the variance of the scalar. And τ is the turbulent time scale, which is defined by

$$\tau = \frac{\kappa}{\varepsilon}, \quad \kappa = \frac{\overline{u_r{}^2} + \overline{u_\theta{}^2} + \overline{u_x{}^2}}{2}, \tag{6}$$

where κ is the turbulent kinetic energy, ε is the dissipation rate of κ. In this model, the scalar attendant on the particle is determined not randomly but deterministicaly.

Modified Curl's model. In the modified Curl's model[9][10], the probability in which the nth particle mixes with other partcles is given by

$$P^{(n)} = \frac{3C_\phi dt}{\tau\left(x^{(n)}(t)\right)}, \tag{7}$$

where C_ϕ is the empirical constant. When the mixing occurs with this probability, the pair to the nth particle is selected (here, the mth particle) and the concentrations of two particles are replaced by the following equations,

$$\Gamma^{(n)}(t+dt) = (1-\beta)\Gamma^{(n)}(t) + \frac{1}{2}\beta\Big(\Gamma^{(n)}(t) + \Gamma^{(m)}(t)\Big), \tag{8a}$$

$$\Gamma^{(m)}(t+dt) = (1-\beta)\Gamma^{(m)}(t) + \frac{1}{2}\beta\Big(\Gamma^{(n)}(t) + \Gamma^{(m)}(t)\Big), \tag{8b}$$

where β is a parameter to control the degree of mixing in a particle pair and a random variable uniformly distributed on the interval $[0,1]$. And the scalar attendant on the unselected particle is not changed. Because mixing occurs locally in the physical space, the nearest particle is selected as the paired particle.

Here, the characteristics of the above two mixing models are summarized, and the reason that thse models are adopted in this research is explained. In the Dopazo's model, the scalar (concentration) of all the particles in the cell changes deterministically with the probability 1, and as shown in (5) the increments of the scalar are proportional to the difference between the scalar of particle and the ensemble average of the scalar over particles within each cell. On the other hand, the modified Curl's model gives the mixing rule of a particle pair which includes some randomness, that is, the change of the concentration of the selected paired particles has occured with a finite probability given by (7), and the model has also the randomness in the distribution of the concentration among the paired particles after mixing as seen in (8a) and (8b), so that the modified Curl's model can be expected to have the larger effect of uniformalization of the concetration than the Dopazo's model. In the future numerical prediction of the various diffusion fields it must be useful to investigate how the difference of these mixing models influences the statistical property of the scalar field, in particular, the change of the PDF profile. That is the reason we adopted the above two different models in this research. It should be noted here that more recently other mixing models have been suggested, e.g., a binomial Langevin model[16] and a mapping closure model[17]. However, these new models are more complicated in comparison with the Dopazo's model and the modified Curl's model. The simplicity of the model is a very important factor for the application of the model to the engineering problems. The use of the new models is our future subject.

4 Simulation conditions

The subject of the present simulation is the oil fog diffusion field by Becker et al.[14]. The simulation conditions are adjusted to those of the experiments by Becker et al.. In the experiments, the measurements of the concentration field of the point source plume of oil fog which is injected from the center of the pipe in a fully developed turbulent pipe flow were made. The fog injector's inner diameter is 2.16 mm, and the outer diameter is 2.77 mm. In the actual experiments, the measurement points were fixed and the injector was moved, then the concentrations were measured at the several downstream cross sections from the injector's exit (the plume source). On the other hand, we fixed the plume source at the origin and calculated concentrations at the same downstream distances from the source as those of experiments. Further, Becker et al. performed the experiments for several values of $\overline{U_c}$, but in this study we chose only one case of these experiments as the simulation subject: the case of $\overline{U_c}$=61 m/s (Re=684,000).

The conditions of the simulation are as follows.

inner radius of the pipe	: $R = 0.1005$ m
cross-sectional mean velocity	: $U_{av} = 52.38$ m/s
kinematic viscosity	: $\nu = 1.54 \times 10^{-5}$ m^2/s
Reynolds number	: $\mathrm{Re} = 2RU_{av}/\nu = 684,000$
friction velocity	: $u_\tau = 2.15$ m/s
Kolmogorov constant	: $C_0 = 2.0$
boundary condition	: absorptive wall

5 Simulation results

5.1 Verification of the velocity field

Here, we made the numerical verifications of the consistency condition[6] up to the second-order moment and the thermodynamic constraint[7] for the calculated velocity field. Since the subject of calculation is a fully developed turbulence field and the velocity field is independent of the azimuthal-direction, only the radial movements of the particles were calculated in this verification. Thus, the increment of the particle movement in the r direction dr^* during the time increment dt is calculated by

$$dr^* = U_r^* dt. \tag{9}$$

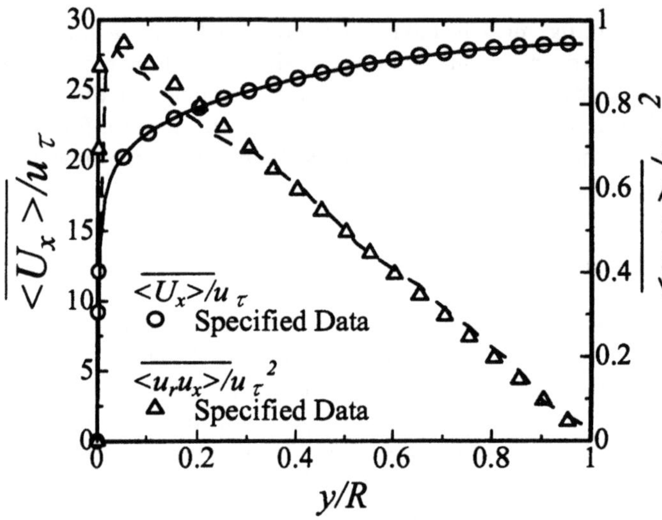

Fig. 2. Radial distributions of the mean velocity and Reynolds shear stress

The total number of particles is $N_t = 20,000$, the time increment is $dt = 4.67 \times 10^{-6}$ sec for one step, and the total number of time steps is 5,000, which corresponds to the real time of 2.335×10^{-2} sec. In order to calculate the statistics of velocity field, we picked up the positions and instantaneous velocities of particles at every 50-time step and take the ensemble average over particles within some spatially discretized cells, which are distributed in the radial direction. The total number of cells is 40 and the width of the kth cell $\Delta r^{(k)}$ from the center of the pipe is given by

$$\Delta r^{(k)}/R = -a(k-1) + 0.04, \tag{10}$$

where the constant a is 0.00076923 which is chosen so that the summation of the width of each cell becomes R. In the following, the simulation results are shown.

Figure 2 shows the radial distributions of the time averages of the main streamwise mean velocity $\langle U_x \rangle$ and the Reynolds shear stress $\langle u_r u_x \rangle$. In the figure, the open circle denotes the specified data of the mean velocity estimated by (3), and the triangle shows the specified data of the Reynolds shear stress estimated by (4). We find good agreement between the inputted data and the simulation results.

Figure 3 shows the radial distributions of the time averages of the r.m.s velocity fluctuations $\sigma_i = \sqrt{\langle u_i^2 \rangle}$ $(i = r, \theta, x)$, together with the experimental data by Laufer denoted by the symbol. We find that the inputted experimental data are well reproduced by the simulation.

From the above results (Fig.2 and Fig.3), we conclude that the velocity field simulated by the present model can reproduce well the prescribed data up to the second-order moment although the input of the third-order moment is omitted. This gives us some practical background to rely on the simulation of the scalar diffusion problem shown in the following section.

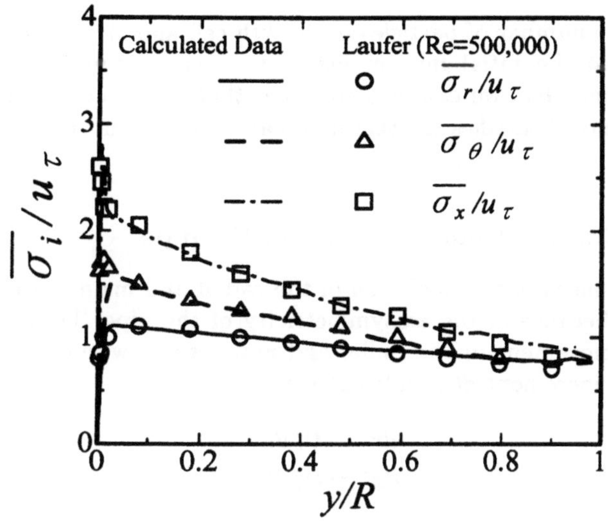

Fig. 3. Radial distributions of the r.m.s velocity fluctuations

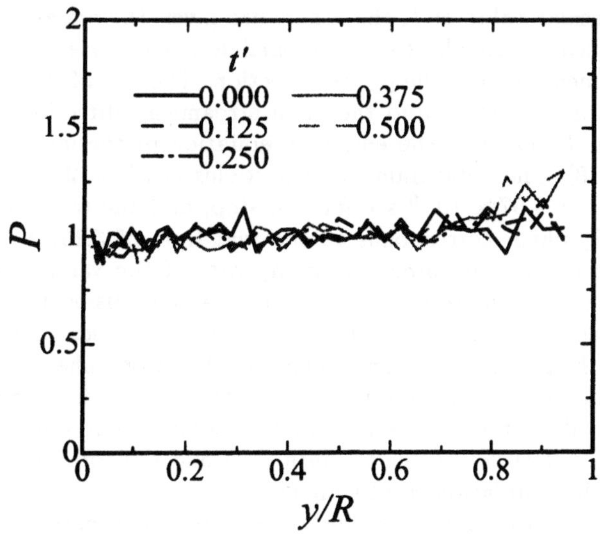

Fig. 4. Time change of the radial distributions of the number density of particles

Figure 4 shows the time change of the radial distribution of the number density of particles in each cell at time step which is a multiple of 0.125 of non-dimensional time $t' = u_\tau t/R$. The number density of particles in the kth cell is defined by

$$P^{(k)} = \frac{R^2 n^{(k)}}{2r^{(k)} \Delta r^{(k)} N}, \tag{11}$$

where $n^{(k)}$ is the number of particles in the kth cell and $\Delta r^{(k)}$ is given by (10). As shown in Fig.4, the initial uniform distribution is almost unchanged with time in the simulation. Thus we can also conclude that the velocity field simulated by the present model satisfies the thermodynamic constraint, i.e., the continuity condition[7].

5.2 Scalar diffusion from the center of the pipe

Here, the two-dimensional calculation of the radial and main streamwise direction was made because of the axisymmetricity of the pipe. Thus, the distance of the particle movement in the radial and main streamwise direction dr^*, dx^* during the time increment dt is calculated by

$$dr^* = U_r^* dt, \tag{12a}$$
$$dx^* = U_x^* dt. \tag{12b}$$

First, at the initial time the particles are distributed uniformly over the area of the calculation ($r = 0 \sim R, x = 0 \sim 7.0R$) and the particles within the source are given the scalar value of 1, the others are given the scalar value of 0. The change of the scalar attendant on each particle is calculated by the molecular mixing model mentioned in the previous section. The size of the source is $0.04R$ in both the radial direction and the main streamwise direction. This size was determined on the basis of the empirical equation of the mixing length given by Nikuradse[18]. The total number of particles is $N_t = 2,000,000$, the time increment is $dt = 4.67 \times 10^{-6}$ sec for one step, and the total number of time steps is 11,000, which corresponds to the real time of 5.137×10^{-2} sec. The parameter C_ϕ, which determines the decay rate of the variance of the scalar, is 7.5. In order to calculate the statistics of scalar diffusion field, we take the ensemble average over particles within each spatially discretized cell which is distributed in the radial and main streamwise direction. The cells of the radial direction are the same as in the verification of the velocity field (See (10)). With regard to the cells of the main streamwise direction, their widths are constant with $\Delta x^{(m)}/R = 0.04$ (m means the mth cell) and the total number is 175. In the following, the simulation results are shown.

Figure 5 shows the radial profiles of the mean concentration. In this figure, the ordinate is normalized by the maximum of mean concentration in each cross-section $\overline{\Gamma}_{max}$ and the abscissa by concentration half-radius of plume $r_{1/2}$. The lines show the simulation results by two molecular mixing models (i.e., the Dopazo's deterministic model and the modified Curl's model) and the symbols show experimental data by Becker et al.[14]. From this figure, it is found that both simulation results agree with experimental data as a whole.

Figure 6 shows the radial profiles of the r.m.s value of concentration fluctuations. In this figure, the lines show the simulation results by two molecular mixing models and the symbols show the experimental data by Becker et al.. The ordinate is normalized by the maximum of mean concentration in each cross-section $\overline{\Gamma}_{max}$ and the abscissa by $r_{1/2}$. From the figure, it is found that

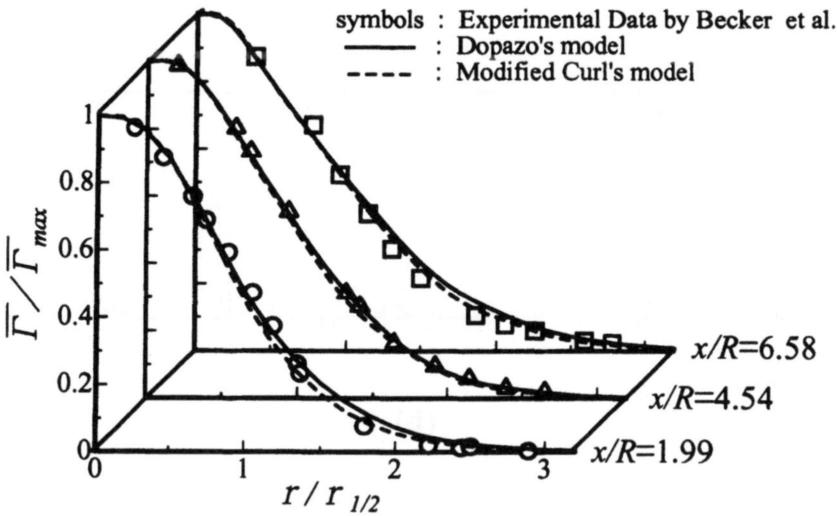

Fig. 5. Radial profiles of the mean concentration

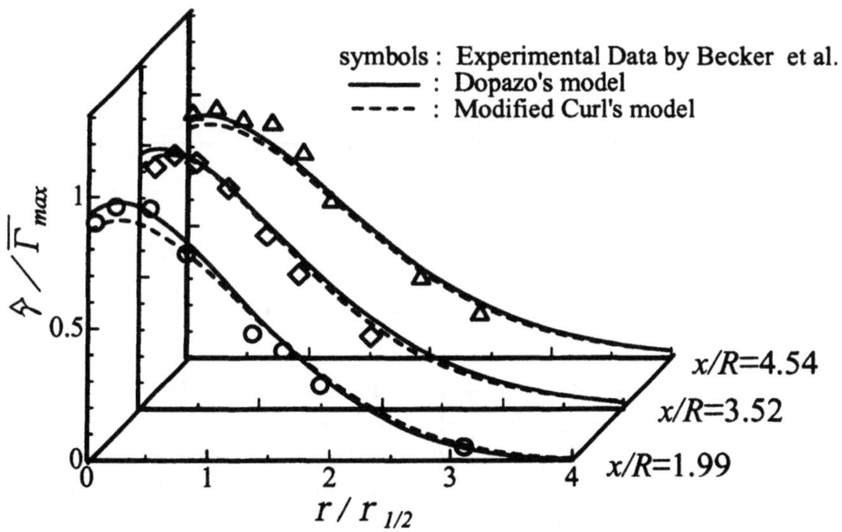

Fig. 6. Radial profiles of the r.m.s value of concentration fluctuations

the difference by the mixing models is not so large, and the whole shape of any simulated distribution almost agrees with the experimental one.

Figure 7 shows the main streamwise variation of the scalar PDF on the centerline of the pipe. Figs.(a),(b) show the simulation results by the Dopazo's deterministic model and the modified Curl's model, respectively. In these figures, the range [0, 1] which the scalar value can take is divided into one hundred pieces. From these figures, it is found that the modified Curl's model has the

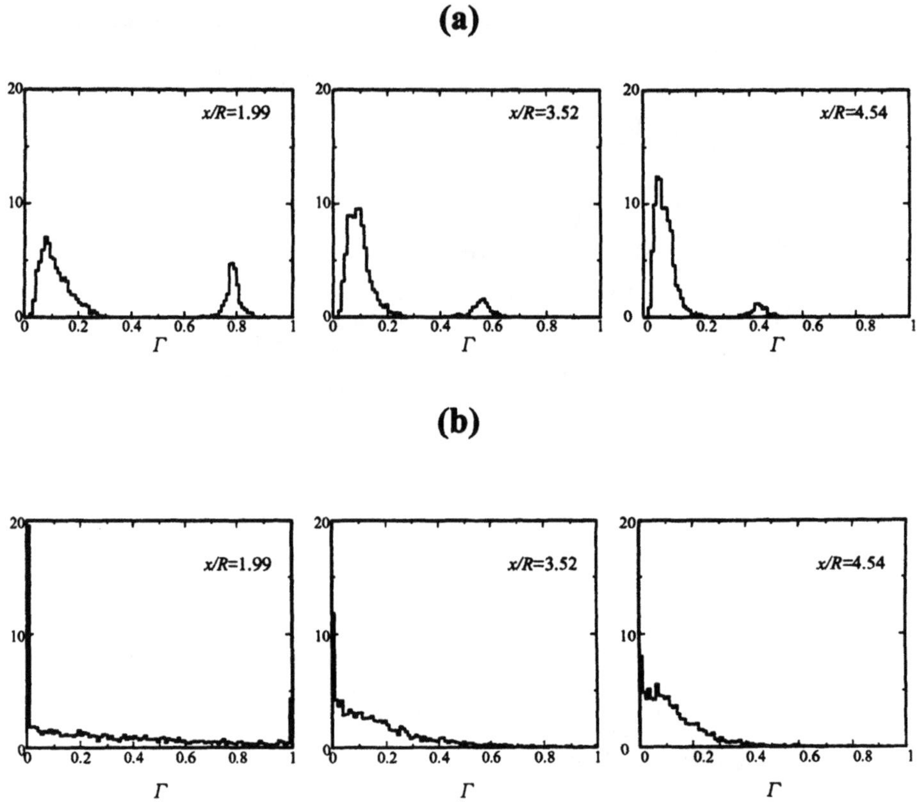

Fig. 7. Downstream variation of the scalar PDF profile

(**a**) Dopazo's model(**b**) Modified Curl's model

larger effect of the molecular mixing than the Dopazo's model: the PDF profile by the modified Curl's model shows the two large spikes near $\Gamma = 0$ and $\Gamma = 1$ at the upstream region (at $x/R = 1.99$), but as going to the downstream direction the uniformalization of the concetration (molecular mixing) has proceeded rapidly, then at the downstream region (at $x/R = 4.54$) the PDF shows the monotonously decreasing profile from the peak near $\Gamma = 0$ to the small value at the large concentration region, while in the PDF profile by the Dopazo's model the two peaks remain at the downstream region because of the less effect of molecular mixing. This difference of molecular mixing effect between the two models seems to be worth consideration to make the prediction of other diffusion fields by the PDF method in the future. Further, although the figures are not shown here, it has been ascertained that both models gives almost the same downstream variations of the mean concentration and the r.m.s. value of concentration (i.e., the first-order and the second-order moment of the PDF). Consequently it is found that both mixing models can express well the characteristics of the mixing process at least up to the second-order moment of the scalar PDF in the plume.

6 Conclusions

The axisymmetric point source plume in a fully developed turbulent pipe flow is studied numerically by the Lagrangian PDF method. The Lagrangian velocities of stochastic particles are modeled by a generalized Langevin equation expressed in the cylindrical coordinates. For the molecular mixing, the Dopazo's deterministic model and the modified Curl's model are used and the simulation results are compared with the experimental data by Becker et al.. The conclusions obtained in the present study are summarized as follows.

1. In the verification of the velocity field, it was confirmed that the present model (i.e., the generalized Langevin model in the cylindrical coordinates) can reproduce well the prescribed data up to the second-order moment of the velocity. It was also ascertained that the simulated velocity field satisfies thermodynamic constraint (i.e., continuity condition).

2. In the simulations of the diffusion field from the source at the center of the pipe, the radial profiles of the mean concentration and concentration fluctuation r.m.s value by two molecular mixing models show good agreement with experimental data as a whole.

3. From the downstream variations of the scalar pdf profiles, it was found that the modified Curl's model has the larger mixing effect than the Dopazo's model, but the downstream variations of the scalar PDF profiles by both mixing models can characterize well the mixing process of the scalar plume from the point source on the center line of the pipe.

From the above, it is confirmed that the present simulation method is valid to predict the distributions of the moments of PDF at least up to the second-order for the axisymmetric plume in the pipe flow.

References

1. S.B. Pope: Prog. Energy Combust. Sci., **11**, 119(1985)
2. S.B. Pope: Annu. Rev. Fluid Mech., **26**, 23(1994)
3. Y.Sakai, T.Kubo, I.Nakamura: 'Stochastic Simulation of a Reactive-Scalar Mixing Layer in a Grid Turbulence by the PDF Method'. In: *Proceedings of the 3rd ASME/JSME Joint Fluids Engineering Conference, San Francisco, California, July 18-23, 1999*, CD-ROM paper No.7771(1999) pp.1-8
4. Y. Sakai, I.Nakamura, H.Tsunoda, K.Hanabusa: JSME International J., B**39**-4, 667(1996)
5. D.C. Haworth, S.B. Pope: Phys. Fluids, **29**-2, 387(1986)
6. S.B. Pope: Phys. Fluids, **30**-8 , 2374(1987)

7. B.L. Sawford: Phys. Fluids, **29**-11, 3582(1986)
8. C. Dopazo: Phys. Fluids, **18**-4, 397(1975)
9. C. Dopazo: Phys. Fluids, **22**-1, 20(1979)
10. J. Janicka, W. Kollmann: Non-Equilib. Thermodyn., **4**, 47(1979)
11. R.L. Curl: A.I.Ch.E. Journal, **9**-2, 175(1963)
12. H. Tennekes, J.L. Lumley: *A First Course in Turbulence* (The MIT Press, New York 1972) p.162
13. J.O. Hinze: *Turbulence*, 2nd edn. (McGraw-Hill, New York 1975) p.621
14. H.A. Becker, R.E. Rosensweig, J.R. Gwozdz: A.I.Ch.E. Journal, **12**-5, 964(1966)
15. J. Laufer: 'The Structure of Turbulence in Fully Developed Pipe Flow', NACA Tech. Report No.1174(1954)
16. L. Valiño, C. Dopazo: Phys. Fluids A, **3**-12, pp.3034-3037(1991)
17. H. Chen, S. Chen, R.H. Kraichnan: Phys. Rev. Lett., **63-24**, pp.2657-2660(1989)
18. H. Schlichting: *Boundary-Layer Theory*, 7th edn. (McGraw-Hill, New York 1979) pp.604-606

A Hybrid RANS/LES Calculation of Turbulent Channel Flow

Fujihiro Hamba

Institute of Industrial Science, University of Tokyo, Meguro-ku, Tokyo 153-8505, Japan

Abstract. LES of wall-bounded flows with the no-slip boundary conditions is limited to moderate Reynolds numbers; wall modeling is necessary for high-Reynolds-number flows. One of approaches to better wall modeling is the hybrid RANS/LES modeling. In this study a hybrid modeling is applied to a channel flow calculation; the k–ε RANS model is used near the wall whereas the one-equation subgrid-scale model is adopted for the LES away from the wall. A mismatch of the mean velocity profile appears between the RANS and LES regions. To improve the velocity profile two values of the velocity component are defined at the boundary between the two regions. As a result the mismatch is removed and better velocity profiles are obtained.

1 Introduction

Turbulent flows at high Reynolds numbers contain various eddies at a wide range of length scale. It is very difficult to solve all eddy motions. In engineering and geophysics mean fluid motions are often of interest more than detailed motions. The transport equation for the mean velocity $\langle \boldsymbol{u} \rangle$ can be derived by applying some average to the Navier–Stokes equation. In this process the Reynolds stress $\langle \boldsymbol{u}'\boldsymbol{u}' \rangle$ appears where $\boldsymbol{u}'$ is the velocity fluctuation. The value of the Reynolds stress needs to be obtained to close the mean velocity equation; this is so-called turbulence modeling. There are two types of turbulence modeling: Reynolds-averaged Navier–Stokes (RANS) equations and large eddy simulation (LES). In RANS the ensemble average is applied. If the turbulent field is statistically steady the ensemble average is equivalent to the time average. Moreover, if the turbulent field is statistically homogeneous in some directions in space the mean quantities depend on the coordinate in the other directions only. The computational cost is much less than that of unsteady three-dimensional problem. Instead, complicated expressions are often required for modeling the Reynolds stress. They may vary for different kind of turbulent flows; universal models have not been developed yet. On the other hand, the grid average is adopted in LES; only the grid-scale (GS) velocity is directly solved whereas the effect of the subgrid-scale (SGS) velocity is modeled. Since unsteady three-dimensional calculation is needed to obtain the GS velocity field, its computational cost is large. However, the SGS velocity field consists of eddies smaller than the grid cell; it is expected that motions of such small eddies are universal and the SGS modeling can be simple compared to RANS.

LES was previously applied to simple flows for basic turbulence research. Due to the rapid development of computers it can now be used to simulate complex flows for practical engineering problems. However, it is still very difficult to simulate wall-bounded flows at high Reynolds numbers with the no-slip boundary conditions. In general the length scale of turbulent eddy is proportional to the distance from the wall. For an accurate LES the grid width near the wall should be narrow enough in both the wall-normal and wall-parallel directions; many grid points are required to resolve the velocity field (Fig. 1(a)). For LES at high Reynolds numbers artificial wall boundary conditions need to be imposed. Wall shear stress is evaluated at the first off-wall cells with algebraic expressions (Fig. 1(b)). For example, in the LES of the atmospheric boundary layer the Monin–Obukhov relationship is used as the wall boundary conditions

$$\frac{\partial \langle u \rangle}{\partial y} = \frac{u^+}{\kappa y} \Phi \left(\frac{y}{L} \right) \text{ at } y = y_1 , \tag{1}$$

where u^+ is the friction velocity, κ is the Kármán constant, and L is the Obukhov length [5]. The nondimensional function Φ equals unity for neutrally stable boundary layer; the boundary condition corresponds to the logarithmic law. The function Φ differs from unity and depends on y/L for stable or unstable boundary layer; it reflects the buoyancy effect on turbulence.

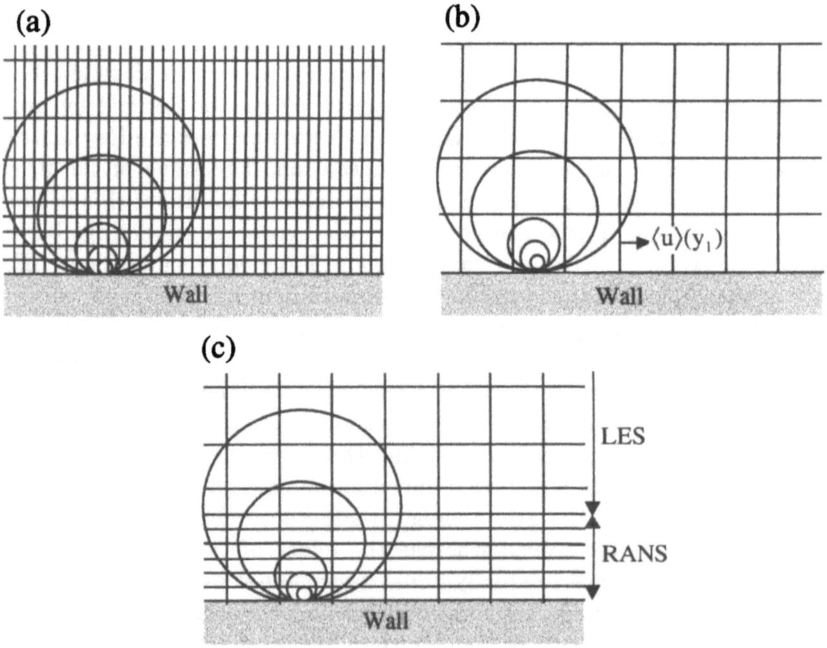

Fig. 1. Schematic drawings of grid systems for (**a**) no-slip conditions, (**b**) wall-law conditions, and (**c**) hybrid simulations. Circles represent turbulent eddies

Boundary conditions with some wall laws are also used in LES of engineering flows. However, it was shown that commonly used boundary conditions are not accurate enough for high Reynolds number flows and better wall modeling is required [8]. One of the approaches to better wall modeling is the hybrid RANS/LES modeling. Near the wall a RANS model is used because the wall-parallel grid width is wider than the turbulent length scale (Fig. 1(c)). Away from the wall the simulation can be switched to LES because the grid is fine enough compared to turbulent eddies. For example, Nikitin et al. [9] used the detached eddy simulation to calculate channel flows; they modified the Spalart–Allmaras model by changing the model length scale from the wall distance to the grid width. Although they were able to treat flows with a friction Reynolds number Re_τ as high as 80 000, the mean velocity profiles show a mismatch between the RANS and LES regions. Davidson and Peng [1] carried out hybrid simulations of a channel flow and a flow around a two-dimensional hill. They adopted the k–ε model for RANS near the wall and the one-equation SGS model for LES away from the wall. The mean velocity profiles of channel flow do not agree well with the logarithmic law; the profiles are too steep. Hamba [3] also carried out a hybrid simulation of a channel flow using the k–ε model and a one-equation SGS model. This simulation is different from the above simulations in that the position of the RANS and LES was exchanged. Nevertheless, a mismatch appeared between the RANS and LES regions.

Although different turbulence models were used, the three simulations show a common defect, a mismatch of the mean velocity profile between the RANS and LES regions. The model length scale were switched from the turbulent integral scale to the grid width; the resulting rapid change of the eddy viscosity seems to cause a steep velocity gradient. In this work, we examine the reason for such a mismatch in the hybrid simulation of a channel flow. We propose a new scheme to remove it and to obtain better velocity profiles.

2 Mismatch of Velocity Profiles Between Two Regions

In this section, to reproduce a mismatched profile and to examine its reason we apply a simple model for both RANS and LES. We treat a channel flow at $Re_\tau = 590$. The size of the computational domain is given by $L_x \times L_y \times L_z = 2\pi \times 2 \times \pi$ where the subscript x denotes the streamwise direction, y the wall-normal direction, and z the spanwise direction. The number of grid points is $64 \times 64 \times 64$. The equations for the GS velocity $(\bar{u}_x, \bar{u}_y, \bar{u}_z)[= (\bar{u}, \bar{v}, \bar{w})]$ are given by

$$\frac{D\bar{u}_i}{Dt} \equiv \left(\frac{\partial}{\partial t} + \bar{u}_j \frac{\partial}{\partial x_j}\right) \bar{u}_i = -\frac{\partial}{\partial x_j}\overline{u'_j u'_i} - \frac{\partial \bar{p}}{\partial x_i} + \nu \frac{\partial^2 \bar{u}_i}{\partial x_j \partial x_j} + f_i \, , \tag{2}$$

$$\frac{\partial \bar{u}_i}{\partial x_i} = 0 \, , \tag{3}$$

$$\overline{u_i' u_j'} = \frac{2}{3} k \delta_{ij} - \nu_s \left(\frac{\partial \bar{u}_i}{\partial x_j} + \frac{\partial \bar{u}_j}{\partial x_i} \right) , \tag{4}$$

where $\overline{u_i' u_j'}$ is the SGS stress, $\bar{p}$ is the GS pressure, ν is the molecular viscosity, f_i is the external force, k is the SGS kinetic energy, and ν_s is the SGS eddy viscosity. For the eddy viscosity the Smagorinsky model is used [6]

$$\nu_s = (C_s \Delta)^2 \sqrt{\frac{1}{2} \left(\frac{\partial \bar{u}_i}{\partial x_j} + \frac{\partial \bar{u}_j}{\partial x_i} \right)^2} , \tag{5}$$

where $C_s (= 0.1)$ is the model constant and Δ is the filter width. When the grid width h $[= (h_x h_y h_z)^{1/3}]$ is adopted as the filter width, the simulation becomes a typical LES. On the other hand, if the integral length scale ℓ is introduced as the filter width, then the simulation can be considered RANS; the Smagorinsky model becomes a kind of mixing length model. The integral scale ℓ for the channel flow at $Re_\tau = 590$ can be obtained in advance from the DNS data by [7] as follows. First we define the average $\langle \ \rangle$ in the x–z plane and in time; the velocity is decomposed as $\bar{u}_i = \langle \bar{u}_i \rangle + \bar{u}_i''$. If the integral scale is adopted as the filter width then the SGS viscosity ν_s is so large that the fluctuation $\bar{u}_i''$ becomes very small; the velocity can be approximated by $\bar{u}_i = \langle \bar{u}_i \rangle$. The SGS viscosity can be written as

$$\nu_s = (C_s \ell)^2 \left| \frac{\partial \langle \bar{u} \rangle}{\partial y} \right| . \tag{6}$$

Using the profiles of the mean velocity $\langle u \rangle$ and the Reynolds shear stress $\langle u'' v'' \rangle$ given bythe DNS, we can evaluate the integral scale as

$$\ell^2 = -\langle u'' v'' \rangle \Big/ C_s^2 \left| \frac{\partial \langle u \rangle}{\partial y} \right| \frac{\partial \langle u \rangle}{\partial y} . \tag{7}$$

Using the two length scales we carry out hybrid simulations in the following two cases:

$$\text{Case 1:} \quad \Delta = \begin{cases} \ell & \text{for } 0 < y^+ < 210 , \\ h & \text{for } 210 \le y^+ < 590 , \end{cases} \tag{8}$$

$$\text{Case 2:} \quad \Delta = \begin{cases} h & \text{for } 0 < y^+ < 210 , \\ \ell & \text{for } 210 \le y^+ < 590 . \end{cases} \tag{9}$$

In Case 1 the RANS model is used near the wall whereas LES is carried out away from the wall. This case corresponds to the wall-modeling approach by [9] and [1]. In Case 2 the position of LES and RANS is exchanged; this is the same as [3]. In this case the Van Driest damping function $f_\nu = 1 - \exp(-y^+/A_\nu)$ (where $A_\nu = 25$) is also used near the wall. Except for the filter width Δ and the damping function the numerical method are the same for both cases.

Figure 2(a) shows the mean velocity profile $\langle \bar{u} \rangle$ for Case 1. Solid line denotes the DNS result by [7]. Vertical line at $y^+ = 210$ denotes the position where

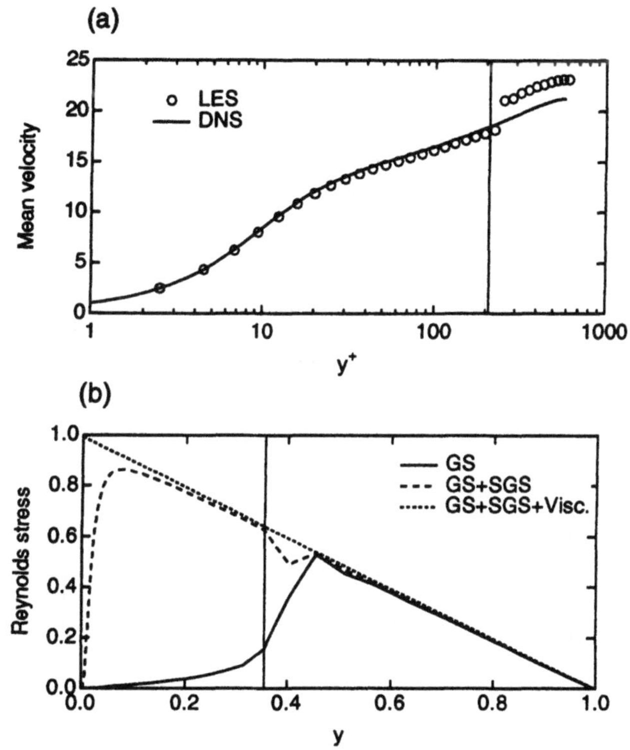

Fig. 2. Profiles of (**a**) mean velocity and (**b**) Reynolds stress for Case 1

the filter width Δ is switched from ℓ to h. A mismatch of the mean velocity is clearly seen in this hybrid simulation. Figure 2(b) shows the profile of the Reynolds shear stress for Case 1: the GS part $\langle \bar{u}''\bar{v}'' \rangle$, the SGS part $\langle \overline{u'v'} \rangle$, and the viscous part $-\nu\partial\langle \bar{u} \rangle/\partial y$. In the LES region at $y^+ > 210$ the GS part dominates as a typical LES. In the RANS region at $y^+ < 210$ the GS part decreases to very small values and the SGS part dominates. Figures 3(a) and 3(b) show the profiles of the mean velocity and the Reynolds shear stress for Case 2, respectively. Although the position of the LES and RANS is different from that for Case 1, a similar mismatch is seen in the mean velocity profile. The profiles of the Reynolds shear stress in each region in Fig. 3(b) are similar to those in the corresponding region in Fig. 2(b).

A simple hybrid simulation with the Smagorinsky model reproduces a mismatch of the velocity profiles; it commonly appears in hybrid simulations. This suggests that the mismatch is not directly related to the details of turbulence modeling. No matter which model is used the GS velocity is governed by (2)–(4). The value of ν_s is always very large in the RANS region compared to that in the LES region. In this simulation the peak value of ν_s is about 50 times as large as ν in the RANS region whereas ν_s is comparable to ν in the LES region. If there

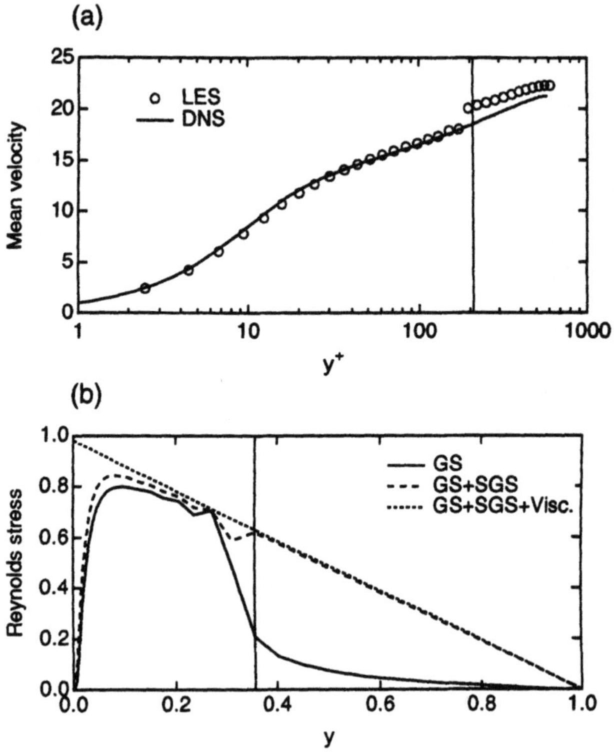

Fig. 3. Profiles of (**a**) mean velocity and (**b**) Reynolds stress for Case 2

is a fluid whose viscosity drastically changes in the two regions, such a steep velocity gradient at the boundary may be a real physical phenomena. However, in the hybrid simulation the real viscosity is not changed but the length scale for the averaging is different in the two regions. This result means that a simulation simply by changing the eddy viscosity in the two regions is not enough for combining the two models appropriately.

3 Model Equations and New Schemes

In this section we propose a new scheme for the velocity equations to reduce the mismatch. The temporal evolution of the GS velocity is obtained by solving the discrete forms of (2) and (3). In this work the second-order central finite-difference scheme is used in a staggered grid. Figure 4 shows the grid cells and the velocity components near the boundary between the RANS and LES regions. Only the $x-y$ plane is shown for simplicity; the subscript j denotes the number of grid cell in the y direction. It is assumed that Δ is set to $5h$ for the cells in the RANS. The superscript such as $5h$ denotes the filter width.

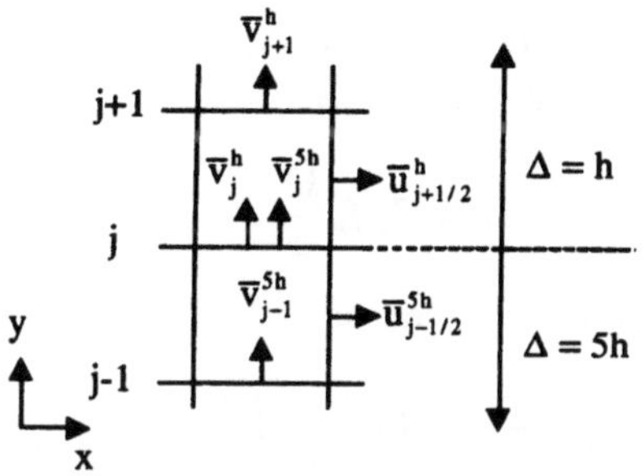

Fig. 4. Grid cells and velocity components defined in new scheme

The continuity equation in the new scheme can be given by

$$\left.\frac{\partial \bar{u}^h}{\partial x}\right|_{j+1/2} + \frac{\bar{v}^h_{j+1} - \bar{v}^h_j}{h_y} + \left.\frac{\partial \bar{w}^h}{\partial z}\right|_{j+1/2} = 0\,, \tag{10}$$

$$\left.\frac{\partial \bar{u}^{5h}}{\partial x}\right|_{j-1/2} + \frac{\bar{v}^{5h}_{j} - \bar{v}^{5h}_{j-1}}{h_y} + \left.\frac{\partial \bar{w}^{5h}}{\partial z}\right|_{j-1/2} = 0\,, \tag{11}$$

for the two cells, respectively. In previous schemes the same velocity component is used for $\bar{v}^h_j$ in (10) and $\bar{v}^{5h}_j$ in (11). If $\bar{v}^h_j$ is adopted in (11) some inconsistency appears; the mass flux in the y direction is associated with a different filter width from other directions. In this work in order to remove this inconsistency we introduce two values of the velocity components at the boundary between the RANS and LES regions. The value of $\bar{v}^h_j$ in (10) is obtained by solving its transport equation. The other value $\bar{v}^{5h}_j$ in (11) is obtained from the following relationship

$$\bar{v}^{5h}_j = \widehat{\bar{v}^h_j}\,, \tag{12}$$

where $\hat{}$ denotes the additional filtering whose width is $\hat{\Delta} = 2\sqrt{6}h (= \sqrt{5^2 - 1^2}h)$. By introducing the two values we can consistently evaluate the continuity equation at the both cells. Similarly, the convection terms in the GS velocity equation are given by

$$\frac{\partial \bar{u}^h_{j+1/2}}{\partial t} = -\frac{1}{h_y}(\cdots - \bar{u}^h_j \bar{v}^h_j) + \cdots\,, \tag{13}$$

$$\frac{\partial \bar{u}^{5h}_{j-1/2}}{\partial t} = -\frac{1}{h_y}(\bar{u}^h_j \bar{v}^h_j - \cdots) + \cdots\,. \tag{14}$$

In (13) and (14) different forms of the momentum flux are adopted; inconsistency about the momentum flux is also removed. It is expected that the velocity fluctuations are evaluated appropriately and therefore the artificial steep velocity gradient is reduced.

Next, we explain the turbulence models for RANS and LES and the method of combining them. Here, we treat the k–ε model and the one-equation SGS model. The equations for the GS velocity are the same as (2)–(4). The eddy viscosity is modeled in terms of k and ε as

$$\nu_t = C_\nu f_\nu \frac{k^2}{\varepsilon} \, . \tag{15}$$

The transport equation for k is written as

$$\frac{Dk}{Dt} = -\overline{u'_i u'_j}\frac{\partial \bar{u}_i}{\partial x_j} - \varepsilon + \frac{\partial}{\partial x_i}\left[\left(\frac{\nu_t}{\sigma_k} + \nu\right)\frac{\partial k}{\partial x_i}\right] , \tag{16}$$

where $\sigma_k = 1$. The equations for the velocity and the kinetic energy explained above are common in RANS and LES. The treatment of the dissipation rate ε is different. In the k–ε model the transport equation for ε_R is solved

$$\frac{D\varepsilon_R}{Dt} = -C_{\varepsilon 1}\frac{\varepsilon_R}{k}\overline{u'_i u'_j}\frac{\partial \bar{u}_i}{\partial x_j} - C_{\varepsilon 2} f_\varepsilon \frac{\varepsilon_R^2}{k} + \frac{\partial}{\partial x_i}\left[\left(\frac{\nu_t}{\sigma_\varepsilon} + \nu\right)\frac{\partial \varepsilon_R}{\partial x_i}\right] , \tag{17}$$

where $C_{\varepsilon 1} = 1.44$, $C_{\varepsilon 2} = 1.92$, $\sigma_\varepsilon = 1.3$, and f_ε is the Van Driest damping function ($A_\varepsilon = 9.5$). On the other hand, in the one-equation SGS model the dissipation rate ε_L is algebraically expressed in terms of k and Δ as

$$\varepsilon_L = \frac{k^{3/2}}{C_\Delta \Delta} , \tag{18}$$

where $C_\Delta = 0.61$. The one-equation SGS model of this type was shown to give results almost same as the Smagorinsky model for channel flows and mixing layers [10,4,2]. In this work we set a buffer region at $y_A < y < y_B$ to connect the RANS and LES regions smoothly. We introduce parameter α to obtain the general expression for ε as

$$\varepsilon = \alpha\varepsilon_R + (1-\alpha)\varepsilon_L, \quad 0 \le \alpha \le 1 \, . \tag{19}$$

The profile of α is shown in Fig. 5. In the case of $\alpha = 1$ at $y < y_A$ the simulation becomes RANS whereas in the case of $\alpha = 0$ at $y > y_B$ LES is carried out.

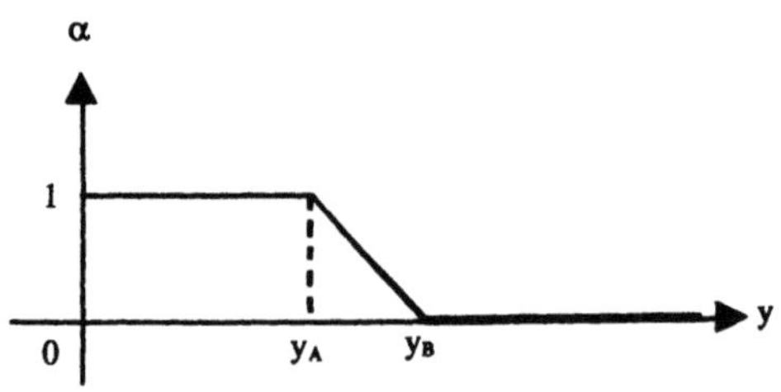

Fig. 5. Profile of parameter α

We carry out simulations of a channel flow for three cases. The details for each case are given by Table 1. At $y_A < y < y_B$ four cells are set to connect the

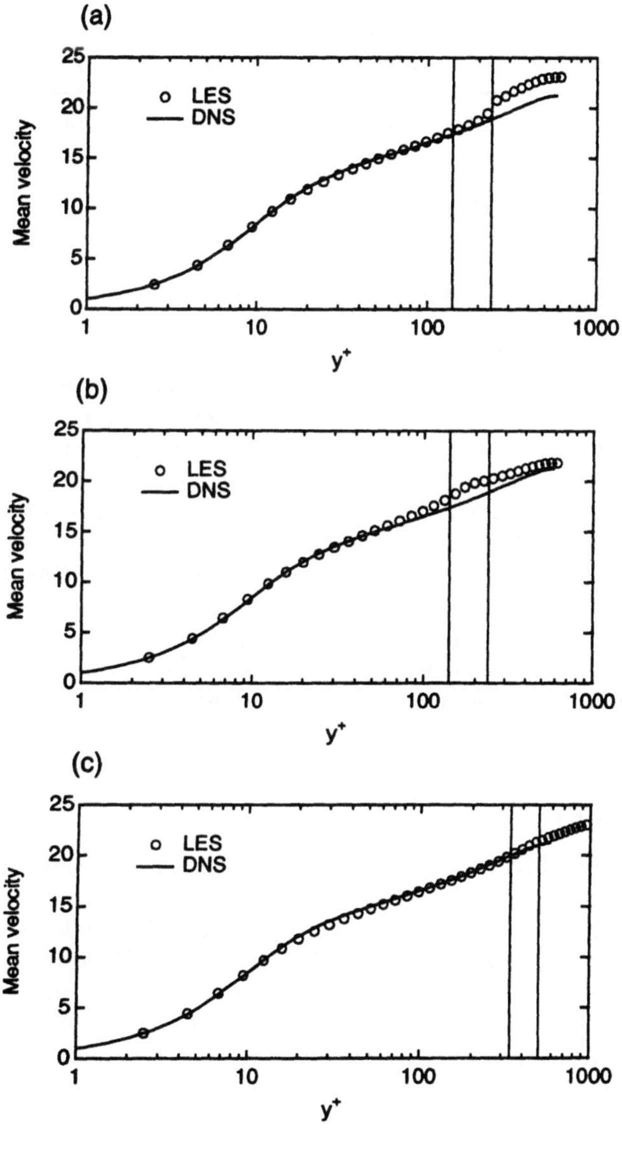

Fig. 6. Mean velocity profiles for (**a**) Case 3, (**b**) Case 4, and (**c**) Case 5

Table 1. Parameters for three cases

Case	Re_τ	Grid point	Scheme	y_A^+	y_B^+
3	590	64 × 64 × 64	Previous	140	238
4	590	64 × 64 × 64	New	140	238
5	1140	64 × 82 × 64	New	338	501

two regions smoothly. In Cases 4 and 5 we apply the new scheme to the velocity components at the four cells.

Figure 6(a) shows the mean velocity profile for Case 3. Two vertical lines denote the position of y_A and y_B. Since the previous scheme is used in Case 3 a mismatch is seen near $y = y_B$ similar to the results in the previous section. Figure 6(c) shows the mean velocity profile for Case 4. Although a region with a steep gradient is still seen, its position is shifted from $y = y_B$ to $y = y_A$. Moreover, the value of the mean velocity at the center line ($y^+ = 590$) decreases; it is close to the DNS data. We found that the steep gradient in this case is not directly related to the connection of the two models. It is the problem in the RANS region only; the boundary $y = y_A$ is too close to the wall. To improve the profile we increase the Reynolds number and shift the boundary $y = y_A$ away from the wall in Case 5. As is seen in the mean velocity profile in Fig. 6(c) a region with a steep gradient disappears and the velocity profile is connected smoothly. This better profile shows that the proposed new scheme is useful.

4 Conclusions

In previous hybrid RANS/LES simulations a mismatch of the mean velocity profile is seen near the boundary between the RANS and LES regions. In this work this mismatched profile is reproduced using a simple SGS model where the filter width is changed from the integral length scale to the grid width. It is suggested that the mismatch is due to the rapid change in the eddy viscosity common in hybrid simulations. A new scheme is proposed to remove the mismatch; two values of the velocity components are introduced at the boundary between the two regions. The inconsistency in the continuity equation and in the convection terms in the GS velocity equation is removed. Using this scheme a hybrid simulation with the k–ε model and the one-equation SGS model is carried out for a channel flow. A mismatch disappears and better velocity profiles are obtained. This scheme is expected to be useful to develop better hybrid simulations for wall modeling of LES.

References

1. L. Davidson, S.H. Peng: 'A hybrid LES-RANS model based on a one-equation SGS model and a two-equation K–ω model'. In: *Proc. 2nd Int. Symp. on Turbulence and Shear Flow Phenomena, 2001*, **2**, pp. 175–180.

2. F. Hamba: J. Phys. Soc. Jpn. **56**, 2721 (1987)
3. F. Hamba: Theoret. Comput. Fluid Dynamics **14**, 323 (2001)
4. K. Horiuti: J. Phys. Soc. Jpn. **54**, 2855 (1985)
5. C.H. Moeng: J. Atmos. Sci. **41**, 2052 (1984)
6. P. Moin, J. Kim: J. Fluid Mech. **118**, 341 (1982)
7. R.D. Moser, J. Kim, N.N. Mansour: Phys. Fluids **11**, 943 (1999)
8. F. Nicoud, J.S. Baggett, P. Moin, W. Cabot: Phys. Fluids **13**, 2968 (2001)
9. N.V. Nikitin, F. Nicoud, B. Wasistho, K.D. Squires, P.R. Spalart: Phys. Fluids **12**, 1629 (2000)
10. A. Yoshizawa, K. Horiuti: J. Phys. Soc. Jpn. **54**, 2834 (1985)

Anisotropy versus Universality in Shear Flow Turbulence

Yoshiyuki Tsuji

Department of Energy Engineerig and Science, Nagoya University, Nagoya city, Aichi, Japan, 464-8603

Abstract. The local isotropy hypothesis presented by Kolmogorov seems to work well as a good approximation depending on the nature of large-scale anisotropy. We discuss how the large-scale anisotropy penetrates the small scales by investigating the anisotropic spectrum measured in the rough wall boundary layers.

1 Introduction

According to the local isotropy hypothesis[1,2], the small-scale velocity fluctuations are to be universal in any kind of turbulent flow. This is one of the key assumptions in turbulence phenomena. At this stage, the question is not whether this assumption is correct or not, but seems to be how the large-scale anisotropy lost its information as the scales become small. In what condition can we expect the universality in small-scale statistics?

Actually, in the laboratory experiments, the large-scale anisotropy caused by the boundary or the initial conditions never disappear, or the finite size of apparatus and small Reynolds number restrict the flow conditions. The anisotropy effect on the inertial-range quantities has been directly checked by several researches so far[5,6]. Recent progress was made by the fusion rule or the SO(3) decomposition[3,4], in which the exact separation of isotropic and anisotropic fluctuations are obtained. In the formula, the tensorial nth order structure functions are necessary and the velocity structure functions appear as sums over all these contributions with different exponents characterizing the scaling factors[3,4]. This is the powerful tool for understanding the isotropic/anisotropic condition in the flow while the nth order structure functions are computed. The conventional experiments don't allow us to get the whole velocity field, then the only indirect fits of the superposition of many anisotropic contribution are possible.

We have reported the anisotropic effect on the small-scale statistics in the shear flow[7]. The focus was brought into the condition how the local isotropy is restored independent of large-scale anisotropy or how the small-scale statistics are affected by large-scale motions. Most of these analyses were based on investigating the velocity structure functions. Here, this study deals with the same problem by analyzing the ansisotropic energy spectra obtained in the rough wall boundary layers. Recently, an excellent result was reported by Ishihara et al. that the scaling formula for the anisotropic spectrum in uniform shear flow was analytical derived[8]. A few experimental evidences for supporting their results are also presented.

2 Local Isotropy in Small Scales

Local isotropy can be made sure by several classical techniques both in the dissipation and the inertial range. It is noted that the "locally isotropic turbulence is not simply an extension of the isotropic turbulence, but the former has both broader and narrower meaning than the latter concept"[18]. It is due to the original definition of local isotropy presented by Kolmogorov[1] that prevent us from disentangling. It is convenient to explain this issue briefly to readers.

2.1 Dissipation Range Isotropy

The concept of isotropy was originally given by G. I. Taylor, which is the mathematical idealization and is, at best, convenient for the approximate description of certain types of turbulent flows[18]. The isotropic turbulence is based on the fact that all the statistical measures of the flow must be invariant to reflections and rotations about all axes. If we use the character $(u, v, w) = (u_1, u_2, u_3)$ for velocity fluctuations, and $(x, y, z) = (x_1, x_2, x_3)$ for coordinate axis as the case may be, a convenient way quantifying the isotropy is via the ratios,

$$Z_i = 2 \cdot \left\langle (\partial u_1/\partial x_1)^2 \right\rangle \Big/ \left\langle (\partial u_j/\partial x_k)^2 \right\rangle , \tag{1}$$

where $(i, j, k) = (1,2,1)$, $(2,3,1)$, $(3,1,2)$, $(4,1,3)$, respectively. Each of these ratios is equal to unity in isotropic flow, but this condition is hardly possible to be realized in conventional laboratory experiments. The isotropy ensures that the energy dissipation rate per unit mass is reduced to

$$\langle \varepsilon \rangle \equiv \nu \left[\left\langle \frac{\partial u_i}{\partial x_j} \frac{\partial u_i}{\partial x_j} \right\rangle + \left\langle \frac{\partial u_i}{\partial x_j} \frac{\partial u_j}{\partial x_i} \right\rangle \right] = 15\nu \left\langle \left(\frac{\partial u}{\partial x} \right)^2 \right\rangle . \tag{2}$$

Kolmogorov subsequently used the idea of isotropic turbulence as the basis for more general concept of locally isotropic turbulence which corresponds more closely to turbulent flows encountered in practice. There are three hypotheses stated by Kolmogorov in 1941 about the smallest eddy size that is responsible for dissipating the energy and the statistically universal state of it (this is abbreviated as **K41**, for convenience [1]). The first hypothesis concerns the isotropy of the small-scale motions. In general, the large scale eddies are anisotropic and are affected by the initial or boundary conditions of the flow. He argued that this anisotropy is lost in the cascade process, by which energy is transferred to successively smaller and smaller eddies. The **Local Isotropy Hypothesis** is restated as, *At sufficiently high Reynolds number, the small-scale turbulent motions are statistically isotropic*[9]. More precise statements of the local isotropy was presented in terms of N-point velocity distributions in the four dimensional space (location and time)[1,18]. The turbulence is locally homogeneous in the domain **G**, if for every fixed N and the location $\boldsymbol{x}^{(n)}$, $n = 1, 2, \cdots, N$, the N-point probability density function of velocity difference is independent of the origin $\boldsymbol{x}^{(0)}$. And it is locally isotropic within the domain **G**, if it is locally homogeneous and in addition if the joint probability of N-point is invariant with respect to rotation and reflections of the coordinate axes.

It is impossible, however, to check the N-point probability density within domain **G** for practical purpose. Then, how can we know whether the flow is in the condition of local isotropy or not. At the early stage, Eq. (1) was adopted as the criterion to judge the local isotropy, but most of the laboratory experiments revealed that each ratio was larger than unity. The typical results in the moderate Reynolds numbers are summarized in refs[29,19]. Following these experimental evidences, George et al. proposed an alternative description of turbulence, namely, *locally axisymmetric turbulence*, which is defined as the locally invariant to rotations about the preferred axis. It requires that $Z_1 = Z_2$ and $Z_3 = Z_4$ for the derivative ratios defined above with no restriction on their values[19]. This description is widely satisfied in laboratory experiments.

We assume that the local isotropy is basically satisfied if the first and/or second similarity hypothesis, mentioned in the next section, is realized. To a good approximation, turbulence becomes locally isotropic in any kind of flow with sufficiently large Reynolds numbers.

2.2 Similarity Hypothesis

In the course of energy cascade, all information of the geometry of large-scale eddies determined by the initial and the boundary conditions are lost. As a consequence, the statistics of the small-scale motions are in a sense universal. In **K41** this universal state depends on two dominant processes, that is, the energy transfer to the successively smaller scales and the viscous dissipation. The important parameters are the rate at which the small scale receive energy from the large scales and the kinetic viscosity ν. The energy transfer rate determines the energy dissipation rate $\langle\varepsilon\rangle$, so these are nearly equal. If the statistically universal state of the small scale is determined by ν and $\langle\varepsilon\rangle$, this is restated as, **Kolmogorov's First Similarity Hypothesis**: *In every turbulent flow at sufficiently high Reynolds number, the statistics of the small-scale motions have a universal form that is uniquely determined by* ν *and* $\langle\varepsilon\rangle$[9]. It is usually called the universal equilibrium range where the first hypothesis is approved. There are unique length, velocity, and time scales that can be formed by the given two parameters; $\eta \equiv (\nu^3/\langle\varepsilon\rangle)^{1/4}$, $u_\eta \equiv (\nu\langle\varepsilon\rangle)^{1/4}$, $\tau_\eta \equiv (\nu/\langle\varepsilon\rangle)^{1/2}$. These are called Kolmogorov scales.

We consider the statistical characteristics of the velocity differences between two points at a fixed time,

$$\Delta u_r = u(x+r) - u(x)\,, \tag{3}$$

where u is the x-component of the velocity vector with separation distance r[m] measured along x. For any positive integer n, dimensional analysis then yields the result that

$$\langle [\Delta u_r]^n \rangle = u_\eta{}^n f_n(r/\eta)\,. \tag{4}$$

The first similarity hypothesis says that the function $f_n(r/\eta)$ is independent of Reynolds number and large scale anisotropy.

If the one-dimensional energy dissipation spectrum is defined by

$$D^{(1)}(k_1) = k_1^2 \cdot E_{11}(k_1) \quad , \quad D^{(2)}(k_1) = k_1^2 \cdot E_{22}(k_1)\,, \tag{5}$$

where the wave number k_1 is obtained by the relation $k_1 = 2\pi f/U$ with time frequency f[Hz] and local mean velocity U[m/s], the ratio Z_1 is re-expressed as

$$Z_1 = 2 \cdot \int_0^\infty D^{(1)}(k_1)dk_1 \Big/ \int_0^\infty D^{(2)}(k_1)dk_1\,. \tag{6}$$

Therefore, not only the small scale but also the large-scale contribution is contained in Z_i. This is the significant reason for $Z_1 \neq 1$ in laboratory experiments or low Reynolds number flows[37]. According to the first similarity hypothesis, the dissipation spectra should have the universal form, thus, the alternative criteria for local isotropy is suggested.

The dissipation spectra indicate the maximum value around some wave number k_p. That is,

$$dD^{(1)}(k_1)/dk_1\Big|_{k_1=k_p^{(1)}} = 0 \quad , \quad dD^{(2)}(k_1)/dk_1\Big|_{k_1=k_p^{(2)}} = 0\,, \tag{7}$$

where $k_p^{(1)}\eta \simeq 0.1$ and $k_p^{(2)}\eta \simeq 0.14$ have been confirmed empirically so far [32]. Dissipation spectra are believed to decay exponentially in the range $0.5 \leq k_1\eta$ like the form $D^{(i)}(k) \propto \exp(-\beta^{(i)}k\eta)$. We investigated these values as the requirement for evaluating the isotropic state in the dissipation range. Both the ratios $k_p^{(2)}/k_p^{(1)}$ and $\beta^{(2)}/\beta^{(1)}$ were examined, because of avoiding the insufficient accuracy of Kolmogorov scale η derived by the mean dissipation rate, they approached the expected values $k_p^{(2)}/k_p^{(1)} \simeq 1.4$ and $\beta^{(2)}/\beta^{(1)} \simeq 1$, respectively, as the Reynolds number increases even at $S^* = 0.1$ (the definition of S^* is given in section 3)[7]. This might be the first signs that the small scales are becoming isotropic in strong mean shear flow. The same tendency was obtained by Saddoughi et al. in which $S^* \simeq 0.01$ and the local isotropy is sufficiently satisfied[30].

2.3 Inertial Range isotropy & Anisotropy

Kolmogorov argued that all information about the geometry of large eddies is lost and the small scale statistics are similar in every turbulent flow. This is mentioned in the first similarity hypothesis. When the Reynolds number becomes so high that the internal length scale η is very small in comparison with the external length scale L. In this kind of turbulence, we can isolate an extensive subrange of motions much smaller than L, but much greater than η. The dominant process in the subrange is the inertial transfer of energy to small-scale disturbances without any conversion of energy into heat. Therefore, the statistical property in this range should not depend on the viscosity[18]. These considerations are restated by the second similarity hypothesis, **Kolmogorov's Second Similarity Hypothesis**: *In every turbulent flow at sufficiently high Reynolds number, the statistics of motions of scale ℓ in the range $L \gg \ell \gg \eta$ have a universal form that is uniquely determined by $\langle\varepsilon\rangle$, independent of ν*[9]. This range is called the inertial subrange since the inertial forces play the main role. The second hypotheses means that the function f_n in Eq. (4) have to assume a form that will suppress ν. Elementary manipulation yield the result that

$$\langle[\Delta u_r]^n\rangle = K_n \cdot (\langle\varepsilon\rangle \cdot r)^{\zeta_n} \quad , \quad \zeta_n = n/3, \tag{8}$$

where coefficient K_n is some suitable constant. With the expectation that K_n is universal, Eq. (8) expresses the global universality of turbulence in the inertial range. For the homogeneous and isotropic flow, the third-order structure function is exactly connected to the second one by Navier-Stokes eq. In the steady state this relation is simply written as follows,

$$\langle\Delta u_r^3\rangle = -4/5 \cdot r \cdot \langle\varepsilon\rangle + 6\nu \frac{\partial\langle\Delta u_r^2\rangle}{\partial r}. \tag{9}$$

Usually the second term in the right-hand side is negligible, this leads to the Kolmogorov's 4/5 law;

$$\langle\Delta u_r^3\rangle = -4/5 \cdot r \cdot \langle\varepsilon\rangle . \tag{10}$$

The relation might be a useful check for the local isotropy in the inertial range.

The velocity structure function is equivalent to the spectral representation in the case of $n = 2$;

$$E_{11}(k_1) = K_2' \cdot \langle\varepsilon\rangle \cdot k_1^{-5/3} , \tag{11}$$

where K_2' is to be universal in high-Reynolds number turbulence. The spectral form is widely used as the isotropy check. The transverse spectrum can be computed from the longitudinal one[18],

$$E_{22}^*(k_1) = E_{33}^*(k_1) = \frac{1}{2} \cdot E_{11}(k_1) \cdot \left(1 - \frac{d \log E_{11}(k_1)}{d \log k_1}\right) , \tag{12}$$

where the superscript $*$ indicates the isotropic value. When the ratio between the real measured spectra E_{22}, E_{33} and E_{22}^*, E_{33}^* are plotted against the wave number, the isotropic condition is $E_{22}^*/E_{22} = E_{33}^*/E_{33} = 1$. It is noticed here that these three different criteria do not always give the consensus but they have different level of interpretation unless the flow is exactly isotropic. This point is emphasized especially for the experimental works.

If the local isotropy is not achieved in the inertial range, the above similarity relations, Eq. (9), Eq. (11), and Eq. (12), are not realized. This might be because of the anisotropic effect in the large scale motions. In the turbulent boundary layers, the anisotropy in the transverse is more significant than that of stream wise direction. So the spectrum E_{22} has little power-law region, nor the structure function does in moderate Reynolds numbers. This may be because the largest scale motions are limited by the distance from the wall. The transverse component shares the unusual effect in the energy transfer process. In this point of view, the cross spectrum E_{12} defined by

$$-\langle uv\rangle = \int_0^{+\infty} -E_{12}(k_1)dk_1\,, \tag{13}$$

might have key information of large scale anisotropy. The anisotropic condition becomes clear by computing the correlation spectrum,

$$R_{12}(k_1) = \left|E_{12}(k_1)\right| \Big/ \sqrt{E_{11}(k_1)\cdot E_{22}(k_1)}\,, \tag{14}$$

which becomes zero for isotropic wave-number range. In this report the large scale anisotropy is investigated by means of the cross spectrum, and discuss the universality in the shear flow.

2.4 Anisotropy in Energy Containing Range

One of the convenient ways to check the anisotropic condition is through using the symmetric, traceless tensors whose elements vanish in isotropic flow. Anisotropy in the energy-containing scale is quantified using the tensor,

$$b_{ij} = \langle u_i u_j\rangle \,/\, \langle u_k u_k\rangle - \delta_{ij}/3\,, \tag{15}$$

in which δ is the Kronecker delta and summation over repeated indices is implied. The combination of second and third principal invariants, $II_b = b_{ij}b_{ji}$, $III_b = b_{ij}b_{jk}b_{ki}$, are useful for representing the anisotropy of velocity-carrying motions. At any point and time, II_b and III_b are determined by the Reynolds stresses. The result plotted as a point in the plane (II_b, III_b) indicates some special state of the Reynods-stress tensor[24]. As the eigenvalues of tensor b_{ij} relate to the invariants II_b, III_b,

$$II_b = -(\lambda_1^2 + \lambda_1\lambda_2 + \lambda_2^2) \quad , \quad III_b = -\lambda_1\lambda_2(\lambda_1 + \lambda_2)\,, \tag{16}$$

it may be possible to characterize the Reynolds stress condition by the eigenvalues. In the isotropic state, $II_b = III_b = 0$ so all eigenvalues are zero.

3 Characteristic Scales in Shear Flow

In this section, both the typical length and velocity scales in the shear flow are summarized, and the scaling form of $E_{12}(k_1)$ is described. A brief review on the anisotropic spectrum is presented in the last subsection.

3.1 Scaling of Ansiotropic Spectrum

The mean shear rate S causes the turbulence anisotropic, or in other words, the shear-stress spectrum is zero in the isotropic flow. There are a few parameters to evaluate the shear effect. In a simple shear flow, for instance, the non-dimensional parameter

$$\hat{S} \equiv Sq^2/\langle\varepsilon\rangle \quad , \quad S = \partial U/\partial y\,, \tag{17}$$

is defined, where $q^2 = u^2 + v^2 + w^2$ is twice the turbulent kinetic energy per unit mass. This is the ratio of the large-eddy time scale $q^2/\langle\varepsilon\rangle$ to the time scale of mean deformation S^{-1}, that characterizes the effects of mean strain rate on the energy containing turbulence scale. Lee et al. obtained the distribution of $\hat{S}$ in a simulated channel flow, and found that $\hat{S}$ reaches the maximum value around $y^+ \simeq 10$ and then decrease toward the outer region[27]. On the other hand, Corrsin proposed the criterion to think about the local isotropy in the shear flow[28]. The local isotropy is expected to be achieved in the wavenumber range in which the inertial transfer and viscous dissipation are important. If $\tau(k_1)$ is the characteristic time scale of motions of wave number k_1, then the influence of the mean shear S is characterized by the non-dimensional parameter $S\cdot\tau(k_1)$. If $S\cdot\tau(k_1)$ is small, then so also the level of anisotropy created by the mean shear might be small. Onsager used dimensional arguments to define the characteristic inertial transfer time per stage as $\tau(k_1) \equiv 1/\left[{k_1}^3\cdot E_{11}(k_1)\right]$. Then in the inertial subrange, this time scale becomes $\tau(k_1) = ({k_1}^2\cdot\langle\varepsilon\rangle)^{-1/3}$. Following these discussions, the criterion for isotropy at wave number k_1 is

$$S\cdot\tau(k_1) = Sk_1^{-2/3}\langle\varepsilon\rangle^{-1/3} \ll 1\,. \tag{18}$$

this criterion is re-expressed as follows with the length scale L_s defined by

$$k_1L_s \gg 1 \quad , \quad L_s \equiv \langle\varepsilon\rangle^{1/2}S^{-3/2}\,. \tag{19}$$

In the dissipation range, Kolmogorov time scale τ_η is the appropriate scale and hence, as first suggested by Corrsin, a criterion for the isotropy of the smallest scales is mentioned as follows,

$$S^* \equiv S\cdot\tau_\eta \ll 1\,. \tag{20}$$

Local isotropy is, therefore, achieved for $S^* \ll 1$ in the wavenumber range of $k_1 \gg L_S^{-1}$.

In the simple shear flow there is the only significant mean velocity gradient. The anisotropy is evident in the Reynolds stress and in view of Eq. (13) it is important to ascertain the contribution to $-\langle uv\rangle$ from the various scale of motions. The dominant contribution to $-\langle uv\rangle$ is from the wavenumbers in the energy containing range. Accorging to Lumley's discussion[22,23], the dimensional analysis then yields

$$E_{12}(k_1)/\left[u_s^2L_s\right] = E_{12}^*(k_1L_s)\,, \tag{21}$$

where E_{12}^* is a non-dimensional function and the velocity scale u_s is

$$u_s \equiv (\langle\varepsilon\rangle/S)^{1/2}\,. \tag{22}$$

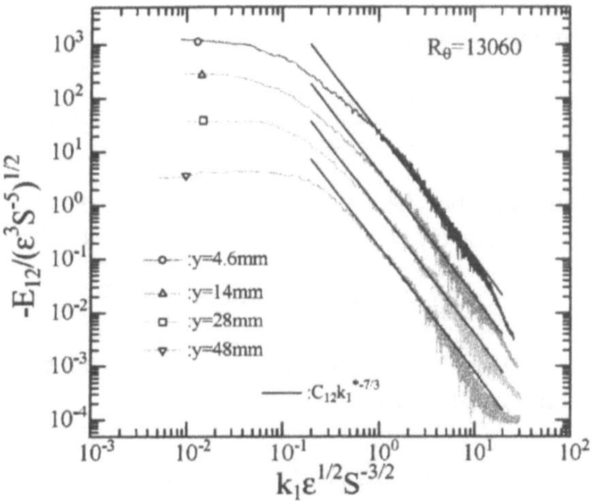

Fig. 1. Anisotropic spectrum measured in this rough wall boundary layer at $R_\theta = 13060$. y indicates the distance from the rough wall. The solid line is the scaling form of Eg. (23). The graphs □, △, and ○ are multiplied by 10, 10^2, and 10^3, respectively.

The linearity of E_{12} with S then determines E_{12}^*,

$$E_{12}^*(k_1^*) = -C_{12} \cdot (k_1^*)^{-7/3} \quad , \quad k_1^* = k_1 L_s \, , \tag{23}$$

where C_{12} is believed to be constant. There are few researches which discuss the criterion Eq. (19) and Eq. (20) qualitatively. Antonia et al. suggested that Eq. (20) may be relaxed to $S^* \leq 0.2$ for the small scales to be isotropic[34]. Saddoughi et al. reported that one decade of inertial subrange with truly negligible shear-stress cospectral density requires S^* not more than about 0.01[30]. And the $-7/3$ power-law scaling is confirmed in $1 \leq k_1 L_s \leq 10$ for their experimental condition. However, it is still open whether the coefficient C_{12} is universal or not, and in what wave-number ranges universal scaling of E_{12} is achieved. This uncertainty is mostly due to the lack of experimental and numerical studies on this topic. In Fig. 1 we show the typical examples of E_{12} measured in this study. Equation (23) seems to work well but the power-law exponents depart slightly from $-7/3$. Detailed discussions are given in the later sections.

3.2 Simple Shear Flow and Boundary Layer

In a simple uniform shear flow, with assuming the energy production P equals to the dissipation $\langle\varepsilon\rangle$, and the Townsend's structure parameter,

$$a_1 \equiv -\langle uv \rangle / \mathbf{u_r} \quad , \quad \mathbf{u_r} = \left[u_r^2 + v_r^2 + w_r^2\right]/2 \, , \tag{24}$$

is constant, where lower subscript r means a root mean square value, we have the simple relation

$$S \cdot \mathbf{u_r} / \langle\varepsilon\rangle = P/(\langle\varepsilon\rangle a_1) = 1/a_1 \, . \tag{25}$$

The normalized mean velocity gradient is

$$S^* = S \cdot \tau_\eta = \frac{\langle\varepsilon\rangle}{\mathbf{u_r}} \cdot \frac{P}{\langle\varepsilon\rangle\, a_1} \left(\frac{\nu}{\langle\varepsilon\rangle}\right)^{1/2} = \frac{1}{a_1} \cdot (Re_L)^{-1/2} \,. \tag{26}$$

where the Reynolds number $Re_L \equiv \mathbf{u_r}^{1/2} \cdot L/\nu$ and $L \equiv \mathbf{u_r}^{3/2}/\langle\varepsilon\rangle$. If the local isotropy is achieved, with $\langle\varepsilon\rangle = 15\nu u_r^2/\lambda^2$ and $\mathbf{u_r} = 3 \cdot u_r^2/2$,

$$R_\lambda \equiv u_r \cdot \lambda/\nu = (20 \cdot Re_L/3)^{1/2} \,, \tag{27}$$

here, the scale λ is called Taylor's micro scale. So the typical characteristic parameters are simply reduced to

$$S^* \simeq \frac{1}{a_1}\sqrt{\frac{20}{3}} \cdot R_\lambda^{-1} \propto R_\lambda^{-1} \,, \tag{28}$$

$$L_s = S^{-3/2} \langle\varepsilon\rangle^{1/2} \simeq a_1^{3/2} \cdot L \quad , \quad u_s = (\langle\varepsilon\rangle/S)^{1/2} \simeq \sqrt{3a_1/2} \cdot u_r \,. \tag{29}$$

In a boundary layer, however, the mean velocity gradient is not constant, but it is represented as a function of the distance from the wall. There is a region called log-law where the mean velocity profile is given by

$$U/u_* = 1/\kappa \cdot \log_e(y^+) + C \,, \tag{30}$$

here u_* is a friction velocity and y^+ is the non-dimensional quantity defined by $y^+ \equiv u_* \cdot y/\nu$. The coefficient κ and C are believed to be universal constant. In the two-dimensional boundary layers the energy production is dominated by the term; $-\langle uv\rangle \cdot dU/dy$, and the structure parameter is well approximated by $a_1 \simeq 0.26$ because the total Reynolds stress is constant; $-\langle uv\rangle \simeq u_*^2$[35]. Thus the relation $S^* \propto R_\lambda^{-1}$ in Eq. (28) is still satisfied in the log-law region and the characteristic scales are re-expressed as

$$S^* = S\,(\nu/\langle\varepsilon\rangle)^{1/2} \propto R_\lambda^{-1} \simeq (\kappa y^+)^{-1/2} \,, \tag{31}$$

$$L_s = S^{-3/2} \langle\varepsilon\rangle^{1/2} \simeq \kappa \cdot u_* \cdot y^+ \quad , \quad u_s = (\langle\varepsilon\rangle/S)^{1/2} \simeq u_* \,. \tag{32}$$

The relation between S^* and R_λ is plotted in Fig. 2. The data was measured in the two-dimensional rough wall boundary layer with no-pressure gradient (experimental conditions are given in section 4, or see the ref[7]). In the low Reynolds number range $R_\lambda \leq 200$, the predicted relation Eq. (28) is not well satisfied. It is emphasized that, as the length scale L_s is a function of the distance from the wall, the large scale motions are restricted by the boundary condition of the wall. This is the significant difference from the uniform shear flow. We aim to reveal the shear effect on the small-scale statistics, however not only the mean shear but also the boundary effect is contained in the flow field. It is not clear whether this non-local interaction might be essential or not.

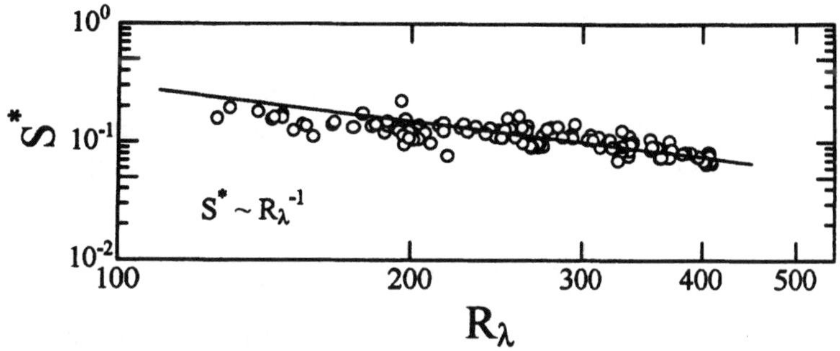

Fig. 2. Shear parameter is plotted as a function of Reynolds number. The solid line indicates the relation Eq (28); $S^* \propto R_\lambda{}^{-1}$

3.3 Review of Anisotropic Spectrum E_{12}

The anisotropic spectrum was measured by several researchers as a pioneering work in 1950s. Klebanoff pointed out the shear effect penetrates deeper and deeper into the dissipation range in the boundary layer[20]. Laufer measured it in the turbulent channel flow and Tani et al. for an axially symmetric turbulent jet. In all these cases, the spectral density $E_{12}(k_1)$ was quite large for small wave numbers, but as k_1 increases, the spectrum falls much more rapidly than E_{11} and E_{22}. It follows that the large-scale turbulence components are definitely anisotropic. Marusic et. al discussed the scaling form of E_{12} in boundary layers. The basic idea is coming from the attached eddy hypothesis of Townsend as extended by Perry et al.[36]. The results indicate a region of inner-flow scaling collapse for mid-range wave numbers. At low wave numbers it would seem that neither an inner-flow nor outer-flow scaling exist, while at high wave-numbers, good collapse occurs using Kolmogorov scaling. The computed anisotropic spectra show qualitative agreement with experimental results[21].

The field measurements were performed over the ocean and the terrain. Pond et al. collected the velocity, temperature, and moisture data from the floating instrument platform at San Diego in 1969[15]. The three components of velocity data were measured by the ultrasonic anemometer and also the cup anemometer is reliable for the mean speed measurements. Although the shear spectrum was measured, the slope seems to be slower than $-7/3$, and also the data points are widely scattered. Caughey et al. mounted the wind and temperature sensor on a 32m tower on the terrain slope at Minesota[14]. Three-component sonic anemometers and fine platinum wire thermometers were used but the sampling interval might to be a second order, therefore the time resolution was not enough to make sure the clear power-law relation. Kaimal et al. presented the form of cross spectrum; $E_{12}(k_1)/u_*^2 \propto G(z/L)k_1^{-7/3}$, where u_* is a friction velocity and z is a distance from the ground. They determined the function G empirically, and made sure that all cross spectra are reduced to a single universal curves in the inertial range. Although the data analysis is limited to $k_1 \leq 1/\ell$, where

ℓ is (the sonic path length) 0.2m, clear power-law relation was confirmed[17]. Wyngaard et al. discussed on the scaling of E_{12} based on the Monin-Obukhov similarity law, which postulate that over a horizontally homogeneous surface the turbulence structure is determined by the surface heat flux, height from the surface, the buoyancy parameter, and the surface shear stress[13]. In the inertial range the following relation is predicted.

$$E_{12}(k_1) = -C_W \cdot S \langle \varepsilon \rangle^{1/3} k_1^{-7/3}, \tag{33}$$

where C_W is a function of Richardson number. On the Taylor's hypothesis Wyngaard generalized Lumley's model and calculate how large wave-number spectral properties are disturbed by convection velocity fluctuations[12]. The inertial range spectra are affected relatively little, but the serious spectral distortion is found at large wave numbers.

Spalart reported the numerical simulation of zero-pressure gradient boundary layers. The Reynolds number based on the momentum thickness and the free-stream wise velocity is $R_\theta = 1410$. The spectrum is normalized by the friction velocity u_* and the distance from the wall y as $E_{12}/(yu_*^2)$ v.s. $k_1 y$, which was following the Perry's Λ model. There is no comment on the scaling exponent but pointed out the k_1^{-1} behavior of E_{12} in the low-wave number range[11]. Three-dimensional flow driven by the large-scale constant force $f = F \cdot cos(x)$ in a period box was numerically simulated by Borue et al. with hyperviscosity[16]. At large scale the flow is anisotropic but for the large Reynolds number the flow is isotropic at small scales. The local isotropy was systematically studied by measuring longitudinal and transverse components of the energy spectra and the cross correlation spectra of velocities. The anisotropic spectrum is well scaled by the expected form with exponent $-7/3$. Its numerical constant is $0.2 \leq C_{12} \leq 0.4$. They also commented on the wave number of onset of local isotropy, which should depend on the local mean shear, however the onset wave number is about $k_1^* = 15$ independently of whether the region have high or low shear. Thus, concluded that the onset of local isotropy seems to be a global property of turbulence at least for such large scale intermittent open systems as Kolmogorov flow.

In the theoretical framework, Lumley presented the scaling form of E_{12} in shear flow, which is briefly explained in section 3[22,23]. Durbin et al. showed that the hypothesis of local isotropy is implausible in the presence of significant mean rates of strain by examining the transport equation of energy dissipation tensor, that was found to have the direct effect of mean strain[26]. The dynamical equations for Reynolds stress spectra was constructed by Canuto et al., which was solved analytically[25]. Ishihara et al. derived the velocity correlation spectrum tensor analytically in the inertial subrange of homogeneous shear flow. The anisotropic spectrum is uniquely determined by the rate of strain tensor of the mean flow, the rate of energy dissipation of per unit mass, and the two non-dimensional constants[8].

To sum up this brief review, it may be said that much less is revealed on the anisotropic spectrum, scaling exponent or the spectrum constant. This is

Table 1. Characteristics of the boundary layers. The experiments are performed several times at each unit Reynolds number U_0/ν. U_0 is the free-stream velocity and ν is kinetic viscosity. u_* is the friction velocity. The momentum thickness is defined as $\theta \equiv \int_0^\delta U(y)/U_0(1 - U(y)/U_0)dy$. The Reynolds number is defined as $R_\theta \equiv U_0\theta/\nu$ and R_λ is based on the Taylor micro scale. In the table R_λ is the averaged value in the log-law region.

	$U_0[m/s]$	$\theta[mm]$	$u_*[m/s]$	R_θ	R_λ	probe
case1	7.65	9.34	0.533	4925	210	I,X,L
case2	10.80	9.47	0.749	6986	280	I,X
case3	13.96	8.99	0.967	8537	340	I,X,L
case4	17.14	9.40	1.184	10910	370	I,X
case5	19.55	9.52	1.348	13060	420	I,X

due to the difficulty of measuring the two-component velocity fluctuation with sufficient resolutions in high Reynolds numbers. As the theoretical treatment has just started, both the experimental and numerical information are required for theoretical studies.

4 Experimental Condition

It is convenient for readers to explain the experimental condition. In a wind tunnel with a test section $0.32 \times 1.06m$ in area and $2.6m$ in length, a typical two-dimensional turbulent boundary layer is generated under the condition of zero-pressure gradient. Most of the data are measured at $1900mm$ downstream from the leading edge by using I-type and X-type probes, and also by the rake made of 24ch I-type probes. The former experimental data were used for computing the energy spectrum $E_{12}(k_1)$, and the latter is for the anisotropic tensor $E_{11}^{12}(k_1)$. The probe's sensitive region is made of tungsten wire whose diameter is $3.1\mu m$ and $0.5mm$ in length, that is operated by a constant temperature anemometer. Velocity signal is sampled during 80 seconds by 12-bit A/D converter at 100kHz.

The turbulent boundary layer developed over the flat working section on which a grain of sand was attached. The roughness elements have a height of $k_s = 4.8mm$ on average and they are spread all over the bakelite plate as densely as possible. A schematic view of rough wall and a photo of roughness elements are shown in Fig. 3. The flow is adjusted to U_0/ν = 5.27, 7.38, 9.49, 11.60, 13.71 $\times$ 10^5[1/m], respectively. Here, U_0 is a free-stream mean velocity and ν is a kinetic viscosity. Considering the unit length, U_0/ν is a kind of Reynolds number, and the measurements are made several times with matching the flow to each of these conditions.

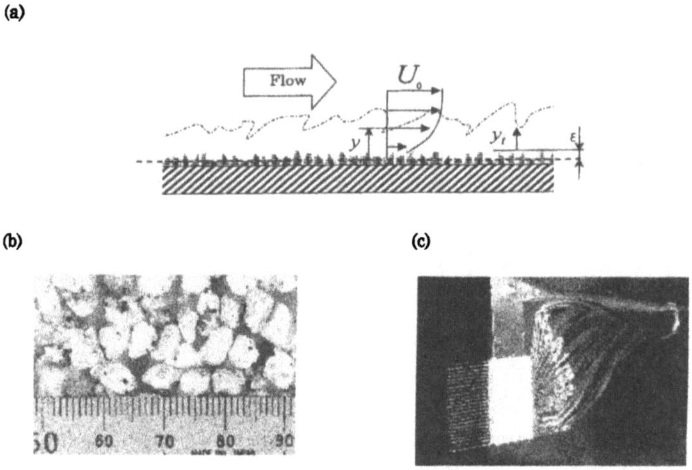

Fig. 3. (a) Schematic view of flow field on rough wall boundary layers. (b) Sand grain used as the roughness element with $k_s = 4.8mm$ height on average. (c) A rake made of 24ch I-type probes. Distance between each probe is set at 1 mm.

5 Anisotropic Spectrum over the Rough-wall Boundary layers

Experimental results are summarized in this section. The anisotropic spectrum is measured within the log-law region at several flow conditions. The focus is on the exam of universal scaling as mentioned by Eq. (23) from the point of local isotropy, and also on presenting the experimental evidence for supporting the novel representation of anisotropic spectrum presented by Ishihara et al. [8].

5.1 Scaling Region and Exponent of $E_{12}(k_1)$

Following the discussions in section 3, the anisptropic spectrum has the scaling form Eq. (23) in the wave-number range until the local isotropy starts, to be exact. If the local isotropy is achieved in $k_1 \ll L_s^{-1}$, in what wave-number ranges the anisotropic spectrum should have the universal scaling. The smallest scale corresponds to the onset of local isotropy, therefore this is given by the point where the correlation spectrum $R_{12}(k_1)$ reaches zero for the first time. We define this wave number as k_τ^1, that is, $R_{12} \simeq 0$ in $k_\tau^1 \leq k_1$. The wave number k_τ^1 is normalized by the typical length scale L_s and η, respectively in Fig. 4 as a function of Reynolds number. The value $k_\tau^1 L_s$ is slightly increasing as the Reynolds number increases, and it is up to 4 for $R_\lambda = 420$. It seems to increase for higher Reynolds number. On the other hand, in the unit of Kolmogorov scale, it decreases as R_λ increases. This means that the local isotropic region gradually extends into the low-wave number range. The wave number k_p, where the energy

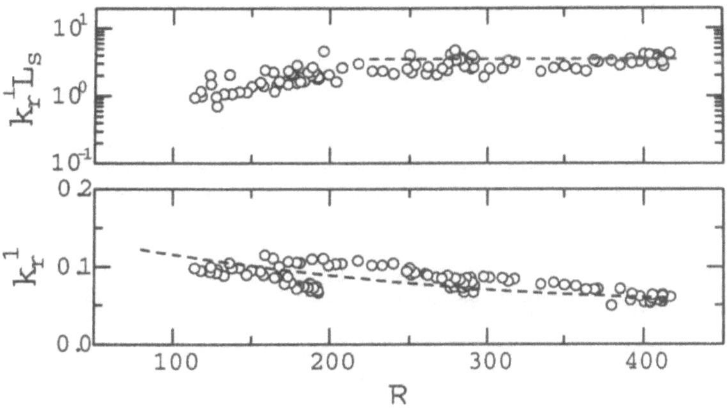

Fig. 4. The onset wave number of local isotropy, which is defined here by the relation $R_{12} \simeq 0$ in $k_r^1 \leq k_1$ is plotted as a function of Reynolds numbers.

dissipation is significantly excited (see Eq. (7)) and empirically $k_p \simeq 0.1/\eta$, is larger than k_r^1, then the dissipation range satisfies the local isotropic state even in the low Reynolds number.

The largest scale for expecting the universal scaling of E_{12} is not clear. It might be a strong function of the flow state, such as initial and boundary conditions, also it is affected by the large energy containing scales. Here we think about the scale k_r^0 that is defined by the wave number for contributing most to the Reynolds shear stress. In the log-log coordinate the relation

$$-\langle uv \rangle = \int_0^{+\infty} -k_1 \cdot E_{12}(k_1)\, d\ln(k_1)\,, \tag{34}$$

makes us know that k_r^0 is equal to the peak position of $E_{12}(k_1)$. We have checked the spectra E_{12} at many locations in the boundary layers but the power-law relation was confirmed in the smaller scale than $1/k_r^0$. Thus it is reasonable $1/k_r^0$ is not the appropriate scale for the scaling region however it is one of the typical scales in shear flow. There is no general consensus of the scaling region at this stage. We empirically suggest here that $0.8 \leq k_1 L_s \leq k_r^1 L_s$ is plausible for the scaling range in these experimental conditions. This is decided by checking lots of data in several flow conditions.

In the next step, we fit the scaling formula, see also Eq. (23),

$$E_{12}^* = -C_{12} \cdot k_1^{*r}\,, \tag{35}$$

within the range $[0.8, k_r^{1*}]$ and figured out the coefficient C_{12} and the exponent r. The subscript $*$ indicates the normalized quantity by the typical velocity and length scales, u_s and L_s. It is noticed that both C_{12} and r slightly depend on the scaling region as far as the Reynolds number is small. The scaling exponents are plotted in Figs. 5 and 6 as a function of k_r^0/k_r^1 and S^*, respectively. The former is equivalent to the Reynolds number and the latter is indicating the effect of

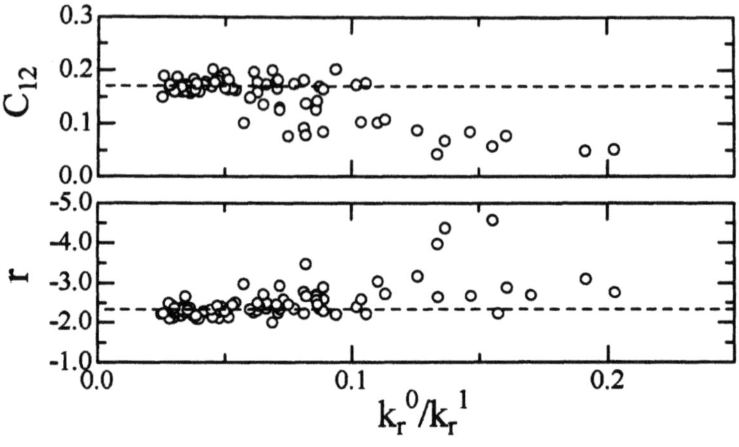

Fig. 5. Spectral constant and scaling exponent defined in Eq. (23) are plotted as a function of k_r^0/k_r^1.

mean shear. The exponent reaches the predicted value 7/3 in $k_r^0/k_r^1 \leq 0.05$, but the large scatter remains in $0.1 \leq k_r^0/k_r^1$. On the other hand, it is the condition $S^* \simeq 0.1$ that the exponent r approaches the constant. The mean shear effect seems to be less important for the exponent r, because the state $S^* = 0.1$ indicates the very strong anisotropic condition, that is far from the local isotropy in the inertial range. We have examined so far the scaling exponents of the velocity increment Δu_r (see Eq. (8)), and reported that the exponents ζ_n are significantly affected by the mean shear rate. Even at $S^* \simeq 0.05$, the ζ_n depart from the isotropic values[7]. So we are surprising why r is so robust for the mean shear rate.

The spectral constant C_{12} is also plotted in Figs. 5 and 6, which converge on 0.17 in the range $k_r^0/k_r^1 \leq 0.05$. There remains large scatter for small Reynolds numbers. But they seems to be decreasing as k_r^0/k_r^1 increases. The mean shear effect is negligible in $S^* \leq 0.1$, that is similar to the tendency of exponent r.

5.2 Anisotropic Spectrum Tensor $E_{11}^{12}(k_1)$

In the inertial subrange the energy spectrum tensor is expected to be divided into the isotropic and the anisotropic part if the Reynolds number is sufficiently large or, in other wards, $S^* \ll 1$. The former part obeys the **K41** scaling and the latter is the deviation coming from the larges scale anisotropic motions. In the case of homogeneous shear flow, Ishihara el al. derived the analytical results representing the anisotropic spectrum and also the spectral constant C_{12} is uniquely determined by two independent parameters, A and B. This is due to the assumption that the direct coupling between the velocity fluctuating field and the mean flow is less dependent on the small scale fluctuation associated by the Kolmogorov scales. According to their formula, spectral constant in Eq. (23)

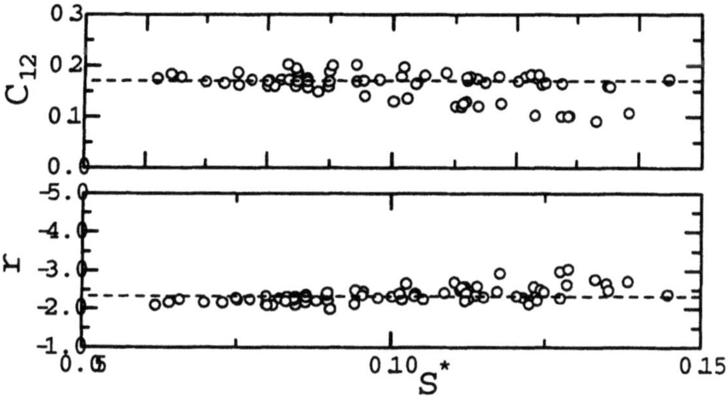

Fig. 6. Spectral constant and scaling exponent defined in Eq. (23) are plotted as a function of S^*.

is given as

$$C_{12} = -36\pi \cdot (-33A + 7B)/1729\,, \tag{36}$$

that should be universal like the Kolmogorov constant in $E_{11}(k_1)$, if these A and B are independent of flow conditions. This is really an attractive theme to be revealed in this research field. But in order to fix these parameters we need at least one another independent relation represented by A and B. It is the anisotropic component E_{11}^{12} defined as follows that can be measurable in conventional experiments.

$$\langle \partial_1 u_1(x) \partial_2 u_1(x + r) \rangle \equiv \int_0^{+\infty} k_1^2 \cdot E_{11}^{12}(k_1) \cdot e^{ik_1 r} dk_1\,, \tag{37}$$

here ∂_i indicates $\partial/\partial x_i$, and the component is scaled by the following relation,

$$E_{11}^{12*}(k_1^*) = -C_{11}^{12} \cdot k_1^{*r_2}\,, \tag{38}$$

$$C_{11}^{12} = 432\pi \cdot (-2A + B)/1729 \quad , \quad r_2 = -7/3\,. \tag{39}$$

Anisotropic component measured in this experiment is plotted in Fig. 7. The power-law relation seems to be reasonable as Ishihara et al. have already checked in their DNS data.

It is noticed that the scaling region is extending into the small wave number range. The lower wave number is down to $O(10^{-1})$ and the higher one is equal to that of E_{12}. The component E_{11}^{12} is principally based on the smallest scales, that is the derivative $\partial_1 u_1$ and $\partial_2 u_1$. On the other hand, spectrum E_{12} is affected by the moderate velocity fluctuations. Thus, we assume that the scaling region of E_{12} and E_{11}^{12} might be close to each other but does not match exactly as far as the Reynolds number is small. Because we have known the similar situation when computing the intermittency exponent defined by $\langle \varepsilon(x)\varepsilon(x + r)\rangle \propto r^{-\mu}$, for instance. The energy dissipation $\varepsilon(x)$ is computed by the velocity derivatives

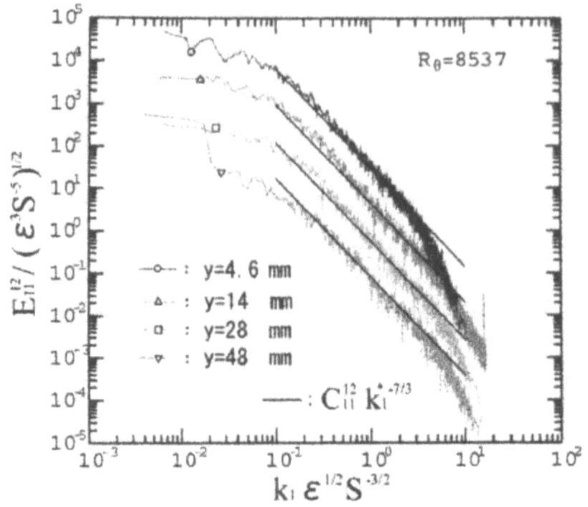

Fig. 7. Anisotropic spectrum measured in the rough wall boundary layer at $R_\theta = 8537$. y indicates the distance from the rough wall. The solid line is the scaling form of Eg. (38). The graphs □, △, and ○ are multiplied by 10, 10^2, and 10^3, respectively.

and the above relation is hoped to be satisfied in the inertial subrange. However, if the inertial subrange is fixed by Eq. (8), the scaling region of μ is extending into the larger scales. We are not sure whether this is caused by the experimental error due to the derivative quantity or representing the physical background, but at this stage we should keep in mind this experimental evidence.

The exponent r_2 and the constant C_{11}^{12} is plotted in Fig. 8 as a function of k_r^0/k_r^1. The exponents converge on 7/3 in $k_r^0/k_r^1 \leq 0.1$, which support well the theoretical prediction of Eqs. (38),(39). The coefficient scatters in wide range due to the difficulty of numerical computation of derivative quantity. We can't figure out the clear dependence of C_{11}^{12} on k_r^0/k_r^1, or the Reynolds number, in the series of this experiment. It can be evaluated as $C_{11}^{12} \simeq -0.08$ in $k_r^0/k_r^1 \leq 0.1$ on average. With the result of Fig. 5, the coefficients are evaluated as

$$A \simeq -0.17 \quad , \quad B \simeq -0.45\,, \tag{40}$$

within the range of $0.05 \leq k_r^0/k_r^1 \leq 0.1$, respectively. But the large uncertainty might be inevitably contained especially in the value of B. Ishihara et al. obtained $A = -0.16 \pm 0.03$ and $B = -0.40 \pm 0.06$, respectively, in the uniform shear flow at $R_\lambda = 284$, and commented the possibility of Reynolds number dependence of A and B. They predicted the asymptotic values are obtained in the large Reynolds numbers.

5.3 Go to the Large Reynolds-Number Flow

It is a challenging theme to figure out the anisotropic feature in the small scale fluctuations in turbulence. We need the large Reynolds number flow for that purpose. It is shown in Fig. 9 that the anisotropic spectrum $E_{12}(k_1)$ was measured

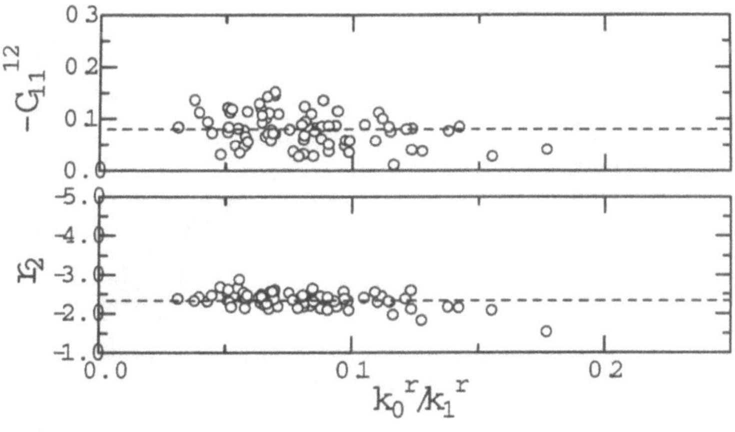

Fig. 8. Spectral constant and scaling exponent defined in Eq. (23) are plotted as a function of k_r^0/k_r^1.

in the atmospheric turbulence. The Reynolds number based on the Taylor micro scale is $R_\lambda \simeq 5000$[33]. There is a clear scaling range in more than one-decade. This encourages us to give further analysis on the constant A and B predicted by Ishihara et al.

In the atmospheric turbulence, we can't control the flow condition and also the typical eddy scale is significantly large. These prevent us from getting the mean velocity gradient S or the scaling plot of Eq. (23) and Eq. (38), accurately. However, there is still an possibility to promote this project. Adopting the another anisotropic tensor, which is independent of E_{12},

$$E_{33}^{12*}(k_1^*) = -C_{33}^{12} \cdot k_1^{*r_3}, \tag{41}$$

$$C_{33}^{12} = 216\pi(A+3B)/1729 \quad , \quad r_3 = -7/3, \tag{42}$$

the ratios E_{11}^{12}/E_{12} and E_{33}^{12}/E_{12} are quantified within the appropriate scaling range. The former is equal to C_{11}^{12}/C_{12} and the latter is C_{33}^{12}/C_{12} subject to $r_2 = r_3 = -7/3$. As is discussed in section 3, C_{12} is less dependent on R_λ but $C_{12} \simeq 0.17$. If the coefficients C_{11}^{12} and C_{33}^{12} can be approximated as

$$C_{11}^{12'} = 0.17 \cdot \left(E_{11}^{12}\Big/E_{12}\right) \quad , \quad C_{33}^{12'} = 0.17 \cdot \left(E_{33}^{12}\Big/E_{12}\right), \tag{43}$$

in the appropriate scaling range, the constants are given as follows,

$$A = \frac{19}{26\pi}\left(2C_{33}^{12'} - 3C_{11}^{12'}\right) \quad , \quad B = \frac{19}{26\pi}\left(C_{11}^{12'} + 4C_{33}^{12'}\right). \tag{44}$$

The purpose is to reveal the asymptotic values of A and B in the condition of $R_\lambda = O(10^4)$. The experiments are under planning and hopefully the results will be reported in the near future. If these data are useful for the theoretical works, we are very pleasure.

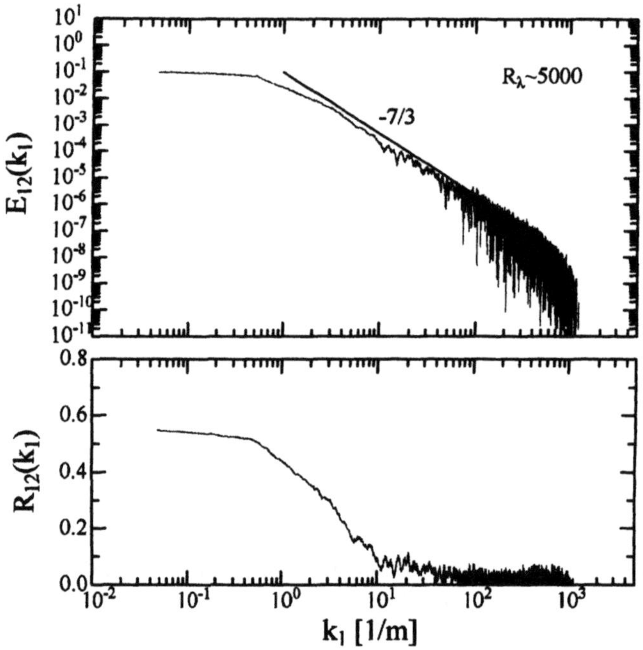

Fig. 9. Anisotropic spectrum measured in the atmospheric boundary layer at $R_\lambda \simeq 5000$. And the spectral correlation defined by Eq. (14). The solid line is the relation; $E_{12} \propto k_1{}^{-7/3}$.

6 Conclusions

The effect of large-scale anisotropy on the small scales statistics is discussed by analyzing the experimental data measured in the rough-wall boundary layers. The discussions are focused on the scaling form of anisotropic spectrum and also its scaling range. The results are summarized as follows.

- The anisotropic spectrum $E_{12}(k_1)$ can be well scaled by the typical length scale L_s and the velocity scale u_s within the range of $0.8 \leq k_1 L_s \leq k_r^1 L_s$. The large scale limit is slightly depending on Reynolds number and it is $k_r^1 L_s \simeq 4$ at $R_\lambda = 420$. The power-law exponents converge on $-7/3$ and the spectral constant $C_{12} \simeq 0.17$ in the range of $k_r^0/k_r^1 \leq 0.05$ or $S^* \leq 0.1$.
- The anisotropic tensor component $E_{11}^{12}(k_1)$ has the scaling form of Eq. (38). The spectral constant C_{11}^{12} is evaluated as -0.08 but there remains large scatter even in $k_r^0/k_r^1 \leq 0.1$. The scaling region is extending smaller wave-number range than that of E_{12}.
- The constants A and B, which determine the spectral constant C_{12} and C_{11}^{12} uniquely, are -0.17 and -0.45, respectively, within the range of $k_r^0/k_r^1 \leq 0.05$ or $S^* \leq 0.1$ in this experiment.

7 Acknowledgements

The author is grateful to Prof. Y.Kaneda and Prof. T.Gotoh for various advices and encouragements at the symposium. Useful discussions with Dr. T.Ishihara and Dr. K.Yoshida are much appreciated. Also thank Prof. K.R.Sreenivasan for his helpful comments on the first version of the manuscript.

References

1. A.N.Kolmogorov, The local structure of turbulence in incompressible viscous fluid for very large Reynolds numbers, Dokl. Akad, Nauk SSSR, **30**, (1941)[reprinted in Proc. R. Soc. Lond. A, **434**, (1991), pp9-13.]
2. A.N.Kolmogorov, Dissipation of energy in the locally isotropic turbulence, Dokl. Akad, Nauk SSSR, **32**, (1941)[reprinted in Proc. R. Soc. Lond. A, **434**, (1991), pp15-17.]
3. I.Arad, B.Dhruva, S.Kurien, V.S.L'vov, I.Procassia, and K.R.Sreenivasan, Extraction of anisotropic contribution in turbulent flows, Phys. Rev. Lett., **81**-24, (1998), pp5330-5334.
4. I.Arad, V.S.L'vov, and I.Procaccia, Correlation functions in isotropic and anisotropic turbulence: The role of the symmetry group, Phys. Rev. E, **59**-6, (1999), pp.6753-6765.
5. K.R.Sreenivasan and G.Stolovitzky, Statistical Dependence of Inertila Range Properties on Large Scales in a High-Reynolds-Number Shear Flow, Physical Rev. Lett., **77**, (1996), pp.2218-2221.
6. K.R.Sreenivasan and B.Dhruva, Is There Scaling in High-Reynolds-Number Turbulence?, Progress of Theoretical Physics Suppul., **130**, (1998), pp.103-120.
7. Y.Tsuji, Large Scale Anisotropy and Small Scale Universality over Rough Wall Turbulent Boundary Layers, submitting. (2001)
8. T.Ishihara, K.Yoshida and Y.Kaneda, Anisotropic Velocity Correlation Spectrum at Small Scales in a Homogeneous Turbulent Shear Flow, Phys. Rev. Lett., **88**, (2002), 154501.
9. S.B.Pope, *Turbulent Flows*, Cambridge University Press, (2000).
10. M. M. Zdravkovich, *Flow Around Circular Cylinder*, Oxford University Press, (1997).
11. P. R. Spalart, Direct Simulation of a Turbulent Boundary Layer up to $R_\theta = 1410$, J. Fluid Mech., **187**, pp.61-98, (1988).
12. J. C. Wyngaard and S. .F. Clifford, Taylor's Hypothesis and High-Frequeny Turbulence Spectra, Journal of the Atomospheric Sciences, vol.34, pp.922-929, (1977)

13. J. C. Wyngaard and O. R. Cote, Cospectral Similarity in the Atmospheric Surface Layer, Quart. J. R. Met. Soc., vol.98, pp.590-603, (1972).
14. S. J. Caughey, J. C. Wyngaard, and J. C. Kaimal, Turbulence in theEvolving Stable Boundary Layer, Journal of the Atmospheric Sciences, vol.36, pp.1041-1052, (1979).
15. S. Pond, G.T. Phelps, J. E. Paquin, G. McBean, and R. W. Stewart, Measurements of the Turbulent Fluxes of Momentum, Moisture and Sensible Heat over the Ocean, Journal of the Atmospheric Sciences, vol.28, pp.901-917, (1971).
16. V. Borue and S. A. Orszag, Numerical Study of Three-dimensional Kolmogorov Flow, J. Fluid Mech., vol.306, pp.293-323, (1996).
17. J. G. Kaimal, J.C. Wyngaard, Y, Izumi, and O. R. Cote, Spectral Characteristics of Surface-layer Turbulence, Quart. J. R. Met. Soc., vol.98, pp.563-589, (1972).
18. A.S.Monin and A.M.Yaglom, *Statistical Fluid Mechanics*, M.I.T. Press, (1975).
19. W.George and H.J.Hussein, Locally axisymmetric turbulence, J. Fluid Mech., **233**, (1991), pp.1-23.
20. P.S.Klebanoff, Characteristics of Turbulence in a Boundary Layer with Zero Pressure Gradient, National Advisory Committee for Aeronautics, Technical Note **3178** (1954).
21. I.Marusic, J.D.Li and A.E.Perry, A Study of the Reynolds-Shear-Stress Spectra in Zero-Pressure-Gradient Boundary Layers, 10th Australasian Fluid Mechanics Conference, University of Melbourne, 11-15 Dec, (1989), pp1.5-1.8.
22. J.L.Lumley, Interpretation of Time Spectra measured in High-intensity Shear Flows, Phys. Fluids, **8**, p.1065-1062, (1965)
23. J.L.Lumley, Similarity and the Turbulent Energy Spectrum, Phys. Fluids, **10**, p.855-858, (1967)
24. J.L.Lumley and G.R.Newman, The Return to Isotropy of Homogeneous Turbulence, J. Fluid Mech., **82**, pp.161-178, (1977).
25. V.M. Canuto, A Dynamical Model for Turbulence I,II,III, Phys. Fluids, **8**, pp.571-613, (1996).
26. P.A.Durbin and C.G.Speziale, Local Anisotropy in Strained Turbulence at High Reynolds Numbers, J. Fluids Engineering, **113**, pp.707-709, (1991).
27. M.Lee, J.Kim and P.Moin, Structure of turbulence at High Shear Rate, J. Fluid Mech., **216**, pp.561-583, (1990).
28. S. Corrsin, On local Isotropy in Turbulent Shear Flow, NACA R & M 58B11, (1958).
29. L.W.Browne, R.A.Antonia and D.A.Shah, Turbulent Energy Dissipation in a Wake, J. Fluid Mech., **179**, pp.307-326, (1987).
30. S.G.Saddoughi and S.V.Veeravalli, Local isotropy in turbulent boundary layers at high Reynolds number, J. Fluid Mech., **268**, (1994), pp.333-372.
31. S.G.Saddoughi, Local isotropy in complex turbulent boundary layers at high Reynolds number, J. Fluid Mech., **348**, (1997), pp.201-245.
32. Y.Tsuji, Peak position of dissipation spectrum in turbulent boundary layers, Phys. Rev. E, **59**, (1999), pp.7235-7238.
33. Y.Tsuji and B.Dhruva, Intermittency feature of shear stress fluctuation in high-Reynolds-number turbulence, Physics of Fluids, **11**, (1999), pp.3017-3025.
34. R.A.Antonia and J.Kim, Isotropy of Small-scale Turbulence, Proc. Summer Program of the Center for Turbulence Research, Stanford, (1992).
35. H.H.Fernholz and P.J.Finley, The Incompressible Zero-Pressure-Gradient Turbulent Boundary Layer: An Assessment of the Data, Prog. Aerospace Sci., **32**, 245 (1996).
36. A.E.Perry and J.D.Li, Experimental Support for the Attached-Eddy Hypothesis in Zero-Pressure-Gradient Turbulent Boundary Layers, J. Fluid Mech., **218**, 405 (1990).
37. J.Schumacher, K.R.Sreenivasan, and P.K.Yeung, Derivative moments in turbulent shear flow, submitting,(2002).

LES Study on the Very Large-Scale Structures of Wall-Bounded Turbulence and an Effect of Thermal Stratification

Makoto Tsubokura and Tetsuro Tamura

Tokyo Institute of Technology, Nagatsuta, Midoriku, Yokohamashi 226-8502, Japan

Abstract. The objective of this work is to investigate the very large-scale flow structures observed in or above the logarithmic layer of a wall-bounded turbulence and the effect of thermal stratification on these structures. Large Eddy Simulation (LES) is adopted as a numerical method in this work. We have found that the large-scale structures are similar to the streak-like structures observed in the vicinity of the wall but their size is hundreds times larger than the fine-scale structures. The large structures are also strongly affected by the thermal stratification and suppressed at a certain weakly stable stratification (bulk *Ri* of about 0.1) tested here, contrarily to the fine-scale streak structures being rather insensitive to the thermal stratification.

1 Introduction

For a couple of last decades, much effort has been devoted to the study of the fine-scale organized structures of turbulence in the vicinity of the wall. One of the reasons of the high attention is that such information is indispensable to control the wall-bounded turbulence in the engineering sense. On the other hand, it is known that there exist much larger structures in or above the logarithmic layer of wall-bounded turbulence. The most famous one among them is a turbulent bulge first observed in a turbulent boundary layer (e.g., [9]). It is known that the turbulent bulges, which have the scale in the same order of magnitude as the boundary layer thickness (δ), have steeply inclined leading fronts and slowly sloped backs (e.g., [3]). These structures of the size comparable to the boundary layer thickness are sometimes called as *large-scale motion* (LSM, or δ-scale motion). So-called hairpin vortex models (e.g., [13]) are frequently used to explain the organized eddy motions of the wall turbulence, and the bulges in the turbulent boundary layer are supposed to be explained as a set of packets generated by the hairpin vortex. Recently Meinhart and Adrian [11] further studied turbulent boundary layer using Particle Image Velocimetry (PIV) and found that there exist long growing zones of relatively uniform low streamwise momentum in the logarithmic layer. Zhou et al. [17] explained from the DNS results of low-Reynolds-number channel flow that the uniform low momentum zones are generated by the streamwise alignment of the hairpin vortices in packets that generate new packets as they go downstream. This streamwise uniform low momentum zones seem to extend well over the boundary layer thickness and is called as *very large-scale motion* (VLSM) in the work of Kim and Adrian [7]. In fact Kim and Adrian found from the hot-film measurement of fully developed turbulent pipe flow that the maximum value of streamwise premultiplied spectrum of the streamwise velocity is 12-14 times the pipe radius. Even though they did not find or visualize definite low momentum zones in their experiment, they conjecture that the maximum of the streamwise spectrum observed at the very large wavelength is associated with the uniform low momentum zones having observed in the turbulent boundary layer, and the same wall model as Zhou et al. [17] is applied to the fully developed turbulence pipe flow. More recently Adrian et al. [1] studied turbulent boundary layer more intensively using PIV and provided the evidence which quantitatively support the existence of hairpin vortices in low to moderate-Reynolds-number boundary layer. But The hypothesis that hairpin vortex packets produce the uniform low momentum zones have not yet been fully proved especially at the high Reynolds number wall-turbulence. Nor there is conclusive proof that the low momentum zones observed in the turbulent boundary layer are the common feature in all wall-turbulence such as pipe or channel flow.

In this work, we would like to define the large-scale structures as the organized flow structures whose size is of the order of or larger than the largest characteristic vertical scale such as the boundary layer thickness or the channel-width. Also they should exist in or around the logarithmic layer as the common feature to wall-turbulence. A study on such large-scale structures is very important when we focus on intermittent or gust phenomena at a surface layer of an atmospheric boundary layer. In addition, considering the fact that Reynolds number of the wall-bounded flow in the engineering problem is very high, information on large-scale structures is also valuable for the actual turbulence control of engineering flow.

Another important topic concerning the large-scale structures is the effect of thermal stratification on these structures. Such an effect on turbulence is very essential in the geophysical sense, because the atmospheric boundary layer is directly exposed by surface heating or cooling due to the diurnal cycle of sunlight. Especially when we consider the atmospheric boundary layer as the very large Reynolds number wall-turbulence, large-scale structures will certainly play an important role in the transfer mechanism of the whole atmospheric boundary layer. But few attempts have so far been made on the investigation of the effect of thermal stratification on such structures.

Accordingly the objectives of this study are as follows. Firstly we would like to see how large-scale structures are reproduced numerically at the logarithmic layer. Especially, interaction of the large-scale structures with the fine-scale structures in the vicinity of the wall (about which we have much knowledge) is our main focus here. Secondly we would like to investigate the effect of thermal stratification on these structures.

There are two important requirements for the analysis of these large-scale structures: one is that three-dimensional analysis is required to extract these structures from flow fields, and the other is that a sufficiently high Reynolds number condition is indispensable for the existence of the large-scale structures at the well large logarithmic region. Considering these two points, Large Eddy Simulation (LES) is adopted as a simulation method in this work.

2 Numerical Methods

2.1 Governing equations and SGS models

We suppose fluid is incompressible and Newtonian, and buoyancy term is modeled by Boussinesq approximation. The governing equations of LES adopted in this study is obtained by spatially filtering the corresponding continuity, momentum and heat transfer equations:

$$\frac{\partial \bar{u}_i}{\partial x_i} = 0, \tag{1}$$

$$\frac{\partial \bar{u}_i}{\partial t} + \frac{\partial \bar{u}_i \bar{u}_j}{\partial x_j} = -\frac{\partial \bar{p}}{\partial x_i} + \frac{1}{Re_\tau}\frac{\partial^2 \bar{u}_i}{\partial x_j \partial x_j} + Ri_\tau \bar{\theta} - \frac{\partial}{\partial x_j}\tau_{ij}, \tag{2}$$

$$\frac{\partial \bar{\theta}}{\partial t} + \frac{\partial \bar{\theta}\bar{u}_j}{\partial x_j} = \frac{1}{Re_\tau Pr}\frac{\partial^2 \bar{\theta}}{\partial x_j \partial x_j} - \frac{\partial}{\partial x_j} q_j, \tag{3}$$

where Re_τ, Ri_τ and Pr are Reynolds, Richardson and molecular Prandtl numbers respectively. We set molecular Prandlt number as 0.7 in this work. Definition of other two parameters will be mentioned in the next section. Here *overbar* denotes the *grid* filtering operation. The last terms on the right of (2) and (3) are the *subgrid*-scale (SGS) stress and heat flux terms to be modeled:

$$\tau_{ij} = \overline{u_i u_j} - \bar{u}_i \bar{u}_j, \tag{4}$$

$$q_j = \overline{q u_j} - \bar{q}\bar{u}_j. \tag{5}$$

In this study these two terms are modeled under the isotropic eddy viscosity assumption:

$$\tau_{ij} - \frac{1}{3}\delta_{ij}\tau_{kk} = -2C\frac{k}{3|\bar{S}|}\bar{S}_{ij}, \tag{6}$$

$$q_j = -C_\theta \frac{k}{3|\bar{S}|}\frac{\partial \bar{\theta}}{\partial x_j}, \tag{7}$$

where $\bar{S}$ is a magnitude of the strain rate tensor, $\bar{S}_{ij} = (\partial \bar{u}_i/\partial x_j + \partial \bar{u}_j/\partial x_i)/2$, and is given as $\bar{S} = \sqrt{2\bar{S}_{ij}\bar{S}_{ij}}$. We would like to note that these forms are different from the famous Smagorinsky's model, and that the SGS turbulence energy, k, is explicitly included in the eddy viscosity or diffusion coefficient. The model coefficients, C and C_θ, included in (6) and (7) are determined following the dynamic procedure [4] [10]. In the dynamic procedure, so-called *subtest*-scale (STS) stress, $T_{ij} = \widetilde{\overline{u_i u_j}} - \tilde{\bar{u}}_i \tilde{\bar{u}}_j$, and corresponding heat flux also must be modeled. We would like to note that *overtilde* denotes the *test* filtering operation required in the dynamic procedure. Here the STS stress is modeled on the analogy of the SGS model given in (6) as follows,

$$T_{ij} - \frac{1}{3}\delta_{ij}T_{kk} = -2C\frac{K}{3|\bar{\tilde{\bar{S}}}|}\bar{\tilde{\bar{S}}}_{ij}, \tag{8}$$

where K is the STS turbulence energy. The SGS and STS energy in (6) and (8) are modeled considering the consistency of the numerical (finite difference) error in the dynamic procedure:

$$k = \overline{\bar{u}_k \bar{u}_k} - \bar{\bar{u}}_k \bar{\bar{u}}_k, \tag{9}$$

$$K = \widetilde{\overline{\bar{u}_k \bar{u}_k}} - \tilde{\bar{\bar{u}}}_k \tilde{\bar{\bar{u}}}_k. \tag{10}$$

The notable features of these SGS models are as follows. Firstly, this model was derived without the assumption of the local equilibrium state of the SGS energy production and the dissipation, which is supposed in the Smagorinsky's model. Therefore, in point of the limitation caused by the physical assumption during the derivation, we can say that this model is more generous in the application of the model to any turbulence. Secondly, because of the consistency of the numerical error in the dynamic procedure, this model shows less sensitivity to the discrete *test* filtering operation than the usual dynamic procedure using Smagorinsky's model. For the detail of the derivation of these models and their fundamental property in LES, please refer to [16] [14] [15]. Because we use Finite Difference Method (FDM) for discretization in this study to reduce required CPU time for numerical simulation, this feature is very important.

2.2 Analysis models

The simplified, or modeled open channel flow, in which no-slip and free-slip conditions are supposed on the lower and upper walls respectively, is adopted as one of the typical wall-bounded turbulence in this study. Hereafter streamwise, normal-wall and spanwise directions are given as x, y and z respectively. The flow field is considered to be periodic for streamwise and spanwise directions, and is driven by the uniform pressure gradient force acting for the streamwise direction to obtain the fully developed turbulence state. The Reynolds number defined by the kinetic viscosity (ν), friction velocity (u_τ) at the lower wall and channel-width (δ), is given as $Re_\tau = u_\tau\delta/\nu = 10000$ to ensure the sufficiently large logarithmic layer for the large motions to be exist. One of the main objectives in this study is to investigate the interaction between well-known fine-scale structures such as streaks in the vicinity of the wall and the large-scale structures. Therefore minimum grid resolution is determined in consideration of the reproduction of the fine-scale structures, and $\Delta x^+ = 82$ and $\Delta z^+ = 41$ for streamwise and spanwise directions are finally adopted (streak structures are known to exist with the interval of 100 for streamwise direction, and with the streamwise size of 1000 in wall unit). By using the 192 uniform grids for these two directions, the analysis region is given as $0.5\pi\delta$ and $0.25\pi\delta$ respectively. For the normal-wall direction, 128 non-uniform grids are arranged. We understand that this analysis region adopted in this study might be too small to observe *overall* features of the large-scale structures, which remains in the future work as the other important topics. As regards the thermal stratification, we test two states of neutral and stable in this study. When the bulk Richardson number is defined as $Ri_\tau = \beta g \Delta T \delta / u_\tau^2$, where β, g and ΔT are the volumetric expansion coefficient, gravitational acceleration and temperature difference between the wall respectively; 0 and 100 for neutral and stable states are considered. The Ri_τ of 100 corresponds to about 0.1 when defined by the velocity on the upper wall, which is categorized as the weakly stable state.

2.3 Discretization

Governing equations are discretized on the staggered grid system based on the fully conservative finite difference scheme developed recently [12]. The fourth-order accuracy is considered for all spatial derivatives except for the SGS term in the governing equations. It should be noted here that the fourth order accuracy is determined from a compromise between the higher- order requirement for SGS stress not to be dominated by numerical errors and the lower-order requirement for non-liner term not to be contaminated by the aliasing error (e.g., [5]). The third-order Runge-Kutta method is basically adopted as the time marching method, and the second derivative for the normal-wall direction included in the viscous term is only treated semi-implicitly using the Crank-Nicolson method for the tolerance of time increment in the numerical simulation. MAC method is used for the velocity-pressure coupling and the corresponding pressure Poisson equation is solved by the discrete FFT method for the periodic direction while the septa-diagonal method is adopted for the normal-wall direction.

2.4 LES accuracy

The fundamental accuracy of the numerical method adopted in this study can be validated by the first and second order turbulent statistics. Figure 1 indicate the mean velocity (left) and the GS turbulent intensity (right) profiles at the two states of neutral and stable stratification. For references, experimental results [2] obtained from the spatially developing boundary layer at the $Re_\tau = 19400$ are also shown. The mean velocity profile at the neutral state shows good agreement with the empirical equation at the logarithmic layer as well as the viscous sublayer. The profile at the stable state also shows almost the same value at or below the logarithmic layer, while it deviates from the neutral state above $y^+ = 1000$. This is one of the typical effects of thermal stratification observed in the wall-turbulence. As regards the GS turbulent intensity profiles, qualitative agreement between the LES results at the neutral state and experimental data is observed especially at the logarithmic region below $y^+ = 1000$. Strictly speaking, open channel flow and spatially developing boundary layer is different in point that former is driven by the constant pressure gradient while the latter

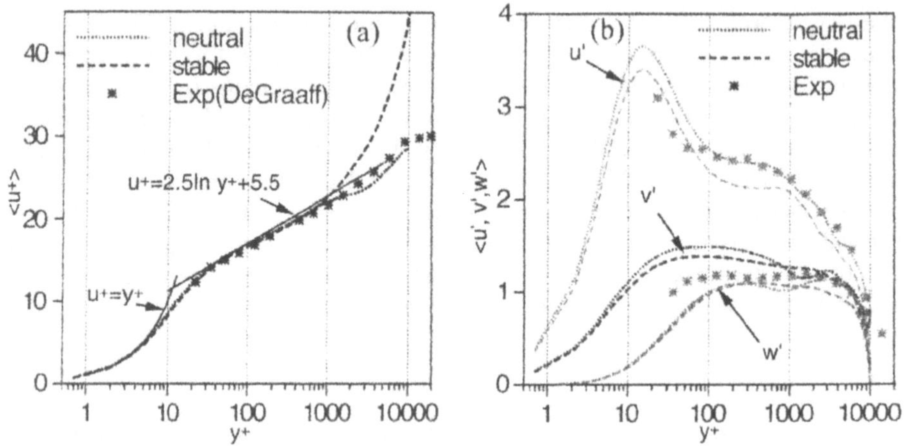

Fig. 1. (a)Mean velocity and (b)GS turbulent intensity profiles.

ideally shows zero pressure gradient for the streamwise direction. But considering the fact that both flows are categorized in the wall-bounded turbulence which shows the common characteristic logarithmic layer in the mean velocity profiles, and that the good correlation of turbulent statistics can be found in or below the logarithmic layer; we can say that the qualitative agreement of LES of open channel flow with the experimental data of spatially developing boundary layer is satisfactory for the validation of the numerical method. At this weakly stable condition tested here, the difference between neutral and stable states at the logarithmic region is not noticeable and is within ten percent, even though stable state shows a little smaller value.

3 Results

3.1 Power Spectra

Pre-multiplied power spectra of fluctuating velocity components are investigated in this section to estimate the flow structures at different spatial scales. When $\Phi_{uu}(\kappa)$ is defined as the power spectrum of fluctuating component of velocity u, so-called pre-multiplied power spectra is given as $\kappa\Phi_{uu}(\kappa)$, where κ is the wave number defined as $\kappa = 1/\lambda$ and λ is the wavelength. Figure 2 indicate spanwise pre-multiplied power spectra of streamwise (left) and spanwise (right) fluctuating velocity components at different heights from the wall. It should be reminded that the horizontal axis indicates a wavelength (not a wave number as is usually used). Jiménez [6] also treated the power spectra in the same manner to study large-scale motions in channel flow. In the vicinity of the wall ($y^+ = 10$), peak values of both Φ_{uu} and Φ_{ww} is observed around $\lambda_z^+ = 200$. This peak position corresponds to the mean space between the streak structures observed at this region. Comparing with the value observed in DNS [8], we are afraid that

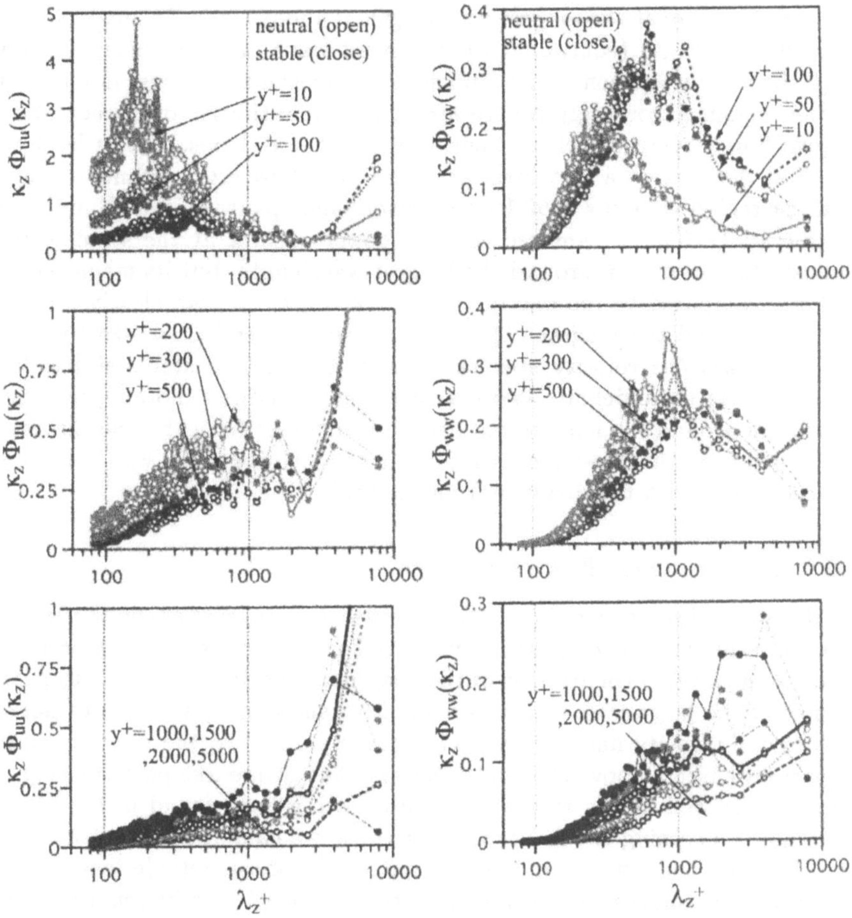

Fig. 2. Pre-multiplied power spectra (for spanwise direction) of streamwise (left) and spanwise (right) velocity fluctuation.

the value in LES is about twice larger estimated. We have already observed this overestimation of fine-scale structures in LES and concluded that this is caused as the fundamental defect of the isotropic eddy viscosity model [14]. This overestimation will be mitigated by adopting finer grid resolutions, but in this study we regard this overestimation as permissible for the study of interaction between small scale in the vicinity of the wall and large-scale structures (hundreds times larger than the small scale structures!) in or above the logarithmic layer. It is interesting to note that fundamental difference cannot be found on the small scale structures between the neutral and stable state in Fig. 2, even though their amplitude in stable state show a little smaller value than those in a neutral state. The most notable feature at or below the logarithmic layer (y^+

is less than 1000) is that the pre-multiplied power spectra at a neutral state show a steep rise at the longest wavelength of about 8000. The maximum at the longest wavelength (equivalent to the lowest wave number) generally indicates that the analysis region is not sufficient to capture the largest motion existing in the region. Accordingly we are afraid that quantitative discussion about the largest flow structure such as their size is rather dubious. But at least we can say that there exists a very large-scale motion whose size amounts to the whole analysis region (10470×7850 for streamwise and spanwise in wall-unit) adopted in this study. The streamwise velocity power spectra at the stable state also indicate the maximum around the longest wavelength, but its magnitude is not so large compared to the one in the neutral state. We can clearly observe two maxima around the longest ($\lambda_z^+ = 8000$) and shorter ($\lambda_z^+ = 1000$) wavelength in the streamwise velocity spectra at both neutral and stable states. We can suppose from these definite two maxima that the largest motion producing the maximum around the longest wavelength is different from the ones having been developed from the fine-scale structures in the vicinity of the wall and grown up in accordance with the distance from the wall.

3.2 Instantaneous flow structures

In the preceding section, we have observed the suggestion of the large-scale structures in the power spectra of velocity fluctuation especially in or above the logarithmic layer. Visualization of the instantaneous flow field will ensure the existence of the large-scale structures. Figure 3 show images of the instantaneous streamwise velocity fluctuation on the plane parallel to the wall ($x - z$ plane) at various heights above the wall. The very fine stripe-like patterns at $y^+ = 10$ indicate the well-known streak structures having been found in the vicinity of the wall. Concerning the fine-scale structures, we cannot see the distinct difference between the neutral and stable states. The most notable feature of these instantaneous images can be found at the region y^+ is between 100 and 1000 (within the logarithmic layer), where we can observe the streak-like low speed flow region elongated for the streamwise direction. These large streak-like structures can be seen in both neutral and stable states and their size amounts to several thousands in wall unit (hundreds times larger than the fine-scale streak structures). In the neutral state, large-scale structures are more clear and enhanced and their streamwise size extends well over the entire analysis region. We can suppose that the large-scale structures in the neutral state in Fig. 3 induce the maximum of the streamwise velocity spectra for spanwise direction at the longest wavelength observed in the preceding section. On the other hand in the stable state, their size seems to be smaller than that in the neutral state and captured generally by the analysis region. At the region $y^+ = 5000$ (well above the logarithmic layer), we cannot see the distinct large-scale structures in both neutral and stable states.

The same images of the streamwise velocity fluctuation on $y - z$ plane are indicated in Figure 4 to investigate the structure of the large motion for the normal-wall direction. The existence of the large-scale motion is also ensured

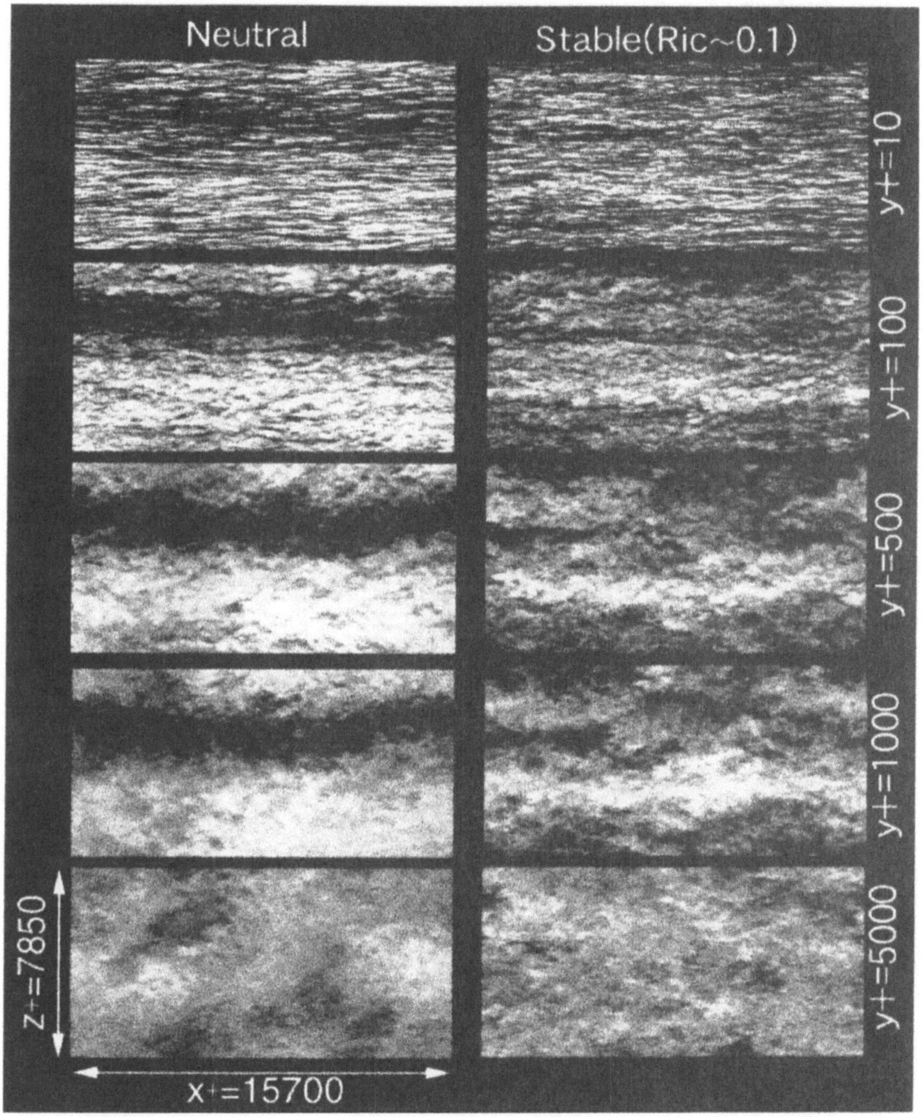

Fig. 3. Images of streamwise velocity fluctuation on $x - z$ plane (parallel to the wall). (white: $u'/u_\tau = 6.0$; black: $u'/u_\tau = -6.0$, left: neutral; right: stable condition, flow from right to left)

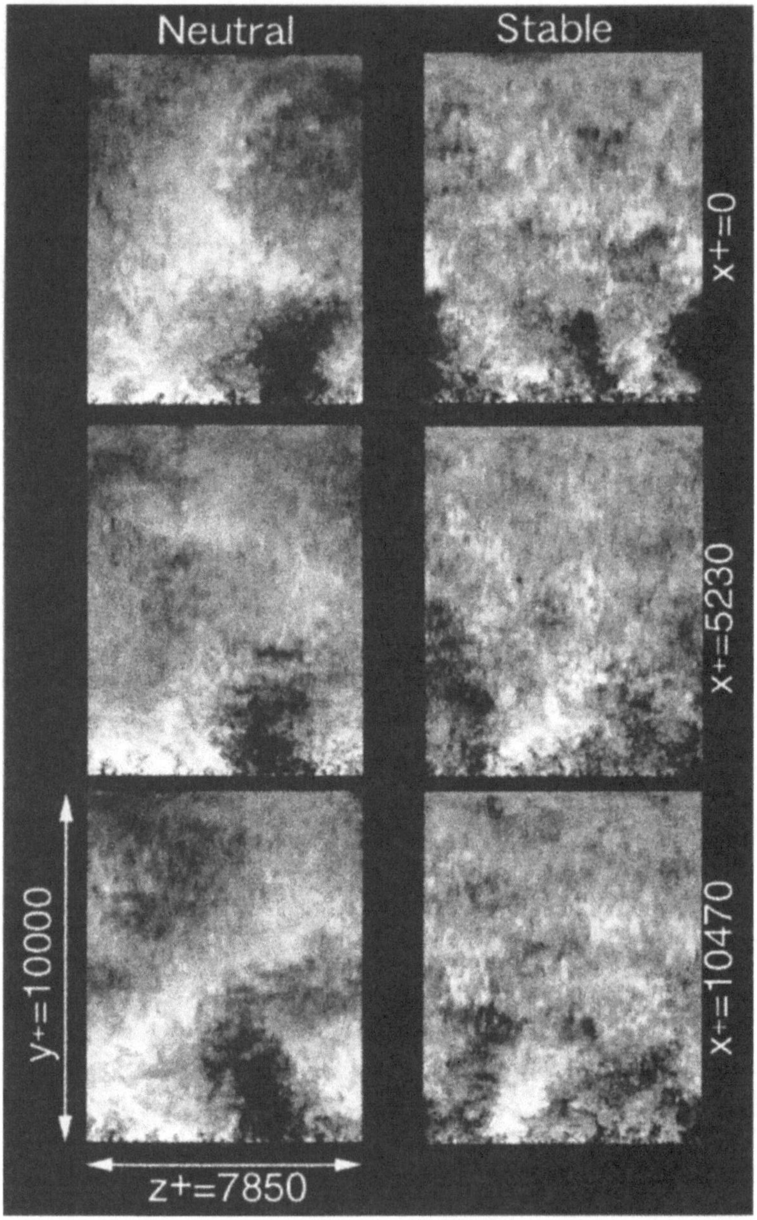

Fig. 4. Images of streamwise velocity fluctuation on $y - z$ plane (parallel to the wall). (white: $u'/u_\tau = 6.0$; black: $u'/u_\tau = -6.0$, left: neutral; right: stable condition, flow from this side to other side)

and their location is found to range from the vicinity of the wall to around $y^+ = 4000$ in both thermal stratification. In the neutral state, we can see only one big low-speed region at all three locations for the streamwise direction, which also means that the large motion extends over the entire region. On the other hand in the stable state, about three big low-speed regions can be identified, but they are rather ambiguous and their streamwise correlation is weak compared with the neutral case. We can recognize how large their size is by comparing them with the fine-scale streaks indicated as small black dots in the vicinity of the wall. The distinct separation as regards the scale of these structures seems to be supposed, and no obvious motions of the intermediate scale between them can be found.

Finally we would like to observe three-dimensional iso-surface images of the low-pressure and the low-speed regions, which is shown in Figs. 5(neutral state) and 6(stable state). The characteristic difference between neutral and stable states can be found in the large-scale motion extracted as the low-pressure region. Generally speaking, the low-pressure region in the stable state is restricted within about one-fourth of whole analysis domain from the wall (equivalent to $y^+ \leq 2500$), while that in the neutral state extends to half of the domain. We can suppose that the large organized motion in the stable state is suppressed by the effect of stable stratification and their motion is restricted in the region where turbulence caused by strong shear is dominant. It is interesting to note here that, even though we cannot see noticeable difference of the first and second order statistics between the neutral and stable states at the logarithmic layer region (see Fig. 1), we can identify the clear difference of the large-scale motions between these two stratified states. The common feature of these large-scale motions as the low-pressure region is that they exist intermittently especially for the streamwise direction and their space is in the same order of magnitude as the channel width (δ). Adrian et al. [1] related the uniform low momentum zones observed in the turbulent boundary layer [11] with the streamwise alignment of the hairpin vortices in packets. At a moment we are afraid that we cannot find a specific clue as to whether the intermittent motion of the low-pressure region observed in Figs. 5 and 6 corresponds to the turbulent bulges as a set of hairpin packets. Nor we can find any definite indication to relate the low-speed flow zone with the large motion visualized as the low-pressure iso-surface.

4 Summary and Discussion

The very large-scale flow structures observed at or above the logarithmic layer of open channel flow have been investigated using the visualized images extracted from the LES results, and the effect of the thermal stratification on these large-scale motions has been considered. We have found that the large-scale structures were visualized as the low-speed flow zones at the logarithmic layer. The structures are similar to the fine-scale streaks observed in the vicinity of the wall, but their size is hundreds times larger than the fine-scale structures. The large structures were also found to be affected by the thermal stratification. In the

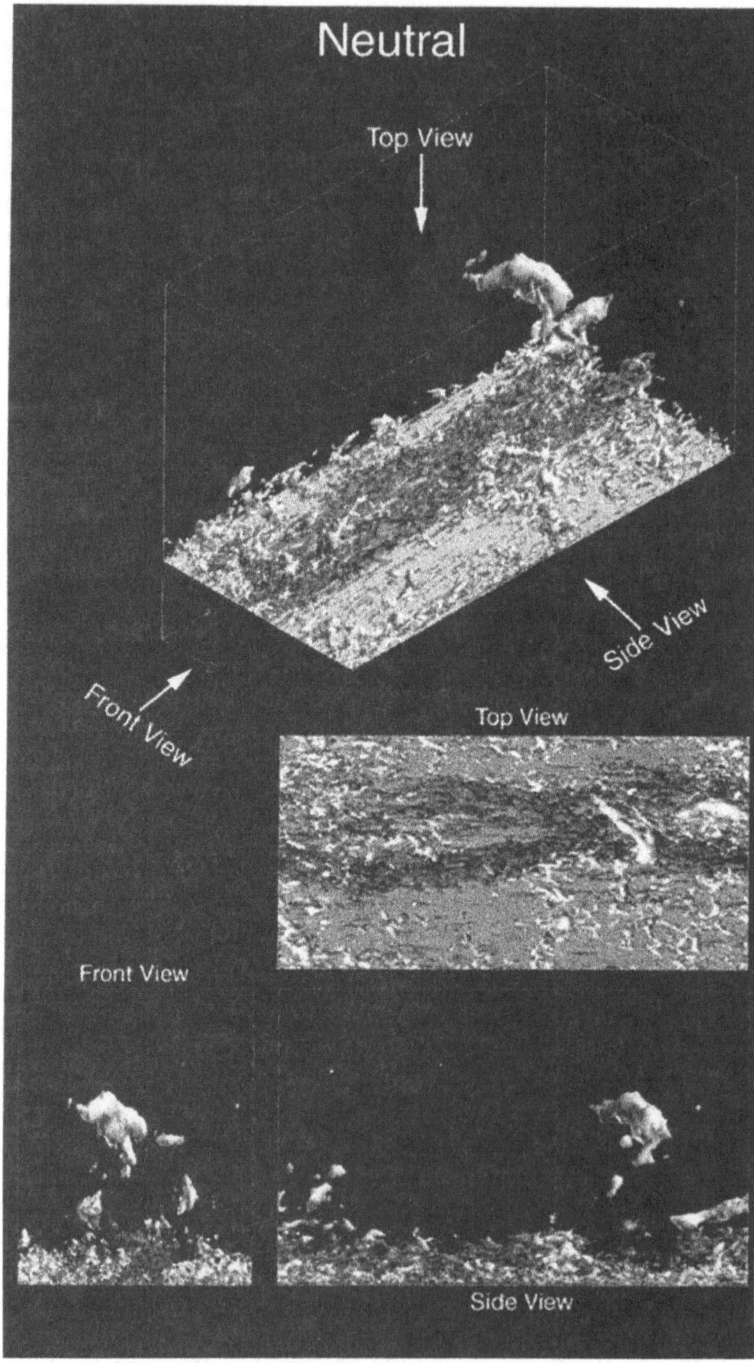

Fig. 5. Three-dimensional iso-surface images of low-pressure (light) and low-speed (dark) flow regions at the *neutral* state. ($p' = -3.0$, $u'/u_\tau = -4.0$)

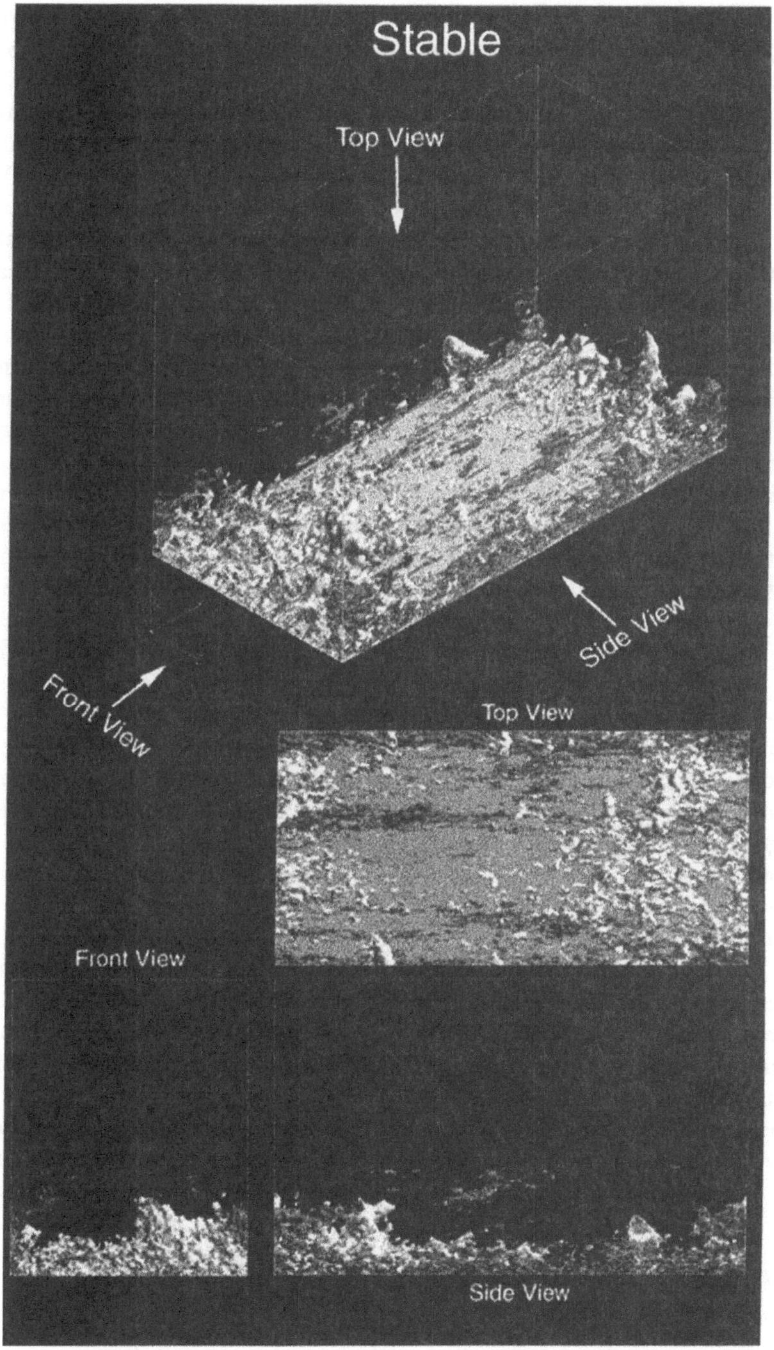

Fig. 6. Three-dimensional iso-surface images of low-pressure (light) and low-speed (dark) flow regions at the *stable* state. ($p' = -3.0$, $u'/u_\tau = -4.0$)

stable condition, they were weakened and their sizes were reduced, contrarily to the fine-scale streak structures being rather insensitive to the stratified condition tested here.

Regarding the kinematic mechanisms or the physical models explaining the uniform low streamwise momentum zones observed in this study, similar large-scale motions are also reported in turbulent boundary layer using PIV technique [11]. Recent physical model [1] of wall-bounded turbulence explains this uniform momentum zones as a result of hairpin vortex packets (with the size equivalent to the boundary layer thickness) aligning coherently for streamwise direction and the low momentum flow in the lower part of each packet fitting together with each other. Kim and Adrian [7] also support this hypothesis in the fully developed pipe flow. We could certainly observe the similar intermittent low-pressure region as the hairpin packets for streamwise direction, in the sense that the interval of the low-pressure region is in the same order of magnitude as the reported hairpin packets. But we are afraid that we could not identify any distinct hairpin vortex or vortex packets in our visualized images, nor we could find any clear organized eddy motion at the logarithmic layer except for the low momentum zones. Therefore at the moment we cannot say definitively as to whether our visualized images support the hypothesis explaining the mechanism of producing the low momentum zones. One reason of such uncertainty in the large eddy identification probably is the inappropriate visualization technique. In this study, we have visualized large eddy motions as the iso-surface of low-pressure region. But large-scale motions are usually contaminated by the surrounding small-scale turbulence eddy especially at the high Reynolds number case conducted here, and their shapes are distorted very much. Development of the more sophisticated vortex identification technique is required for the study of large eddy motions at the high Reynolds number flow.

However we also would like to bear in mind that there still is a possibility to propose other new physical models for the wall-bounded turbulence which explain the mechanism of producing the low-momentum zones observed in this study. We understand that investigation of the scaling parameter on these large-scale motions (inner-layer scaling or outer-layer scaling?) will be a important clue for the new physical models. But because we have only one Reynolds number case with two different stratified conditions at the moment, data are not sufficient to inquire further into the topic.

Anyhow this work has just begun and many topics including the development of new visualization technique, the whole kinematics using the vast analysis region and the scaling of these large-scale motions are left in our future works.

References

1. R. J. Adrian, C. D. Meinhart and C. D. Tomkins: J. Fluid Mech. **422**, 1 (2000)
2. D. B. DeGraaff and J. K. Eaton: J. Fluid Mech. **422**, 319 (2000)
3. R. E. Falco: Phys Fluids **20**, S124 (1977)
4. M. Germano, U. Piomelli, P. Moin and W. H. Cabot: Phys. Fluids **A3**, 1760 (1991)
5. S. Ghosal: J. Comp. Phys **125**, 187 (1996)

6. J. Jiménes: Ann. Research Briefs, Center for Turbulence Research, 137 (1998)
7. K. C. Kim and R. J. Adrian: Phys Fluids **11**, 417 (1999)
8. J. Kim, P. Moin and R. Moser: J. Fluid Mech. **177**, 133 (1987)
9. L. S. G. Kovasznay, V. Kibbens and R. F. Blackwelder: J. Fluid Mech. **41**, 283 (1970)
10. D. K. Lilly: Phys. Fluids **A4**, 633 (1992)
11. C. D. Meinhart and R. J. Adrian: Phys. Fluids **7**, 694 (1995)
12. Y. Morinishi, T. S. Lund, O. V. Vasilyev and P. Moin: J. Comput. Phys. **143**, 90 (1998)
13. A. E. Perry and M. S. Chong: J. Fluid Mech. **119**, 173 (1982)
14. M. Tsubokura, T. Kobayashi and N. Taniguchi: JSME Int. J. Ser. B **44**, 487 (2001)
15. M. Tsubokura: Phys. Fluids **13**, 500 (2001)
16. A. Yoshizawa, M. Tsubokura, T. Kobayashi and N. Taniguchi: Phys. Fluids **8**, 2254 (1996)
17. J. Zhou, R. J. Adrian, S. Balachandar and T. M. Kendall: J. Fluid Mech. **387**, 353 (1999)

Part III

Statistical Theory of Turbulence and LES Modeling

High Resolution DNS of Incompressible Homogeneous Forced Turbulence –Time Dependence of the Statistics–

Takashi Ishihara and Yukio Kaneda

Nagoya University, Nagoya 464-8603, JAPAN

Abstract. Results of direct numerical simulation (DNS) of incompressible homogeneous and isotropic forced turbulence with grid points up to 1024^3 are presented with emphasis on the time dependence of the turbulence statistics. The Taylor micro scale Reynolds number at a quasi-stationary state is 283 in the DNS with 1024^3 grid points, and the energy spectrum exhibits an universal equilibrium range. The results suggest that for $R_\lambda > 280$ or so, the energy dissipation rate normalized by the integral length scale and root-mean-square velocity is almost independent of R_λ and is $0.44 - 0.45$.

1 Introduction

Direct numerical simulation (DNS) of turbulence provides us with detailed data of turbulence statistics free from experimental uncertainties, under well controlled conditions. DNS can therefore be a powerful means not only for direct applications for practical problems, but also for the understanding of basic features of turbulence dynamics. In DNS for the latter purpose, a key role is played by the idea that the turbulence statistics at sufficiently small scales are universal in the sense that they are insensitive to the details of the initial and boundary conditions, and the external forcing at large scales, provided that the Reynolds number is sufficiently high. This idea is in accordance with the Kolmogorov hypothesis [1], and has been supported by experiments and simulations. This idea also implies that for the understanding of the universal feature of small scale statistics in turbulence, it is desirable to simulate turbulence at Reynolds number as high as possible, and that such features may be captured by DNS under simple boundary and forcing conditions. It is then reasonable to use very simple boundary conditions such as periodic boundary conditions, particularly because the number of the degrees of freedom to be treated in DNS of turbulence increases rapidly with the Reynolds number so that the use of complicated boundary conditions would result in heavier computation.

There have been in fact extensive studies of turbulence based on DNS using such simple conditions, as seen in Fig.1, which shows a rough sketch (of course not complete) of the history of the DNS of homogeneous isotropic incompressible turbulence obeying the Navier-Stokes equations under periodic boundary conditions. Since the pioneering work by Orszag in 1969[2], in DNS of which the number of grid points was 32^3, the number has been increasing approximately exponentially in time, and now becomes over 1024^3, as seen in Fig. 1 [3–7].

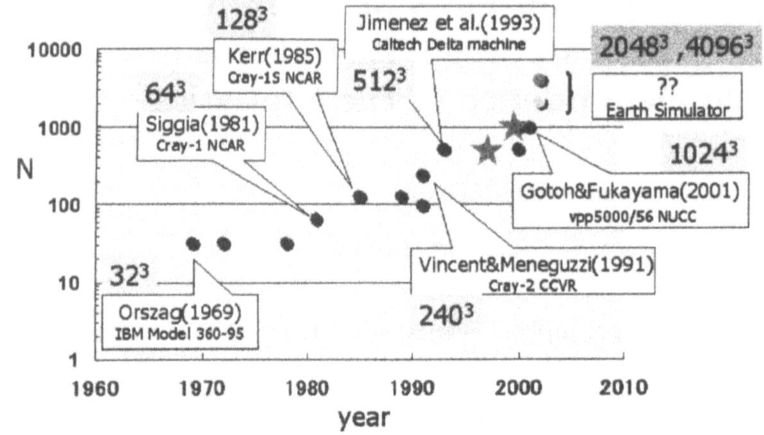

Fig. 1. History of DNS of incompressible homogeneous and isotropic turbulence under periodic boundary conditions. Star denotes the position of our DNS.

It is to be recalled here that in DNS of turbulence, the field is generally simulated by solving the equation of motion as an initial value problems. The filed cannot be stationary if there is no external forcing, because turbulence is a dissipative system. Even if there is external forcing, it cannot be stationary in a rigorous sense. If one could continue the simulation for sufficiently long time, then one may expect to realize stationary or quasi-stationary state. However, it is difficult in practice to simulate the DNS for very long time, particularly because the DNS of turbulence is generally quite expensive. This is particularly so in the DNS of turbulence at high Reynolds number in which the number of the degrees of freedom to be solved is enormous. One must stop the DNS at a certain finite time. But little seems to be known, for example, on the effect of the finiteness of the simulation time, how long the initial transient stage is, how strong the effect of non-stationarity is, and so on, in DNS of apparently "stationary" turbulence under external forcing.

In this paper, we report some results of our recent DNS of forced turbulence with the number of grid points up to 1024^3, where particular attention is paid on the effect of non-stationarity or the transient effect. In the following, we first review the method of our DNS in §2, and then present (1) time series of turbulence characteristics, (2) energy spectrum at a statistically quasi-stationary state, and (3) time series of the energy spectra in §3.

2 Numerical method

We consider the motion of an incompressible fluid of unit density which obeys the forced Navier Stokes equations

$$\frac{\partial}{\partial t}\mathbf{u} + (\mathbf{u}\cdot\nabla)\mathbf{u} = -\nabla p + \nu\Delta\mathbf{u} + \mathbf{f}, \qquad (1)$$

Table 1. DNS parameters and turbulence characteristics at $t = 10$ for Run256 and Run512, and at $t = 5.5$ for Run1024.

	R_λ	k_{max}	$\Delta t(\times 10^{-3})$	$\nu(\times 10^{-3})$	E	ϵ	L	λ	η
Run256	94	121	1.0	2.0	0.5	0.0936	1.10	0.327	0.0171
Run512	164	241	1.0	0.763	0.5	0.0811	1.26	0.217	0.00860
Run1024	283	483	0.625	0.289	0.5	0.0727	1.21	0.141	0.00427

$$\nabla \cdot \mathbf{u} = 0 \tag{2}$$

under periodic boundary conditions with fundamental periodic box of size 2π in each direction of the Cartesian coordinates, so that the minimum wavenumber is unity. Here $\mathbf{u}$, p, ν, and $\mathbf{f}$ denote velocity, pressure, kinematic viscosity, and external force which satisfies $\nabla \cdot \mathbf{f} = 0$. We solve (1) using an alias-free spectral method for spatial differentiation and a fourth order Runge-Kutta method for time advancing.

A negative viscosity is used to maintain quasi-steady turbulence; in wavenumber space the forcing is given by $\hat{\mathbf{f}}(\mathbf{k}) = -c\hat{\mathbf{u}}(\mathbf{k})$, where $\hat{\mathbf{f}}$ and $\hat{\mathbf{u}}$ are Fourier coefficient of $\mathbf{f}$ and $\mathbf{u}$, respectively, and the value of c is set non-zero only in the wavenumber range $k < 2.5$ and is adjusted every time step so as to keep the total kinematic energy constant ($\langle \mathbf{u} \cdot \mathbf{u} \rangle /2 = 0.5$). The kinematic viscosity is so chosen that $k_{max}\eta \sim 2$ in the statistically (quasi-)steady state, where k_{max} is the maximum wavenumber retained in the simulation and η is the Kolmogorov length scale. In our largest DNS (hereafter called Run1024), the number of grid points is 1024^3 and $k_{max} = 483$.

Prior to carrying out Run1024, we had performed DNS's with 256^3 and 512^3 grid points (hereafter they are referred to as Run256 and Run512, respectively) to study R_λ-dependence of the turbulence statistics and also to generate initial conditions for the higher-resolution DNS's. The characteristic parameters of the DNS are listed in Table 1. The initial condition for Run256 is a random solenoidal velocity field with the energy spectrum $E(k) \propto k^4 \exp(-k^2/2)$, and the initial fields of Run512 and Run1024 are the same as those of Run256 and Run512 at $t = 10(\sim 5T)$, respectively, where T is the eddy turnover time (see §3 for its definition). The kinematic viscosities of Run 512 and Run1024 are smaller than those of Run256 and Run512, respectively. Run1024 was performed up to $\sim 2.53T$ It took 8,800 time steps.

3 Numerical results

3.1 Time dependence of one-point statistics

Turbulence can be characterized various statistics. Among them are the so-called one-point statistics or the integrated quantities defined as follows:

the total kinematic energy $E = \frac{1}{2}\langle \mathbf{u}\cdot\mathbf{u}\rangle = \frac{3}{2}u'^2 = \int_0^{k_{\max}} E(k)dk,$

the mean rate of energy dissipation per unit mass $\epsilon = 2\nu \int_0^{k_{\max}} k^2 E(k)dk,$

the integral length scale $L = \frac{\pi}{2u'^2}\int_0^{k_{\max}} k^{-1}E(k)dk,$

the Taylor microscale $\lambda = \left(\frac{15\nu u'^2}{\epsilon}\right)^{1/2},$

the Kolmogorov length scale $\eta = \left(\frac{\nu^3}{\epsilon}\right)^{1/4},$

the eddy turnover time $T = \frac{L}{u'},$

the Kolmogorov time scale $\tau_\eta = \left(\frac{\nu}{\epsilon}\right)^{1/2},$

the Taylor microscale Reynolds number $R_\lambda = \frac{u'\lambda}{\nu},$

the energy dissipation rate scaled on L and u' $a = \frac{\epsilon L}{u'^3}.$

Since the total kinematic energy $E = 3u'^2/2$ is kept constant ($= 0.5$) in our DNS, it follows that (i) λ, η, τ_η, and R_λ are completely determined by ν and ϵ, (ii) T is proportional to L, and (iii) $a \equiv \epsilon L/u'^3$ is determined by ϵ and L. These facts suggest that the time dependence/independence of ϵ and L can be a good measure showing the degree of nonstationarity/stationarity of the simulated field. Note that ϵ and L depend on the energy spectrum, so that their time dependence reflects that of the energy spectrum. With these in mind, we starts with observing the time dependence of ϵ and L and then study that of the other one-point statistics, in the following.

Figure 2 shows the evolution of the mean rate of energy dissipation ϵ per unit mass. It is seen in Fig. 2 that, irrespectively of the differences between the run conditions, all the three curves show that ϵ increases initially, attains its peak value, and then decreases to attain a local minimum. The curve of Run512 shows that the time dependence of ϵ after the local minimum is weak, and the other two curves suggest that it is also the case for Run1024 and Run256. The state at which $\epsilon \approx$const. is not realized before the time of the local minimum of ϵ. The time at which ϵ takes the peak or the local minimum is moved forward in time for DNS with higher resolution. This suggests that the time necessary for the turbulence to attain its statistically steady state is shorter in DNS with higher Reynolds number.

Fig. 3 shows the time dependence of the integral length scale L. Since $u' \sim 1/\sqrt{3}$ in our DNS's, T is obtained by multiplying the values of L in Fig. 3 by $\sqrt{3}$. It is seen that L fluctuate slowly in time around a constant (≈ 1.2). Since the main contribution to the value of L is from the energy containing range where an artificial forcing acts, the behavior of L may be directly influenced by the nature of the forcing used in the DNS. Note that the time dependence of L in Run1024 is the weakest among the three runs.

Fig. 4 shows the time dependence of the characteristic length scales λ and η. Since $\lambda \propto \epsilon^{-1/2}$ and $\eta \propto \epsilon^{-1/4}$, the values of λ and η reflect the time dependence of ϵ. They are seen to be almost constant after the time at which ϵ takes the local minimum (in that period ϵ is almost constant).

Fig. 5 shows the time dependence of the length-scale ratios L/λ and λ/η. Both the ratios eventually become almost constant, but it is seen that reflecting the behavior of L, the ratio L/λ varies with time even after the time at which ϵ takes the local minimum. The ratio L/λ is a measure of the width of the inertial subrange. It is seen that L/λ is larger for DNS with higher Reynolds number, but $L/\lambda \approx 8$ for Run1024 is still quite small. Fig. 6 shows that the time dependence of the length-scale ratio L/η is relatively smaller than that of the time-scale ratio T/τ_η, where T/τ_η is given by $T/\tau_\eta = \sqrt{15}L/\lambda$, and L/λ is shown in Fig. 5.

Fig. 7 shows that R_λ initially decreases and becomes almost constant after the time at which ϵ takes the local minimum. Note that the value of R_λ is relatively high at an initial short transient period, but the ratios, L/λ and T/τ_η, are smaller than those at the later time. This fact gives a warning that in order to estimate R_λ-dependence of statistical quantities properly, one need simulate the turbulence field for sufficiently long time, so that a statistically stationarity in the sense $\epsilon \approx$const. is achieved.

Fig. 8 shows the time dependence of $a = \epsilon L/u'^3$. The behavior of a is similar to that of ϵ in Fig. 2, but even if $\epsilon \approx$const. is approximately satisfied, a fluctuates

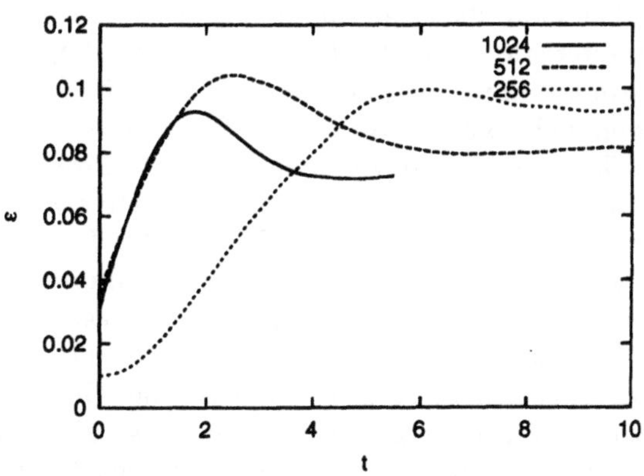

Fig. 2. Time dependence of ϵ in Run1024, Run512 and Run256.

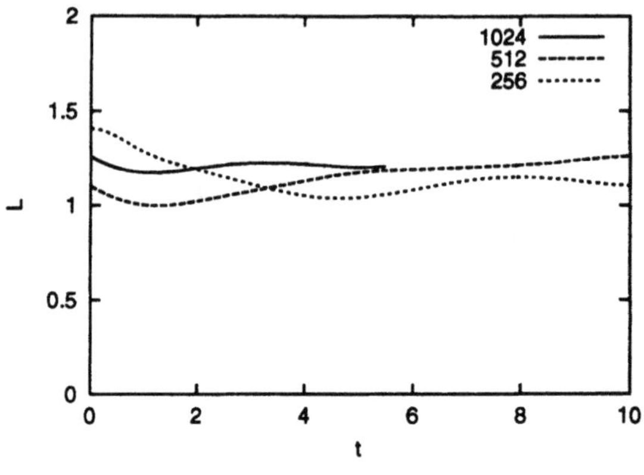

Fig. 3. Time dependence of the integral length scale, L. The eddy turnover time, T, is obtained by multiplying the values of L by $\sqrt{3}$.

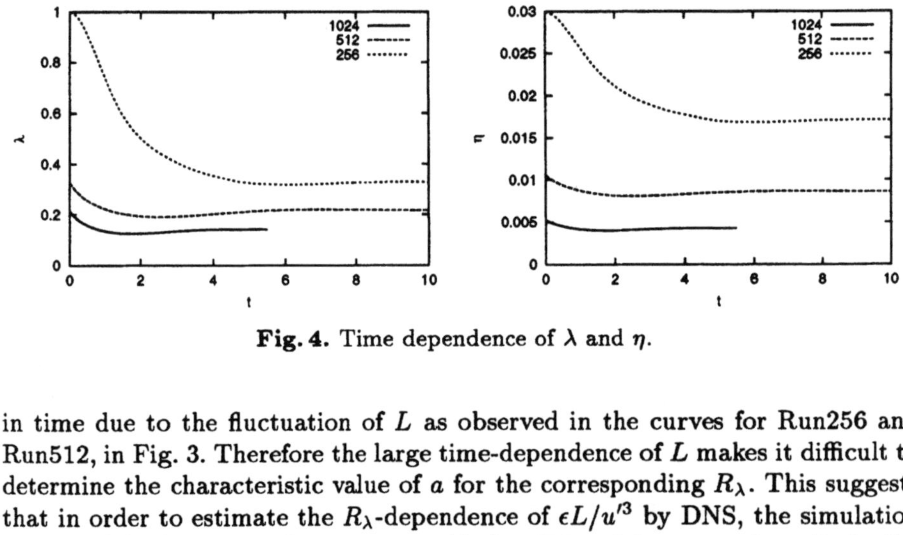

Fig. 4. Time dependence of λ and η.

in time due to the fluctuation of L as observed in the curves for Run256 and Run512, in Fig. 3. Therefore the large time-dependence of L makes it difficult to determine the characteristic value of a for the corresponding R_λ. This suggests that in order to estimate the R_λ-dependence of $\epsilon L/u'^3$ by DNS, the simulation time need be long enough so as to avoid the effect of the nonstaionarity in the values of L and ϵ. This is particularly true in DNS with lower resolution such as our Run256 and Run512. Since the variance of L is relatively small for Run1024, it is easier to determine the characteristic value of a for $R_\lambda = 283$.

In Fig. 9, we plot, as a function of R_λ, the first minimum value of $\epsilon L/u'^3$ after $\epsilon L/u'^3$ attains its peak value, and compare the results with recent data from Gotoh *et al.* [8] as well as previous DNS data [6,8–11], which are listed in Ref. [12]. The data, especially those from our DNS and Gotoh *et al.*[8], suggest that $\epsilon L/u'^3$ approaches to a constant with the increase of R_λ, and it is almost constant ($= 0.44 - 0.45$) for $R_\lambda > 280$ or so. The consideration in the above paragraph suggests that the data for $R_\lambda < 200$ are scattered presumably because of insufficient length of simulation time.

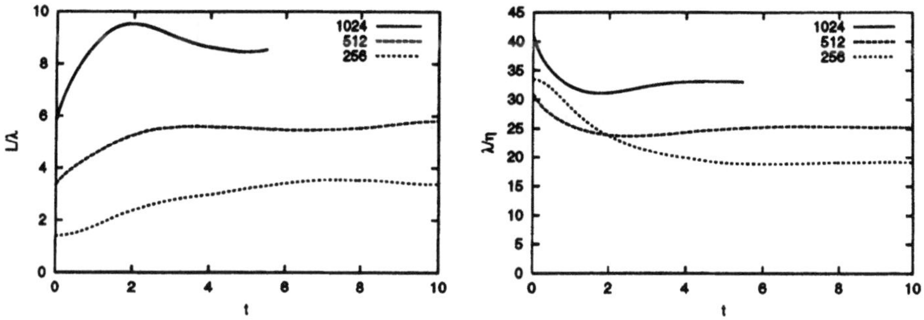

Fig. 5. Time dependence of the ratios L/λ and λ/η.

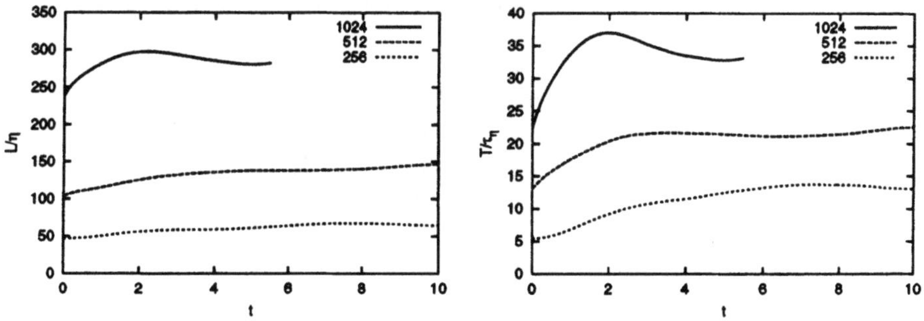

Fig. 6. Time dependence of the ratios L/η and T/τ_η.

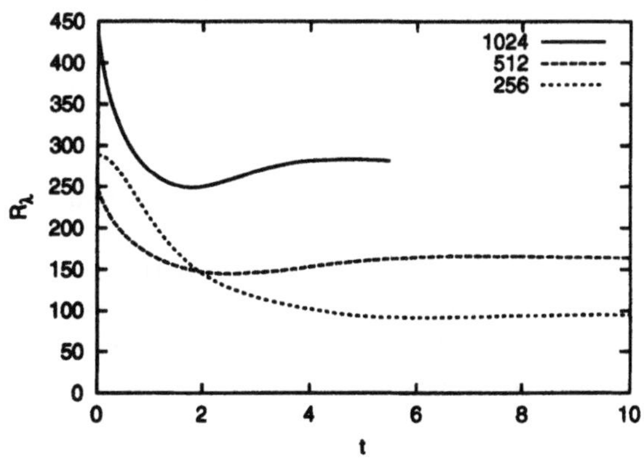

Fig. 7. Time dependence of R_λ.

It may be interesting to note here that
(i) in spite of the similarity of our forcing to that of Jimenez *et al.*[6], the values of $\epsilon L/u'^3$ for $R_\lambda < 200$ are different from each other; $\epsilon L/u'^3 = 0.49 - 0.57$ for

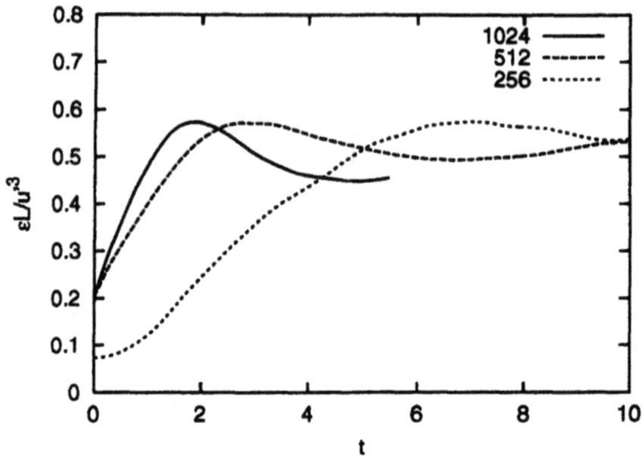

Fig. 8. Time dependence of the normalized rate of energy dissipation $\epsilon L/u'^3$.

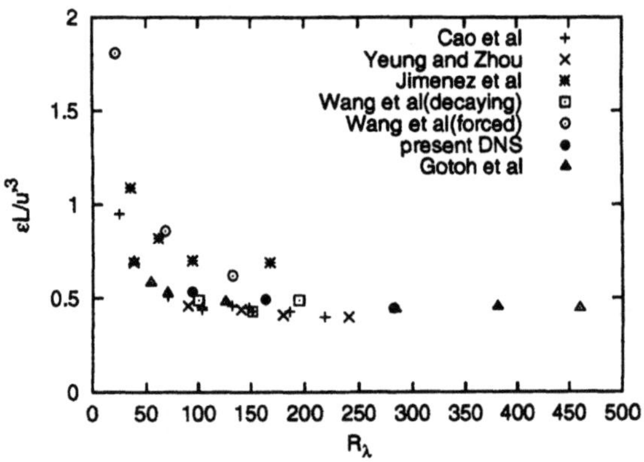

Fig. 9. $\epsilon L/u'^3$ vs. R_λ.

$R_\lambda = 164$ in our DNS, while $\epsilon L/u'^3 = 0.69$ for $R_\lambda = 168$ in DNS by Jimenez *et al.*,
(ii) on the other hand, in spite of the difference between our forcing and that in Gotoh *et al.*[8], the values of $\epsilon L/u'^3$ are almost the same to each other for $R_\lambda > 280$; $\epsilon L/u'^3 = 0.447$ for $R_\lambda = 283$ in our DNS, while $\epsilon L/u'^3 = 0.442$ for $R_\lambda = 284$ in the DNS by Gotoh *et al.*.

3.2 Energy spectrum at a statistically quasi-stationary state

Figure 10 shows the normalized energy spectra

$$\epsilon^{-1/4}\nu^{-5/4}E(k)$$

obtained from snapshots of Run256 ($t = 10$), Run512 ($t = 10$), and Run1024 ($t = 5.5$), as functions of the normalized wavenumber $k\eta$. It can be seen that

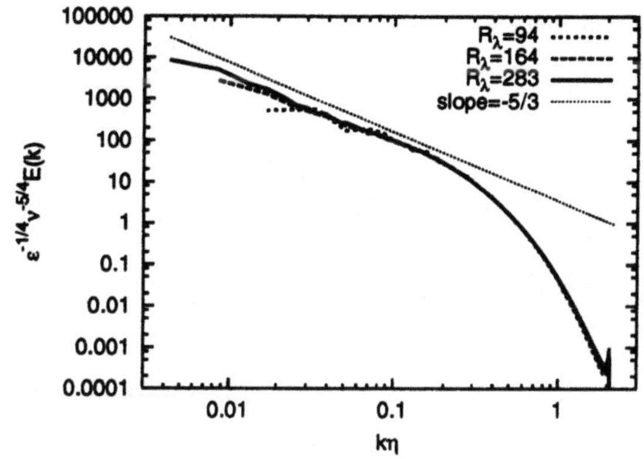

Fig. 10. Normalized energy spectrum $\epsilon^{-1/4}\nu^{-5/4}E(k)$ at $t = 10$ of Run256 and Run512, and at $t = 5.5$ of Run1024.

irrespectively of R_λ the normalized energy spectra overlap each other at high wavenumbers.

Figure 11 and 12 respectively show the comparisons of the compensated energy spectrum defined by

$$\epsilon^{-2/3}k^{5/3}E(k)$$

and normalized energy flux defined by

$$\frac{\Pi(k)}{\epsilon} = \frac{1}{\epsilon}\int_k^\infty T(k')dk',$$

at different R_λ, where $T(k)$ is the energy transfer function. These plots are obtained from the same fields as in Fig. 10. In contrast to Fig.10, the difference between the three spectra is clearly visible in Fig.11.

Figure 12 shows that there exists a plateau, at $k\eta \approx 0.06$ for $R_\lambda = 94$, at $0.02 < k\eta < 0.07$ for $R_\lambda = 167$ and at $0.01 < k\eta < 0.06$ for $R_\lambda = 283$, where $\Pi(k)$ is nearly independent of k. It is seen that the value of $\Pi(k)$ at the plateau agrees well with the value of ϵ for $R_\lambda = 283$, while the agreement is not so good for $R_\lambda = 94$ and 164.

In Fig. 11, $k^{5/3}E(k)/\epsilon^{2/3}$ is seen to be nearly constant in the plateau range for the case of $R_\lambda = 283$. On the other hand, the plateau range of $\Pi(k)$ for $R_\lambda = 94$ and 164 is very small, and it is difficult to find in Fig. 11 a scaling range, where the compensated energy spectrum is constant independent of k. In Fig. 11, we can also observe bumps, each of which has a peak at $k\eta \approx 0.15$ irrespectively of R_λ. This kind of bump has been also observed in experiments, e.g. [13]. Note that the bump is at the wavenumber range $k\eta > 0.07$, where $\Pi(k)$ is a decreasing function of k, and the inertial subrange lies in $k\eta < 0.07$. This fact suggests that in order to obtain an inertial subrange we have to resolve modes with wavenumbers much smaller than $k\eta \approx 0.07$.

The Kolmogorov constant C_K defined by

$$E(k) = C_K\epsilon^{2/3}k^{-5/3} \tag{3}$$

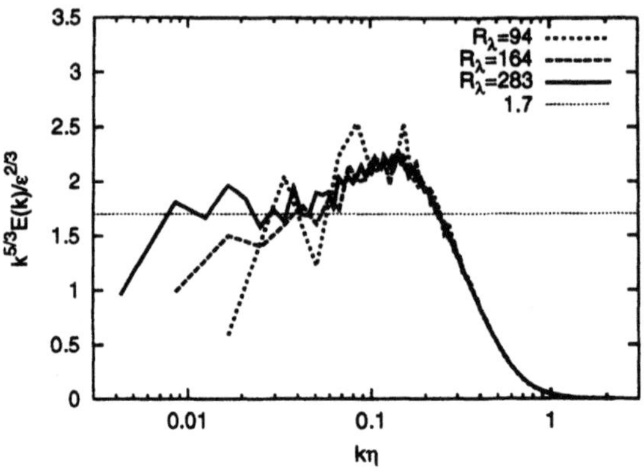

Fig. 11. Compensated energy spectrum $\epsilon^{-2/3}k^{5/3}E(k)$ at $t = 10$ of Run256 and Run512, and at $t = 5.5$ of Run1024.

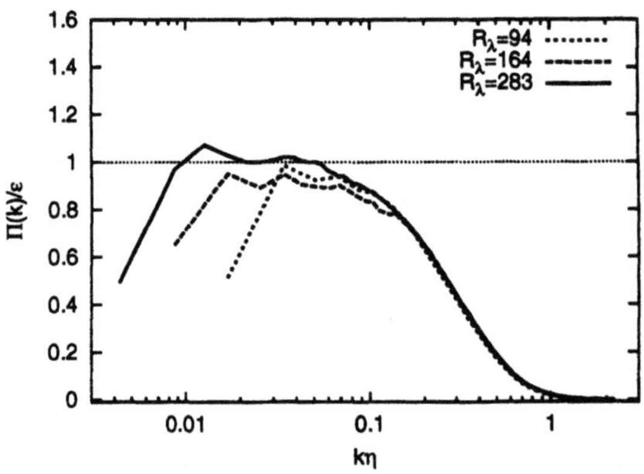

Fig. 12. Normalized energy flux $\Pi(k)/\epsilon$ at $t = 10$ of Run256 and Run512, and at $t = 5.5$ of Run1024. The dotted straight line denotes $\Pi(k)/\epsilon = 1$.

in the inertial subrange can be estimated from the plot for $R_\lambda = 283$ at $0.01 < k\eta < 0.06$ in Fig. 11. We then obtain $C_K \approx 1.7$. This value 1.7 is consistent with previous experiments [14], spectral closure theories [15,16] and recent numerical simulations, [10,8] whose forcing scheme and the value of $k_{\max}\eta (\approx 1)$ are different from those in our DNS. These suggest that the Kolmogorov constant of the inertial subrange are insensitive to the detail of such flow conditions as forcing at large scale and wavenumber truncation at small scale. However it is also to be remembered that some statistical quantities such as high order structure functions in the inertial subrange are sensitive to the value of $k_{\max}\eta$ (see analysis in Ref. [17]).

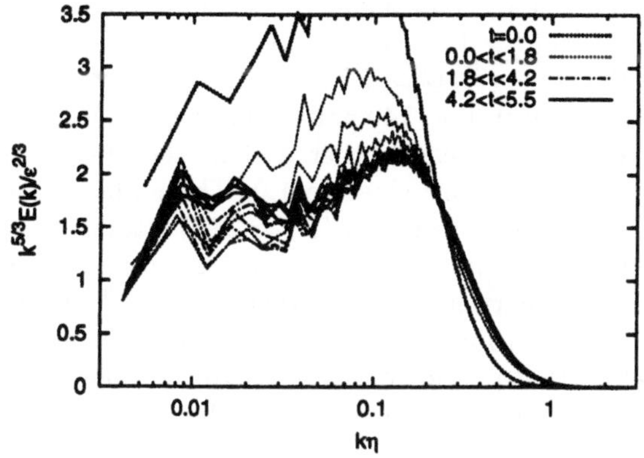

Fig. 13. Time series of compensated energy spectrum $\epsilon^{-2/3}k^{5/3}E(k)$ in Run1024.

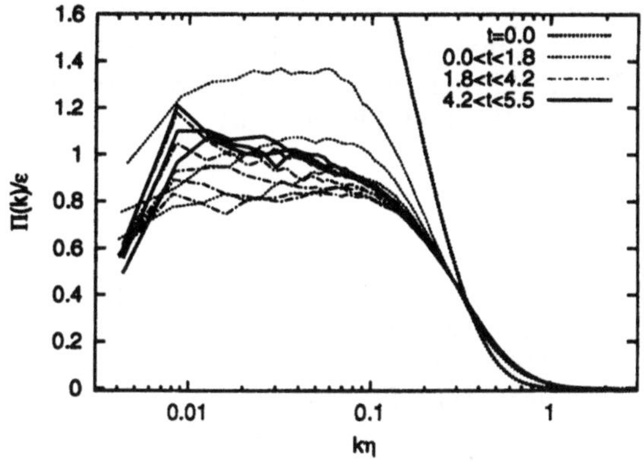

Fig. 14. Time series of normalized energy flux $\Pi(k)/\epsilon$ in Run1024.

3.3 Time dependence of the energy spectrum

In the present DNS, the total energy E is kept constant so that the behavior of ϵ and L determines that of the other one-point statistics discussed in §3.1. It was shown in §3.1 that (i) L depends on time, but its dependence is weak, and (ii) the time dependence of L is weaker for larger R_λ. Therefore we study here the time dependence of spectra in the DNS with the highest R_λ, i.e., Run1024, with taking account of the time dependence of ϵ. Regarding the evolution of ϵ, it was seen in Fig.2 that there are at least three characteristic stages;
(i) for $0.0 < t < 1.8$, ϵ increases sharply with t,
(ii) for $1.8 < t < 4.2$, ϵ decreases gradually, and
(iii) for $4.2 < t$, ϵ is almost constant.

Figures 13 and 14 respectively show the compensated energy spectrum and the energy flux function normalized by ϵ at various times from Run1024. It can be seen in Fig. 13 that in the stage (i) the compensated spectrum as well as the normalize energy flux are not stationary in the entire wavenumber range, while in the stage (ii) and (iii) ($t > 1.8$), the energy spectra at $k\eta > 0.1$ agree well with each other and at the stage (iii) ($t > 4.2$), the energy spectra at $0.02 < k\eta < 0.06$ are close to 1.7. It is seen in Fig. 14 that in the stage (ii), the values of $\Pi(k)/\epsilon$ at $k\eta > 0.1$ agree well with each other and in the stage (iii), the values of $\Pi(k)/\epsilon$ at $0.02 < k\eta < 0.06$ are close to 1.0. These observations suggest that the turbulent velocity field by Run1024 ($R_\lambda = 283$) attains its quasi-statistically steady state (at least, regarding the energy spectrum and the energy flux) and exhibits an equilibrium inertial subrange with the $k^{-5/3}$ energy spectrum after $t = 4.2(\approx 2T)$. One needs to be careful not to misread the energy spectra at the transient stage before that period to be an equilibrium one; the confusion might then result in, for example, an overestimate of the Kolmogorov constant.

4 Acknowledgement

The authors are grateful to Professor K. R. Sreenivasan for stimulating discussions which encourage our making plots in Fig. 9. This work was supported by "Research for the Future" Program of the Japan Society for the Promotion of Science under the project JSPS-RFTF97P01101.

References

1. A. N. Kolmogorov: Dokl. Akad. Nauk SSSR **30** 9 (1941)
2. S. A. Orszag: Phys. Fluids, **12** II-250 (1969)
3. E. D. Siggia: J. Fluid Mech. **107** 375 (1981)
4. R. M. Kerr: J. Fluid Mech. **153** 31 (1985)
5. A. Vincent, M. Meneguzzi; J. Fluid Mech. **225** 1 (1991)
6. J. Jimenez, A. A. Wray, P. G. Saffman, R. S. Rogallo:J. Fluid Mech. **255** 65 (1993)
7. T. Gotoh, D. Fukayama: Phys. Rev. Lett. **86** 3775 (2001)
8. T. Gotoh, D. Fukayama, T. Nakano: Phys. Fluids **14**, 1065 (2002)
9. N. Cao, S. Chen, G. D. Doolen: Phys. Fluids **11** 2235 (1999)
10. P. K. Yeung, Y. Zhou: Phys. Rev. E **56**, 1746 (1997)
11. L. P. Wang, S. Chen, J. G. Brasseur, J. C. Wyngaard: J. Fluid Mech. **309**, 113 (1996)
12. K. R. Sreenivasan: Phys. Fluids **10**, 528 (1998)
13. K. R. Saddoughi, S. V. Veeravalli: J. Fluid Mech. **268**, 333 (1994)
14. K. R. Sreenivasan: Phys. Fluids **7**, 2778 (1995)
15. R. H. Kraichnan: Phys. Fluids **8**, 575 (1965)
16. Y. Kaneda: J. Fluid Mech. **107**, 131 (1981)
17. Y. Yamazaki, T. Ishihara, Y. Kaneda: J. Phys. Soc. Jpn. **71**, 777 (2002)

Subgrid Models for Two-Dimensional Turbulence based on Lagrangian Spectral Theory

Toshiyuki Gotoh[1], Isao Kakui[1], and Yukio Kaneda[2]

[1] Department of System Engineering, Nagoya Institute of Technology, Showa-ku, Gokiso, Nagoya, 466-8555, JAPAN,

[2] Depratment of Computational Science and Engineering, Graduate School of Engineering, Nagoya University, Chikusa-ku, Nagoya,464-8603, JAPAN

Abstract. An LES subgrid model in the enstrophy cascading range of two-dimensional turbulence is proposed. The effects of the subgrid scales are modeled by a Langevin-type equation containing an eddy damping factor and a random force. The eddy damping and forcing are expressed as functionals of the energy spectrum using the Markovianized Lagrangian renormalized approximation (MLRA), a Lagrangian spectral theory without *ad hoc* parameters. It is found that the enstrophy spectrum predicted by MLRA-LES ($N = 128^2$) agrees well with the DNS data ($N = 1024^2$) for forced turbulence. The growth of the error spectrum predicted by the MLRA-LES is also found to be very similar to that of the DNS. Some simplifications of the eddy damping factor are proposed for application to global-scale flow prediction. One model is found to give predictions of the enstrophy spectrum and error growth which are comparable to the MLRA-LES results.

1 Introduction

Large Eddy Simulation (LES) for turbulent flow has become increasingly important in atmospheric, oceanographic and engineering research [1]. In LES, the coarse-grained equations obtained by projecting the Navier-Stokes flow onto a grid-scale flow are integrated numerically. In many applications of LES, the effects of the subgrid scale (SGS) fluid motions are expressed by a Smagorinsky-type eddy viscosity. However, as shown in non-equilibrium statistical mechanics, projection of a large number of degrees of freedom corresponding to fast modes onto a smaller number of retained degrees of freedom corresponding to slow modes produces two effects: the decay of memory (or damping) and random forcing, which are related through the so-called 'fluctuation dissipation' theorem [2]. An analogy between Brownian motion of a small particle in a quiescent fluid and the random motion of grid-scale eddies is very attractive and useful for tubulence modeling; but since the turbulent flow is far from thermal equilibrium, we must use the analogy with great care.

Kraichnan has discussed a model representation of the DIA equations which contains a nonlinear eddy damping factor and a random force [3]. Inspired by this model, Bertoglio derived a set of equations for LES in which the effects of SGS motions were represented by an eddy damping factor plus a random force. This model is referred as a Langevin model in this paper [4]. The forcing and damping are related so that the correlation function satisfies the EDQNM spectral evolution equation. A similar approach has been taken in Chasnov [5]. The random force is considered as back-scatter, which represents the backward transfer of the energy from the subgrid scales to the grid scales.

Although there are two unknowns in the Langevin-type equation for the SGS, the eddy damping factor and the random force intensity, the condition used by Bertoglio [4] and Chasnov [5] to determine them was the energy equation alone. This means that there is no unique way to determine the eddy damping factor and the random force spectrum. They determined the two factors as follows: the nonlinear energy transfer function in the EDQNM spectral equation consists of source and drain terms; the former determines the spectrum of the random force, and the latter determines the eddy damping factor.

Here we present a set of Langevin-type SGS equations which is determined in a more theoretically consistent way. The guiding principle in our study is to use the MLRA spectral equation for LES modeling, because although the LRA and MLRA both yield good results for the energy spectrum in two and three dimensions without any *ad hoc* parameters, the MLRA is easier to use than the LRA [6–10]. In order to determine the two factors more rigorously, we consider the evolution of the errors arising from insufficient knowledge of the fluid motion below the grid scale $1/k_c$. By requiring that the evolution of the spectra of the energy and error predicted by the Langevin-type SGS equation is statistically the same as that predicted by the MLRA equations, we determine the two factors. The LES equations obtained by this method is integrated and compared with DNS results.

2 Basic equations in two dimensions

We consider two-dimensional homogeneous isotropic turbulence in a steady state. The basic equation for an incompressible fluid in an unbounded domain is

$$\frac{\partial \boldsymbol{u}}{\partial t} + \boldsymbol{u} \cdot \nabla \boldsymbol{u} = -\nabla p + \nu \nabla^2 \boldsymbol{u} + \boldsymbol{f}, \quad \nabla \cdot \boldsymbol{u} = 0, \tag{1}$$

where $\rho = 1$ is assumed for simplicity, and $\boldsymbol{f}$ is a solenoidal external force acting at low wavenumbers which maintains the turbulence. For later use, we also write the vorticity equation:

$$\left(\frac{\partial}{\partial t} + \nu k^2\right) \omega(\boldsymbol{k}, t) = -\frac{1}{2} \sum_{\boldsymbol{p},\boldsymbol{q}}^{\Delta} \left(\frac{1}{p^2} - \frac{1}{q^2}\right) (\boldsymbol{p} \times \boldsymbol{q}) \omega(\boldsymbol{p}, t) \omega(\boldsymbol{q}, t) + \boldsymbol{f}_{\omega}. \tag{2}$$

3 MLRA equations for two-dimensional turbulence

3.1 Energy spectrum

A set of Lagrangian renormalized approximation (LRA) equations for turbulence is derived by considering the internal dynamics of the fluid motion in the Lagrangian framework. A detailed derivation is found in Kaneda [6]. In the analysis of the LRA equations, it is found that the LRA equations (reffered to as MLRA hereafter) obtained by Markovian modification of the fluctuation-dissipation relation, replacing $Q^L(k,t,s) = G^L(k,t,s)Q(k,s,s)$ by $Q^L(k,t,s) = G^L(k,t,s)Q(k,t,t)$ for $t \geq s$, preserve the realizability of the LRA equations and are numerically more stable than the LRA equations [6,7]. Here, Q^L and G^L are the Lagrangian velocity covariance and response functions, respectively. The Markovian modification is exact when the turbulence is statistically steady, and is a good approximation when the characteristic time of the energy spectrum is much longer than that of the internal dynamics.

A set of MLRA equations for two-dimensional homogeneous isotropic turbulence is as follows [9,10];

$$\left(\frac{\partial}{\partial t} + 2\nu k^2\right) Q(k,t) = B(k,t), \tag{3}$$

$$\left(\frac{\partial}{\partial \tau} + \nu k^2 + \eta(k,t)\right) G(k,t,\tau) = 0, \quad \tau > 0, \tag{4}$$

$$G(k,t,\tau = 0) = 1, \tag{5}$$

where

$$B(k,t) = \iint_\Delta dpdq\ b_{kpq}\theta_{kpq}(t)Q(q,t)[Q(p,t) - Q(k,t)], \tag{6}$$

$$\theta_{kpq}(t) = \int_0^\infty G(k,t,\tau)G(p,t,\tau)G(q,t,\tau)d\tau, \tag{7}$$

$$\eta(k,t,\tau) = k\int_0^\infty dq\ q^2 J(q/k)Q(q,t)\int_0^\infty G(q,t,\tau)d\tau, \tag{8}$$

$$J(x) = J(1/x) = \frac{\pi}{2}\begin{cases} 3x - x^3, & x \leq 1 \\ 3/x - 1/x^3, & x > 1 \end{cases} \tag{9}$$

$$b_{kpq} = 4k^{-2}(k^2 - q^2)(p^2 - q^2)(1 - x^2)^{1/2}. \tag{10}$$

Assuming $G(k,\tau) = \exp\left[-(\nu k^2 + \eta(k))\tau\right]$ we obtain an approximate formula for θ_{kpq} as

$$\theta_{kpq} = \left(\nu(k^2 + p^2 + q^2) + \eta(k) + \eta(p) + \eta(q)\right)^{-1}, \tag{11}$$

$$\eta(k) \approx \left(\frac{3}{2}\int_0^k q^2 E(q)dq\right)^{1/2}. \tag{12}$$

3.2 Error spectrum

The error field is defined in terms of the difference between the Eulerian velocity fields of pairs of flows chosen from statistically identical ensembles [11,12];

$$\delta \boldsymbol{u}(\boldsymbol{x},t) = \boldsymbol{u}^{(1)}(\boldsymbol{x},t) - \boldsymbol{u}^{(2)}(\boldsymbol{x},t). \tag{13}$$

Both fields are solutions of the Navier-Stokes equation with the same external forces and boundary conditions, statistically homogeneous and isotropic, and have zero means and identical energy spectra. The relevant single-time two-point spectra are defined as

$$Q(k,t) = \left\langle u_i^{(n)}(\boldsymbol{k},t)u_i^{(n)}(-\boldsymbol{k},t)\right\rangle, \tag{14}$$

$$W(k,t) = \left\langle u_i^{(m)}(\boldsymbol{k},t)u_i^{(n)}(-\boldsymbol{k},t)\right\rangle, \tag{15}$$

$$\Delta(k,t) = \frac{1}{2}\left\langle \delta u_i^{(m)}(\boldsymbol{k},t)\delta u_i^{(n)}(-\boldsymbol{k},t)\right\rangle = Q(k,t) - W(k,t), \tag{16}$$

where m and n are 1 or 2.

The equations for W and Δ are also given by the MLRA as follows;

$$\left(\frac{\partial}{\partial t} + 2\nu k^2\right) W(k,t) = B_W(k,t), \tag{17}$$

$$\left(\frac{\partial}{\partial t} + 2\nu k^2\right) \Delta(k,t) = B_\Delta(k,t), \tag{18}$$

$$B_W(k,t) = \iint_\Delta dpdq\ b_{kpq}\theta_{kpq}(t)Q(q,t)[W(p,t) - W(k,t)] - B_X(k,t), \tag{19}$$

$$B_\Delta(k,t) = \iint_\Delta dpdq\ b_{kpq}\theta_{kpq}(t)Q(q,t)[\Delta(p,t) - \Delta(k,t)] + B_X(k,t), \tag{20}$$

$$B_X(k,t) = \iint_\Delta dpdq\ b_{kpq}\theta_{kpq}(t)W(p,t)\Delta(q,t). \tag{21}$$

4 A Langevin equation model for two-dimensional turbulence

We introduce a projection such that for a Fourier mode $A(\boldsymbol{k})$

$$A(\boldsymbol{k}) = \mathcal{P}A + \mathcal{P}'A \equiv A^<(\boldsymbol{k}) + A^>(\boldsymbol{k}), \tag{22}$$

$$\mathcal{P}A \equiv \begin{cases} A(\boldsymbol{k}) & |\boldsymbol{k}| \le k_m \\ 0 & |\boldsymbol{k}| > k_m \end{cases}, \quad \mathcal{P}' = 1 - \mathcal{P}. \tag{23}$$

Correpsponding to this we have a decomposition of A in physical space

$$A(\boldsymbol{x}) = A^<(\boldsymbol{x}) + A^>(\boldsymbol{x}), \tag{24}$$

$$A^<(\boldsymbol{x}) = \int A^<(\boldsymbol{k})e^{i\boldsymbol{k}\cdot\boldsymbol{x}}d\boldsymbol{k}, \quad A^>(\boldsymbol{x}) = \int A^>(\boldsymbol{k})e^{i\boldsymbol{k}\cdot\boldsymbol{x}}d\boldsymbol{k}, \tag{25}$$

The velocity field $\boldsymbol{u}$ and the pressure p are decomposed as

$$\boldsymbol{u} = \boldsymbol{u}^< + \boldsymbol{u}^>, \quad p = p^< + p^>. \tag{26}$$

The equation for $\boldsymbol{u}^<$ is obtained as

$$\begin{aligned} \frac{\partial \boldsymbol{u}^<}{\partial t} &= -\mathcal{P}[\boldsymbol{u}\cdot\nabla\boldsymbol{u}] - \nabla p^< + \nu\nabla^2\boldsymbol{u}^< \\ &= -\boldsymbol{u}^<\cdot\nabla\boldsymbol{u}^< - \nabla p^< + \nu\nabla^2\boldsymbol{u}^< + \boldsymbol{q}^<, \end{aligned} \tag{27}$$

$$\nabla\cdot\boldsymbol{u}^< = 0, \tag{28}$$

where $\boldsymbol{q}^<$ represents the grid-subgrid interaction:

$$\boldsymbol{q}^< = \boldsymbol{u}^<\cdot\nabla\boldsymbol{u}^< - \mathcal{P}[\boldsymbol{u}\cdot\nabla\boldsymbol{u}] \tag{29}$$

The effects of the subgrid scales are assumed to be expressed in terms of the eddy damping factor and the random force as

$$\boldsymbol{q}^<(\boldsymbol{k},t) = -\mu_e(k|k_c)k^2\boldsymbol{u}^<(\boldsymbol{k},t) + \boldsymbol{f}^<(\boldsymbol{k},t), \tag{30}$$

where the random force $\boldsymbol{f}^<(\boldsymbol{k},t)$ is white in time and obeys Gaussian statistics with zero mean and variance

$$\langle \boldsymbol{f}^<(\boldsymbol{k},t)\rangle = 0, \qquad \langle f_i^<(\boldsymbol{k},t)f_j^<(-\boldsymbol{k},s)\rangle = 2\delta(t-s)D(k,t|k_c). \tag{31}$$

The problem is to find the two functions $\mu_e(k|k_c)$ and $D(k|k_c)$ and we need two conditions to determine them. Bertoglio and Chasnov have determined $\mu_e(k|k_c)$ and $D(k|k_c)$ by making the eddy damping factor correspond to the term proportional to $Q(k)$ in the right hand side of Eq.(6) and $D(k|k_c)$ correspond to the first term of the integrand of Eq.(6). More precisely, they are given by those terms but with the integration domain restricted to p and/or q higher than k_c [4,5].

Instead, we consider two different conditions: one condition is the requirement that the evolution of the energy spectrum according to the SGS model is identical to that given by the spectral theory, the MLRA. Another condition is the requirement that the evolution of the error spectrum is identical to that computed by the MLRA. The evolution equations for the spectra Q^{LES}, W^{LES}, and Δ^{LES} are

$$\left(\frac{\partial}{\partial t} + 2\nu k^2\right) Q^{LES}(k,t) = B^{LES}(k,t), \tag{32}$$

$$\left(\frac{\partial}{\partial t} + 2\nu k^2\right) W^{LES}(k,t) = B_W^{LES}(k,t), \tag{33}$$

$$\left(\frac{\partial}{\partial t} + 2\nu k^2\right) \Delta^{LES}(k,t) = B_\Delta^{LES}(k,t), \tag{34}$$

and

$$B^{LES}(k,t) = 2Real\left\langle u^{<} \cdot M \sum u^{<} u^{<} \right\rangle - 2\mu_e(k|k_c)k^2 Q^{LES} + 2D(k|k_c), \quad (35)$$

$$B_W^{LES}(k,t) = Real\left\langle u^{<,(2)} \cdot M(k) \sum u^{<,(1)} u^{<,(1)} \right\rangle$$
$$+Real\left\langle u^{<,(1)} \cdot M(k) \sum u^{<,(2)} u^{<,(2)} \right\rangle - 2\mu_e(k|k_c)k^2 W^{LES}, \quad (36)$$

$$B_\Delta^{LES}(k,t) = 2Real\left\langle (u^{<,(1)} - u^{<,(2)}) \right.$$
$$\left. \cdot M(k) \sum (u^{<,(1)} u^{<,(1)} - u^{<,(2)} u^{<,(2)}) \right\rangle$$
$$-2\mu_e(k|k_c)k^2 \Delta^{LES} + 2D(k|k_c). \quad (37)$$

Applying the MLRA procedure to the third order moments in Eqs.(35), (36), and (37), we obtain the closed set of equations for the right hand sides of Eqs.(32), (33), and (34)

$$B^{LES}(k,t) = \iint_{\Delta_1} dpdq\ b_{kpq} \hat{\theta}_{kpq}(t) Q^{LES}(q,t)[Q^{LES}(p,t) - Q^{LES}(k,t)], \quad (38)$$

$$B_W^{LES}(k,t) = \iint_{\Delta_1} dpdq\ b_{kpq} \hat{\theta}_{kpq}(t) Q^{LES}(q,t)[W^{LES}(p,t) - W^{LES}(k,t)]$$
$$-B_X^{LES}(k,t), \quad (39)$$

$$B_\Delta^{LES}(k,t) = \iint_{\Delta_1} dpdq\ b_{kpq} \hat{\theta}_{kpq}(t) Q^{LES}(q,t)[\Delta^{LES}(p,t) - \Delta^{LES}(k,t)]$$
$$+B_X^{LES}(k,t), \quad (40)$$

$$B_X^{LES}(k,t) = \iint_{\Delta_1} dpdq\ b_{kpq} \hat{\theta}_{kpq}(t) W^{LES}(p,t) \Delta^{LES}(q,t) \quad (41)$$

where

$$\iint_{\Delta_1} dpdq = \int_0^{k_c} \int_{|p-k|}^{min(p+k,k_c)} . \quad (42)$$

Note that the triple relaxation factor $\hat{\theta}_{kpq}(t)$ is not given by Eq.(11) but by

$$\hat{\theta}_{kpq} = \left(\nu(k^2 + p^2 + q^2) + \eta(k|k_c) + \eta(p|k_c) + \eta(q|k_c)\right.$$
$$\left. +\mu_e(k|k_c)k^2 + \mu_e(p|k_c)p^2 + \mu_e(q|k_c)q^2)\right)^{-1}, \quad (43)$$

where the factor $\eta(k|k_c)$ is

$$\eta(k|k_c) = k \int_0^{k_c} dq\ q^2 J(q/k) Q(q,t) \int_0^\infty G(q,t,\tau) d\tau$$
$$\approx \frac{3}{2} \left(\int_0^k dq\ q^2 E(q) + k^2 \int_k^{k_c} dq\ E(q) \right) \int_0^\infty G(q,t,\tau) d\tau$$
$$\approx \left(\frac{3}{2} \int_0^{k_c} q^2 E(q) dq \right)^{1/2} . \quad (44)$$

In passing from the second to the last line, we have used the fact that $\mu_e(k|k_c)$ is very small for $k \ll k_c$ but rises quickly when $(k_c - k)/k_c \ll 1$. Since the present Langevin model mimics the effects of the SGS at the level of second order statistics, there are differences between the modeled effects and those in the real Navier-Stokes dynamics, especially for $k \lesssim k_c$; however, when $k \ll k_c$, the difference should be small enough.

The two requirements above are equivalent to demanding

$$
\begin{aligned}
&Q^{LES}(k,t) = Q^{MLRA}(k,t), \quad W^{LES}(k,t) = W^{MLRA}(k,t), \\
&\Delta^{LES}(k,t) = \Delta^{MLRA}(k,t)
\end{aligned} \tag{45}
$$

for $k \le k_c$, where the quantities with superscript ^{MLRA} are those given by the solution of the MLRA equations. Then we obtain the equations for $k \le k_c$

$$B^{LES}(k,t) - 2\mu_e(k|k_c)k^2 Q^{LES} + 2D(k|k_c) = B(k,t), \tag{46}$$

$$B_{\Delta}^{LES}(k,t) - 2\mu_e(k|k_c)k^2 \Delta^{LES} + 2D(k|k_c) = B_{\Delta}(k,t). \tag{47}$$

By Eqs.(45), we drop the superscripts ${}^{LES}, {}^{MLRA}$ for Q, W, and Δ in what follows.

Since the fluid motion below the grid scale is not resolved in LES, the details of the fluid motion is lost, but with the help of the MLRA spectral theory we can infer how the spectrum behaves in the range of wavenumbers higher than k_c. Suppose that all knowledge of the velocity field except the energy spectrum is lost for wavenumbers greater than k_c. Then we have

$$W(k,t) = 0, \quad \Delta(k,t) = Q(k,t) \qquad \text{for } k > k_c. \tag{48}$$

When the domain of the (p,q) integrals in $B(k,t), B_W(k,t)$, and $B_{\Delta}(k,t)$ is split into the two parts Δ_1 and Δ_2, where Δ_2 is the domain such that p and/or q are larger than k_c, the contributions from terms containing W in Δ_2 domain vanishes, for example,

$$\iint_{\Delta_2} dpdq\; b_{kpq}\hat{\theta}_{kpq}(t) W(p,t) W(q,t) = 0. \tag{49}$$

Subtracting Eq.(46) from Eq.(47) and using Eqs.(48) and (49), and retaining the lowest order terms in Δ, we obtain

$$
\begin{aligned}
\mu_e(k|k_c)k^2 = &-\frac{1}{2Q(k,t)} \iint_{\Delta_1} dpdq\; b_{kpq}(\theta_{kpq}(t) - \hat{\theta}_{kpq}(t)) \\
&\qquad\qquad \times Q(q,t)[Q(p,t) - Q(k,t)] \\
&+ \frac{1}{2} \iint_{\Delta_2} dpdq\; b_{kpq}\theta_{kpq}(t) Q(q,t).
\end{aligned} \tag{50}
$$

Similarly by adding Eq.(46) to Eq.(47), we have the spectrum of the random force

$$D(k|k_c) = \frac{1}{2} \iint_{\Delta_2} dpdq\; b_{kpq}\theta_{kpq}(t) Q(p,t) Q(q,t). \tag{51}$$

Note that at the lowest order there is no contribution to the random force from the wavenumbers in Δ_1.

The present LES equations differ in two important respects from those of the previous studies. First, the triple relaxation factor is affected by the subgrid scale modeling. When the difference between θ_{kpq} and $\hat{\theta}_{kpq}$ is neglected, the expression for the eddy damping factor (50) reduces to that given by Bertoglio and by Chasnov [4,5]. Second, the present LES model can compute the error growth by computing the difference between pairs of velocity fields generated by two different sequences of random forces $\boldsymbol{f}^{<}(\boldsymbol{k},t)$. This means that the LES using Langevin-type modeling can compute both the average energy spectrum and a measure of predictability. Computation of a predictability measure has not been considered previously in the context of LES.

The eddy damping factor is obtained by solving Eq. (50) iteratively, but we consider only the zero-th order iteration for μ_e in what follows: neglecting the first term of (50), the eddy damping factor is

$$\mu_e(k|k_c)k^2 = \frac{1}{2}\iint_{\Delta_2} dpdq\ b_{kpq}\theta_{kpq}(t)Q(q,t). \tag{52}$$

The results obtained by considering corrections to μ_e will be reported elsewhere.

5 LES computation and comparison with DNS

We consider LES in the enstrophy cascading range of two-dimensional steady turbulence. First, DNS of the Navier-Stokes equation (2) is done by using the pseudo-spectral method in space and the 4-th order Runge-Kutta-Gill method for time advancing [13]. The external random force is assumed to be white in time and to obey Gaussian statistics with spectral support in a low wavenumeber band below $k = 10$. An additional damping factor effective only for wavenumbers below $k = 3$ is also introduced to prevent the inverse cascade of energy, so that the turbulence has an enstrophy cascade range. Total energy, total enstrophy, and Reynolds number are defined by

$$E = \frac{1}{2}\langle \boldsymbol{u}^2\rangle = \bar{u}^2, \qquad \Omega = \frac{1}{2}\langle \omega^2\rangle, \qquad R_L = \frac{\bar{u}L}{\nu} \tag{53}$$

respectively, where $L = 1/k_L = \bar{u}/\eta^{1/3}$ is the integral scale and η is the enstrophy dissipation rate. The numerical parameters and data are listed in Table 1.

It is useful to see how the enstrophy spectrum evolves when the subgrid model is neglected. Figure 1 shows the evolution of the enstrophy spectrum computed

Table 1. DNS parameters and statistical quantities

N	k_{max}	ν	R_L	E	Ω	η
1024^2	483	1.04×10^{-4}	1.48×10^{4}	3.40	163	11.9

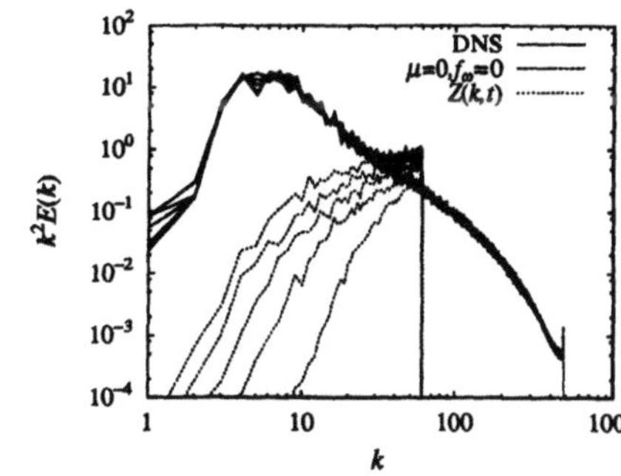

Fig. 1. Comparison of the enstrophy spectrum. Solid line: DNS, dashed line: DNS with cut off at $k = k_c$, dotted line: $Z(k,t)$. Lines are plotted at each one eddy turnover time.

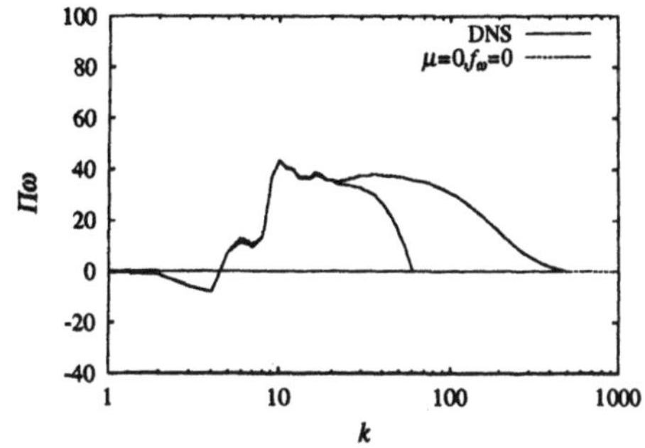

Fig. 2. Comparison of $\Pi_\omega(k,t)$ by DNS and by DNS with cut off.

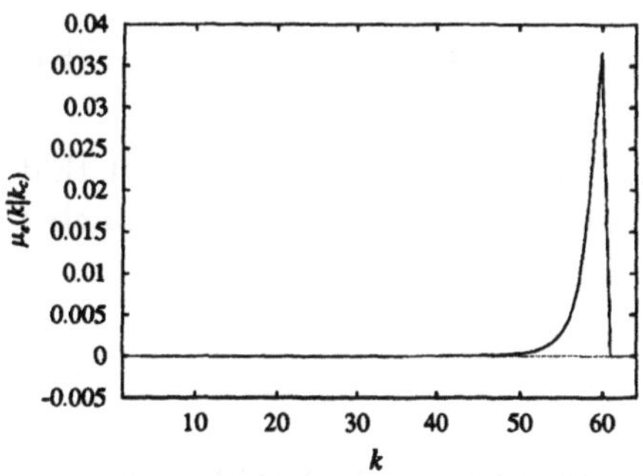

Fig. 3. Eddy damping factor.

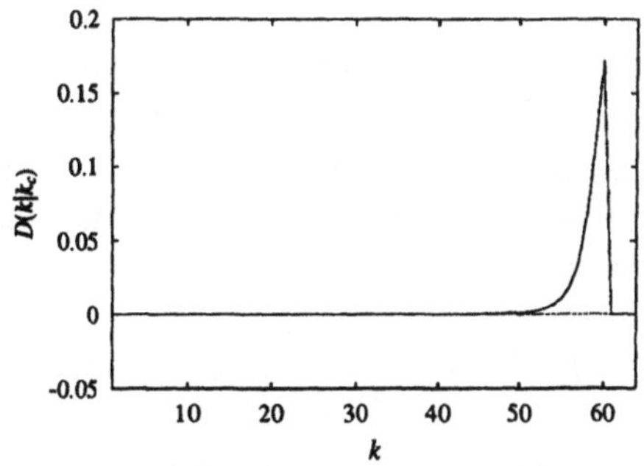

Fig. 4. Random force spectrum.

by DNS with truncation of all Fourier modes with wavenumbers higher than k_c. The spectrum rises near k_c and penetrates into lower wavenumber range as time goes on. The enstrophy transfer flux Π_ω begins to decay at about $k = 20$ as seen in Fig. 2. The resolution in the LES becomes worse at higher wavenumebers.

Since the DNS resolution is insufficient to reach the asymptotic energy spectrum $E(k) \propto k^{-3}[\ln(k/k_0)]^{-1/3}$, we cannot use the asymptotic spectrum to evaluate $\mu_e(k|k_c)$ and $D(k|k_c)$. Instead, when the DNS reaches a steady state, the energy spectrum obtained by the DNS is substituted into formulas (51) and (52). The results are shown in Figs. 3 and 4. The eddy damping factor increases rapidly as k tends to k_c, while it tends to a small negative constant when $k \ll k_c$. The random force spectrum also has a sharp rise near k_c and approaches zero for $k \ll k_c$. The LES equation is identical to the DNS equation in the low wavenumber range: it has the same damping factor to kill the inverse cascade and the same external force. For LES, the initial value of the vorticity field was taken from that of the DNS for wavevectors below k_c. Figure 5 compares the evolution of the enstrophy spectrum after one eddy turnover time. Also included is the

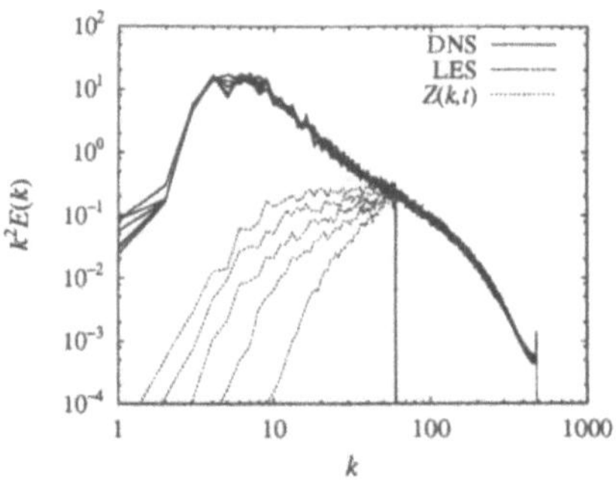

Fig. 5. Comparison of the enstrophy spectrum. Solid line: DNS, dashed line: LES, dotted line: $Z(k,t)$. Lines are plotted at each one eddy turnover time.

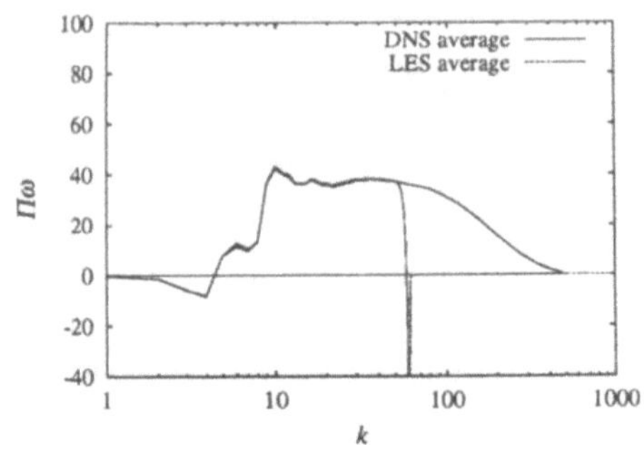

Fig. 6. Comparison of the time averaged $\Pi(k,t)$.

spectrum of the vorticity difference between LES and DNS:

$$Z(k,t) = \sum_{shell} \left\langle |\omega^{LES}(\boldsymbol{k},t) - \omega^{DNS}(\boldsymbol{k},t)|^2 \right\rangle. \tag{54}$$

The agreement of the enstrophy spectrum predicted by the present LES with that of DNS is very satisfactory. The enstrophy transfer flux functions in wavenumber space, defined as

$$\Pi^{DNS}(k) = -\int_0^k k'^2 T(k') dk', \tag{55}$$

$$\Pi^{LES}(k) = -\int_0^k \left[k'^2 T^{<}(k') - 2\mu_e(k'|k_c) k^4 E(k') + 2\pi k'^3 D(k'|k_c)\right] dk' \tag{56}$$

for DNS and LES respectively are plotted in Fig. 6. Again agreement between LES and DNS is excellent.

Figure 7 compares the vorticity contours predicted by LES and DNS at about 5 eddy turnover times. In the plot of the DNS vorticity, the Fourier components larger than k_c are set to zero to permit comparison with the LES which has a cutoff at k_c. General features of the two figures are (1) the vorticity patterns are very similar to each other; in particular, the position and intensity of strong vorticity regions resemble each other, but (2) small scales differ: despite the cutoff at k_c, the DNS contours have a finer structure than the LES contours. The difference is partly due to the random force in the LES that is destructive of the scales near $1/k_c$.

One important advantage of using Langevin-type modeling for the subgrid scales is that the present LES can compute the error growth, a measure of the predictability of the LES computation, which should be compared to that of the DNS. Suppose that the velocity field is known for grid scales larger than $1/k_c$ but the only information known about scales below $1/k_c$ is the energy spectrum [11,12]. The DNS initial condition is such that the velocity vectors with wavevectors $k \leq k_c$ are the same as those of DNS, while those for $k > k_c$

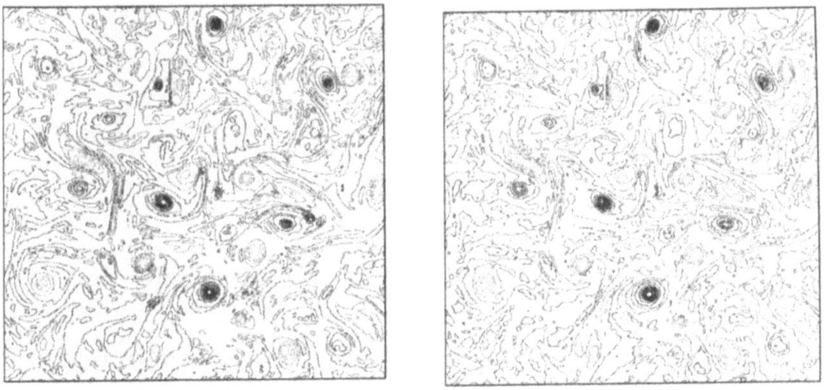

Fig. 7. Comparison of the vorticity contours by the DNS and LES at about five eddy turnover time. All the Fourier components higher than k_c in DNS are discurded. (a) DNS, (b) LES.

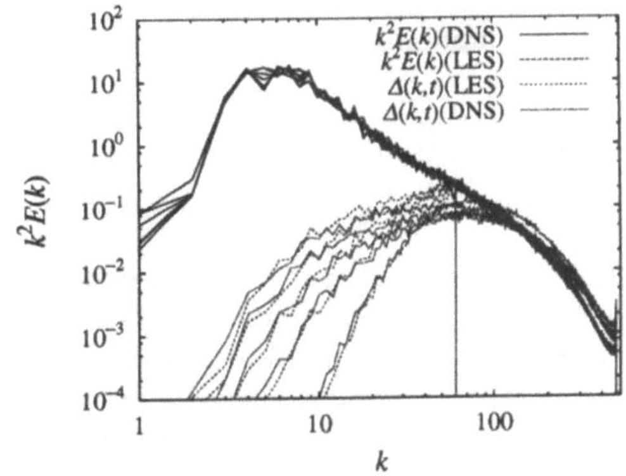

Fig. 8. Comparison of the evolution of the error spectrum $\Delta(k,t)$. Solid line: DNS, dashed line: LES. Lines are plotted at each one eddy turnover time.

are given by Gaussian random solenoidal vectors with the same spectrum as the DNS. Then the error given by the difference between pairs of realizations of the velocity field is computed by DNS. On the other hand, in the present LES, the error is defined by two velocity fields with two different realizations of the random forces representing the effects of the subgrid scales, so that the error growth found from the present LES can be compared to that of the DNS. Figure 8 compares the growth of the error spectrum. Although there are differences between DNS and LES near k_c, the evolution of $\Delta(k,t)$ predicted by the present LES agrees well with that predicted by DNS. In this sense, the present LES successfully computed the enstrophy spectrum as well as the predictability error.

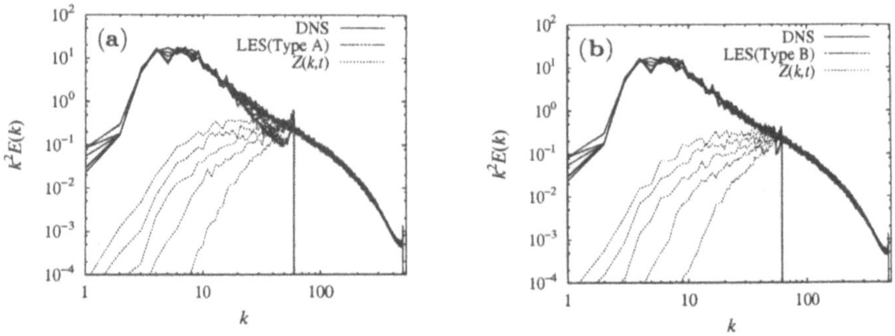

Fig. 9. Comparison of $k^2E(k,t)$ by LES's with two types of the eddy damping factor A and B at about five eddy turnover time. (a) Type A (b) Type B.

6 Simplification of the eddy damping factor

Although the eddy damping factor (52) and the random force (51) work well, their analytical expressions are not suitable for practical applications, which require a simpler, more convenient formula [14,15]. Here we consider some possible simplifications of the eddy damping factor. Two types of simplification have been considered:

$$\text{Type A: } \mu_e^A(k|k_c) = \mu_e(0|k_c) + ak^2, \quad (57)$$
$$\text{Type B: } \mu_e^B(k|k_c) = \mu_e(0|k_c) + bH(k - k_1), \quad (58)$$

where $H(x)$ is the Heaviside function, and a and b are constants to be determined, and k_1 is set to be 54 in the present study. The type A eddy damping factor implies 4th order hyperviscosity in physical space because $\mu_e(k|k_c)k^2 = ak^4$. Type B is only constant for wavenumbers near k_c; otherwise it is zero, reflecting the fact that $\mu_e(k|k_c)$ from (52) rises quickly near k_c. Constants a and b are determined by the relation

$$\int_0^{k_c} \mu_e^\alpha(k|k_c)k^4E(k)dk = \int_0^{k_c} \mu_e(k|k_c)k^4E(k)dk, \quad (59)$$

where α equals A or B. This states that the total amount of the enstrophy extracted by $\mu_e^\alpha(k|k_c)$ is equal to that extracted by $\mu_e(k|k_c)$, the MLRA eddy damping.

Figure 9 compares the enstrophy spectrum obtained from two LES runs with the DNS results. The enstrophy spectrum predicted by the type A model deviates from the DNS curves for $k > k_c/4$. On the other hand, the type B enstrophy spectrum closely follows the DNS enstrophy spectrum up to k near k_c. Despite the simplicity of formula (58), its performance is quite satisfactory. When the vorticity contour in Fig. 10 is compared to those of MLRA-LES and DNS in Fig. 10, agreement is satisfactory.

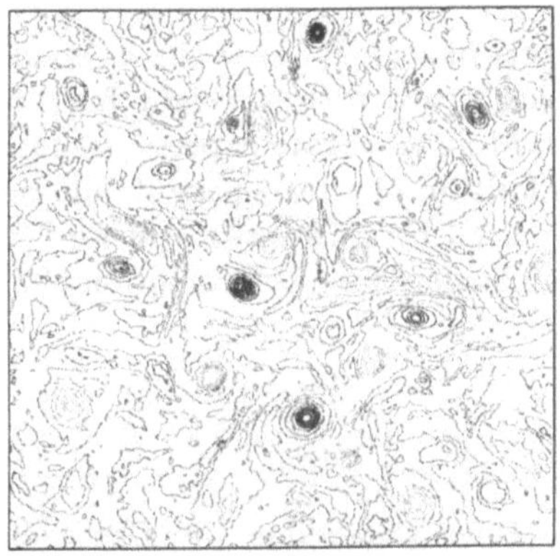

Fig. 10. Comparison of the vorticity contours by Type B LES at about five eddy turnover time.

7 Discussion and summary

We have studied a Langevin-type subgrid model in the two-dimensional enstrophy cascading range. By requiring that the evolution of the energy and error spectra predicted by LES be the same as that of MLRA, the eddy damping factor and random force intensity were obtained. Since the dynamics is altered by subgrid-scale modeling, the interaction among the wavenumber components below the cut off wavenumber differs from that of the Navier-Stokes equation, which means the triple relaxation time $\hat{\theta}_{kpq}(t)$ is not equal to that computed by MLRA in the case of no wavenumber cut off. To obtain the simplest model, the difference has been neglected in the present study.

It should be stressed that the eddy damping factor $\mu_e(k|k_c)$ is not the eddy viscosity which is defined by $\nu_e(k|k_c) = -T(k|k_c)/(2k^2E(k))$. When an asymptotic analysis is done for $(k_c - k)/k_c \ll 1$ at very high Reynolds number, it can be shown that both $\mu_e(k|k_c)$ and $D(k|k_c)$ diverge, but their net contribution to enstrophy transfer appears as a cusp and remains finite at k_c.

Comparison between the LES and DNS results show that the enstrophy spectrum and enstrophy transfer flux agree well with the DNS data. This is partly due to the fact that in the computation of $\mu_e(k|k_c)$ and $D(k|k_c)$, the energy spectrum for $k > k_c$ obtained by DNS is used.

One important feature of the present LES is that the error growth in the LES computation, a measure of predictability, can be computed by the LES itself. It was found that the error growth agrees well with that of DNS. This means that the present LES is capable of computing the evolution of the average energy spectrum with assurance of its predictability.

Also considered was a simplification of the eddy damping factor for practical applications. One simplified form is obtained by setting $\mu_e(k|k_c)$ to a constant over a range of wavenumbers near k_c, to zero elsewhere, and by determining the constant to yield the same enstrophy sink strength as that computed by

MLRA. The finite support of $\mu_e(k|k_c)$ near the k_c boundary implies a non-local eddy viscosity in physical space with a weight function, meaning an increase of computational effort. More elaborate expressions for both the eddy damping factor and the random force in physical space should be explored.

This work was supported by a Grant-in-Aid for "Research for the Future" Program from the Japan Society for the Promotion of Science under the project JSPS-RFTF97P01101.

References

1. S. B. Pope: *Turbulent Flows.* Cambridge. UK: Cambridge Univ. Press. (2000).
2. R. Kubo, M. Toda, and N. Hashitsume: *Statistical Phyasics II, Non-equilibrium statistical mechanics.* Springer-Verlag (1991).
3. R. H. Kraichnan: Eddy viscosity in two and three dimensions. J. Atmos. Sci. **33** 1521 (1976).
4. J. P. Bertoglio: A stochastic subgrid model for sheared turbulence. *Macroscopic modelling of Turbulent Flows* ed. by U. Frisch and J.B. Keller. pp.100 Springer (1984).
5. J. R. Chasnov: Simulation of the Kolmogorov inertial subrange using an improved subgrid model. Phys. Fluids **A 3** 188 (1991).
6. Y. Kaneda: Renormalized expansions in the theory of turbulence with the use of the Lagrangian position function. J. Fluid Mech. **107** 131 (1981).
7. T. Gotoh, Y. Kaneda and N. Bekki: Numerical integration of the Lagrangian renormalized approximation. J. Phys. Soc. Jpn., **57** 866 (1988).
8. Y. Kaneda: Inertial range structure of turbulent velocity and scalar fields in a Lagrangian renormalized approximation. Phys. Fluids **29** 701 (1986).
9. Y. Kaneda: Inertial range of two-dimensional turbulence in a Lagrangian renormalized approximation. Phys. Fluids **30** 2672 (1987).
10. T. Gotoh: Passive scalar diffusion in two dimensional turbulence in the Lagrangian renormalized approximation. J. Phys. Soc. Jpn, **58** 2365 (1989).
11. C. E. Leith: Atmospheric predictability and two-dimensional turbulence. J. Atmos. Sci. **28** 145 (1971).
12. C. E. Leith and R. H. Kraichnan: Predictability of turbulent flows. J. Atmos. Sci. **29** 1041 (1972).
13. T. Gotoh: Energy spectrum in the inertial and dissipation ranges of two-dimensional steady turbulence. Phys. Rev. E **57** 2984 (1998).
14. C. Basdevant, M. L essieur, and R. Sadourny: Subgrid-scale modeling of enstrophy transfer in two-dimensional turbulence. J. Atmos. Sci. **35** 1028 (1978).
15. O. Métais and M. Lessieur: Spectral simulation of isotropic and stably stratified turbulence. J. Fluid Mech. **239** 157 (1992).

LES Modelings based on the Lagrangian Renormalized Approximation

Kyo Yoshida, Takashi Ishihara, Daishi Fujita, Tomomichi Yamahira, and Yukio Kaneda

Graduate School of Engineering, Department of Computational Science and Engineering, Graduate School of Engineering, Nagoya University, Chikusa-ku, Nagoya 464-8603, Japan

Abstract. A systematic way for formulating large-eddy simulation (LES) models based on the Lagrangian renormalized approximation is presented. A spectral LES to simulate the energy spectrum and a probabilistic LES to simulate both the energy spectrum and the error spectrum are formulated for 3-dimensional turbulence and 2-dimensional turbulence with the inverse energy cascading range. The derived models are free from any ad hoc parameters. The models are verified using high resolution numerical simulations with 512^3 grid points for 3-dimension and 1024^3 for 2-dimension.

1 Introduction

Environmental and industrial flows, in general, have huge degrees of freedoms and wide ranges of time and length scales. There are practical needs to solve problems concerning such flows; to predict their time evolution with a given initial condition, to obtain their mean profiles averaged over long time period, etc. . With recent development of the computer hardware, the numerical simulations are acquiring the potential to cope with such problems. However, the most powerful supercomputers today still can not take account of the all degrees of freedoms involved in such flow systems. Therefore the number of degrees of freedoms must be reduced by some mean in the simulation . One of such methods is a large-eddy simulation (LES) in which small scale eddies which fall into subgrid scale are modeled, while large scale eddies are directly solved. Note that the sub-grid scale eddies whose individual information is lost are strongly coupled with the resolved scale eddies in general turbulent flows. Therefore one can hardly expect LES to simulate the individual flows exactly at resolved scales over a long time period. LES may only simulate some averaged quantities, or in other words statistical quantities, where the average may be taken over a spatial domain, a time period or some ensemble, depending on the flow under consideration. What statistical quantities LES can simulate should depend on the details of the modeling.

In this paper, a systematic way of developing LES models based on the Lagrangian renormalized approximation (LRA) is presented. The LRA is the simplest 2-point closures among few ones free from any ad hoc parameters and gives the proper Kolmogorov energy spectrum $E(k) = K_o\epsilon^{2/3}k^{-5/3}$ as its prediction up to the constant factor ($K_o = 1.72$) [1,2]. Here, k is the wavenumber and

ϵ is the energy dissipation rate per unit mass. The guiding principle to construct the LES models is the universality assumption, that is, although the large scale eddies of turbulence are strongly influenced by external conditions (geometry, type of boundary conditions and forcing), the statistics of small-eddies at sufficiently high Reynolds number are insensitive to the details of these external conditions. The assumption is widely supported by experiments and observations. The LES models are derived for when subgrid scales are in the universal scale range. The statistical quantities to simulate are specified first and then the form of the LES model is determined accordingly. The model parameters are estimated systematically and without any ad hoc parameters.

In Section 2, a spectral LES model which aims at simulating the energy spectrum is formulated and the result of the LES is compared with that of a high resolution direct numerical simulation (DNS). In Section 3, the concept of an error due to the uncertainty in the subgrid scales is introduced and a probabilistic LES (PLES) model which aims at not only simulating the energy spectrum but also estimating the error is formulated. The PLES result is compared with the corresponding result of high resolution DNS. In Section 4, LES and PLES models are formulated for 2-dimensional turbulence with inverse energy cascading scale range. In Section 5, further application of the present method is discussed.

2 Spectral LES model

We consider 3-dimensional statistically homogeneous turbulent velocity field which obeys the incompressible Navier-Stokes equations. The equations are given in spectral form by

$$\left(\frac{\partial}{\partial t}+\nu k^2\right)u_i(\boldsymbol{k})=M_{imn}(\boldsymbol{k})\sum_{\boldsymbol{k}=\boldsymbol{p}+\boldsymbol{q}}u_m(\boldsymbol{p})u_n(\boldsymbol{q})+f_i(\boldsymbol{k},t), \tag{1}$$

$$k_iu_i(\boldsymbol{k})=0 \tag{2}$$

with

$$M_{imn}(\boldsymbol{k})=-\frac{\mathrm{i}}{2}[k_mP_{in}(\boldsymbol{k})+k_nP_{im}(\boldsymbol{k})],\quad P_{ij}(\boldsymbol{k})=\delta_{ij}-\frac{k_ik_j}{k^2}. \tag{3}$$

Here, $\boldsymbol{u}(\boldsymbol{k}),\nu$, and $\boldsymbol{f}(\boldsymbol{k})$ are a velocity field, kinematic viscosity and an external solenoidal force field respectively. Here and hereafter, we use the summation convention for repeated indices.

Throughout this paper, we consider the sharp spectral filter. The filtered field $\boldsymbol{u}^<$ and the subgrid field $\boldsymbol{u}^>$ in wavevector space are defined by

$$\boldsymbol{u}^<(\boldsymbol{k})=\begin{cases}\boldsymbol{u}(\boldsymbol{k}) & k<k_c\\ \boldsymbol{0} & k\geq k_c\end{cases},\qquad \boldsymbol{u}^>(\boldsymbol{k})=\begin{cases}\boldsymbol{0}, & k<k_c\\ \boldsymbol{u}(\boldsymbol{k}) & k\geq k_c\end{cases}, \tag{4}$$

where k_c is a cut-off wavenumber. The model field in LES, which we will denote by $\tilde{\boldsymbol{u}}$, is supposed to simulate $\boldsymbol{u}^<$ in some statistical sense. In this section, we formulate a LES model equation which simulates the energy spectrum. The energy spectrum of the velocity field $\boldsymbol{u}$ is defined by

$$E(k) = \frac{1}{2} \sum_{k-1/2<|\boldsymbol{k}'|<k+1/2} \langle \boldsymbol{u}(\boldsymbol{k}') \cdot \boldsymbol{u}(-\boldsymbol{k}') \rangle, \tag{5}$$

with $\langle\rangle$ denoting a statistical average over an ensemble.

In order to simulate this quantity, we consider a LES model equation of the form

$$\left(\frac{\partial}{\partial t} + [\nu + \nu_e(k|k_c)]k^2 \right) \tilde{u}_i(\boldsymbol{k}) = M_{imn}(\boldsymbol{k}) \sum_{\boldsymbol{k}=\boldsymbol{p}+\boldsymbol{q}} \tilde{u}_m(\boldsymbol{p})\tilde{u}_n(\boldsymbol{q}) + f_i(\boldsymbol{k},t), \tag{6}$$

for $k < k_c$ and $\tilde{\boldsymbol{u}} = \boldsymbol{0}$ for $k \geq k_c$. Here k_c is the cut-off wavenumber and $\nu_e(k|k_c)$ is an eddy viscosity which depends on k. This form of LES model was proposed by Chollet and Lesieur [3].

We determine the eddy viscosity $\nu_e(k|k_c)$ using the Lagrangian renormalized approximation (LRA). The LRA provides a set of closed equations for Lagrangian 2-points 2-time correlation function and the averaged Lagrangian response function. See Refs. [1,2] for the details. We put $\boldsymbol{f} = \boldsymbol{0}$ in the formulation of the closure equations for simplicity. However, it can be checked that the present analysis is also valid in the presence of the external forcing when it is concentrated at large scales, i.e. $\boldsymbol{f}^< = \boldsymbol{f}$. We assume that the turbulence is statistically quasi stationary in the sense that the characteristic time of the energy spectrum $E(k,t)$ is much larger than that of the averaged Lagrangian response function $G(k,t,s)$ with k in the inertial subrange. The assumption validates a Markovian approximation and the LRA equations for isotropic turbulence are given as follows.

$$\left(\frac{\partial}{\partial t} + 2\nu k^2 \right) E(k) = T(k), \tag{7}$$

$$\left(\frac{\partial}{\partial t} + \nu k^2 + \eta(k,t,s) \right) G(k,t,s) = 0, \tag{8}$$

$$G(k,t,t) = 1, \tag{9}$$

where

$$T(k) = \int\!\!\int_\triangle dpdq\; k^3 pq b_{kpq} \theta_{kpq} q^{-2} E(q)[p^{-2}E(p) - k^{-2}E(k)], \tag{10}$$

$$\eta(k,t,s) = \frac{1}{2} \int\!\!\int_\triangle dpdq\; kpq \int_s^t ds' d_{kpq} G(q,t,s') q^{-2} E(q), \tag{11}$$

$$\theta_{kpq} = \int_{-\infty}^t ds\; G(k,t,s)G(p,t,s)G(q,t,s), \tag{12}$$

$$\int\!\!\int_\triangle dpdq = \int_0^\infty dp \int_{|p-k|}^{p+k} dq. \tag{13}$$

The geometric factors are given by

$$b_{kpq} = \frac{p}{k}(xy + z^3), \qquad d_{kpq} = (1 - y^2)(1 - z^2). \tag{14}$$

Here x, y, z are cosines of the interior angles opposite the triangle sides k, p, q respectively.

The LRA closure equations for the model (6) are given similarly. Here and hereafter, quantities associated with the model field $\tilde{u}$ will be denoted by $\tilde{\cdot}$.

$$\left(\frac{\partial}{\partial t} + 2\nu k^2\right) \tilde{E}(k) = \tilde{T}(k|k_c), \tag{15}$$

$$\left(\frac{\partial}{\partial t} + [\nu + \nu_e(k|k_c)]k^2 + \tilde{\eta}(k|k_c, t, s)\right) \tilde{G}(k, t, s) = 0, \tag{16}$$

$$\tilde{G}(k, t, t) = 1, \tag{17}$$

where

$$\tilde{T}(k|k_c) = \int\int_{\Delta_1} dpdq\; k^3 pq b_{kpq} \tilde{\theta}_{kpq} q^{-2} \tilde{E}(q)[p^{-2}\tilde{E}(p) - k^{-2}\tilde{E}(k)], \tag{18}$$

$$\tilde{\eta}(k|k_c, t, s) = \frac{1}{2}\int\int_{\Delta} dpdq\; kpq \int_s^t ds' d_{kpq} q^{-2} \tilde{G}(q, t, s') \tilde{E}(q, s'), \tag{19}$$

$$\tilde{\theta}_{kpq} = \int_{-\infty}^t \tilde{G}(k, t, s)\tilde{G}(p, t, s)\tilde{G}(q, t, s), \tag{20}$$

$$\int\int_{\Delta_1} dpdq = \int_0^{k_c} dp \int_{|p-k|}^{\min(p+k, k_c)} dq. \tag{21}$$

Note that the energy spectrum $E^<(k)$ of the filtered field $u^<$ is identical to $E(k)$ for $k < k_c$ and $E^<(k) = 0$ for $k \geq k_c$ since the sharp spectral filter is used. The requirement for the model is stated as follows.

Let the energy spectrum $\tilde{E}(k)$ of the model field simulate $E^<(k)$:

$$\tilde{E}(k) = E(k), \qquad \frac{\partial}{\partial t}\tilde{E}(k) = \frac{\partial}{\partial t}E(k), \qquad \text{for } k < k_c. \tag{22}$$

Substituting (22) into the closure equations (15)–(21) leads to

$$\nu_e(k|k_c) = -\frac{T(k) - \tilde{T}(k)}{2k^2 E(k)}. \tag{23}$$

Note that $\tilde{T}(k)$ depends on $\nu_e(k|k_c)$ through $\tilde{G}(k, t, s)$. The eddy viscosity $\nu_e(k|k_c)$ may be numerically estimated by iteration. Let

$$\tilde{G}^{(0)}(k, t, s) = G(k, t, s) \tag{24}$$

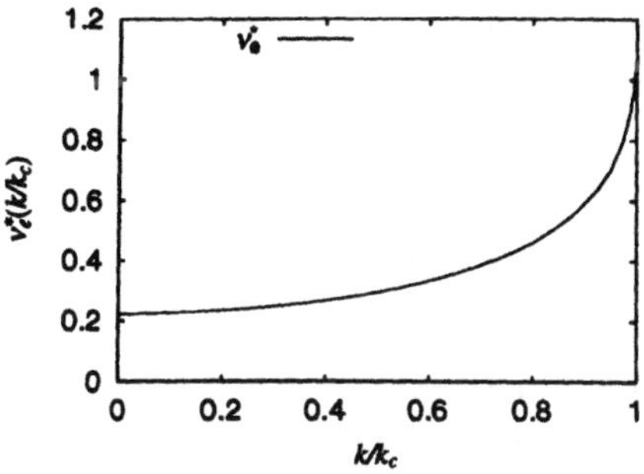

Fig. 1. Normalized eddy viscosity ν_e^*

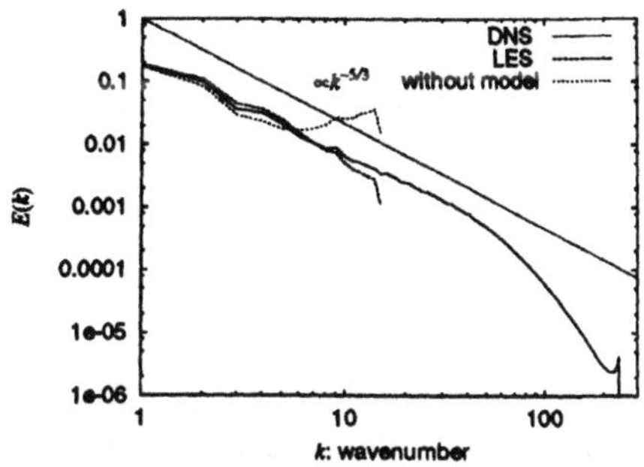

Fig. 2. The energy spectrum $E(k)$ of DNS and $\tilde{E}(k)$ of LES at $t = 2.25$.

be the 0-th iteration. The 0-th iteration of ν_e is obtained by substituting (24) into (23):

$$\nu_e^{(0)}(k|k_c) = \frac{1}{2E(k)} \int\int_{\Delta_2} dpdq\ kpqb_{kpq}\theta_{kpq}q^{-2}E(q)[k^{-2}E(k) - p^{-2}E(p)], \quad (25)$$

where the integral domain $\Delta_2 = \Delta - \Delta_1$. This formula coincide with the Kraichnan's definition of the eddy viscosity [4].

We consider the case when k_c is in the inertial subrange where energy spectrum is given by the universal form

$$E(k) = K_o\epsilon^{2/3}k^{-5/3}, \quad (26)$$

with K_o being the Kolmogorov constant. Although the form (26) may not hold for $k \gg k_c$ and $k \ll k_c$, we may substitute the form (26) into the closure equations when the inertial range is broad since p, q-integrals converges at both limits: p or $q \to 0$ and p or $q \to \infty$ and the contribution from the both limits is small. The kinematic viscosity ν is set to 0 since its contribution is small in the inertial subrange. The equation (8) with (9) and (11) is then solved numerically for $G(k, t, s)$ and a further analysis yields $K_o = 1.72$ [2].

The eddy viscosity may be normalized as

$$\nu_e^{(0)}(k|k_c) = \epsilon^{1/3}k_c^{4/3}\nu_e^*(k/k_c), \quad (27)$$

where $\nu_e^*(\cdot)$ is a dimensionless function. The normalized eddy viscosity ν_e^* was numerically computed from the LRA and is given in Fig. 1.

The response function $\tilde{G}^{(1)}$ of the next iteration is obtained by solving (16) with ν_e substituted by $\nu_e^{(0)}$. Let us compare the order of the second and the third terms in the l.h.s. of (16) at the stage of the 0-th iteration. Since $\nu_e^{(0)}(k|k_c)k^2 \sim C_0\epsilon^{1/3}k_c^{-4/3}k^2$ and $\eta(k,t,s) \sim C_1\epsilon^{1/3}k^{2/3}$ where the orders of C_0 and C_1 are unity, the ratio $\nu_e^{(0)}(k|k_c)k^2/\eta(k,t,s) \sim (k/k_c)^{4/3} < 1$. Therefore the difference between $\tilde{G}^{(0)}$ and $\tilde{G}^{(1)}$ may be small and the 0-th order iteration $\nu_e^{(0)}$ is expected to be an appropriate approximation already, especially for small

Table 1. The characteristic quantities of the 512^3 DNS field. R_λ: Taylor microscale Reynolds number, E: total energy, ϵ: energy dissipation rate, L_0: integral length scale, λ: Taylor microscale, η: Kolmogorov microscale

R_λ	E	ϵ	L_0	λ	η
284	0.500	7.15×10^{-2}	1.20	0.142	4.29×10^{-3}

k_c. In the following of this paper, $\nu_e^{(0)}$ is used in the place of ν_e in the model equation (6). The analysis of the further iteration is not given in this paper.

The energy dissipation rate ϵ in (27) must be evaluated within the model. In the present context, ϵ is interpreted as the energy transfer from the model field $\tilde{\boldsymbol{u}}$ to the subgrid field $\boldsymbol{u}^>$. Therefore the model value $\tilde{\epsilon}$ may be related to the eddy viscosity ν_e as

$$\tilde{\epsilon} = \int_{k<k_c} d\boldsymbol{k}\; 2\nu_e(k|k_c)k^2 \boldsymbol{u}(\boldsymbol{k}) \cdot \boldsymbol{u}(-\boldsymbol{k}). \tag{28}$$

Note that $\nu_e(k|k_c)$ in (28) depends on ϵ. Since ϵ is a spatially averaged quantity and expected to vary slowly with time, we use (28) to determine the value of ϵ at the next time step, namely,

$$\tilde{\epsilon}(t+\Delta t) = \int_{k<k_c} d\boldsymbol{k}\; 2\nu_e(k|k_c,t)k^2 \boldsymbol{u}(\boldsymbol{k}) \cdot \boldsymbol{u}(-\boldsymbol{k}), \tag{29}$$

$$\nu_e(k|k_c,t) = [\tilde{\epsilon}(t)]^{1/3} k_c^{-4/3} \nu_e^* \left(\frac{k}{k_c}\right), \tag{30}$$

where Δt is the time step interval. The complete set of the LES model equations is given by (6), (29) and (30).

The present LES model was tested by comparing the result with that of corresponding high resolution DNS. The DNS and the LES were performed in the same $(2\pi)^3$-periodic boundary box domain. The numbers of grid points are 512^3 and 32^3 for the DNS and the LES respectively. It was confirmed that the present DNS gives almost the same energy spectrum as that with 1024^3 grid points and the same viscosity [5,6]. Prior to the present DNS and LES, another DNS with 512^3 grid points was performed to simulate almost statistically stationary forced isotropic turbulence. The initial condition of the present DNS was given by a statistically almost stationary state of the prior DNS. The characteristic quantities of the initial field are listed in Table 1. The initial condition of the LES was generated by cutting off the high wavenumber modes $k > k_c$ of the DNS initial field. The cut-off wavenumber $k_c = 15$ was set to be in the region where the energy spectrum approximately obeys $k^{-5/3}$-power law. The external force $\boldsymbol{f}$ was given in the form of a negative friction restricted to small wavenumber modes:

$$\boldsymbol{f}(\boldsymbol{k}) = \begin{cases} \gamma \boldsymbol{u}(\boldsymbol{k}), & k < 2.5, \\ 0, & \text{otherwise}. \end{cases} \tag{31}$$

The coefficient γ was so chosen at each time step that the total energy is constant in the DNS, while it was fixed to the typical DNS value 0.136 in the LES. The kinematic viscosity $\nu = 2.89 \times 10^{-4}$ was same for the DNS and the LES. A 4-th order Runge-Kutta method were used for time advancing with $\Delta t = 1.25 \times 10^{-3}$ both in the DNS and the LES. An alias free spectral method was used in both the DNS and the LES. The initial value of $\tilde{\epsilon}$ was set to a small quantity 1×10^{-12}.

After a short initial transient time interval, the value of $\tilde{\epsilon}$ reaches the order of the DNS value of ϵ. The energy spectra $E(k)$ and $\tilde{E}(k)$ of the DNS and the LES respectively are almost stationary. Simulated $E(k)$ and $\tilde{E}(k)$ at $t = 2.25 = 1.08T$ are given in Fig. 2, where $T = L_0/\langle u_1^2\rangle^{1/2}$ is the eddy turn over time. The energy spectrum $\tilde{E}(k)$ is in good agreement with $E(k)$. This means that the energy transfer from the filtered field to the subgrid field is appropriately represented by the present model. The energy spectrum of a 32^3-grid-point simulation without model (i.e. $\nu_e(k) = 0$) at the same time is also shown in Fig. 2 for comparison. The energy accumulates in the wavenumber region near k_c in the simulation without model since it is not transfered to the subgrid scale.

3 Probabilistic LES

In many practical applications of numerical simulations, it is required to simulate not only the statistical averages such as the energy spectrum but also the individual realization of the flow. Such applications include for example the weather forecast. Although the energy spectrum can be simulated by the spectral LES as we have seen in Section 2, there is no guarantee that the individual realization of the real flow or the high resolution DNS can be simulated by the LES. Generally speaking, it is impossible to simulate the individual realization for a long time period in LES since the small eddies whose individual information is lost in LES are strongly coupled with the large eddies.

Let us consider two initial conditions $\boldsymbol{u}^{(1)}$ and $\boldsymbol{u}^{(2)}$ of isotropic turbulence with the identical filtered field, $\boldsymbol{u}^{(1)<} = \boldsymbol{u}^{(2)<}$, and different subgrid fields, $\boldsymbol{u}^{(1)>} \neq \boldsymbol{u}^{(2)>}$. As the flows evolve with time, it is expected that the two fields become different not only in the subgrid scales but also in the resolved scales, i.e. $\boldsymbol{u}^{(1)<} \neq \boldsymbol{u}^{(2)<}$. We will see later in this section that this is the case as a result from DNS. Let us assume in the following that the two fields $\boldsymbol{u}^{(\alpha)}(\boldsymbol{k}), \alpha = 1, 2$ are statistically identical and therefore we omit the index α for the statistical quantities such as $E(k)$ and $T(k)$. Following Leith and Kraichnan's concept [7], we introduce the error spectrum $\Delta(k)$ which is the half of the energy spectrum of the difference field $\delta\boldsymbol{u} = \boldsymbol{u}^{(1)} - \boldsymbol{u}^{(2)}$:

$$\Delta(k) = \frac{1}{4} \sum_{k-1/2<|\boldsymbol{k}'|<k+1/2} \langle \delta\boldsymbol{u}(\boldsymbol{k}') \cdot \delta\boldsymbol{u}(-\boldsymbol{k}') \rangle. \tag{32}$$

If the two fields $\boldsymbol{u}^{(1)}$ and $\boldsymbol{u}^{(2)}$ are completely correlated at the scale $1/k$, then $\Delta(k) = 0$ and if completely uncorrelated, $\Delta(k) = E(k)$. The error spectrum shows the limitation of the LES predictability of the individual realization, since if $\Delta(k,t) = E(k,t)$ at a certain scale $1/k$ and a certain time t, then there is no way that the LES can simulate $\boldsymbol{u}^{(1)}(\boldsymbol{k})$ or $\boldsymbol{u}^{(2)}(\boldsymbol{k})$ without further subgrid scale information at the initial time. If not only the energy spectrum but also the error spectrum is simulated in LES, then the LES can estimate its own reliability of the prediction. In this section, we aim to construct a LES model which can simulate both the energy spectrum and the error spectrum.

We employ a Langevin type model:

$$\left(\frac{\partial}{\partial t} + [\nu + \mu_e(k|k_c)]k^2\right) \tilde{u}_i^{(\alpha)}(\boldsymbol{k}) = M_{imn}(\boldsymbol{k}) \sum_{\boldsymbol{k}=\boldsymbol{p}+\boldsymbol{q}} \tilde{u}_m^{(\alpha)}(\boldsymbol{p})\tilde{u}_n^{(\alpha)}(\boldsymbol{q}) + f_i(\boldsymbol{k},t) + (f_e^{(\alpha)})_i(\boldsymbol{k}|k_c,t), \quad \alpha = 1,2 \tag{33}$$

as the LES model for the present purpose. Two realizations are necessary for this LES and α denotes the index of the realizations. Here, $\boldsymbol{f}_e$ is a statistically isotropic random forcing delta-correlated in time with a given correlation spectrum

$$F(k|k_c,t) = 4\pi k^2 \int_{-\infty}^{t} ds \langle \boldsymbol{f}_e^{(\alpha)}(\boldsymbol{k}|k_c,t) \cdot \boldsymbol{f}_e^{(\alpha)}(-\boldsymbol{k}|k_c,s)\rangle. \tag{34}$$

The random forcing $\boldsymbol{f}_e$ of different realizations are set to be uncorrelated. We call the model (33) the probabilistic LES (PLES) model to distinguish it from the model formulated in Section 2.

The problem is how to determine the functions $\nu_e(k|k_c)$ and $F_e(k|k_c)$. In this paper, the functions are determined from the explicit requirement that the model should simulate the energy spectrum and the error spectrum by using the LRA. In previous studies of PLES by Bertoglio [8] and Chasnov [9], the functions are determined in an intuitive manner by using the eddy damped quasi normal (EDQNM) closure and such requirement is not considered. The difference between the LRA and the EDQNM is discussed in Section 5.

The LRA equation for $\Delta(k)$ is given by

$$\left(\frac{\partial}{\partial t} + 2\nu k^2\right) \Delta(k) = S(k), \tag{35}$$

where

$$S(k) = \int\int_{\triangle} dpdq\, k^3 pq b_{kpq} \theta_{kpq} \times \{q^{-2}E(q)[p^{-2}\Delta(p) - k^{-2}\Delta(k)] + q^{-2}W(q)p^{-2}\Delta(p)\}, \tag{36}$$

with $W(k) = E(k) - \Delta(k)$.

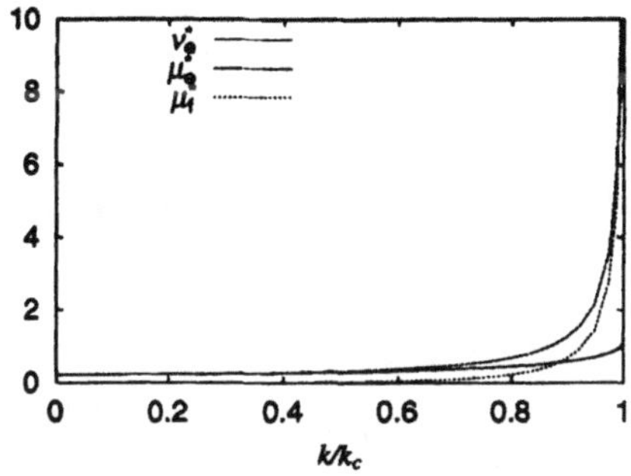

Fig. 3. Normalized eddy viscosity μ_e^* and eddy forcing μ_f^*.

The LRA closure equations for the model (33) are given as follows.

$$\left(\frac{\partial}{\partial t} + 2[\nu + \mu_e(k|k_c)]k^2\right) \tilde{E}(k,t) = \tilde{T}(k) + F_e(k|k_c), \tag{37}$$

$$\left(\frac{\partial}{\partial t} + 2[\nu + \mu_e(k|k_c)]k^2\right) \tilde{\Delta}(k,t) = \tilde{S}(k) + F_e(k|k_c), \tag{38}$$

$$\left(\frac{\partial}{\partial t} + [\nu + \mu_e(k|k_c)]k^2 + \tilde{\eta}(k|k_c,t,s)\right) \tilde{G}(k,t,s) = 0, \tag{39}$$

$$\tilde{G}(k,t,t) = 1, \tag{40}$$

where

$$\tilde{T}(k) = \int\!\!\int_{\Delta_1} dpdq\; k^3pqb_{kpq}\tilde{\theta}_{kpq}q^{-2}\tilde{E}(q)[p^{-2}\tilde{E}(p) - k^{-2}\tilde{E}(k)], \tag{41}$$

$$\tilde{S}(k) = \int\!\!\int_{\Delta_1} dpdq\; k^3pqb_{kpq}\tilde{\theta}_{kpq} \times\{q^{-2}\tilde{E}(q)[p^{-2}\tilde{\Delta}(p) - k^{-2}\tilde{\Delta}(k)] + q^{-2}W(q)p^{-2}\tilde{\Delta}(p)\}, \tag{42}$$

$$\tilde{\eta}(k,t,s) = \frac{1}{2}\int\!\!\int_{\Delta_1} dpdq\; kpq \int_s^t ds' d_{kpq}q^{-2}\tilde{G}(q,t,s')\tilde{E}(q). \tag{43}$$

Let $\Delta^<(k)$ be defined by the half of the energy spectrum of $\delta\boldsymbol{u}^< = \boldsymbol{u}^{(1)<} - \boldsymbol{u}^{<(2)}$. In a similar way as in Section 2, the requirement for the model is stated as follows.

Let the energy spectrum $\tilde{E}(k)$ and the error spectrum $\tilde{\Delta}(k)$ of the model field simulate $E^<(k)$ and $\Delta^<(k)$ respectively:

$$\tilde{E}(k) = E(k), \qquad \frac{\partial}{\partial t}\tilde{E}(k) = \frac{\partial}{\partial t}E(k),$$

$$\tilde{\Delta}(k) = \Delta(k), \qquad \frac{\partial}{\partial t}\tilde{\Delta}(k) = \frac{\partial}{\partial t}\Delta(k), \qquad \text{for } k < k_c. \tag{44}$$

Here, we assume that the two fields are uncorrelated at the subgrid scales, $\Delta(k) = E(k), k > k_c$. Then the above requirement (44) together with (7)-(13) and (35)-(43) leads to

$$\mu_e(k|k_c) = \int\int_{\Delta_2} dpdq\ k^{-1}pqb_{kpq}\theta_{kpq}q^{-2}E(q) + \frac{1}{W(k)}\int\int_{\Delta_1} dpdq\ kpqb_{kpq} \times(\theta_{kpq} - \tilde{\theta}_{kpq})q^{-2}E(q)[k^{-2}W(k) - p^{-2}W(p)], \tag{45}$$

$$F_e(k|k_c) = \int\int_{\Delta_2} dpdq\ k^3pq\theta_{kpq}q^{-2}E(q)p^{-2}E(p) + \frac{1}{W(k)}\int\int_{\Delta_1} dpdq\ kpqb_{kpq}(\theta_{kpq} - \tilde{\theta}_{kpq}) \times[2q^{-2}E(q)p^{-2}\Delta(p)E(k) - q^{-2}p^{-2}E(p)\Delta(k) - q^{-2}\Delta(q)p^{-2}\Delta(p)E(k)]. \tag{46}$$

Here we use the approximation $\tilde{G}(k) = G(k)$ as in Section 2. Then (45) and (46) reduce to

$$\mu_e(k|k_c) = \int\int_{\Delta_2} dpdq\ k^{-1}pqb_{kpq}\theta_{kpq}q^{-2}E(q), \tag{47}$$

$$F_e(k|k_c) = \int\int_{\Delta_2} dpdq\ k^3pq\theta_{kpq}q^{-2}E(q)p^{-2}E(p). \tag{48}$$

We assume that k_c is in the inertial subrange. Similarly in Section 2, μ_e and F_e may be normalized to the dimensionless functions μ_e^* and μ_f^* as

$$\mu_e(k|k_c) = \epsilon^{1/3}k_c^{4/3}\mu_e^*(k/k_c), \qquad F_e(k|k_c) = 2K_o\epsilon k_c^{5/3}\mu_f^*(k/k_c). \tag{49}$$

Note that $\mu_e^* - \mu_f^* = \nu_e^*$. The normalized functions μ_e^* and μ_f^* are numerically computed and shown in Fig. 3. n

Correction of $\mu_e(k)$ and $F_e(k)$ may be obtained by an iteration method with some extra assumptions. The similar method for 2-dimensional turbulence with the enstrophy cascading range is used by Gotoh *et al.* [10]. In this paper, we use (47) and (48) as a first approximation and their correction is briefly discussed at the end of this section.

The model value of $\tilde{\epsilon}$ is simulated in a similar way as in Section 2. In PLES, $\tilde{\epsilon}$ may be evaluated as the difference between the energy transfer from the model

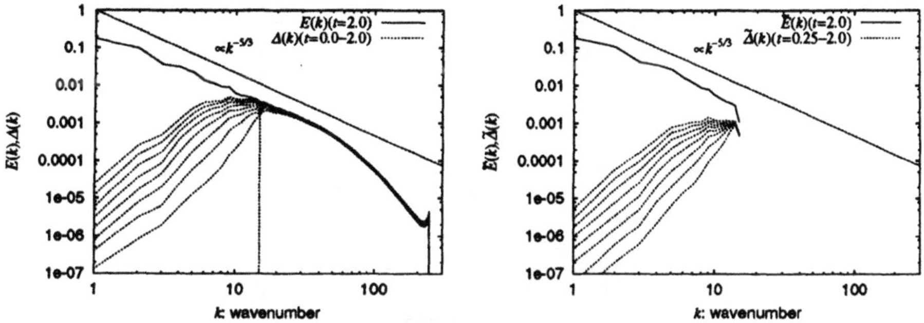

Fig. 4. Energy spectrum $E(k)$ and error spectra $\Delta(k)$ of the DNS (left) and $\tilde{E}(k)$ and $\tilde{\Delta}(k)$ of PLES (right). The error spectra are plotted with the time interval 0.25.

field to the subgrid scale field due to μ_e and the energy input due to $\boldsymbol{f}_e$:

$$\tilde{\epsilon}(t+\Delta t) = \int_{k<k_c} d\boldsymbol{k}\ [2\mu_e(k|k_c,t)k^2\boldsymbol{u}(\boldsymbol{k})\cdot\boldsymbol{u}(-\boldsymbol{k}) - \frac{1}{2}\boldsymbol{f}_e(\boldsymbol{k}|k_c)\cdot\boldsymbol{f}_e(-\boldsymbol{k}|k_c)\Delta t], \tag{50}$$

$$\mu_e(k|k_c,t) = [\tilde{\epsilon}(t)]^{1/3}k_c^{-4/3}\mu_e^*\left(\frac{k}{k_c}\right), \tag{51}$$

$$f_e(k|k_c,t) = 2K_o\tilde{\epsilon}(t)k_c^{5/3}\mu_f^*\left(\frac{k}{k_c}\right), \tag{52}$$

where Δt is the time step interval. The equations (33),(50),(51) and (52) form the complete set of PLES model equations.

The PLES model was tested by comparing the result with that of corresponding high resolution DNS. Two DNS were performed with different initial conditions. Let $\boldsymbol{u}^{(1)}$ and $\boldsymbol{u}^{(2)}$ denote the velocity fields in these simulations. The initial condition of $\boldsymbol{u}^{(1)}$ is same as that of the DNS in Section 2. The initial condition of $\boldsymbol{u}^{(2)}$ was generated by copying $\boldsymbol{u}^{(1)}(\boldsymbol{k})$ for the modes with $k < k_c$ and phase-randomizing $\boldsymbol{u}^{(1)}(\boldsymbol{k})$ for the modes $k > k_c$, where the amplitudes were maintained. Consequently $\Delta(k) = 0$ $(k < k_c)$ and $\Delta(k) = E(k)$ $(k \geq k_c)$ holds at $t = 0$. The wavenumber $k_c = 15$ is set to be in the wavenumber range where the initial energy spectrum exhibits nearly $k^{-5/3}$-power law. The same k_c was used as the cut-off wavenumber in the PLES. The initial condition of $\tilde{\boldsymbol{u}}^{(1)}$ and $\tilde{\boldsymbol{u}}^{(2)}$ in the PLES were identical and generated by cutting off the high wavenumber modes $k > k_c$ of $\boldsymbol{u}^{(1)}$. The random force $\boldsymbol{f}_e$ was imposed at every time step with no correlation between different time steps. All the other run conditions of the DNS and the PLES are the same as those of the DNS and the LES in Section 2.

The simulated energy spectra $E^{(1)}(k)$ and $E^{(2)}(k)$ in DNS are almost identical and stationary. The time development of $\Delta(k)$ is shown in the left figure in Fig.4. It is observed that the error due to the uncertainty of the small scale propagates from small scales to larger scales. The behavior of $\Delta(k)$ would worth detailed analysis. However the subject is beyond the scope of this paper. The right figure in Fig.4 shows $\tilde{E}(k)$ and $\tilde{\Delta}(k)$ simulated in the PLES. The qualitative

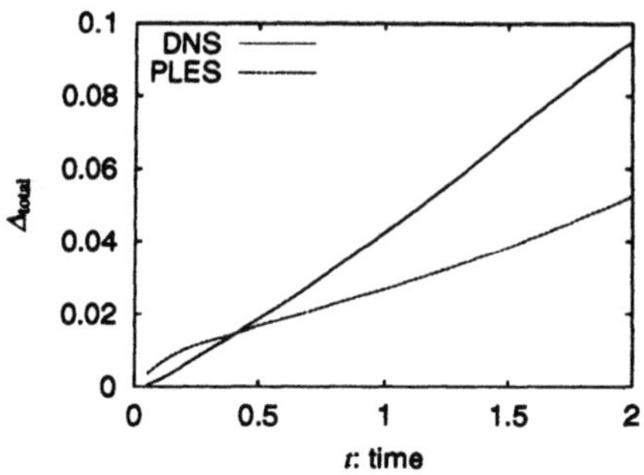

Fig. 5. Total error Δ_{total} of DNS and PLES

behavior of $\tilde{\Delta}(k)$ is in agreement with that of $\Delta(k)$. A quantitative comparison was made for the total error defined by $\Delta_{\text{total}} = \int_0^{k_c} dk\Delta(k)$. The total errors normalized by the the energy of the filtered field $E^< = \int_0^{k_c} dkE(k)$ of the DNS and the LES are given in Fig. 5. The PLES underestimates the total error $\tilde{\Delta}_{\text{total}}$ approximately by a factor of 2 in comparison with Δ_{total} in the DNS.

The correction of $\mu(k|k_c)$ by an iteration method will reduce $\mu(k|k_c)$ at $k \sim k_c$ since $\tilde{G}(k,t,s)$ decays faster in time $(t - s)$ than $G(k,t,s)$ due to the eddy viscosity $\mu_e(k|k_c)$ and therefore $\tilde{\theta}_{kpq} < \theta_{kpq}$. The correction of $F_e(k|k_c)$ may be small if $\Delta(k) \ll E(k)$ for $k < k_c$. The error spectrum $\tilde{\Delta}(k)$ in the PLES would increase by the correction. However, its significance is questionable since the asymptotic behavior of $\mu_e^*(x) \sim C(1-x)^{-2/3}, x \to 1$ where C is a constant is not modified by the correction. The reason of the gap in $\Delta(k)$ and $\tilde{\Delta}(k)$ is left to be examined in the future.

4 Application to 2-dimensional turbulence with inverse energy cascading range

4.1 Model equations

It is possible to formulate the LES and PLES models for 2-dimensional turbulence using the LRA in a similar way as in 3-dimensional turbulence. There are two possible inertial scaling ranges in isotropic 2-dimensional turbulence, which are the inverse energy cascading range with $k^{-5/3}$ energy spectrum and the enstrophy cascading range with k^{-3} energy spectrum. The LES and PLES models are formulated in this paper for the case when the cut-off wavenumber k_c lays in the former range. The model for the case when k_c lays in the latter range is discussed by Gotoh *et al.* in this proceedings [10].

The concept of the model formulation for 2-dimensional turbulence is almost the same as that for 3-dimensional turbulence. The LES and PLES model equations are given by (6) and (33), respectively. By using the same approximation $\tilde{G}(k,t,s) = G(k,t,s)$ as in Sections 2 and 3, $\nu_e^{(2D)}(k|k_e)$, $\mu_e^{(2D)}(k|k_e)$ and

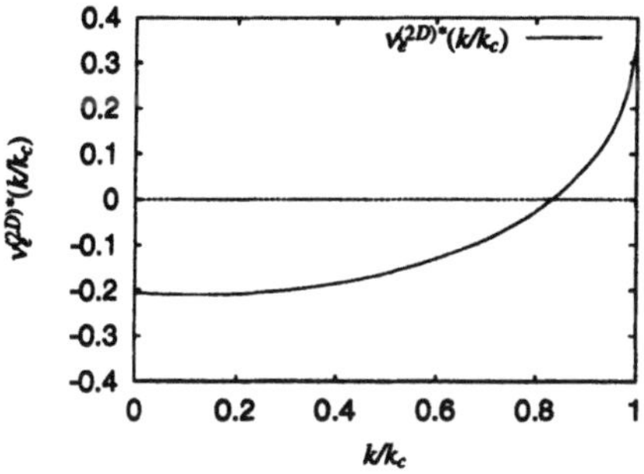

Fig. 6. Normalized eddy viscosity ν_e^* for LES

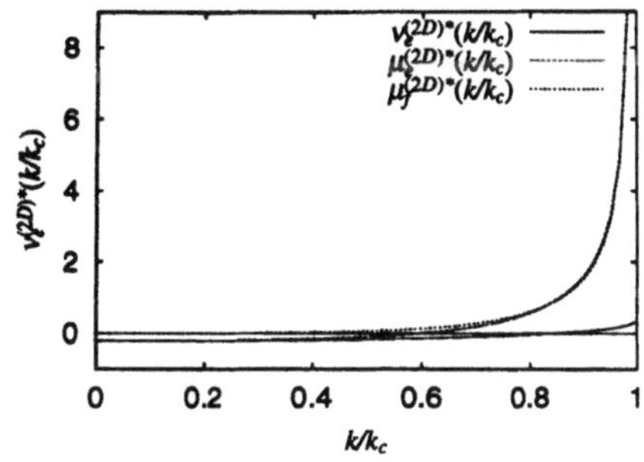

Fig. 7. Normalized eddy viscosity μ_e^* and eddy force μ_f^* for PLES.

$F_e^{(2D)}(k|k_e)$ in the model equations can be estimated by the LRA. (See Ref. [11] for the complete set of the LRA closure equations for the 2-dimensional isotropic turbulence.)

$$\nu_e^{(2D)}(k|k_c) = \frac{1}{2\pi E(k)}\int\int_{\Delta_2} dpdq\ kb_{kpq}^{(2D)}\theta_{kpq} \times q^{-1}E(q)[k^{-1}E(k) - p^{-1}E(p)], \quad (53)$$

$$\mu_e^{(2D)}(k|k_c) = \frac{1}{2\pi}\int\int_{\Delta_2} dpdq\ b_{kpq}^{(2D)}\theta_{kpq}q^{-1}E(q), \quad (54)$$

$$F_e(k|k_c) = \frac{1}{\pi}\int\int_{\Delta_2} dpdq\ k^3 b_{kpq}^{(2D)}\theta_{kpq}p^{-1}E(p)q^{-1}E(q), \quad (55)$$

where θ_{kpq} is defined by (12) and the response function $G(k,t,s)$ in 2-dimensional turbulence which satisfies (8) with

$$\eta(k,t,s) = \frac{1}{\pi}\int\int_{\Delta} dpdq\ k^2 \int_s^t ds' d_{kpq}^{(2D)}G(q,t,s')q^{-1}E(q). \quad (56)$$

The geometric factors for 2-dimension are given by

$$b_{kpq}^{(2D)} = 4k^{-4}(p^2 - q^2)(k^2 - q^2)(1 - x^2)^{1/2}, \quad (57)$$

$$d_{kpq}^{(2D)} = 2(1 - y^2)(1 - z^2)(1 - x^2)^{-1/2}. \quad (58)$$

In the inverse energy cascading range, the energy spectrum is given by

$$E(k) = K_o^{(2D)}\epsilon^{2/3}k^{-5/3}. \quad (59)$$

Note that ϵ denote the energy transfer from subgrid-scale field $\boldsymbol{u}^>$ to filtered field $\boldsymbol{u}^<$. The LRA estimates the Kolmogorov constant to be $K_o^{(2D)} = 7.41$. Substituting (59) with the LRA estimate of $K_o^{(2D)}$ in (53), (54) and (55) yields the following expressions:

$$\nu_e^{(2D)}(k|k_c) = \epsilon^{1/3}k_c^{-4/3}\nu_e^{(2D)*}(k/k_c), \quad (60)$$

$$\mu_e^{(2D)}(k|k_c) = \epsilon^{1/3}k_c^{-4/3}\mu_e^{(2D)*}(k/k_c), \quad (61)$$

$$F_e^{(2D)}(k|k_c) = 2K_o^{(2D)}\epsilon k_c^{5/3}\mu_f^{(2D)*}(k/k_c), \quad (62)$$

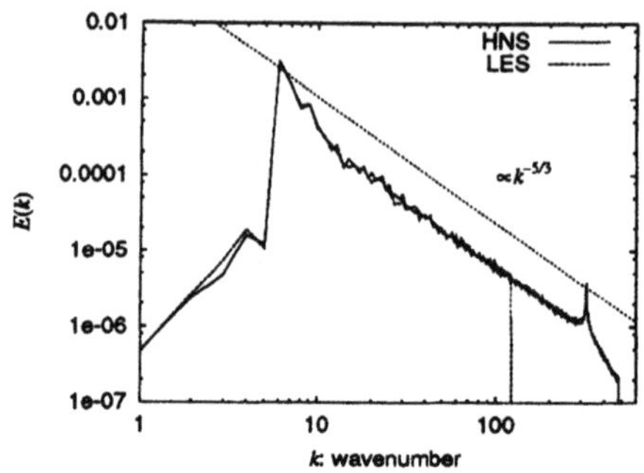

Fig. 8. The energy spectrum $E(k)$ of HNS and $\tilde{E}(k)$ of LES at $t = 20 = 170T$

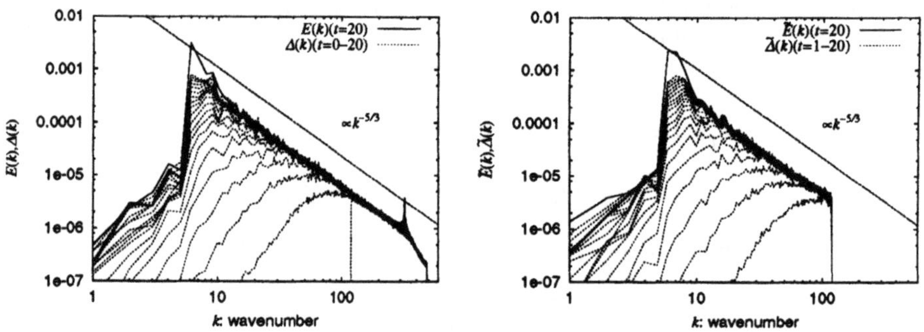

Fig. 9. Energy spectrum $E(k)$ and error spectra $\Delta(k)$ of the HNS (left) and $\tilde{E}(k)$ and $\tilde{\Delta}(k)$ of the PLES (right). The error spectra are plotted with the time interval 1.

where the normalized eddy viscosities $\nu_e^{(2D)*}$ and $\mu_e^{(2D)*}$ and the normalized eddy force $\mu_f^{(2D)*}$ are dimensionless functions. These functions were numerically computed and are shown in Fig. 6. Note that both eddy viscosities $\nu_e^{(2D)}(k|k_c)$ and $\mu_e^{(2D)}(k|k_c)$ are negative for small k/k_c. This means that the eddies whose scales are much larger than $1/k_c$ receive energy from the subgrid scale eddies.

Since the energy is transfered from the subgrid scale to the resolved scale, the transfer ϵ is determined by the subgrid field $\boldsymbol{u}^>$ rather than by the filtered field $\boldsymbol{u}^<$. We assume statistical stationarity in the subgrid scale and put $\nu_e^{(2D)}(k|k_c)$, $\mu_e^{(2D)}(k|k_c)$ and $F_e^{(2D)}(k|k_c)$ to be constant in time. We replace ϵ in (60), (61) and (62) by $[E(k_c)/(K_o^{(2D)})]^{3/2}k_c^{5/2}$ with $E(k_c)$ obtained from the initial data of the high resolution numerical simulation.

The LES and PLES of 2-dimensional turbulence with the inverse energy cascade range were performed with 256^2 grid points ($k_c = 120$) and the results were compared with the corresponding high resolution numerical simulations (HNS) with 1024^4 grid points. By using the similar procedures as those in Section 2 and 3, the initial conditions of HNS, LES and PLES were generate from an almost statistically stationary state of forced 2-dimensional turbulence with the energy inverse cascading range obtained in Ref. [12] ($\beta = 0$ in Series E). See

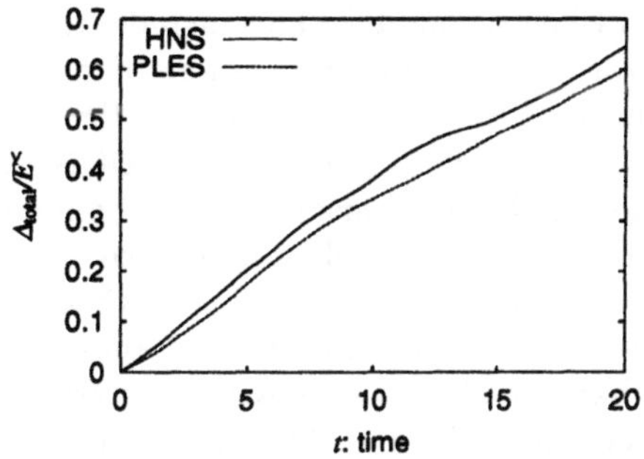

Fig. 10. Total error Δ_{total} of DNS and PLES

the reference for the detailed setting for the HNS. The settings for LES and PLES are the same with the HNS except for the number of grid points, the eddy viscosity and the eddy forcing.

Fig. 8 shows the energy spectrum $\tilde{E}(k)$ of LES compared with $E(k)$ of the higher resolution simulation at $t = 20 = 170T$, where the characteristic time T is defined by the inverse of the r.m.s. of the vorticity field. The agreement of $\tilde{E}(k)$ with $E(k)$ is satisfactory, which means that the energy transfer from the subgrid scales to larger scales are well modeled by the present eddy viscosity $\nu_e^{(2D)}$. Figure 9 shows $E(k)$ and $\Delta(k)$ simulated in the HNS and $\tilde{E}(k)$ and $\tilde{\Delta}(k)$ simulated in the PLES. Figure 10 shows the time evolution of the total error Δ_{total} normalized by $E^{<}$ for both the HNS and LES. From the figures, it can be seen that $\tilde{E}(k)$ and $\tilde{\Delta}(k)$ are qualitatively and quantitatively in good agreement with $E(k)$ and $\Delta(k)$ respectively. The agreement is better than in the 3-dimensional case.

5 Discussion

A systematic way of formulating equations for LES models is given in this paper. Once the statistical quantities to simulate and the type of the LES model are fixed and the universality in the subgrid scales is known, the model parameters are estimated using the LRA without any ad hoc parameters. In this paper, we employed the spectral eddy viscosity model to simulate the energy spectrum and the Langevin type model to simulate both the energy spectrum and the error spectrum. Assumed universality in the subgrid scales is the Kolmogorov energy spectrum for 3- or 2- dimensional isotropic turbulence.

The previous studies of the spectral LES (Chollet and Lesieur [3], Métais and Lesieur [13]) and the PLES (Bertoglio [8], Chasnov [9]) are based on the EDQNM closure. Unlike in the LRA, the response function G is not directly solved and a certain form is assumed for θ_{kpq} in the EDQNM. Consequently, Kolmogorov constant K_o depends on a parameter which can not be determined within the closure.

The advantage of the modelings based on the LRA to those of the EDQNM is that the model parameters are completely derived from the closure. The point may be more elucidated when the model which includes the effect of anisotropy in the subgrid scales are considered. Some applications of the EDQNM to anisotropic turbulence were studied by Cambon [14]. However, the anisotropic effect upon the response function is not known a priori in the EDQNM. On the other hand, the LRA is in principle capable for anisotropic and inhomogeneous turbulence although the computation becomes much complicated.

The next application of the present modeling method would be to formulate a LES model to simulate the mean flow and the anisotropic spectrum in turbulence. Recently, Ishihara *et al.* [15] suggested a certain universal form of the anisotropic energy spectrum in shear turbulence. Such anisotropic energy spectrum can be substituted into the LRA closure equations to estimate the effect of subgrid scale anisotropy to the resolved scale flow. The detailed analysis will be reported elsewhere.

Acknowledgment

The DNS and LES were performed using a Fujitsu VPP5000/56 computer at the Nagoya University Computation Center. The present study was supported by the "Research for the Future" Program of the Japan Society for the Promotion of Science under project JSPS-RFTF97P01101.

References

1. Y. Kaneda: J. Fluid Mech. **107**,131 (1981)
2. Y. Kaneda: Phys. Fluid **29**, 701 (1986)
3. J. P. Chollet and M. Lesieur: J. Atmos. Sci. **38**, 2747 (1981)
4. R. H. Kraichnan: J. Atmos. Sci. **33**, 1521 (1976)
5. Y. Yamazaki, T. Ishihara, and Y. Kaneda: J. Phys. Soc. Jpn. **71**,777 (2002)
6. T. Ishihara and Y. Kaneda: present volume
7. C. E. Leith and R. H. Kraichnan: J. Atmos. Sci. **29**, 1041 (1972)
8. J. P. Bertoglio: 'Stochastic subgrid model for sheared turbulence'. In *Macroscopic Modelling of Turbulent Flows*, ed. by U. Frisch, J. B. Keller, G. C. Papanicolaou, and O. Pironneau, (Berlin Heidelberg New York Tokyo, 1984. Springer-Verlag) pp. 100–119
9. J. R. Chasnov: Phys. Fluids A **3**, 188 (1991)
10. T. Gotoh, I. Kakui, and Y. Kaneda: present volume
11. Y. Kaneda: Phys. Fluid **30**, 2672 (1987) erratum: see the reference in Ishihara and Kaneda, Phys. Fluids, **13**, 1431 (2001)
12. T. Ishihara and Y. Kaneda: Phys. Fluids **13**, 2338 (2001)
13. O. Métais and M. Lesieur: J. Fluid Mech. **239**, 157 (1992)
14. C. Cambon: present volume
15. T. Ishihara, K. Yoshida, and Y. Kaneda: Phys. Rev. Lett. **88**, 154501 (2002)

LES of Stably Stratified Turbulence

Kyo Yoshida, Takashi Ishihara, and Yukio Kaneda

Graduate School of Engineering, Department of Computational Science and Engineering, Graduate School of Engineering, Nagoya University, Chikusa-ku, Nagoya 464-8603, Japan

Abstract. Large-eddy simulations (LES) of forced uniformly stably stratified turbulence with grid points up to 512^3 were performed with the degrees of stratification such that both nonlinear turbulent interactions and internal gravity wave motions play essential roles. The degrees of stratification correspond to those of the ocean turbulence at length scale $O(10)$[m]. It is assumed that at the subgrid scales, the flow is nearly isotropic and that the density fluctuation is approximated by a passive scalar. The LES model equations are derived by using the Lagrangian renormalized approximation which is free from any ad hoc parameter. Energy density of the simulated flow in the wavevector space is strongly anisotropic. The vertical shear spectrum is in qualitative agreement with the measurements in the ocean by Gargett *et al.*. However, the slope of the spectrum in the simulation ($\propto k^{-1/3}$) is shallower than that by the measurement ($\propto k^{-1}$).

1 Introduction

Flows in atmosphere and ocean are markedly anisotropic due to geophysical effects. Among the effects are the separation between the typical scales in horizontal direction ($\sim O(10^3)$[km]) and those in the vertical direction ($\sim O(10^0)$[km]), rotation of the Earth, density stratification, etc. . Although it is widely accepted since the 1941 Kolmogorov theory that the statistics of turbulence at very high Reynolds number and at sufficiently small scales is universal and isotropic, the statistical non-universality and anisotropy prevail in a vast scale range in geophysical turbulences. In ocean turbulence, the effect of stratification is probably the one that penetrate most into the small scales among the geophysical effects. The wavenumber (inverse of the length scale) where the typical time scale of nonlinear eddy interactions and that of the internal gravity waves balance is estimated to be $O(10^0)$[cpm] (e.g. See Ref. [1]). Therefore it is expected that the statistics of ocean turbulence is approximated by that of the universal isotropic turbulence in the wavenumber range $> O(10^0)$[cpm] and is non-universal and anisotropic in the wavenumber range $< O(10^0)$[cpm] depending on individual geophysical conditions.

However, there is evidence for some degree of universality in ocean turbulence not only at wavenumbers larger than $O(10^0)$[cpm] but also at wavenumbers smaller than $O(10^0)$[cpm]. The internal wave statistics up to vertical wavenumber of $O(10^{-1})$[cpm] is well described by the empirical Garret-Munk spectrum [2] and a certain form of spectrum is suggested for the intermediate range $O(10^{-1}) \sim O(10^0)$[cpm] by measurements (See Ref.[1] and references cited therein). The mechanisms underlying this universal spectrum in not well understood.

Numerical simulations should give us some insight into the mechanism. But the degree of freedoms that the computers can handle is limited and the most powerful computer today still can not simulate directly from gross scales (larger than $O(100)$[m]) to micro scales ($\sim O(10^{-2})$[m]) in the ocean turbulence. In order to simulate the scales of the present interest ($> O[10^0]$[m]), the smaller scales should be modeled to reduce the degree of freedom. At sufficient small scales, it is expected that the velocity field is approximated by the isotropic turbulence and that the density fluctuating field by passive scalar.

In this paper, we derive a large eddy simulation (LES) model for passive scalar turbulence using a method based on the Lagrangian renormalized approximation (LRA) which was proposed in Ref. [3]. Then the model is applied to simulate homogeneous stratified turbulence in which the degree of stratification corresponds to that of small scales ($O(10^1)$[m]) in the ocean turbulence.

Besides the fundamental numerical studies of stably stratified turbulence (e.g. Refs [4–6]), the studies along the context of the ocean turbulence include those by Ramsden and Holloway [7], Siegel and Domaradzki [8], Hibiya *et al.* [9] and Carnevale *et al.* [10]. The setting of the present numerical simulations is similar to those by Carnevale *et al.*. The difference from their simulations are the larger number of grid points (up to 512^3), the manner of the forcing at large scales and the use of the LRA eddy viscosity and eddy diffusivity.

In Section 4, the results from the present simulations are compared with the measurements in the ocean by Gargett *et al.* and the simulations by Carnevale *et al.*.

2 Model equations

We consider statistically homogeneous turbulent velocity and density fields in a uniform mean density gradient that satisfy the Boussinesq approximation. The model equations are given in spectral form by

$$\left(\frac{\partial}{\partial t} + (\nu_e(k|k_c) + \nu)k^2\right) u_i(\boldsymbol{k}) + \rho\delta_{i3} = M_{imn}(\boldsymbol{k}) \sum_{\boldsymbol{p}+\boldsymbol{q}=\boldsymbol{k}} u_m(\boldsymbol{p})u_n(\boldsymbol{q}) + F_i^u(\boldsymbol{k}) \quad (1)$$

$$\left(\frac{\partial}{\partial t} + (\kappa_e(k|k_c) + \kappa)k^2\right) \rho(\boldsymbol{k}) - N^2 u_3(\boldsymbol{k}) = K_m(\boldsymbol{k}) \sum_{\boldsymbol{p}+\boldsymbol{q}=\boldsymbol{k}} u_m(\boldsymbol{p})\rho(\boldsymbol{q}) + F^\rho(\boldsymbol{k}) \quad (2)$$

$$k_i u_i(\boldsymbol{k}) = 0, \quad (3)$$

where

$$M_{imn}(\boldsymbol{k}) = -\frac{i}{2}\left[k_m P_{in}(\boldsymbol{k}) + k_n P_{im}(\boldsymbol{k})\right], \quad K_m(\boldsymbol{k}) = -ik_m, \tag{4}$$

$$P_{ij}(\boldsymbol{k}) = \left(\delta_{ij} - \frac{k_i k_j}{k^2}\right). \tag{5}$$

Here, $\boldsymbol{u}$ is the velocity field and $u_i (i = 1, 2, 3)$ denote velocity components on orthonormal basis vectors $(\boldsymbol{i}_1, \boldsymbol{i}_2, \boldsymbol{i}_3)$ where $\boldsymbol{i}_3$ is anti-parallel to the vertical gravitational acceleration $\boldsymbol{g}$; $\rho = g\rho'/\rho_0$ where ρ_0 and ρ' are the reference density and the density deviation from the mean density $\bar{\rho}(x_3)$ respectively; $N = \sqrt{-(g/\rho_0)(d\bar{\rho}/dx_3)}$ is the Brunt-Väisälä frequency which is assumed to be uniform in x_3; ν and κ are the kinematic viscosity coefficient and the molecular diffusivity coefficient, respectively. The external forcing and injection of ρ are denoted by $\boldsymbol{F}^u$ and F^ρ respectively. In the simulations, only the modes with wavenumber smaller than the cut-off wavenumber k_c are solved directly and the effect of subgrid modes $k \geq k_c$ are modeled through wavenumber-dependent eddy viscosity $\nu_e(k|k_c)$ and eddy diffusivity $\kappa_e(k|k_c)$.

In sufficiently high wavenumber range, say $k > k_0$, it is expected that the effect of stratification is small and that the statistics of $\boldsymbol{u}$ and ρ are approximated by that of the velocity field and the passive scalar field, respectively, in the isotropic turbulence. In the following, we consider the case that k_c is located in the range $k_0 < k_c < k_1$, where k_1 is the higher end of the inertial subrange. Let the kinematic energy spectrum, potential energy spectrum and buoyancy flux spectrum be defined by

$$E(k) = \frac{1}{2} \sum_{k-1/2<|\boldsymbol{k}'|<k+1/2} \boldsymbol{u}(\boldsymbol{k}') \cdot \boldsymbol{u}(-\boldsymbol{k}'), \tag{6}$$

$$E_P(k) = \frac{1}{2N^2} \sum_{k-1/2<|\boldsymbol{k}'|<k+1/2} \rho(\boldsymbol{k}')\rho(-\boldsymbol{k}'), \tag{7}$$

$$F_B(k) = - \sum_{k-1/2<|\boldsymbol{k}'|<k+1/2} u_3(\boldsymbol{k}')\rho(-\boldsymbol{k}'), \tag{8}$$

respectively. We assume the inertial-range kinematic energy spectrum

$$E(k) = K_o \epsilon^{2/3} k^{-5/3} \tag{9}$$

for $k_0 < k < k_1$, where ϵ is the kinematic energy dissipation rate and K_o is the Kolmogorov constant. Similarly, the potential energy spectrum is assumed to take the form proportional to the inertial-range passive scalar spectrum:

$$E_P(k) = C_o \epsilon_P \epsilon^{1/3} k^{-5/3} \tag{10}$$

in the same wavenumber range $k_0 < k < k_1$, where ϵ_P is the potential energy dissipation rate and C_o is a constant. Both K_o and C_o are estimated from the Lagrangian renormalized approximation (LRA) to be 1.72 and 0.34, respectively

[11], however the latter value is not used in the subsequent computation. The amplitude of buoyancy flux $F_B(k)$ is assumed to be sufficiently small and can be neglected for $k_0 < k < k_1$.

The method to determine the function $\nu_e(k|k_c)$ by using the LRA is given in Ref.[3]. Here, we determine the function $\kappa_e(k|k_c)$ following the similar method. The equations for the potential energy spectrum may be written as follows,

$$\left(\frac{\partial}{\partial t} + 2\kappa k^2\right) E_P(k) = T_P(k) \tag{11}$$

where $T_P(k)$ represent the potential energy transfer due to the advection. Here, the terms due to the buoyancy, the external force $\boldsymbol{F}^u$ and the injection F^ρ are omitted for simplicity. Such simplification and the following analysis are validated when the forcing $\boldsymbol{F}^u(\boldsymbol{k})$, and the injection $F^\rho(\boldsymbol{k})$ have no subgrid component and the amplitude of $F_B(k)$ is sufficiently small in the wavenumber range $k_0 < k < k_1$. The amplitude of $F_B(k)$ is not necessarily small at low wavenumbers $k \ll k_0$.

Let the quantities associated with the model equations (1) and (2) denoted by $\tilde{\cdot}$. The LRA equation of the potential energy spectrum associated with the model is given by

$$\left(\frac{\partial}{\partial t} + 2[\kappa + \kappa_e(k|k_c)]k^2\right) \tilde{E}_P(k) = \tilde{T}_P(k). \tag{12}$$

where the terms due to $\boldsymbol{F}^u, F^\rho$ and the buoyancy are neglected for the same reason mentioned above. The requirement for the model to simulate the potential energy spectrum:

$$\tilde{E}_P(k) = E_P(k), \qquad \frac{\partial}{\partial t}\tilde{E}_P(k) = \frac{\partial}{\partial t}E_P(k), \tag{13}$$

leads to

$$\kappa_e(k|k_c) = -\frac{T_P(k) - \tilde{T}_P(k|k_c)}{2k^2 E_P(k)}. \tag{14}$$

In the computation of $\nu_e(k|k_c)$ in Ref.[3], the Lagrangian response function of the model velocity field $\tilde{G}(k,t,s)$ was approximated by that of the original velocity field $G(k,t,s)$. Using the same approximation and the similar approximation for the response function of ρ, $\kappa_e(k|k_c)$ is expressed in the following form.

$$\kappa_e(k|k_c) = -\frac{1}{2E_P(k)}\int\!\!\int_{\Delta>} dpdq\ kpq(1-y^2)$$
$$\times[p^{-2}E_P(p,t) - k^{-2}E_P(q,t)]q^{-2}E(q)\int_{-\infty}^{0} ds' G(q,t,s'), \tag{15}$$

where $\int\int_{\Delta>} dpdq = \int_{k_c-k}^{k_c} dp \int_{k_c}^{p+k} dq + \int_{k_c}^{\infty} dp \int_{p-k}^{p+k} dq$, and x, y, z are the cosine of the interial angles opposite the triangle sides k, p, q respectively. (See Ref.

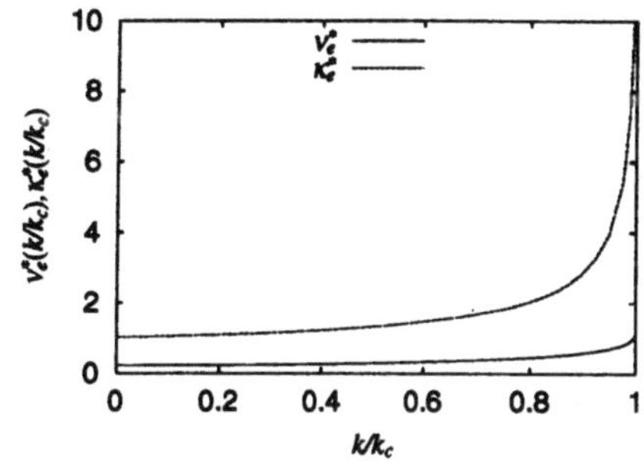

Fig. 1. Normalized eddy viscosity ν_e^* and eddy diffusivity κ_e^*.

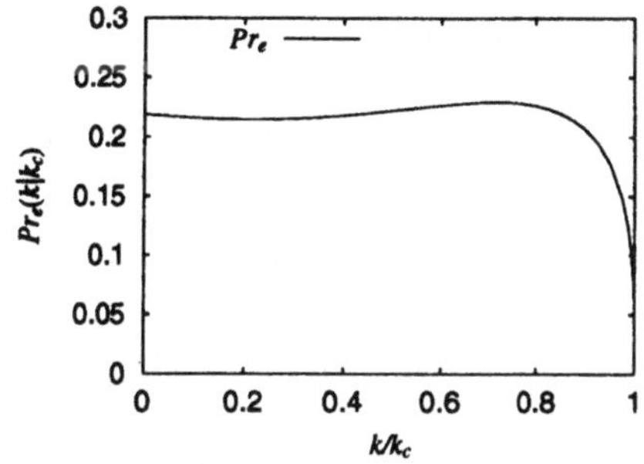

Fig. 2. Eddy Prandtl number $Pr_e(k|k_c)$.

[11] for the detailed derivations of the LRA equations for the passive scalar turbulence.)

When the inertial-range forms of $E(k), E_P(k)$ and $G(k)$ are substituted into (15), $\kappa_e(k|k_c)$ may be normalized as follows,

$$\kappa_e(k|k_c) = \epsilon^{1/3} k_c^{4/3} \kappa_e^*(k/k_c), \tag{16}$$

where $\kappa_e^*(\cdot)$ is a dimensionless function. The normalized eddy diffusivity κ_e^* was numerically computed and is shown in Fig. 1 together with the normalized eddy viscosity ν_e^* defined by $\nu_e(k|k_c) = \epsilon^{1/3} k_c^{4/3} \nu_e^*(k/k_c)$, and computed in Ref.[3].

The eddy diffusivity $\kappa_e(k|k_c)$ is quite larger than the eddy viscosity $\nu_e(k|k_c)$. Unlike $\nu_e^*(k/k_c)$, $\kappa_e^*(k/k_c)$ diverges at $k/k_c = 1$. The asymptotic formula for $x = k/k_c \to 1$ is given by $\kappa_e^*(x) \sim 2.23(1-x)^{-1/3}$. The eddy Prandtl number defined by $Pr_e(k|k_c) = \nu_e(k|k_c)/\kappa_e(k|k_c)$ is given in Fig. 2. The eddy Prandtl number is smaller than 0.23 for the entire range $0 < k < k_c$ and $Pr_e(k|k_c) \to 0$ as $k/k_c \to 1$. The estimate is much smaller than that from the eddy damped quasi normal Markovian (EDQNM) closure, which is $Pr_e(k|k_c) \sim 0.6$. (See Ref. [12] and the references cited therein.) Note that $Pr_e(k|k_c)$ in the EDQNM closure depends on parameters which can not be fixed from the theory alone. On the other hand, the derivation of $Pr_e(k|k_c)$ in the LRA needs no parameterization. However, there is a possibility that the LRA is underestimating Pr_e, or in other words, overestimating κ_e since the Lagrangian response function for the density fluctuation field is not renormalized and do not damp the T_P term in the LRA. The LRA for the passive scalar turbulence possibly improves by changing so-called "representatives" to close the equations. Such an improvement is a theoretical challenge. Although the closure problem of the passive scalar turbulence is hardly settled (not at least to the level of that of the velocity field turbulence), in the following section, we apply the present eddy diffusivity κ_e to LES of stratified turbulence with expectation that qualitative characters of the statistics may be simulated by the LES.

In LES, the kinematic energy dissipation rate ϵ may be interpreted as the kinematic energy transfer from the resolved wavenumber range $k < k_c$ to the subgrid wavenumber range $k \geq k_c$. The kinematic energy dissipation rate ϵ must be evaluated within the model. The model value $\tilde{\epsilon}$ is determined in the simulation by following the method in Ref. [3]:

$$\tilde{\epsilon}(t+\Delta t) = \int_{k<k_c} dk\ 2\nu_e(k|k_c,t)k^2\boldsymbol{u}(\boldsymbol{k})\cdot\boldsymbol{u}(-\boldsymbol{k}), \tag{17}$$

$$\nu_e(k|k_c,t) = [\tilde{\epsilon}(t)]^{1/3}k_c^{-4/3}\nu_e^*\left(\frac{k}{k_c}\right), \tag{18}$$

$$\kappa_e(k|k_c,t) = [\tilde{\epsilon}(t)]^{1/3}k_c^{-4/3}\kappa_e^*\left(\frac{k}{k_c}\right), \tag{19}$$

where Δt is the time step interval.

3 Numerical experiments

The LES based on the model equations (1)–(3) and (17)–(19) were performed in a periodic boundary $(2\pi)^3$-box domain. An alias free spectral method with the cut-off wavenumber k_c was used with a fast Fourier transform. The time was advanced by using a fourth-order Runge-Kutta method. The initial velocity and density fluctuation fields are almost isotropic. The energy is equally partitioned within modes with a wavevector, that is, the kinetic energy is twice as large as the potential energy for every wavevector. The total (kinetic + potential) energy spectrum are concentrated around a peak wave number $k_p = 2$. The kinematic viscosity coefficient ν and the molecular diffusivity coefficient κ are both 0. The initial $\tilde{\epsilon}$ is set to be a small quantity 1×10^{-12}. The external forcing and injection of density fluctuation were put by amplifying the smallest wavevector modes with $1.5 \le k \equiv |\boldsymbol{k}| \le 2.5$:

$$F_i^u(\boldsymbol{k}) = cu_i(\boldsymbol{k}), \qquad F^\rho(\boldsymbol{k}) = c\rho(\boldsymbol{k}), \tag{20}$$

where a time-dependent and wavevector-independent value c is so determined to keep the total energy constant.

The LES were firstly performed with resolution of $128^3(k_c = 60.3)$ up to a certain time when the energy was cascaded to modes with highest wavenumber and the energy spectrum became quasi stationary. Then the simulations were continued with higher resolution of $256^3(k_c = 121)$ until the energy spectrum reached quasi stationary state. Finally, the simulations with resolution of $512^3(k_c = 241)$ were performed using the final states of the 256^3 simulation fields as the initial conditions. The simulations were performed with different Brunt-Väisaälä frequencies, $N = \pi, 2\pi$ and 3π.

The buoyancy wavenumber $k_b = N^{3/2}\epsilon^{-1/2}$ at which the time scale of nonlinear turbulent interactions and that of the internal gravity wave balance are $k_b = 27, 99$ and 231 for the statistically quasi stationary state of the simulations with $N = \pi, 2\pi$ and 3π, respectively. In order not to violate the assumption that the subgrid scale is approximated by the isotropic turbulence, k_b should be at least smaller than $k_c = 241$. Therefore we did not venture the simulation with larger N.

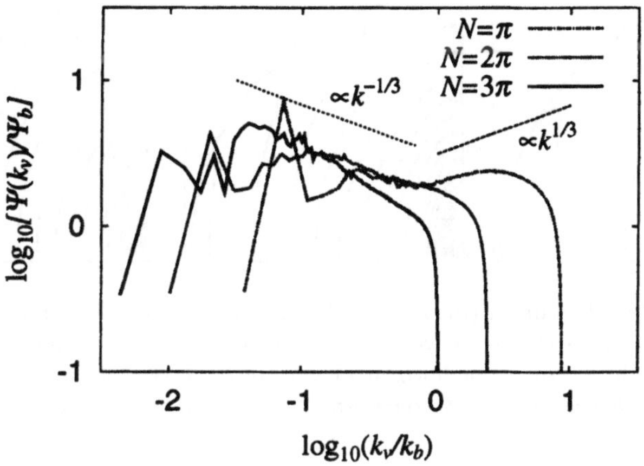

Fig. 3. Normalized vertical shear spectrum of the simulations with different N.

A spectrum of vertical shear in the ocean was obtained from measurements by Gargett *et al.* [1]. The corresponding vertical shear spectrum in the present simulation is defined by

$$\Psi(k_v) = \sum_{k_v - 1/2 \le k'_v < k_v + 1/2} {k'_v}^2 [|u_1(\mathbf{k}')|^2 + |u_2(\mathbf{k}')|^2], \tag{21}$$

where $k_v = |k_3|$ is the vertical wavenumber. The vertical shear spectra $\Psi(k_v)$ of statistically quasi-stationary states in the simulations are given in Fig. 3. The wavenumber and the vertical shear spectrum are normalized by $k_b = N^{3/2}\epsilon^{-1/2}$ and $\Psi_b = N^2/k_b$, respectively in the figure. Since the energy is pumped into the lowest wavenumber range and the statistics in the range directly feels the individual manner of the forcing, the range should be discarded in the analysis of universality. The fall of spectrum from power-law straight line at the highest wavenumber range $k \sim k_c$ is the artifact of the spherical simulation domain $|\boldsymbol{k}| = k < k_c$ in the wavevector space. It is suggested from the intermediate-range spectra of the simulations with different N that there are two scaling ranges. One is $k > k_b$ where the vertical shear spectrum scales as $\Psi(k_v) \propto k_v^{1/3}$. The scaling corresponds to Kolmogorov scaling for the energy spectrum $E(k) \propto k^{-5/3}$ and it is expected that the turbulence is quasi isotropic in this wavenumber range. The change of the slope in the shear spectrum is observed at $k \sim k_b$. In the other scaling range $k < k_b$, the vertical shear spectrum approximately scales as $\Psi(k_v) \propto k_v^{-1/3}$. Comparison of the result with that of the ocean measurements is given in Section 4.

Numerical simulations have the advantage that the detailed flow data are easily obtained in comparison with the measurements in real ocean. One can investigate not only 1-dimensional spectra but also how the energy is distributed in (k_h, k_v) space, where $k_h = \sqrt{k_1^2 + k_2^2}$ is the horizontal wavenumber. Note that for each wavevector $\boldsymbol{k}$, there are three degrees of freedom, two for $\boldsymbol{u}(\boldsymbol{k})$ and one

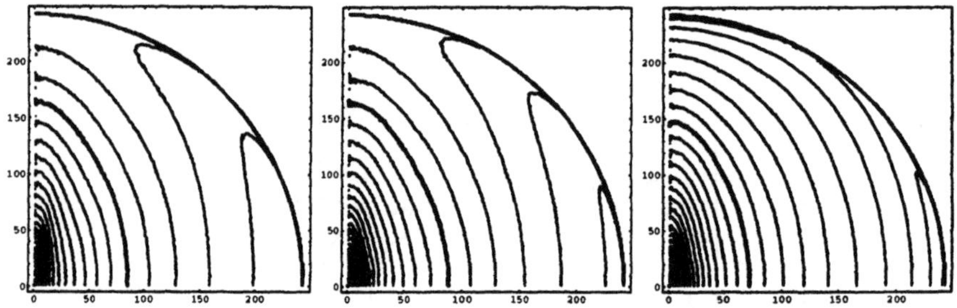

Fig. 4. Energy densities of $E_V(k_v, k_h)$ (left), $E_W(k_v, k_h)$ (middle) and E_P (right) normalized by $k^{-5}N^2$ averaged over wavevectors with same (k_h, k_v) and the time period $9.8 < t \leq 10.6$. The numbers in horizontal and vertical axes show $k_h + 1$ and $k_v + 1$ respectively. The levels of the contours are 2^n where n is a integer. Thick contours show the level 2^0.

for $\rho(\boldsymbol{k})$. A decomposition of $\boldsymbol{u}(\boldsymbol{k})$ is given by

$$\boldsymbol{u}(\boldsymbol{k}) = \phi_1(\boldsymbol{k})\boldsymbol{e}^1(\boldsymbol{k}) + \phi_2(\boldsymbol{k})\boldsymbol{e}^2(\boldsymbol{k}), \tag{22}$$

where

$$\boldsymbol{e}^1(\boldsymbol{k}) = \frac{\boldsymbol{i}_3 \times \boldsymbol{k}}{|\boldsymbol{i}_3 \times \boldsymbol{k}|}, \qquad \boldsymbol{e}^2(\boldsymbol{k}) = \frac{\boldsymbol{k} \times \boldsymbol{e}^1(\boldsymbol{k})}{|\boldsymbol{k} \times \boldsymbol{e}^1(\boldsymbol{k})|} \tag{23}$$

[13]. In the linearized non-viscous non-diffusive system, $\phi_1(\boldsymbol{k})$ does not oscillate in time while $\phi_2(\boldsymbol{k})$ oscillates with the frequency Nk_h/k. In the following, we call $\phi_1(\boldsymbol{k})$ and $\phi_2(\boldsymbol{k})$ vortical component and wave component, respectively, just for convenience. Here, the terminologies 'vortical' and 'wave' should not be taken too literally. The energy density in wavevector space for vortical kinematic component, wave kinematic component and potential component are defined as

$$E_V(\boldsymbol{k}) = \frac{1}{2}|\phi_1(\boldsymbol{k})|^2, \quad E_W(\boldsymbol{k}) = \frac{1}{2}|\phi_2(\boldsymbol{k})|^2, \quad E_P(\boldsymbol{k}) = \frac{1}{2N^2}|\rho(\boldsymbol{k})|^2, \tag{24}$$

respectively. The energy densities may be averaged over $\boldsymbol{k}$ with the same (k_h, k_v) and reduce to functions of (k_h, k_v). In Fig. 4, the averaged energy densities $E_V(k_h, k_v), E_W(k_h, k_v)$, and $E_P(k_h, k_v)$ normalized by $k^{-5}N^2$ are shown for a statistically quasi stationary state averaged over a time interval of 0.8 in the $N = 3\pi$ simulation with 512^3 grid points. The buoyancy wavenumber in the period is $k_b = 231$ which is only slightly smaller than the cut-off wavenumber $k_c = 241$. Therefore anisotropic stratification effect is significant almost over the entire simulation domain $k < k_c$. It is found from Fig. 4 that all three components of the energy are concentrated near the k_v axis. The result is consistent with the previous numerical studies. (See Ref.[14] and the references cited therein.) It is interesting that vortical kinetic component which feels the anisotropic stratification effect not directly but only indirectly through nonlinear interaction also develops to a strongly anisotropic distribution.

4 Discussion

The vertical shear spectrum $\Psi(k_v)$ obtained in the present LES is comparable with that by the measurements in the ocean by Gargett *et al.* [1]. They are in qualitative agreement in the sense that there are two scaling ranges $k_v < k_0$ and $k_v > k_0$ where $\Psi(k_v)$ is a decreasing and increasing function of k_v, respectively and that $k_0 \sim k_b$. But the slope of the spectrum $\Psi(k_v) \propto k_v^{-1/3}$ for $k < k_0$ obtained in the simulation is shallower than the slope $\Psi(k_v) \propto k_v^{-1}$ obtained in Ref. [1]. The spectrum in the LES which was performed by Carnevale *et al.* [10] exhibit somewhat steeper slope in the buoyancy wavenumber range $k < k_b$ than the present LES although its scaling range is quite narrow. The major difference between the present simulation and that in Ref. [10] is the forcing in the small wavenumber range. In the LES of Ref. [10], the energy was supplied to the system by fixing the amplitude of a certain standing wave.

There may be 2 types of scaling in the wavenumber range $k < k_b$ depending on whether local or nonlocal interaction in wavenumber space dominates the dynamics of the scale range. If so, the scaling in the range can be sensitive to the larger scale statistics. In this paper, we refrain from going into a dimensional argument to derive the scaling exponent for the both cases since the strongly anisotropic distribution of the energy in the wavevector space would not permit a simple dimensional analysis. Further detailed data analysis of the energy distribution in (k_h, k_v)-space would give us hints for understanding the mechanism. The analysis is now underway and will be reported elsewhere.

Acknowledgment

The authors are grateful to Drs. G. Holloway, G. F. Carnevale, and T. Hibiya for valuable discussions and helpful comments. The numerical simulations were performed using a Fujitsu VPP5000/56 computer at the Nagoya University Computation Center. The present study was supported by the "Research for the Future" Program of the Japan Society for the Promotion of Science under project JSPS-RFTF97P01101.

References

1. A.E. Gargett, P.J. Hendricks, T.B. Sanford, T.R. Osborn, and A.J. Williams III: J. Phys. Oceanogr. **11**, 1258 (1981)
2. W. Munk. 'Internal wave and small-scale processes'. In: *Evolution of Physical Oceanography, Scientific Surveys in Honour of Henry Stommel*, ed. by B. A. Warren and C. Wunsch (MIT Press, 1981) pp. 264–291
3. K. Yoshida, T. Ishihara, D. Fujita, T. Yamahira, and Y. Kaneda: 'LES modelings based on the Lagrangian renormalized approximation', present volume
4. J. R. Herring and O. Métais: J. Fluid Mech. **202**, 97 (1989)
5. O. Métais and J. R. Herring: J. Fluid Mech. **202**, 117 (1989)
6. O. Métais and M. Lesieur: J. Fluid Mech. **239**, 157 (1992)

7. D. Ramsden and G. Holloway: J. Geophys. Res. **97**, 3659 (1992)
8. D. A. Siegel and J. A. Domaradzki: J. Phys. Oceanogr. **24**, 375 (1994)
9. T. Hibiya, Y. Niwa, and K. Fujiwara: J. Geophys. Res. **103**, 18715 (1998)
10. G. Carnevale, M. Briscolini, and P. Orlandi: J. Fluid Mech. **247**, 205 (2001)
11. Y. Kaneda: Phys. Fluid **29**, 701 (1986)
12. M. Lesieur and R. Rogallo: Phys. Fluids A **1**, 718 (1989)
13. J. R. Herring: Phys. Fluids **17**, 859 (1974)
14. C. Cambon: present volume

The Eulerian Time Correlation Function in Homogeneous Isotropic Turbulence

Robert Rubinstein[1] and Guo-Wei He[2]

[1] Computational Modeling and Simulation Branch Mail Stop 128, NASA Langley Research Center, Hampton, VA 23681-2199, USA
[2] ICASE, Mail Stop 132C, NASA Langley Research Center, Hampton, VA 23681-2199, USA

Abstract. Two general models are proposed for the Eulerian time correlation function in homogeneous isotropic turbulence. The first is based on continued fraction approximations to its Laplace transform, and the second is based on random sweeping by a possibly non-Gaussian velocity field. Both models can give reasonable quantitative agreement with DNS data for moderate time separations over which the time correlation functions at different wavenumbers exhibit a common self-similar form.

1 Introduction

The analysis of time correlations in turbulence begins with Kraichnan's explanation of the inconsistency of Eulerian turbulence closures with Kolmogorov scaling [11]. This work revealed that the dynamic decorrelation mechanisms are distinct for Eulerian and Lagrangian time correlations and that energy transfer in turbulence must be analyzed in terms of Lagrangian quantities.

Since Lagrangian time correlations arise both in the problem of energy transfer in turbulence and in the equally fundamental problem of passive scalar diffusion, they would seem to be more important. Nevertheless, Eulerian properties are relevant in a broad class of problems in which turbulence acts as a time-dependent random medium and properties at fixed locations in space are required. Wave scattering by turbulence is one example; provided the turbulence is not simply frozen during the passage of the wave, analysis of this problem will depend on Eulerian time correlations.

Sound radiation by turbulence is another problem in which Eulerian correlations might be relevant. As applied to homogeneous, isotropic turbulence, Lighthill's theory appears to treat fixed regions of space, not moving volumes of fluid, as sound sources [13]. Since the space-time properties of fixed spatial volumes determine the sound source, Eulerian time correlations are relevant. Since the observer defines a coordinate system at rest, Eulerian time correlations are certainly consistent with Lighthill's theory. More generally, Kaneda has suggested that whereas energy transfer in turbulence depends on Lagrangian time correlations [8](??), momentum transfer depends on Eulerian time correlations, which also suggests the relevance of Eulerian time correlations to sound radiation.

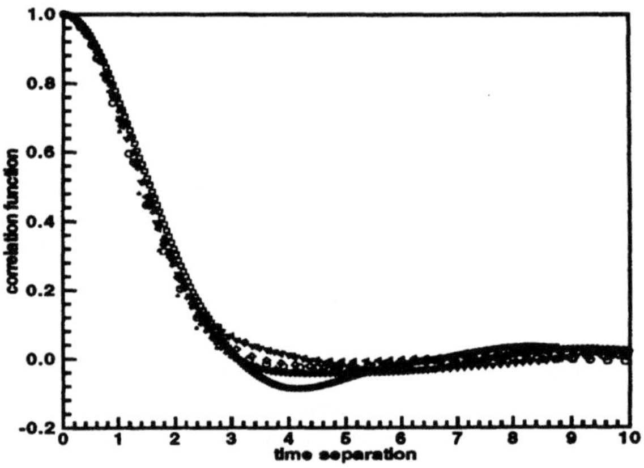

Fig. 1. Time correlation function at various wavenumbers plotted as functions of the similarity variable $\xi = Vk\tau$. The thick curve with the lowest minimum corresponds to a low wavenumber; increasing wavenumber corresponds to increasing $\psi(\xi)$ at $\xi = 4$.

The Eulerian time correlation function $\psi(r,t)$ is defined in homogeneous, isotropic turbulence by

$$\langle \mathbf{u}(\mathbf{x},t) \cdot \mathbf{u}(\mathbf{x}+\mathbf{r},t+\tau)\rangle = Q(r)\ \psi(r,\tau) \tag{1}$$

where

$$Q(r) = \langle \mathbf{u}(\mathbf{x},t) \cdot \mathbf{u}(\mathbf{x}+\mathbf{r},t)\rangle \tag{2}$$

is the single-time correlation function.

The basic fact about Eulerian time correlations is the similarity form for inertial range separations [11,8],

$$\psi(r,\tau) = \psi\left(\frac{V\tau}{r}\right) \tag{3}$$

where

$$V^2 = \langle \mathbf{u}(\mathbf{x},t) \cdot \mathbf{u}(\mathbf{x},t)\rangle \tag{4}$$

is the *rms* velocity fluctuation.

The validity of the similarity form Eq. (3) has been demonstrated repeatedly in the literature [8]. We confirm this similarity form with new obervations summarized in Fig. 1, which shows the time correlation function in Fourier space, $\psi(k,\tau)$ for various wavenumbers k, plotted as functions of the similarity variable $Vk\tau$. The data is taken from He et al. [6] The time correlations for all wavenumbers collapse very well for time separations $Vk\tau < 3$. At longer times, the time correlation functions cross zero and oscillate. These features appear to be quite robust, although our data is not sufficiently well resolved to draw definitive conclusions at very long time separations. This behavior is somewhat in contrast to the data of Kaneda et al. [9], however, we provisionally assume it is correct.

Although the correlation functions at different wavenumbers no longer collapse so well when $Vk\tau > 3$, the time correlation functions for wavenumbers in the range $10 < k < 35$, appear to exhibit a common similarity form even in the neighbourhood of the zero-crossing. We tentatively identify these wavenumbers with an inertial range, although the relatively low-resolution (128^3) DNS does not contain an extended region of unambiguous Kolmogorov scaling.

In this paper, we explore some possible analytical forms for the function ψ. Apart from its purely theoretical interest, there are applications in which the precise analytical form of the time correlation is important. In astophysical applications [14], the total acoustic radiation from stars depends surprisingly sensitively on the assumptions made about the time correlation function. A similar sensitivity to the time correlation function was found by Bertoglio et al. [1].

2 Continued fraction methods

Kaneda et al. curvefit turbulent time correlation functions by rational functions, using the Taylor series approximations and the Padé table [9]. Thus, if for fixed k, we write

$$\psi(k,\tau) = \omega_0(k) + \frac{1}{2!}\omega_2(k)\tau^2 + \frac{1}{4!}\omega_4(k)\tau^4 + \cdots \tag{5}$$

then the Taylor coefficients are evaluated as

$$\begin{aligned}
\omega_0(k) &= \psi(k,0) = 1 \\
\omega_2(k) &= \frac{d^2\psi}{dt^2}(k,0) = -\frac{\langle \dot{\mathbf{u}}(\mathbf{k},t)\cdot\dot{\mathbf{u}}(-\mathbf{k},t)\rangle}{\langle \mathbf{u}(\mathbf{k},t)\mathbf{u}(\mathbf{k},t)\rangle} \\
\omega_4(k) &= \frac{d^4\psi}{dt^4}(k,0) = +\frac{\langle \ddot{\mathbf{u}}(\mathbf{k},t)\ddot{\mathbf{u}}(-\mathbf{k},t)\rangle}{\langle \mathbf{u}(\mathbf{k},t)\mathbf{u}(\mathbf{k},t)\rangle} \\
&\cdots
\end{aligned} \tag{6}$$

The time derivatives on the right side can be replaced by single-time moments of the velocity field by applying the Navier-Stokes equations. The time derivative of order n then is given in terms of a single-time moment of order $2n$. This fact connects the time correlation function to intermittency [4].

This method of calculation leads to very good results. Here, a different method is followed, which leads to a rational approximation of the Laplace transform of the correlation function [12]. In molecular hydrodynamics, the correlation function is expressed as a continued fraction through the Zwanzig-Mori projection operator formalism [2]. Recently, this formalism has been generalized to non-equilibrium conditions by Daems et al. [4].

However, we can also give an elementary account as follows. Define the Laplace transform of the time correlation function as usual by

$$\psi(s) = \int_0^\infty d\tau\, e^{-s\tau}\psi(\tau) \tag{7}$$

and assume the continued fraction representation

$$\psi(s) = \frac{a_0}{s+}\frac{a_2}{s+}\frac{a_4}{s+}\frac{a_6}{s+}\cdots \tag{8}$$

and the Taylor series expansion Eq. (5). Substituting Eq. (5) in Eq. (7),

$$\psi(s) = \frac{1}{s} + \frac{\omega_2}{s^3} + \frac{\omega_4}{s^5} + \cdots \tag{9}$$

Equating the expressions for $\psi(s)$ Eqs. (8) and (9), the Euclidean algorithm gives

$$\begin{aligned} a_2 &= -\omega_2 \\ a_4 &= -\frac{\omega_4}{\omega_2} + \omega_2 \\ a_6 &= -\frac{\omega_4 - \omega_6/\omega_2}{\omega_2 - \omega_4/\omega_2} + \frac{\omega_4}{\omega_2} \end{aligned} \tag{10}$$

An obvious approximation method is finite truncation of this continued fraction [2]. We obtain in this way first the the one-parameter or Markovian model

$$\psi(\tau) = e^{-\omega\xi} \tag{11}$$

and at the next order, the two-parameter model

$$\psi(\tau) = \left(\cos\tilde{\omega}\xi + \frac{\omega}{2\tilde{\omega}}\sin\tilde{\omega}\xi\right)e^{-\omega\xi/2} \tag{12}$$

In Eqs. (11) and (12), $\xi = Vk\tau$, the similarity variable corresponding to Eq. (3)

It should be stressed that an assumption of complete similarity is made in writing Eqs. (11) and (12), namely that

$$\omega_{2n} \sim (Vk)^{2n} \tag{13}$$

This is an assumption of 'normal scaling' for the Taylor coefficients; in view of Eqs. (10), it also implies normal scaling of the coefficients in the continued fraction expansion Eq. (8). This assumption may be satisfactory for Fourier coefficients [3].

The Markovian model is well-known to be incorrect, because it does not have zero slope at $\tau = 0$. But Fig. 1 also suggests that the two-parameter model is qualitatively inadequate, because it predicts equally spaced zeroes of the correlation function. We could proceed to the general three-parameter model, but prefer to consider a special form of this model [2] defined in terms of the *memory function*

$$K(\tau) = \tilde{\omega}^2(1 + \omega\tau)e^{-\omega\tau} \tag{14}$$

by the equation

$$\dot{\psi}(\tau) + \int_0^\tau d\tau' K(\tau - \tau')\psi(\tau') = 0 \tag{15}$$

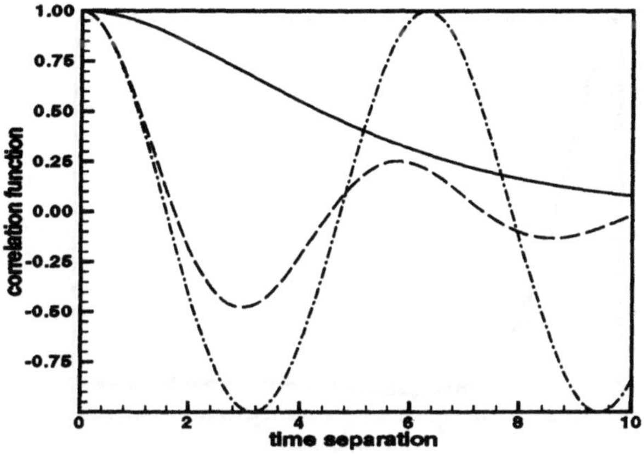

Fig. 2. The two-parameter correlation function of Eqs. (14)–(15) plotted for various values of the ratio $r = \tilde{\omega}/\omega$: $r = 0$ (solid), $r = 1$ (dashed), $r = 10$ (dot-dash). As r increases from zero, the function goes from pure decay to pure oscillation.

Eq. (15) is equivalent to a system of ordinary differential equations, expressed in terms of Laplace transforms as

$$\begin{aligned} &s\psi(s) + \tilde{\omega}^2\phi_1(s) = 1 \\ &s\phi_1(s) + \omega\phi_1(s) - \psi(s) - \omega\phi_2(s) = 0 \\ &s\phi_2(s) + \omega\phi_2(s) - \psi(s) = 0 \end{aligned} \tag{16}$$

which proves more convenient for numerical evaluation.

Since this is a two-parameter model, it is defined by two properties of the correlation function, which we take to be the *microscale*

$$\tau_\mu^2 = -\frac{1}{\ddot{\psi}(0)} \tag{17}$$

and the *integral scale*

$$\tau_M = \int_0^\infty dt\ \psi(t) \tag{18}$$

The connection between these quantities and the model parameters is

$$\tilde{\omega}^2 = \frac{1}{\tau_\mu} \tag{19}$$

$$2\frac{\tilde{\omega}^2}{\omega} = \frac{1}{\tau_M} \tag{20}$$

The generality of this class of functions is evident from the equivalent system Eq. (16), which shows that pure exponential decay occurs in the limit $\tilde{\omega} \approx 0$ and that oscillations occur in the limit $\omega \approx 0$. Intermediate behavior can be

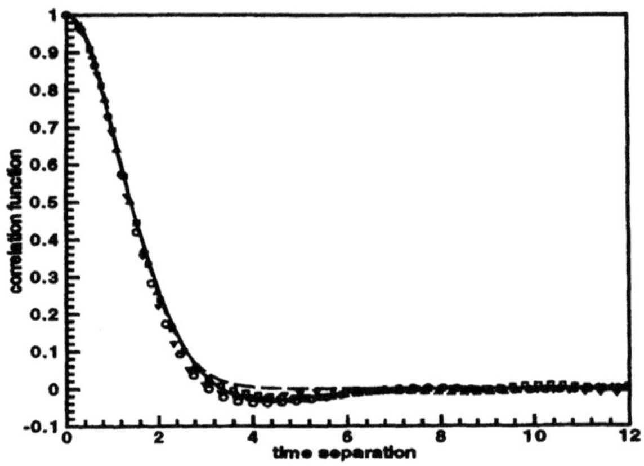

Fig. 3. The correlation function Eqs. (14)–(15) fit to the data of Fig. 1. Symbols are DNS, the solid line is the curvefit, and the dashed line is the best squared exponential curvefit. The fitting was done by trial and error. The region of negative values of the correlation functions is reproduced well by the empirical function.

anticipated between these limits and is confirmed in Fig. 2. Trial and error adjustment of the parameters led to an approximation to the measured Eulerian correlation functions which is shown in Fig. 3. We emphasize that the relations Eqs. (19)–(20) were not used to obtain the curvefit.

This result shows that a good fit to the time correlation function can be obtained in principle knowing only the microscale and integral scale. Nevertheless, the fact remains that neither the equations relating the Taylor coefficients and the model parameters Eq. (10) nor the system of ordinary differential equations for the correlation function Eq. (16) has much apparent physical significance.

3 The random sweeping model

An important theoretical model for the Eulerian time correlation function is provided by the direct interaction approximation [10]. However, the functional form predicted, $\psi(\tau) \sim J_1(\xi)/\xi$ is unrealistically oscillatory.

Another approximation for Eulerian time correlations is based on Kraichnan's model problem for the random sweeping effect [11]. In this problem, a fixed frozen random velocity field $\mathbf{u}(\mathbf{x})$ is advected by a random constant velocity field $\mathbf{v}$. The result is the time dependent field defined by

$$\dot{\mathbf{u}}(\mathbf{k},t) = -i\mathbf{k}\cdot\mathbf{v}\mathbf{u}(\mathbf{k},t) \tag{21}$$

where

$$\mathbf{u}(\mathbf{k},0) = \mathbf{u}(\mathbf{k}) \tag{22}$$

so that

$$\mathbf{u}(\mathbf{k},t) = \exp(i\mathbf{v}\cdot\mathbf{k}t)\mathbf{u}(\mathbf{k}) \tag{23}$$

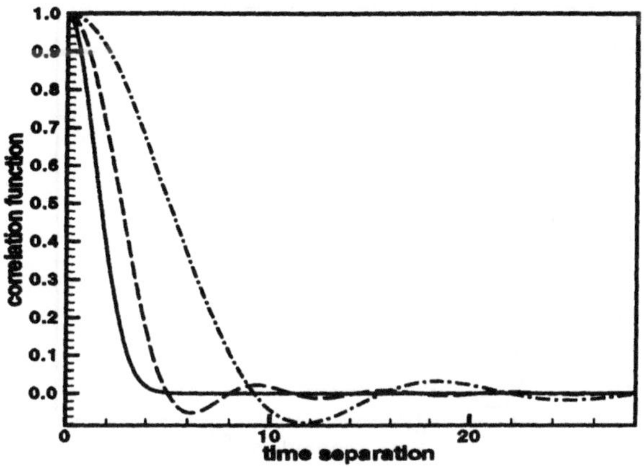

Fig. 4. The one-parameter correlation function defined by Eq. (28) plotted for various values of the parameter a: $a = 100$ (solid), $a = 10$ (dashed), $a = 1$ (dot-dash). The limiting cases are $\psi(\xi) = \exp(-\xi^2)$ when $a = \infty$ and $\psi(\xi) = \sin(\xi)/\xi$ when $a = 0$.

If $\mathbf{v}$ is Gaussian, then

$$\psi(\tau) = \langle \exp(i\mathbf{v}\cdot\mathbf{k}\tau)\rangle = \exp(-V^2\tau^2k^2/2) \tag{24}$$

where

$$V^2 = \langle \mathbf{v}\cdot\mathbf{v}\rangle \tag{25}$$

The time correlations of the actual turbulent velocity field are approximated by the (ensemble-averaged) time correlations of the velocity field defined in Eq. (23) The assumption underlying this model is that the large scales responsible for sweeping in the actual velocity field can be considered independent of the small scales at which the time correlation is evaluated. The first equality in Eq. (24) shows that the pdf is related to the frequency spectrum.

It is evident that this model will not provide a very good fit to the data because the measured correlation functions all take negative values. We consider how this model might be generalized to obtain better agreement. If the pdf of $\mathbf{v}$ is non-Gaussian, then certainly the second equality in Eq. (24) does not apply. There is considerable recent work [5,7] suggesting that the single-point pdf of velocity fluctuations is not exactly Gaussian, but is instead *sub-Gaussian*: that is, the tails are lighter than Gaussian, in contrast to familiar 'intermittency' behavior.

A simple sub-Gaussian is defined by truncating a Gaussian,

$$p(\mathbf{v}) = \begin{cases} b\exp(-v^2/2V^2) & \text{if } v \le a \\ 0 & \text{if } v > a \end{cases}$$

where the standard deviation $\propto V$ defines the relevant sweeping velocity. Limiting cases of the corresponding correlation function are the squared exponential

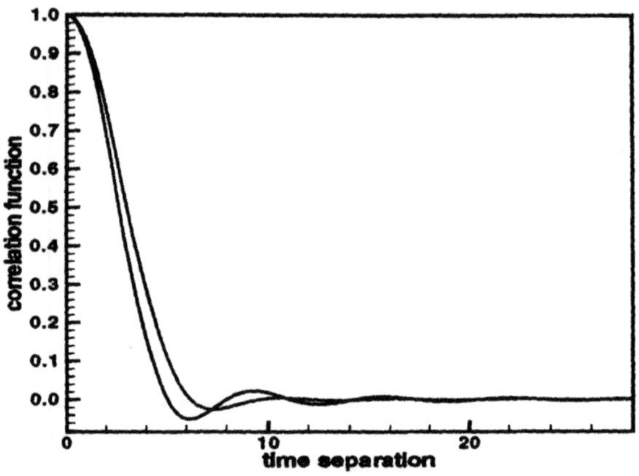

Fig. 5. The one-parameter correlation function defined by Eq. (30) plotted for two values of the parameter α: $\alpha = 100$ (more rapid decay), $\alpha = 10$ (less rapid decay).

Eq. (24) when $a = \infty$ and

$$\psi(\tau) = \frac{\sin(\xi)}{\xi} \tag{27}$$

when $a = 0$. In general, this correlation function satisfies the inhomogeneous differential equation

$$\frac{d\psi}{d\tau} + \frac{1}{2}\tau\psi = -\pi a^2 e^{-a^2} \frac{d}{d\tau} \frac{\sin(a\tau)}{a\tau} \tag{28}$$

Fig. 4 plots this correlation function for various values of the parameter a.

Although these correlation functions look promising, there is little doubt that the underlying model for the pdf is somewhat artificial. Falkovich and Lebedev have attempted to compute the tail of the single-point velocity pdf [5]. The tail is found to depend on the forcing statistics, and is therefore not universal. Simply to illustrate some possibilities, we follow this article and consider a non-Gaussian pdf with an $\exp(-v^4)$ tail. Suppose, to have consistency with near Gaussian behavior for small deviations from zero, that

$$p(\mathbf{v}) = a \exp[-(v^2 + \alpha v^4)/2V^2] \tag{29}$$

Then $\psi(t)$ satisfies

$$\alpha \frac{d}{d\tau}\left[\frac{d^2\psi}{d\tau^2} + \frac{2}{\tau}\frac{d\psi}{d\tau}\right] - \frac{1}{2}\frac{d\psi}{d\tau} - \frac{1}{4}\tau\psi = 0 \tag{30}$$

Note that when $\alpha = 0$, the time correlation function reduces correctly to the squared exponential form corresponding to a Gaussian-distributed sweeping velocity. Two typical forms of this function are shown in Fig. 5.

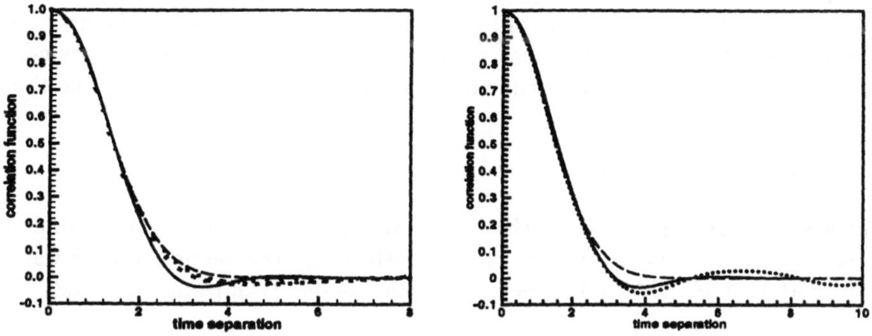

Fig. 6. The correlation function Eq. (30) fit to the data of Fig. 1. Symbols are DNS, the solid line is the curvefit, and the dashed line is the best Gaussian (squared exponential) curvefit. The graph on the left corresponds to relatively high wavenumbers, the graph on the right to a low wavenumber. The Gaussian provides a good fit to the high wavenumber data, but at low wavenumber, the sub-Gaussian provides much better agreement.

We note the existence of asymptotic solutions to Eq. (30)

$$\psi(\tau) \sim \cos(\sqrt{3}\xi^{4/3}/2 + \phi)\exp(-\xi^{4/3}/2) \tag{31}$$

This solution indicates that the correlation function falls off considerably more slowly than $exp(-\xi^2)$ and that the correlation function crosses zero with increasing frequency as $\tau \to \infty$. In this respect, the simpler model Eq. (26) seems more realistic, but the fact is that the available data do not rule Eq. (31) out.

In Fig. 6, this function is compared to the empirical distribution corresponding to a variety of wavenumbers k. Again, the curvefitting was done by trial and error, with no attempt to determine the best fit from numerical properties of the measured correlation function. A best fit squared exponential is included for comparison. This figure shows that the low k correlation function is fit remarkably well by this type of function.

Despite the good fit for the low wavenumber case, these sub-Gaussian models are generally too 'elastic' in comparison to the data: the negative amplitude after the first zero crossing tends to be larger than the data, and it tends to recover too quickly. This raises the question of the accuracy of the data in this range. More accurate measurements should be made in order to make more definitive comparisons.

4 Conclusions

The collapse of the Eulerian time correlation function to a self-similar form based on the sweeping velocity is confirmed by DNS data. For moderate time-separations, the collapse is excellent for all wavenumbers, and continues for larger time separations for intertial range modes. Models for the time correlation function based on continued fraction approximation and on a random sweeping model can both give satisfactory agreement with the data. More accurate measurements of the time correlations at large time separation will be required to refine these models.

Acknowledgments

This work was supported by the National Aeronautics and Space Administration under NASA Contract No. NAS1-97046 while the second author was in residence at ICASE, NASA Langley Research Center, Hampton, VA 23681.

References

1. J.-P. Bertoglio, F. Bataille, and J.-D. Marion (2001), Two-point closures for weakly compressible turbulence, Phys. Fluids **13**, 290.
2. J.-P. Boon and S. Yip (1980), *Molecular Hydrodynamics*, Dover.
3. C. Brun and A. Pumir (2001), Statistics of Fourier modes in a turbulent flow, Phys. Rev. E **63** 056313.
4. D. Daems, S. Grossmann, and V. S. L'vov (1999), Continued fraction representation of temporal multiscaling in turbulence, Phys. Rev. E. **60**, 6656.
5. G. Falkovich and V. Lebedev (1997), Single-point velocity distribution in turbulence, Phys. Rev. Lett. **79**, 4159.
6. G.-W. He, R. Rubinstein, and L.-P. Wang (2001), Effects of eddy viscosity on time correlations in large eddy simulations, ICASE Report 20001-10.
7. J. Jimenez (1998), Turbulent velocity fluctuations need not be Gaussian, J. Fluid Mech. **376** 139.
8. Y. Kaneda (1993), Lagrangian and Eulerian time correlations in turbulence, Phys. Fluids A **5**, 2835.
9. Y. Kaneda, T. Ishihara, and K. Gotoh (1999). Taylor expansions in powers of time of Lagrangian and Eulerian two-point two-time velocity correlations in turbulence, Phys. Fluids **11**.
10. R. H. Kraichnan (1959), The structure of isotropic turbulence at very high Reynolds number, J. Fluid Mech. **5**, 497.
11. R. H. Kraichnan (1964), Kolmogorov's hypotheses and Eulerian turbulence theory, Phys. Fluids **7**, 1723.
12. R. H. Kraichnan (1968), Convergents to infinite series in turbulence theory, Phys. Rev. **174**, 240.
13. M. J. Lighthill (1952), On sound generated aerodynamically, Proc. Roy. Soc. A **211**, 564.
14. P. Ulmschneider, J. Theurer, and Z. E. Musielak (1996), Acoustic wave energy fluxes for late-type stars, Astron. Astrophys. **315**, 212.

Predictability of 3D Isotropic Turbulence –Effect of Data Assimilation–

Kazuki Hayashi, Takashi Ishihara, and Yukio Kaneda

Department of Computational Science and Engineering, Graduate School of Engineering

Abstract. In numerical simulation of turbulent flows, it is usually difficult to obtain exact initial conditions and it is well known that initial uncertainty in high wavenumbers spreads toward large scale through nonlinear dynamics. These imply the limitation of the numerical prediction. We present here a series of numerical experiments to study the effect of data assimilation on the predictability of isotopic 3D turbulence. In the experiment we have two parameters K_c and T to control the data assimilation, where K_c is the maximum of the assimilated wavenumber range and T is the time interval of the assimilation. It is found not only that we can suppress the growth of uncertainty by data assimilation but also that we can reduce it under appropriate conditions on K_c and T. It is also found that there exists a critical K_c^* such that if $K_c < K_c^*$, the uncertainty grows for any T.

1 Introduction

It is widely accepted that the motion of turbulence is unpredictable. For example, on the first page of his textbook on turbulence, Batchelor[1] wrote "... It is a well-known fact that under suitable conditions, which normally amount to a requirement that the kinematic viscosity ν be sufficiently small, some of these motions are such that the velocity at any given time and position in the fluid is not found to be the same when it is measured several times under seemingly identical conditions. ...". In real turbulence, it is difficult to identify or control the initial and boundary conditions as well as the external force in full detail, so that "seemingly" identical conditions are not rigorously identical. They suffer from small uncertainties or differences. The evolution of turbulence motion is sensitive to such inevitable uncertainty. This sensitivity is one of the main origins of the principal difficulty in numerical weather prediction. Small initial uncertainty in small scale may give rise to big uncertainty in large scale as times go on. Such view has been supported by recent extensive studies on chaotic systems.

In this paper, we try in a sense to challenge such a view, i.e., instead of considering the sensitivity of the turbulence to the initial uncertainty, we explore the possibility of its insensitivity under appropriate conditions. For example, consider the numerical weather prediction, for which observational data from satellite may be available. One may then ask whether it is possible or not to improve the prediction or elongate the time for which the prediction is reliable, by using such data appropriately. If it is the case, then one may further ask how much improvement is possible, or what is the best way of the use for the improvement. Although it is generally true that the initial uncertainty or error would grow with time when there is no mechanism of suppression or control of the uncertainty, it is of fundamental as well as practical interest to explore the possibility that the growth of the uncertainty may be suppressed when the uncertainty in large scale is controlled by certain means.

In this respect, it is encouraging that recent numerical experiments by Kreiss and Yestrom demonstrate that the time history of the large scale flows of the solution of the three-dimensional (3D) Navier-Stokes equations generates the small scale[2]. In this paper, we investigate the effect of periodic (in time) data assimilation at large scale on the evolution of uncertainty at small scale in 3D incompressible turbulence under periodic boundary conditions. Particular attention is paid to the role of the time interval of the data assimilation and the scale of eddies to be assimilated.

2 Method of numerical experiments

We consider the motion of 3D incompressible turbulence of unit density obeying the Navier-Stokes equations under periodic boundary conditions in each of the Cartesian co-ordinates with period 2π. In the spectral space, the Navier-Stokes equations may be written as

$$\left(\frac{\partial}{\partial t} + \nu k^2\right)\hat{u}_i(\boldsymbol{k}) = M_{imn}(\boldsymbol{k}) \sum_{\boldsymbol{p}+\boldsymbol{q}=\boldsymbol{k}} \hat{u}_m(\boldsymbol{p})\hat{u}_n(\boldsymbol{q}) + f_i(\boldsymbol{k}), \tag{1}$$

after the elimination of the pressure term by using the incompressibility condition $\boldsymbol{k}\cdot\hat{\boldsymbol{u}}(\boldsymbol{k}) = 0$, where the summation convention is used for the repeated indices, ν the kinematic viscosity, $\boldsymbol{k} = (k_1, k_2, k_3)$ with k_1, k_2 and k_3 being integers, f_i the Fourier transform of the i-th component of the external force, and

$$M_{ijm}(\boldsymbol{k}) = -\frac{i}{2}\left[k_m P_{ij}(\boldsymbol{k}) + k_j P_{im}(\boldsymbol{k})\right], \quad P_{ij}(\boldsymbol{k}) = \delta_{ij} - k_i k_j / k^2.$$

Let $\hat{\boldsymbol{u}}_1(\boldsymbol{k},t)$ and $\hat{\boldsymbol{u}}_2(\boldsymbol{k},t)$ be two velocity fields, where $\hat{\boldsymbol{u}}_1$ evolves according to (1). The initial condition of the second field $\hat{\boldsymbol{u}}_2$ is exactly the same as that of $\hat{\boldsymbol{u}}_1$, but only at small wavenumber range such that the wavenumber k is smaller than a certain wavenumber, say $k_{initial}$. For $k \geq k_{initial}$, $\hat{\boldsymbol{u}}_2(\boldsymbol{k},0)$ is different from $\hat{\boldsymbol{u}}_1(\boldsymbol{k},0)$, but their energy spectra are the same to each other. Thus the two fields are different initially only at large wavenumbers i.e., small scales, and

$$\begin{cases} \hat{\boldsymbol{u}}_1(\boldsymbol{k},0) = \hat{\boldsymbol{u}}_2(\boldsymbol{k},0), & \text{for } k < k_{initial}, \\ \hat{\boldsymbol{u}}_1(\boldsymbol{k},0) \neq \hat{\boldsymbol{u}}_2(\boldsymbol{k},0), & \text{for } k \geq k_{initial}. \end{cases} \tag{2}$$

Starting with such an initial condition, the field $\hat{\boldsymbol{u}}_2$ evolves according to the same equation (1) as for the basis field $\hat{\boldsymbol{u}}_1$ up to time $t = T$, where not only the viscosity but also the external force for the field $\hat{\boldsymbol{u}}_2$ are the same as those for $\hat{\boldsymbol{u}}_1$.

During the evolution up to $t = T$, the difference between the fields $\hat{\boldsymbol{u}}_1$ and $\hat{\boldsymbol{u}}_2$ is not confined to high wave numbers $k > k_{initial}$, but spreads towards smaller wavenumbers $k \leq k_{initial}$.

At $t = T$, we perform data assimilation on large scale, such that the field $\hat{\boldsymbol{u}}_2(\boldsymbol{k}, T)$ is replaced by $\hat{\boldsymbol{u}}_1(\boldsymbol{k}, T)$ for the Fourier-modes with wavenumber smaller than a certain wavenumber, say $K_c(< k_{initial})$, as illustrated in Fig.1. Thus the field $\hat{\boldsymbol{u}}_2$ is renewed and exactly the same as $\hat{\boldsymbol{u}}_1$ at $t = T$, but only at small wavenumbers satisfying $k < K_c$. At large wavenumbers $k \geq K_c$ the two fields remain different from each other.

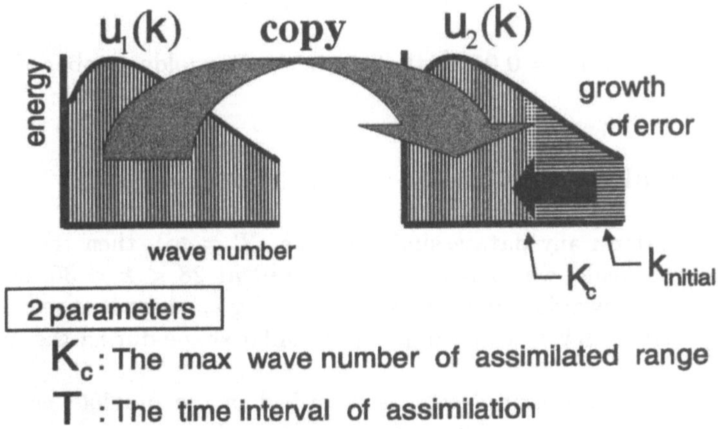

Fig. 1. Method of data assimilation

Then, we let the two fields evolve again according to (1) up to time $t = 2T$, and continue the similar process, i.e., we make the similar data assimilation at $t = 2T, 3T, ...$, and let the two field evolve according to (1) during the time intervals. Our main interest here is the role of the two key parameters K_c and T, characterizing the maximum wavenumber of the assimilation range, and the time interval or the frequency of the data assimilation, respectively.

As a quantitative measure characterizing the difference between the two fields, we use the difference spectrum $\Delta(k, t)$ defined as

$$\Delta(k, t) = \frac{1}{2} \sum_{|\boldsymbol{k}|=k} \delta\hat{\boldsymbol{u}}(\boldsymbol{k}, t) \cdot \delta\hat{\boldsymbol{u}}(-\boldsymbol{k}, t),$$

$$\delta\hat{\boldsymbol{u}}(\boldsymbol{k},t) = \hat{\boldsymbol{u}}_1(\boldsymbol{k},t) - \hat{\boldsymbol{u}}_2(\boldsymbol{k},t),$$

where $\sum_{|\boldsymbol{k}|=k}$ denotes the sum over the wavevector domain $\boldsymbol{k}$ satisfying $k-1/2 < |\boldsymbol{k}| \le k+1/2$.

The two fields $\boldsymbol{u}_1(\boldsymbol{k})$ and $\boldsymbol{u}_2(\boldsymbol{k})$ were simulated by solving (1) using an alias free spectral method with a 4th-order Runge-Kutta method for the time marching. The maximum wavenumber of the retained Fourier modes is 30, whereas the wavenumber $k_{initial}$ was set to 28. In order to keep the turbulence quasi-stationary, we introduced a negative viscosity that is wavevector independent for $2 < k < 3$, and zero outside the range.The value of the negative viscosity is adjusted every time step so as to keep the total energy constant i.e., $< \mathbf{u}_1 \cdot \mathbf{u}_1 >= 1$. The initial energy spectrum $E(k,0)$ is given by

$$E(k,0) \propto k^4 \exp(-\frac{2k^2}{k_p}),$$

where $k_p = 3.0$. We set $\nu = 0.01$, so that the initial Reynolds number of the field $\mathbf{u}_1$ is 53.

3 Numerical results

If we do not perform any data assimilation (i.e., $T = \infty$), then the difference between the two fields, which is initially confined at $28 < k < 30$, in general spreads with time towards smaller k as shown in Fig.2, where it is seen that the difference spectrum $\Delta(k,t)$ keeps almost the similar shape during the spreading process.

If we perform the data assimilation as described in the previous section, the growth or the decrease of the difference depends on the assimilation conditions, especially on T and K_c, as well as the wavenumber k, (see Fig.1). Some quantitative idea on the difference growth may be obtained by the observation of the evolution $\Delta(k_s,t)$ at a fixed the wavenumber $k = k_s$. Although we have measured the difference spectrum at several values of k_s in $4.0 \le k_s \le 25$, the results were found to be not very sensitive to k_s, at least qualitatively and for a certain range of k_s. We therefore show in the following only the results for $k_s = 10$ as a typical example.

Figure 3 shows the evolution of the normalized difference spectrum $\bar{\Delta}(k,t)$ for $K_c = 0, 4, 5, ..., 8$ and $T = 2$, where the normalized spectrum $\bar{\Delta}(k,t)$ is defined as

$$\bar{\Delta}(k,t) = \Delta(k,t)/\{2E_1(k,t)\},$$

in which $E_1(k,t)$ is energy spectrum of $\hat{\boldsymbol{u}}_1(\boldsymbol{k})$ at time t. Figure 3 suggests
(i) the decrease or increase of $\bar{\Delta}(k,t)$ is nearly exponential in t after an initial transient stage, and
(ii) there is a critical value of K_c, say K_c^*, such that $\bar{\Delta}(k,t)$ decreases exponentially with time for $K_c > K_c^*$, while it increases exponentially for $K_c < K_c^*$. These results imply that copying/assimilating the large scale eddies may result

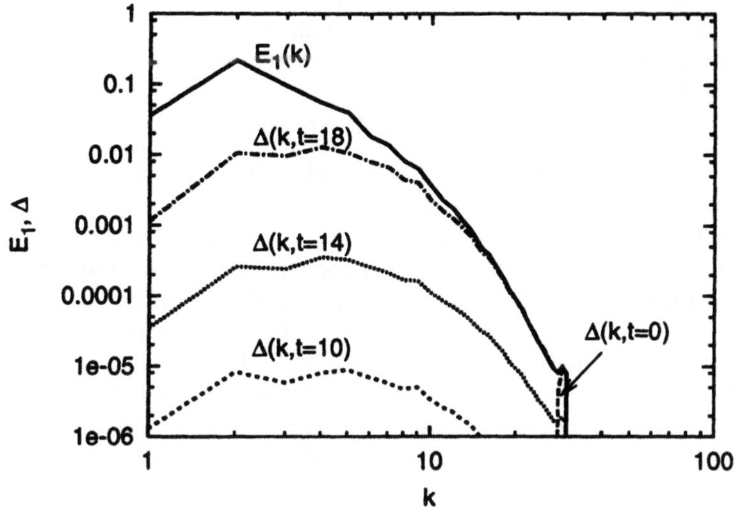

Fig. 2. Energy spectrum $E_1(k,t)$ of $\mathbf{u}_1(k,t)$ and difference spectrum $\Delta(k,t)$ for $T = \infty$.

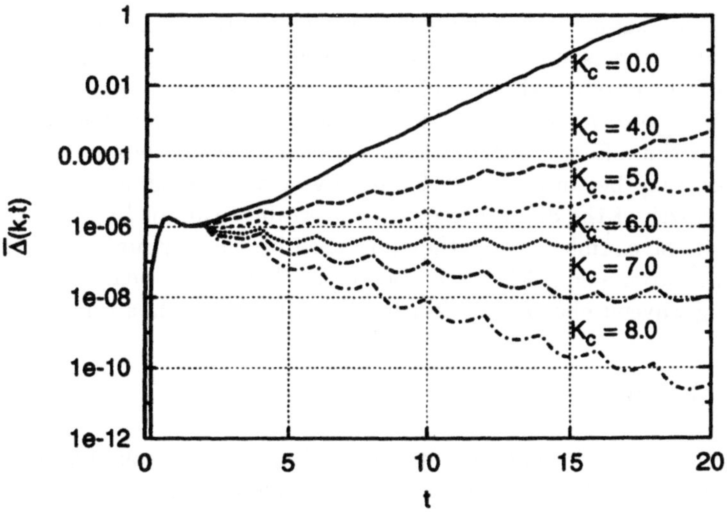

Fig. 3. Evolution of the normalized difference $\bar{\Delta}(k_s,t)$ for $K_c = 0, 4, 5, .., 8$ with $T = 2$.

in the decreasing with time of the difference between the two fields, provided that copy/assimilation time interval T is small enough.

It is tempting to approximate the time dependence of $\bar{\Delta}(k,t)$ in (i) by an exponential form

$$\bar{\Delta}(k_s,t) = C\exp(\alpha t), \tag{3}$$

where C and α are time independent constants. The negativity of α then implies that the difference decreases with time.

A rough estimate of the exponent $\alpha = \alpha(K_c,T)$ may be obtained by fitting the curves to (3). Figure 4 shows the dependence of α thus obtained on K_c and T, whereas Figs. 5 and 6 show the cross sections of the contour of Fig.4 at given T and K_c, respectively. In these figures the symbols are the measured values and the lines show the interpolated values. Figure 5 shows that the exponent α is smaller for larger K_c and smaller T, as would be expected. It also shows that the dependence of α on K_c is weaker for larger T, and that for large T, α is positive in a wide range of K_c. Figure 6 shows that α is larger for larger T and smaller K_c, again as would be expected. The increase of α with T is seen more rapid for smaller T. It is also seen that $\alpha > 0$ for $K_c = 4$ and 5, irrespectively of the value T. This implies that the difference cannot be suppressed however frequent the data assimilation may be, if the assimilated wavenumber range is too small.

In Fig.7 are shown the iso-lines of $\alpha(K_c,T)$ in (K_c,T)-plane, where the K_c-dependence of the iso-line for $\alpha = 0$ is seen to be almost linear, i.e., T increases almost linearly with K_c for large K_c, but it sharply drops near $K_c = K_c^*$ ($K_c^* \sim 4.2$). Figure 7 implies that there is a critical wavenumber K_c^* such that if $K_c < K_c^*$, the difference $\Delta(k_s,t)$ cannot be suppressed however frequent the assimilation may be. In other words, if only very large eddies are assimilated, then the growth of the difference or the uncertainty cannot be suppressed. The assimilation wavenumber range therefore need be wide enough to a certain extent for the suppression. Figure 7 also shows that at K_c near K_c^*, the critical assimilating wavenumber K_c for which $\alpha(K_c,T) = 0$ depends very weakly on T.

4 Conclusion

1. A numerical study is made of the effect of data assimilation of large eddies on the evolution of small scale uncertainty in 3D incompressible turbulence. In particular, the role of two key parameters, the assimilation time interval T and the wavenumber K_c characterizing the maximum assimilation wavenumber, in the growth or the decrease of the uncertainty in small scale is investigated quantitatively. The uncertainty is shown to decrease with time or to be suppressed under appropriate conditions of large eddy data assimilation. This implies the possibility of significantly improving the prediction by appropriate data assimilation at large scale.

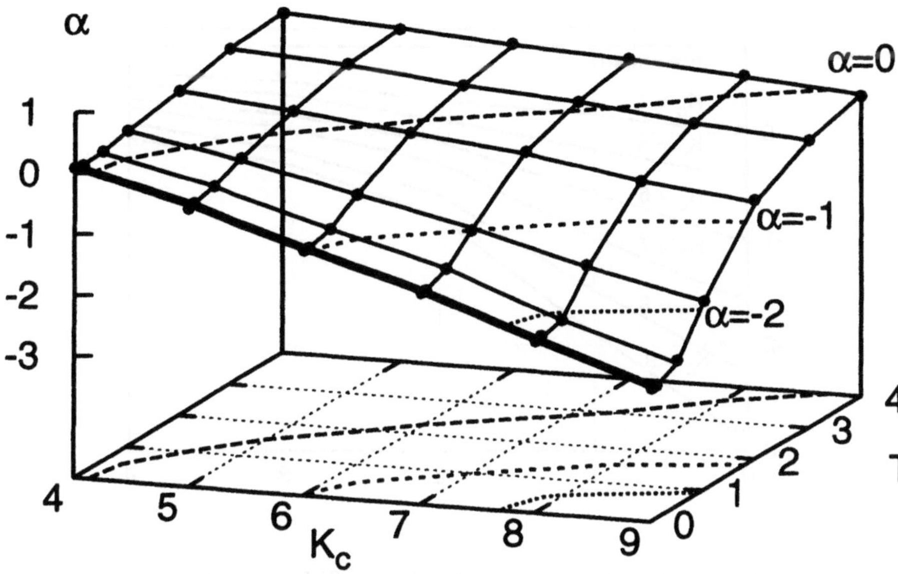

Fig. 4. Dependence of the growth rate α on T and K_c.

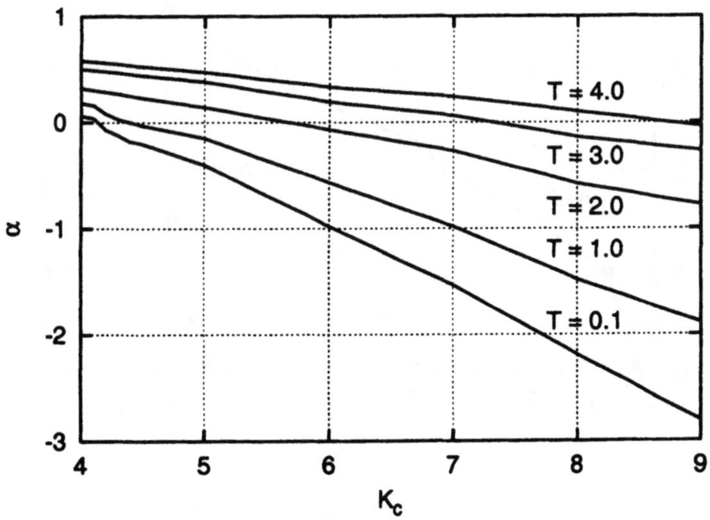

Fig. 5. Growth rate α as a function of K_c for fixed T.

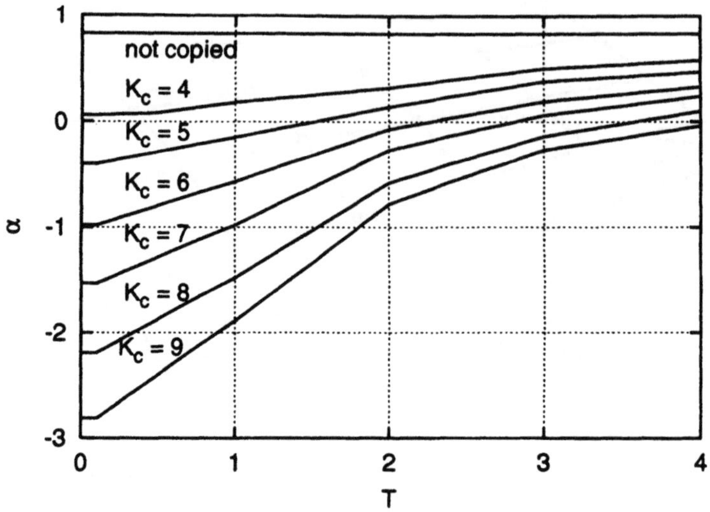

Fig. 6. Growth rate α as a function of T for fixed K_c.

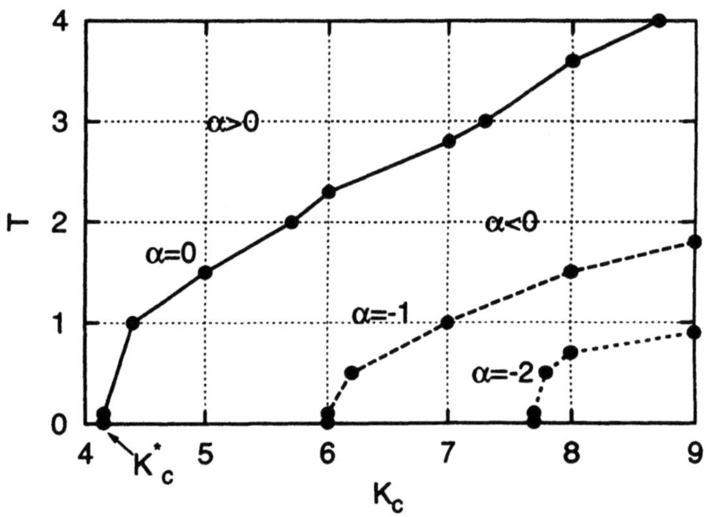

Fig. 7. Iso-lines of α in the (K_c, T) plane.

2. There is a critical wave number K_c^* such that if $K_c < K_c^*$, then the growth rate $\alpha(K_c, T)$ (see (3)) is positive for any T, so that the growth cannot be suppressed, however frequent the assimilation may be.

5 Acknowledgement

This work was supported by "Research for the Future" Program of the Japan Society for the Promotion of Science under the project JSPS-RFTF97P01101.

References

1. G.K. Batchelor: *The theory of homogeneous turbulence* (Cambridge University Press, Cambridge 1953)
2. Heinz-Otto Kreiss, J. Ystrom: A numerical study of the solution to the 3D incompressible Navier-Stokes equations. Research report, UCLA, Los Angeles (1998)

Orthonormal Divergence-Free Wavelet Analysis of Spatial Correlation between Kinetic Energy and Nonlinear Transfer in Turbulence

Keiji Kishida[1] and Keisuke Araki[2]

[1] Research Organization of Science and Engineering, Ritsumeikan Univ., JAPAN
[2] Dept. Assistive and Rehabilitation Engineering, Okayama Univ. Sci., JAPAN

Abstract. Helical Meyer wavelet, an orthonormal divergence-free vector wavelet basis[1] is applied to the analysis of direct numerical simulation (DNS) data of incompressible isotropic turbulence. In the present study we focus our analysis on the correlation between the scale-location wavelet energy spectrum and nonlinear energy transfers relevant to its dynamics. Since helical Meyer wavelets are orthonormal and solenoidal, we can evaluate precisely the magnitude of nonlinear energy transfer between the wavelet modes with retaining the detailed energy balance. It is found that the magnitude of scale-location energy spectrum is positively correlated to those of relevant nonlinear energy transfers, i.e. nonlinear energy transfer actively occurs mainly in such domains that the fluid motions of assigned scale are intense. Batchelor and Townsend conjectured that intermittency of energy distribution in small scales may persist if the amount of energy transferred from a certain scale to smaller scales is determined by the energy density on the spot[2]. Our observation supports that such hypothetical process actually occurs as a result of the Navier-Stokes dynamics.

1 Introduction

Phenomenological theories of intermittency in turblence, whether based on velocity increment or dissipation rate, rely on the postulate that energy transfer process is somewhat multiplicative, that is, kinetic energy of fluid motion is transferred from larger scales to smaller scales in some stochastic and step-by-step manner (see Ref.[3] and references therein). Roughly speaking, a process is called multiplicative if a certain quantity Q_j, which may be the velocity increments with distance $\sim \lambda^j l_0$ where l_0 is integral scale and λ is a certain number, obeys

$$Q_j = Q_0 X_1 X_2 \cdots X_j \tag{1}$$

where X_j is some stochastic variable, which may be the ratio of dissipation rates of the nearest scales. Although quite a few models have flourished, the research of the relation between the vortex dynamics and generation of intermittency on the basis of the Navier-Stokes equation (NSE)

$$\frac{\partial \boldsymbol{u}}{\partial t} + (\boldsymbol{u} \cdot \nabla)\boldsymbol{u} = -\nabla P + \nu \nabla^2 \boldsymbol{u} \tag{2}$$

with incompressibility condition

$$\nabla \cdot \boldsymbol{u} = 0 \tag{3}$$

seems still in a rudimentary state. One of the reasons is, we speculate, the Fourier analytical method, which is a conventional tool to analyze the cascading process, cannot provide the information on location of the fluid motion that is relevant to transfer. Wavelet analysis, on the other hand, has been regarded as a promising method since it provides the information on scale and location simultaneously. Recently we developed an orthonormal divergence-free wavelet and applied it to the analysis of nonlinear transfer in fully developed turbulence[1]. We found that the principal nonlinear interaction is local though some analysis based on the Fourier analysis had suggested non-local interaction is important (e.g. Ref.[4]).

How is intermittency generated? As a scenario to generate and maintain the intermittency in turbulence, Batchelor and Townsend conjectured that, roughly speaking, the amount of energy transfer from an observing scale to nearest smaller scale spatially occurs in such a way that is proportional to the intensity of fluid motion of the observed scale and location. Although their scenario is qualitative and primitive, principal features of multiplicative process are included. In the present study we will survey whether such process is realistic or not in terms of wavelet analysis. Capturing the information on scale and location simultaneously is one of the advantages of wavelet analysis so that one can investigate the following two points: (1) whether the transfer process has characteristics of multiplicative process, (2) how the transfer process is related to the coherent structures.

2 Mathematical preliminaries

In this subsection we summarize mathematical preliminaries: definition and properties of orthonormal divergence-free wavelet, wavelet expansion of the velocity field, wavelet energy spectrum and its application to NSE.

2.1 Helical wavelet

Orthonormal divergence-free wavelet (we call it *helical wavelet* hereafter) on a three-dimensional, periodic box T^3 is obtained by unitary transform of complex helical waves[5], which constitutes an orthonormal basis of divergence free vector field on T^3 or R^3. Helical wavelet is *divergence-free, real-valued* vector field on T^3 which is defined by the following formula[1]:

$$\boldsymbol{\psi}_{j\epsilon \boldsymbol{l} s}(\boldsymbol{x}) := \frac{1}{\sqrt{2^{3j}}} \sum_{\boldsymbol{k}\in Z^3\setminus\{0\}} \widehat{\psi}_\xi\left(\frac{k_x}{2^j}\right)\widehat{\psi}_\eta\left(\frac{k_y}{2^j}\right)\widehat{\psi}_\zeta\left(\frac{k_z}{2^j}\right)$$
$$\times \frac{\boldsymbol{e}_\vartheta(\boldsymbol{k}) + is\boldsymbol{e}_\varphi(\boldsymbol{k})}{\sqrt{2}} \exp\left[2\pi i \boldsymbol{k}\cdot\left(\frac{\boldsymbol{x}}{L} - \frac{\boldsymbol{l}}{2^j}\right)\right]. \tag{4}$$

Indices and notations are summarized in Tables 1 and 2. Roughly speaking,

Table 1. indices of helical wavelet and their implication.

index	range of index	physical implication
j	$\{0, 1, 2, ...\}$	spatial scale, mean wavelength
$\epsilon := \xi + 2\eta + 4\zeta$	$\{1, 2, ..., 7\}$	anisotropy (see. Fig.1)
$\boldsymbol{l} := (l_x, l_y, l_z)$	$l_j \in \{0, 1, 2, ..., 2^j - 1\}$	location
s	$\{1, -1\}$	helicity

Table 2. nomenclature of the definition of helical wavelet.

$\boldsymbol{k}$:	wave number, $\boldsymbol{k} = (k_x, k_y, k_z)$,
$\widehat{\psi}_0(k)$:	Fourier coefficient of scaling function on R,
$\widehat{\psi}_1(k)$:	Fourier coefficient of mother wavelet on R,
ξ, η, ζ :	index for the kind of wavelet, $\xi, \eta, \zeta \in \{0, 1\}$,
$\boldsymbol{e}_\vartheta(\boldsymbol{k}), \boldsymbol{e}_\varphi(\boldsymbol{k})$:	base vector of spherical coordinate system in the ϑ- and φ-directions,
L :	length of each side of periodic box.

the helical wavelet $\psi_{j\epsilon \boldsymbol{l} s}(\boldsymbol{x})$ is such window function that has a window width $\Delta x \sim L/2^j$ and is centered at $\boldsymbol{x} = (L/2^j)\boldsymbol{l}$. In Fig.1 we show isosurfaces of the amplitude of Fourier coefficient of helical wavelet. It is obvious from the figure that we regard that the index ϵ represents *anisotropy*. By defini-

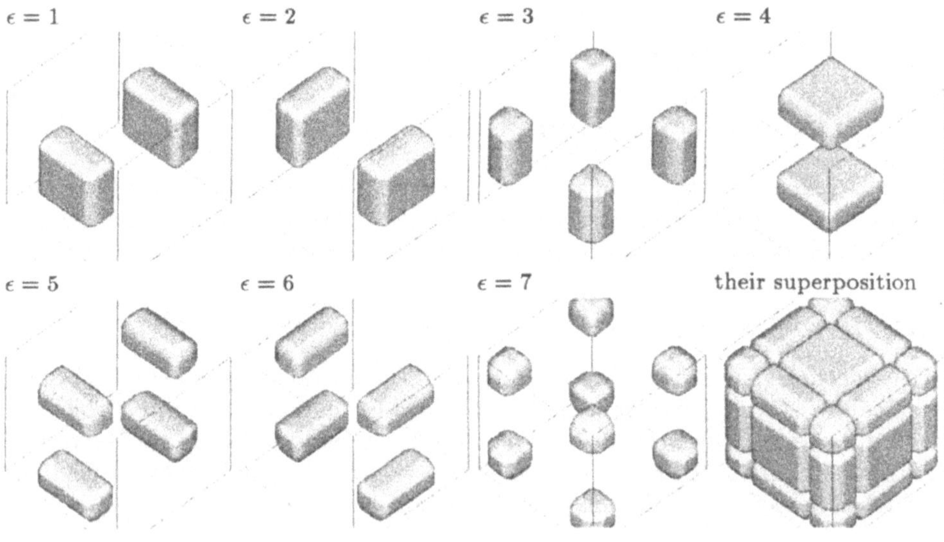

Fig. 1. amplitude of Fourier coefficient of helical wavelets. Threshold of isosurface is 71% of each peak value. Each wavelet covers different direction in Fourier space so that ϵ implies *anisotropy*.

tion helical wavelets constitute an orthonormal basis of L^2 function space. It should be also emphasized that the helical wavelet is *not* Beltrami flow, i.e. $\nabla \times \boldsymbol{\psi}_{j\epsilon l s}(\boldsymbol{x}) \not\parallel \boldsymbol{\psi}_{j\epsilon l s}(\boldsymbol{x})$ in general. This property causes non-trivial nonlinear interaction between two wavelet modes.

2.2 Helical wavelet spectrum of velocity field

Helical wavelet decomposition of the velocity field is given by

$$\boldsymbol{u}(\boldsymbol{x},t) = \sum_{j,\epsilon,\boldsymbol{l},s} \boldsymbol{u}_{j\epsilon\boldsymbol{l}s}(\boldsymbol{x},t), \tag{5}$$

where $\boldsymbol{u}_{j\epsilon\boldsymbol{l}s}(\boldsymbol{x},t)$ is helical wavelet spectrum defined by

$$\boldsymbol{u}_{j\epsilon\boldsymbol{l}s}(\boldsymbol{x},t) := u_{j\epsilon\boldsymbol{l}s}(t)\boldsymbol{\psi}_{j\epsilon\boldsymbol{l}s}(\boldsymbol{x}), \quad u_{j\epsilon\boldsymbol{l}s}(t) := \frac{1}{L^3}\int_{\mathrm{T}^3} \boldsymbol{u}(\boldsymbol{x},t)\cdot\boldsymbol{\psi}_{j\epsilon\boldsymbol{l}s}(\boldsymbol{x})\,\mathrm{d}\boldsymbol{x}. \tag{6}$$

Hereafter we will call $u_{j\epsilon\boldsymbol{l}s}(t)$ *helical wavelet coefficient* of the velocity field. Helical wavelet coefficient is a real number because helical wavelet is real-valued function. Roughly speaking, the spectrum $\boldsymbol{u}_{j\epsilon\boldsymbol{l}s}(\boldsymbol{x},t)$ implies the portion of the flow centered at $\boldsymbol{x} = (L/2^j)\boldsymbol{l}$, with the envelope of the size $\Delta x \sim L/2^j$, with helicity s and anisotropy of type ϵ. There are too many parameters to treat. So we reduce the information on, for example, anisotropy and helicity. In the following analysis we use the scale-location spectrum

$$\boldsymbol{u}_{j\boldsymbol{l}}(\boldsymbol{x},t) := \sum_{\epsilon,s} \boldsymbol{u}_{j\epsilon\boldsymbol{l}s}(\boldsymbol{x},t), \tag{7}$$

and the scale spectrum

$$\boldsymbol{u}_{j}(\boldsymbol{x},t) := \sum_{\epsilon,\boldsymbol{l},s} \boldsymbol{u}_{j\epsilon\boldsymbol{l}s}(\boldsymbol{x},t). \tag{8}$$

It should be noted that they are still orthogonal, i.e. they satisfy

$$\int_{\mathrm{T}^3} \boldsymbol{u}_{j\boldsymbol{l}}(\boldsymbol{x},t)\cdot\boldsymbol{u}_{j'\boldsymbol{l}'}(\boldsymbol{x},t)\,\mathrm{d}\boldsymbol{x} \propto \delta_{jj'}\delta_{l_x l'_x}\delta_{l_y l'_y}\delta_{l_z l'_z}, \tag{9}$$

$$\int_{\mathrm{T}^3} \boldsymbol{u}_{j}(\boldsymbol{x},t)\cdot\boldsymbol{u}_{j'}(\boldsymbol{x},t)\,\mathrm{d}\boldsymbol{x} \propto \delta_{jj'}, \tag{10}$$

respectively, where δ is Kronecker's delta.

2.3 Helical wavelet energy spectra

Scale-anisotropy-location-helicity energy spectrum. Helical wavelet basis is self-dual and normalized so that the kinetic energy of velocity field is straghtforwardly obtained by the square sum of helical wavelet coefficients;

$$E(t) := \frac{1}{2}\int_{\mathrm{T}^3} |\boldsymbol{u}(\boldsymbol{x},t)|^2 \mathrm{d}\boldsymbol{x} = \frac{1}{2}\sum_{j,\epsilon,\boldsymbol{l},s}\int_{\mathrm{T}^3} |\boldsymbol{u}_{j\epsilon\boldsymbol{l}s}(\boldsymbol{x},t)|^2\mathrm{d}\boldsymbol{x} = \frac{1}{2}\sum_{j,\epsilon,\boldsymbol{l},s} u_{j\epsilon\boldsymbol{l}s}(t)^2. \tag{11}$$

Compared with bi-orthogonal wavelet, we regard this fact as an important advantage of the helical wavelet basis for the energy budget analysis of turbulence due to Navier-Stokes dynamics. The quantity $u_{j\epsilon l s}(t)^2/2$ works as the *scale-anisotropy-location-helicity* energy spectrum and one can reduce the information, for example anisotropy and helicity, according to the objective of the survey.

Scale-location energy spectrum. In the present study we focus our attention on the inter-scale nonlinear interaction and its relation to coherent structures. So we reduced the information on anisotropy and helicity. Summing $u_{j\epsilon l s}(t)^2/2$ with respect to ϵ and s, we obtain the *scale-location* energy spectrum defined by the following formula;

$$E_{jl}(t) := \frac{1}{2}\sum_{\epsilon,s} u_{j\epsilon l s}(t)^2. \tag{12}$$

This quantity represents the amplitude of kinetic energy of fluid motion with spatial scale $\sim 2^{-j}L$ located around the center $2^{-j}Ll$. Therefore spatial coherence of the fluid motion for an assigned scale is available by observing spatial distribution of scale-location energy spectrum.

Scale energy spectrum. Furthermore, reducing the information on location of wavelet, we obtain the *scale* energy spectrum

$$E_j(t) := \sum_l E_{jl}(t). \tag{13}$$

We have already discussed the dynamics of scale energy spectrum and found that local nonlinear interaction is dominant in homogeneous, isotropic turbulence[1].

2.4 Helical wavelet representation of the Navier-Stokes equation and energy budget equation

Substituting wavelet expansion Eq.(5) into NSE Eq.(2) and taking inner product with helical wavelet $\psi_{j\epsilon l s}$, one obtain the helical wavelet representation of the NSE

$$\begin{aligned}\frac{\mathrm{d}u_{j\epsilon l s}(t)}{\mathrm{d}t} &= \sum_\lambda\sum_\mu \langle \psi_{j\epsilon l s}|\psi_\lambda|\psi_\mu\rangle\, u_\lambda(t)\, u_\mu(t) \\ &\quad +\nu\sum_\mu \langle \psi_{j\epsilon l s}|\nabla^2\psi_\mu\rangle u_\mu(t)\end{aligned} \tag{14}$$

where λ, μ denotes the indices of helical wavelets and the brackets are given by

$$\langle \boldsymbol{a}|\boldsymbol{b}|\boldsymbol{c}\rangle = -\frac{1}{L^3}\int_{\mathrm{T}^3} \boldsymbol{a}(\boldsymbol{x})\cdot(\boldsymbol{b}(\boldsymbol{x})\cdot\nabla)\,\boldsymbol{c}(\boldsymbol{x})\,\mathrm{d}\boldsymbol{x}, \tag{15}$$

$$\langle \boldsymbol{a}|\boldsymbol{b}\rangle = \frac{1}{L^3}\int_{\mathrm{T}^3} \boldsymbol{a}(\boldsymbol{x})\cdot\boldsymbol{b}(\boldsymbol{x})\,\mathrm{d}\boldsymbol{x}. \tag{16}$$

Pressure term vanishes because helical wavelet is *divergence-free.* Time development of the helical wavelet energy spectrum is obtained by multiplying $u_{j\epsilon l s}(t)$ on both sides of Eq.(14)

$$\frac{\mathrm{d}E_{j\epsilon l s}(t)}{\mathrm{d}t} = \sum_{\lambda}\sum_{\mu}\langle \psi_{j\epsilon l s}|\psi_{\lambda}|\psi_{\mu}\rangle\, u_{j\epsilon l s}(t)\, u_{\lambda}(t)\, u_{\mu}(t) + \nu\sum_{\mu}\langle \psi_{j\epsilon l s}|\nabla^2\psi_{\mu}\rangle u_{j\epsilon l s}(t)\, u_{\mu}(t). \tag{17}$$

We call the integral

$$\langle \psi_{j\epsilon l s}|\psi_{\lambda}|\psi_{\mu}\rangle\, u_{j\epsilon l s}(t)\, u_{\lambda}(t)\, u_{\mu}(t) = \langle u_{j\epsilon l s}|u_{\lambda}|u_{\mu}\rangle \tag{18}$$

wavelet energy transfer hereafter. It is very remarkable that helical wavelet representation of the nonlinear term satisfies detailed energy balance between arbitrary two or three wavelet mode (for details, see Ref.[6]). The definition of wavelet nonlinear transfer may seem naive for some readers, but this form has its own mathematically rigorous foundations. That is, the wavelet energy transfer is the *Levi-Civita connection* on the group of volume preserving diffeomorphism, which is the configuration space of Lagrangian mechanical description of hydrodynamics of an incompressible fluid (see appendix 2 of Ref.[7]).

In the present study we focus our interest on the relation between nonlinear transfer in turbulence and coherent structures so that we reduced the information on anisotropy and helicity by summing up with respect to ϵ and s;

$$\frac{\mathrm{d}E_{jl}(t)}{\mathrm{d}t} = \sum_{\lambda}\sum_{\mu}\langle u_{jl}|\psi_{\lambda}|\psi_{\mu}\rangle\, u_{\lambda}(t)\, u_{\mu}(t) + \nu\sum_{\mu}\langle u_{jl}|\nabla^2\psi_{\mu}\rangle u_{\mu}(t). \tag{19}$$

It should be remarked that these equations do not constitute a dynamical system, but are energy budget equations.

3 Result

We analyzed a shapshot data of the decaying homogeneous isotropic turbulence calculated by Kishiba el al.[8], which is the same data that is analyzed in Ref.[1]. NSE is solved by the pseudospectral method with a polyhedral dealiasing method, periodic boundary conditions ($L = 2\pi$), and 128^3 grid points. The snapshot is taken at $t = 1.0$ in the time unit adopted in Ref.[8]. The Taylor microscale Reynolds number of the snapshot is $R_\lambda = 100$.

3.1 Comparison of Fourier and wavelet energy spectra

It seems a quite interesting attempt to draw simultaneously three different kind of information on energy spectrum (see Fig.1): the Fourier energy spectrum $E(k)$, the wavelet scale energy spectrum E_j, and the probability density function (PDF) of the wavelet scale-location energy spectrum E_{jl}. The Fourier energy spectrum $E(k)$ provides a conventional method to represent energy density of each spatial scale. On the other hand, the PDF of E_{jl} describes statistics of spatial distribution of kinetic energy in an assigned scale, i.e. intermittency of each resolution class. However, the direct comparison of $E(k)$, E_j and E_{jl} has no quantitative significance because they are summation of different number of modes; the mode numbers to obtain $E(k)$, E_j and E_{jl} are $4\pi k^2 \times 2$, $2^{3j} \times 7 \times 2$, and 7×2, respectively. To mollify the difference we normalized the wavelet spectrum E_j and E_{jl} by $\Delta_j = 2^j(3/4\pi)^{1/3}$ and $\Delta_j/2^{3j}$, respectively. The value Δ_j is so carefully chosen that the number of independent Fourier modes included in the interval $\Delta_j \leq |\boldsymbol{k}| \leq 2\Delta_j$ is 14×2^{3j}. As is clearly seen from the figure, the amplitude of the normalized wavelet spectrum E_j/Δ_j and the abscissa of the PDF of $2^{3j}E_{jl}/\Delta_j$ are quantitatively comparable to the amplitude of Fourier energy spectrum $E(k)$. Thus the normalization is found convenient for graphical comparison of the wavelet spectrum with the Fourier one. The PDF of the wavelet scale-location energy spectrum E_{jl} for each j is plotted in the plane perpendicular to the k-$E(k)$ plane.

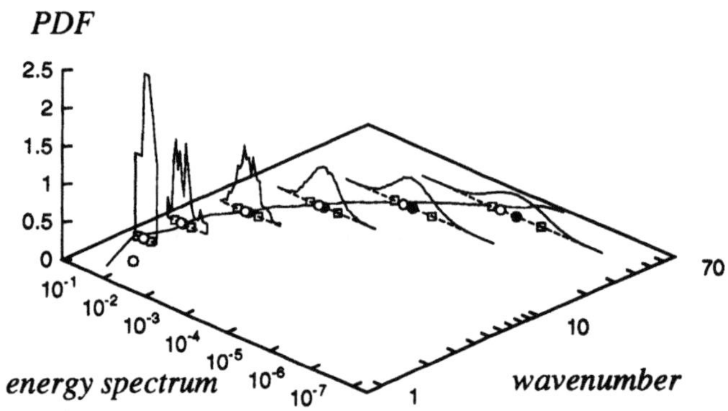

Fig. 2. simultaneous presentation of $E(k)$ (solid line on the bottom plane), E_j(open circle on the bottom plane) and PDF of the ensemble $\{E_{jl}\}$ (solid lines on the vertical planes) with appropriate normalization. Wavelet spectrum is plotted according to the wave number defined by $k_j := \sqrt{\int |\nabla \times \boldsymbol{u}_j|^2 d\boldsymbol{x} / \int |\boldsymbol{u}_j|^2 d\boldsymbol{x}}$. Solid circles and open squares are average and standard deviation of the ensemble $\{\log E_{jl}\}$.

Statistics of E_{jl} provide a measure of spatial variation of energy distribution in an assigned scale. It should be noted that the histogram of E_{jl} in Fig.2 is taken on the logarithmically partitioned bands. The peak of the PDF decreases and its width broadens as the resolution class j increases, i.e. as the spatial resolution becomes finer. This implies that the spatial variation of kinetic energy of the fluid motion become more intermittent as the scale decreases. To verify this tendency quantitatively, we analyzed the volume and energy portions of high energy wavelet sites E_{jl} (see Table 3). Here we tentatively take $\mu_j + \sigma_j$ as a threshold to measure the portions, where μ_j and σ_j are the mean and standard deviation of scale-location wavelet energy $\{E_{jl}\}$ for each j, respectively. Helical wavelet analysis is able to capture the increasing tendency of intermit-

j	energy	volume
2	24.5 %	14.1 %
3	27.2 %	13.3 %
4	33.3 %	12.3 %
5	38.1 %	9.43 %

Table 3. energy and volume portion of energetic sites. Threshold is defined by $\mu_j + \sigma_j$ where μ_j and σ_j are respectively average and standard deviation of the ensemble $\{E_{jl}\}$ for each j.

tency as is found by Fourier analysis. What we would like to stress here is that the decomposition of kinetic energy of fluid into wavelet scale-location spectrum and their statistics provides an immediate measure of intermittency. So we expect that statistical analysis of time development of E_{jl}, i.e. of wavelet energy transfer $\langle \boldsymbol{u}_{jl} | \boldsymbol{u}_{j'\epsilon' l' s'} | \boldsymbol{u}_{j''\epsilon'' l'' s''} \rangle$ may work as a direct, powerful tool to investigate the origin of intermittency due to the Navier-Stokes dynamics without any phenomenological assumption. It should be noted here that the PDF's in Fig.2 are very close to the log-normal distribution for $j > 3$ though we cannot judge whether it is accidental or not.

3.2 Correlation between energy and transfer

Survey of the statistics of all the possible combinations of wavelet energy transfer $\langle \boldsymbol{u}_{jl} | \boldsymbol{u}_{j'\epsilon' l' s'} | \boldsymbol{u}_{j''\epsilon'' l'' s''} \rangle$ is substantially impossible. So we will investigate the statistics of reduced quantities given by

$$\langle \boldsymbol{u}_{jl} | \boldsymbol{u} | \boldsymbol{u}_{j''} \rangle = \sum_{j',\epsilon',l',s'} \sum_{\epsilon'',l'',s''} \langle \boldsymbol{u}_{jl} | \boldsymbol{u}_{j'\epsilon' l' s'} | \boldsymbol{u}_{j''\epsilon'' l'' s''} \rangle . \tag{20}$$

In Fig.3 scatter plots of the pair of wavelet scale-location energy spectrum and absolute value of wavelet energy transfer $(E_{jl}, \left| \langle \boldsymbol{u}_{jl} | \boldsymbol{u} | \boldsymbol{u}_{j-1} \rangle \right|)$ for each wavelet site $\boldsymbol{l}$ are shown. Figures are drawn for the $j = 4$ and 5 cases. It is an interesting finding that wavelet energy transfer is very active for both the forward and backward transfers. (Surely, we must be careful for naive interpretation of wavelet energy transfer. See §4.) So figures are separately drawn according to the sign of wavelet energy transfer. It is verified that the absolute value of wavelet energy transfer positively correlates with scale-location energy spectrum. This

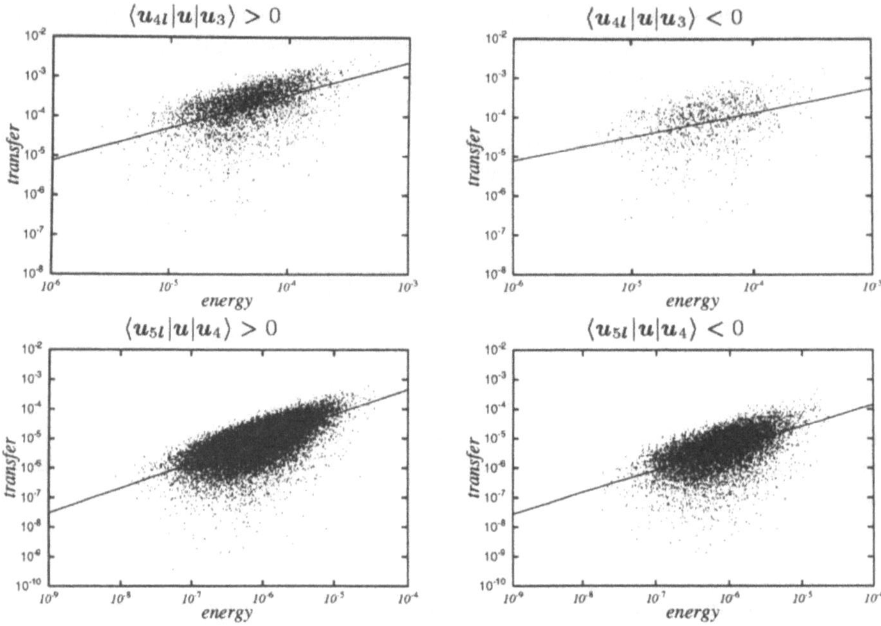

Fig. 3. scatter plot of the pair of scale-location energy spectrum and nonlinear transfer.

positive correlation implies, roughly speaking, that nonlinear energy transfer is active where the amplitude of fluid motion is intense.

It is an interesting feature that the scattered dots seem to obey the same probability density function and the difference between positive and negative nonlinear transfers is due to their population.

In Fig.4 we show the spatial distribution of wavelet energy transfer $\langle \boldsymbol{u}_{jl}|\boldsymbol{u}|\boldsymbol{u}_{j-1}\rangle$. Each ball is centered at $2^{-j+1}\pi \boldsymbol{l}$. The *volume* of each ball is determined to be proportional to the amplitude of transfer $|\langle \boldsymbol{u}_{jl}|\boldsymbol{u}|\boldsymbol{u}_n\rangle|$. The color of each ball denotes the sign of transfer; white denotes positive (i.e., the site acquires energy) and black does negative (i.e., the site losts energy). It should be noted that relatively small transfers (small balls) are not depicted in order to observe the feature of active region. Wire frames denotes the coherent structures, i.e., the clustered domain where the wavelet scale-location energy spectra E_{jl}'s are relatively large.

Two significant features are seen from these figures. Firstly, the large balls tend to locate at the inside and neighborhood of wire frames. That is, the wavelet energy transfers $\langle \boldsymbol{u}_{jl}|\boldsymbol{u}|\boldsymbol{u}_{j-1}\rangle$ tend to have large values on such wavelet sites that the scale-location energy spectrum E_{jl} is large. This is consistent with the result of the scatter-plot analysis. Secondly, both the wavelet energy spectrum E_{jl} and the wavelet energy transfer $\langle \boldsymbol{u}_{jl}|\boldsymbol{u}|\boldsymbol{u}_{j-1}\rangle$ are distributed not in uniformly scattered manner, but in some coherent, clustering manner. This fact implies that intensive fluid motions take place at some coherent regions and they are relevant to the nonlinear transfer in turbulence.

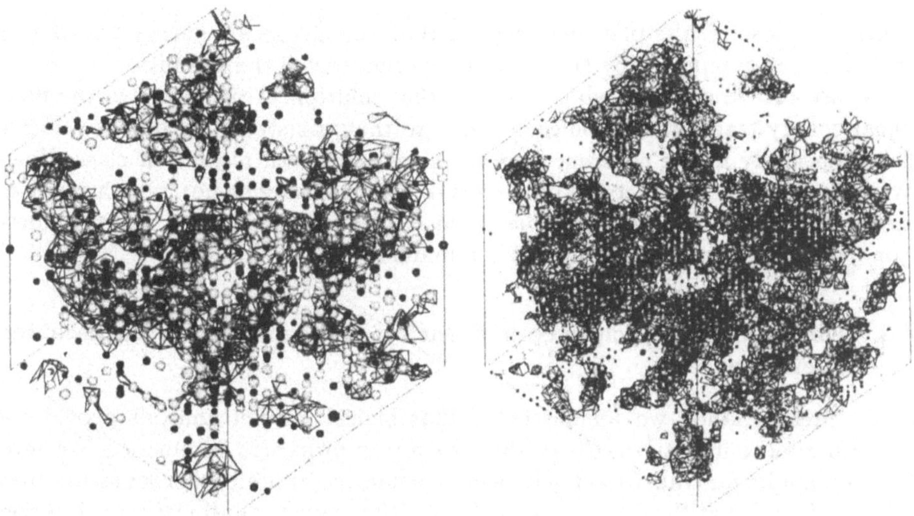

Fig. 4. spatial distribution of nonlinear wavelet transfer $\langle \boldsymbol{u}_{jl}|\boldsymbol{u}|\boldsymbol{u}_{j-1}\rangle$. Left: $j = 4$, right: $j = 5$. Wire frames are the coherent structures of $\boldsymbol{u}_{jl}$. Threshold of wire frames is $\mu_j + \sigma_j$ where μ_j and σ_j are respectively the average and standard deviation of the ensemble $\{E_{jl}\}$ for each j.

4 Discussion

4.1 Batchelor and Townsend scenario of the generation of intermittency and coherent structures

In this section we will discuss the relation between the wavelet nonlinear energy transfer analysis shown in the previous section and the generating process of intermittency in turbulence.

The scenario of the generation and maintenance of intermittency conjectured by Batchelor and Townsend is that, roughly speaking, the amount of energy transfer from an observing scale to nearest smaller scale spatially occurs in such a way that is proportional to the intensity of fluid motion of the observed scale and location. Scatter plot analysis clearly shows that the wavelet energy transfer $\langle \boldsymbol{u}_{jl}|\boldsymbol{u}|\boldsymbol{u}_{j-1}\rangle$ is positively correlated with wavelet scale-location energy spectrum E_{jl}, irrespective of its sign. This statistical result based on NSE revealed that such hypothetical process that is proposed by Batchlor and Townsend is realistic. In other words, the result implies that energy transfer process in turbulence has somewhat multiplicative nature. Thus it is verified quantitatively that generation and maintainance of intermittency is intrinsic in the NSE dynamics.

Spatial distribution analysis of wavelet energy transfer, on the other hand, demonstrates that amplitude of nonlinear transfer is large around the "coherent structures" of $\boldsymbol{u}_j$, i.e. the region where E_{jl}'s have large amplitude and are clustering. (It should be noted here that these coherent structures do not exactly correspond to the worm vortices identified by the isosurfaces of enstrophy density.) These results plausibly suggest that the inter-scale energy transfer in turbulence is dominated by the coherent stuructures of the resolution class.

These two result, as a whole, implies that multiplicative nature of the nonlinear energy transfer in turbulence is due to the geometry of turbulent velocity field, especially to the coherent structures found in each resolution class. However, the problem whether the process is self-maintaining or not, which requires more consideration on the dynamics, especially, on the instability and regeneration cylce of coherent structures, still remains.

4.2 Uncertainty principle consideration and remarks on backscatter phenomena

In the present study we demonstrated that helical wavelet analysis provides a powerful tool to investigate the nonlinear transfer process in turbulence. We verified that nonlinear transfer actively occurs inside and around coherent structures. It is also found that the population of the positive wavelet nonlinear transfers and that of negative ones are of comparable order to each other. Surely, these transfers may simply be able to be interpreted as the forward and backward transfer of energy, and one may conclude that both forward and backward transfer are very active. But we should emphasize that such naive interpretation may be misleading in some cases.

Iima and Toh have applied the wavelet analysis to the nonlinear dynamics of the Burgers equation[9]. They analysed the nonlinear transfer process of a Burgers shock solution and claimed that wavelet analysis shows fictitious oscillations. Though their claim that the such oscillatory appearance is due to "phase" of wavelet function is dubious (because some other possible explanation exists as is discussed below), one should be careful with interpreting the result.

The key idea to understand the origin of such fictitious oscillation is the uncertainty principle of Fourier analysis. We will discuss firstly the nature of wavelet function from Fourier analytical viewpoint. Then we will discuss the consequence of wavelet expansion as a band-pass filter.

As is seen from Fig.1, helical wavelet is such function whose support in the Fourier space is located at the "center" wave number given by $k_c \sim 2^j/L$ and has window width $\Delta k \sim 2^j/L$. Thus the wavelet function $\boldsymbol{\psi}_{jl\epsilon\sigma}$ is zero mean ($\int \boldsymbol{\psi}_{jl\epsilon\sigma} \mathrm{d}\boldsymbol{x} = \mathbf{0}$), spatially oscillates by the wavelength determined by the center wave number $\lambda \sim 1/k_c$, and is localized by the scale determined by the window width in Fourier space $\Delta x \sim 1/\Delta k$. We conclude that the origin of oscillation is not the "phase", but the "center frequency" of Fourier coefficient of wavelet.

Next, we consider the wavelet expansion of velocity field. When we analysed nonlinear transfer, we used two kinds of filtered field; one is the scale-location spectrum given by $\boldsymbol{u}_{jl}(\boldsymbol{x},t) = \sum_{\epsilon,s} \boldsymbol{u}_{j\epsilon ls}(\boldsymbol{x},t)$, and the other is the scale spectrum given by $\boldsymbol{u}_j(\boldsymbol{x},t) = \sum_{\epsilon,l,s} \boldsymbol{u}_{j\epsilon ls}(\boldsymbol{x},t)$. The former has the same spatial characteristics as the wavelet itself. The latter, on the other hand, is not localized. The scale spectrum $\boldsymbol{u}_j(\boldsymbol{x},t)$ is expected to be such a function that oscillates with wavelength $\lambda \sim 1/k_c$ and has an "envelope" which varies on the scale

The scaling law (1) is obtained from a dimensional analysis of $\bar{\epsilon}$ and r. The scaling exponent ζ_p, defined by $S_p(r) \propto r^{\zeta_p}$, increases linearly with p, the normal scaling. The Kolmogorov spectrum, which is the wavenumber equivalent for $p = 2$, is $E(k) = K\bar{\epsilon}^{2/3}k^{-5/3}$, where K is a universal constant. One of the central issues in turbulence studies has been whether the above scaling law holds for all flows [2,3]. The exponent of the inertial range spectrum is now widely accepted to be $-5/3$, with a possible small correction due to intermittency. The Kolmogorov constant is between $K = 1.5$ and 2.0. For higher order moments, however, experiments have shown that the scaling exponents increase more slowly than $p/3$[3]. Thus, when the scale of turbulence becomes small, the fluctuations of the velocity increment become stronger and the statistics of the velocity field deviate from Gaussian.

A phenomenological model that explains why the statistics change as the scale of turbulence decreases was first proposed by Kolmogorov(K62) [4]. In this theory, the locally volume-averaged energy dissipation rate per unit mass is defined as

$$\epsilon_r = \frac{3}{4\pi r^3}\int_{|\boldsymbol{y}|\le r} \epsilon(\boldsymbol{x}+\boldsymbol{y})\mathrm{d}^3\boldsymbol{y} \tag{3}$$

and a distribution for ϵ_r is introduced:

$$P(\epsilon_r) = \frac{1}{\sqrt{2\pi}\sigma_{\ln\epsilon_r}}\exp\left(-\frac{(\ln\epsilon_r - m_r)^2}{2\sigma^2_{\ln\epsilon_r}}\right), \tag{4}$$

$$m_r = \ln\bar{\epsilon} - \frac{1}{2}\sigma^2_{\ln\epsilon_r}, \qquad \sigma^2_{\ln\epsilon_r} = A + \mu\ln(L/r), \tag{5}$$

where μ is a universal positive constant and A is a constant that depends on the flow. The K41 scaling (1) is replaced by the conditional average for a given value of ϵ_r so that $\langle\delta u_r^p|\epsilon_r\rangle \propto \epsilon_r^{p/3}r^{p/3}$, and the total average is obtained by taking the average over the distribution of ϵ_r. This gives

$$S_p(r) = \int_0^\infty \langle\delta u_r^p|\epsilon_r\rangle P(\epsilon_r)\mathrm{d}\epsilon_r \propto \bar{\epsilon}^{p/3}\,(r/L)^{\mu p(p-3)/18}\,. \tag{6}$$

The scaling exponent ζ_p is now quadratic in p.

Many phenomenological theories for the energy cascade towards small scales have been proposed since K62 [2,3]. Of interest are the probability density function of ϵ_r, the structure functions, and the constant μ appearing in the scaling exponents. The one-dimensional surrogate dissipation equation is generally used in experiments,

$$\epsilon'_r(\boldsymbol{x}) = \frac{15\nu}{r}\int_0^r\left(\frac{\partial u_1(\boldsymbol{x}+y\boldsymbol{e}_1)}{\partial x_1}\right)^2\mathrm{d}y \tag{7}$$

and local isotropy is assumed. However, the degree of homogeneity and isotropy at small scales is sometimes questionable, which affects the accuracy of estimates of the energy dissipation rate [5–8,10–12]. Moreover, a one-dimensional surrogate energy dissipation rate is used as a flow parameter; there are very few studies of the statistics of the energy dissipation rate itself [5,9,12]. Praskovski and Oncley measured the PDF for $\ln \epsilon'_r$ experimentally in a wind tunnel at a Taylor microscale Reynolds number of $R_\lambda = 3200$, and in the atmospheric surface layer over Wyoming at $R_\lambda = 12700$. They found that the PDF for $\ln \epsilon'_r$ was close to Gaussian over the inertial range of r at $R_\lambda = 3200$, while the PDF at $R_\lambda = 12700$ decayed faster than the Gaussian PDF [9]. The constant μ was 0.2, and increased very slightly with R_λ.

Direct numerical simulation (DNS) of turbulent flows does not include any models of motion for scales below the grid size, and any quantities such as ϵ_r that are difficult to determine experimentally are computed directly, without deformation of the flow field. Therefore, DNS is a powerful tool for studying turbulence. Wang *et al.* studied statistics of the one-dimensional average of the energy dissipation rate ϵ_{1r} numerically for decaying and forced turbulence using DNS [12]. The inertial range of the energy spectrum was observed only at the lower end of the forced turbulence spectrum at $R_\lambda = 151$. The PDF for $\ln \epsilon_{1r}$ was negatively skewed for all r, and approached a Gaussian PDF with increasing r and R_λ. This was in contrast to the experiments of Praskovski and Oncley [9]. Also, Wang *et al.* reported a value of $\mu = 0.28$.

The turbulence fields obtained in the present DNS have a finite inertial range. Therefore, statistics of the locally volume-averaged energy dissipation rate and the velocity field can be extracted for scales of motion that are between the inertial and dissipation ranges. The present DNS has two advantages over those of previous studies: (1) the study was performed using the ideal conditions of homogeneous isotropic forcing at large scales, which are difficult to realize experimentally, and (2) the data extend into a small but finite portion of the inertial range. Therefore we can examine the statistics of the locally volume-averaged energy dissipation rate more precisely, using its full definition.

2 Numerical simulation

The numerical method is described only briefly here; complete details can be found in Gotoh *et al.* [13,14]. The Navier-Stokes equations were integrated in Fourier space. A solenoidal Gaussian random force that was white in time with a narrow spectrum support in the low wavenumber band was applied to the flow. A pseudo-spectral code was used to compute the convolution sums and the aliasing error was effectively removed. The time integration was performed using the fourth-order Runge Kutta Gill method. The resolution condition $k_{max}\eta > 1$ was satisfied when $R_\lambda = 381$, and $k_{max}\eta$ was 0.96, slightly less than unity, when $R_\lambda = 460$. The statistical averages were calculated using time averages over several large eddy turnover times. The DNS parameters are listed in Table I.

Table 1. DNS parameters and statistical quantities. T^{av}_{eddy} is the length of the time average.

R_λ	N	k_{max}	ν	T^{av}_{eddy}	E	$\bar{\epsilon}$	L	λ	$\eta(\times 10^{-2})$	K
381	1024^3	483	2.80×10^{-4}	4.21	1.74	0.499	1.139	0.0989	0.258	1.63
460	1024^3	483	2.00×10^{-4}	2.14	1.79	0.506	1.150	0.0841	0.199	1.64

3 Results

3.1 Structure functions

The isotropy at 2nd and 3rd order moments was well satisfied for $r \leq L/2$. Also, the energy budget for various scales agreed well with the Kármán-Howarth-Kolmogorov relation

$$\frac{4}{5}\bar{\epsilon}r = -D_{LLL} + 6\nu\frac{\partial D_{LL}}{\partial r} + Z \tag{8}$$

for steady turbulence, where D_{LL} and D_{LLL} are the second and third order longitudinal structure functions, respectively, and $Z(r)$ denotes contributions due to the external force [14]. The inertial range of the 4/5 law was clearly observed, as shown in Fig.1.

In the scaling range, the structure functions are expected to be

$$S_p^L(r) = \langle|\delta u_r|^p\rangle \propto r^{\zeta_p^L}, \quad S_p^T(r) = \langle|\delta v_r|^p\rangle \propto r^{\zeta_p^T}, \tag{9}$$

where $\delta v_r = v(\boldsymbol{x}+r\boldsymbol{e}_1) - v(\boldsymbol{x})$. The structure functions are computed as follows. For a separation vector $r\boldsymbol{e}_i$, $|\delta u_r(\boldsymbol{x}, s)|^p$ is spatially averaged, then averaged across the three directions of the separation vector, and finally averaged over time,

$$\langle|\delta u_r|^p\rangle = \frac{1}{3NT^{av}_{eddy}}\int_t^{t+T^{av}_{eddy}}\left(\sum_{i=1}^{3}\sum_{\boldsymbol{x}}|u(\boldsymbol{x}+r\boldsymbol{e}_i, s) - u(\boldsymbol{x}, s)|^p\right)ds, \tag{10}$$

where $\boldsymbol{e}_i$ is the unit vector in the direction of the x_i axis.

Figures 2 and 3 show $S_p^L(r)$ and $S_p^T(r)$ for various orders at $R_\lambda = 460$. When $\lambda < r < L$, there is a range of finite lengths over which the curves of $S_p^L(r)$ and $S_p^T(r)$ are straight. The scaling exponents ζ_p^L and ζ_p^T were computed by taking the logarithmic derivatives of $S_p^L(r)$ and $S_p^T(r)$; these are summarized in Fig.4. Unlike the Kolmogorov theory, both ζ_p^L and ζ_p^T vary nonlinearly with p. The quantities ζ_p^L and ζ_p^T are almost identical up to the 6th order. At higher orders, however, ζ_p^T is less than ζ_p^L, which indicates that the transverse velocity increment is more intermittent than the longitudinal velocity increment. Due to insufficient statistical convergence of $S_p^L(r)$ and $S_p^T(r)$ for $p > 6$, however, the differences observed at higher orders are not conclusive.

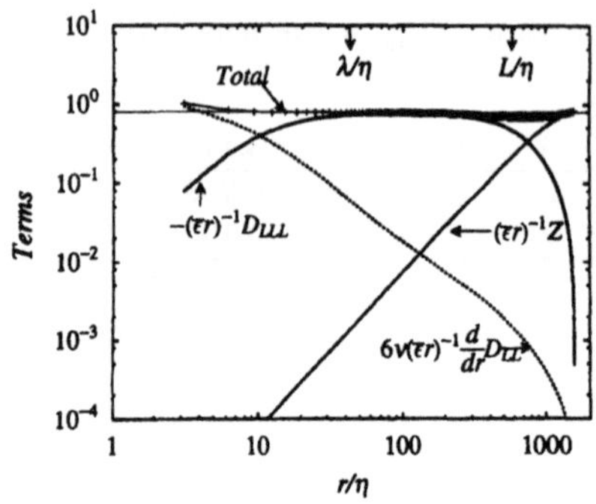

Fig. 1. Terms in the Kármán-Howarth-Kolmogorov equation. $R_\lambda = 381$. D_{LL}, D_{LLL}, and Z denote $S_2^L(r), S_3^L(r)$, and the forcing term, respectively. The thin solid line is 4/5.

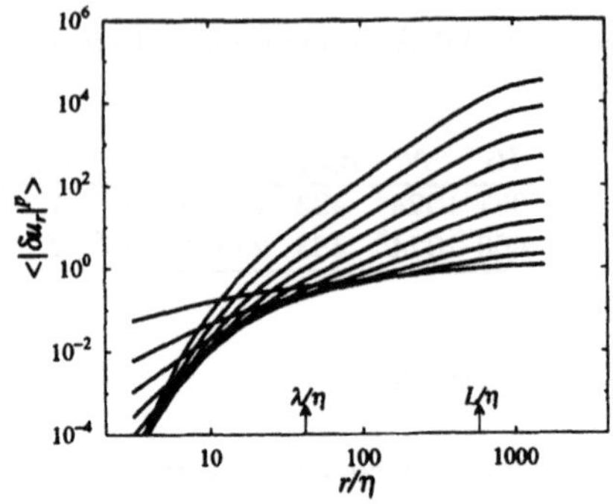

Fig. 2. Plot of $S_p^L(r)$ against r/η for various orders. $R_\lambda = 460$. The curves represent $p = 1, 2, \cdots, 10$ from the lowermost at $r/\eta = L/\eta$.

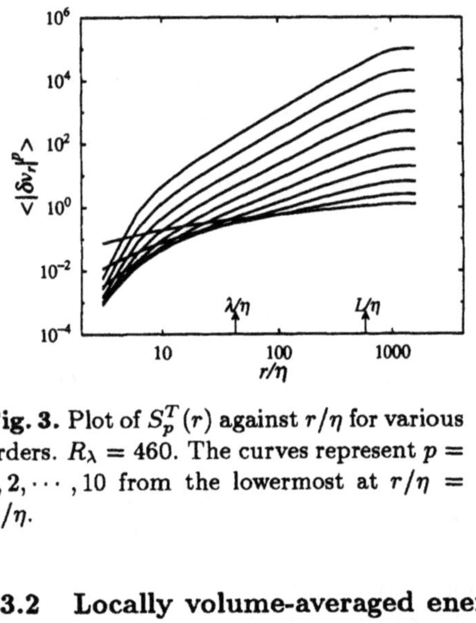

Fig. 3. Plot of $S_p^T(r)$ against r/η for various orders. $R_\lambda = 460$. The curves represent $p = 1, 2, \cdots, 10$ from the lowermost at $r/\eta = L/\eta$.

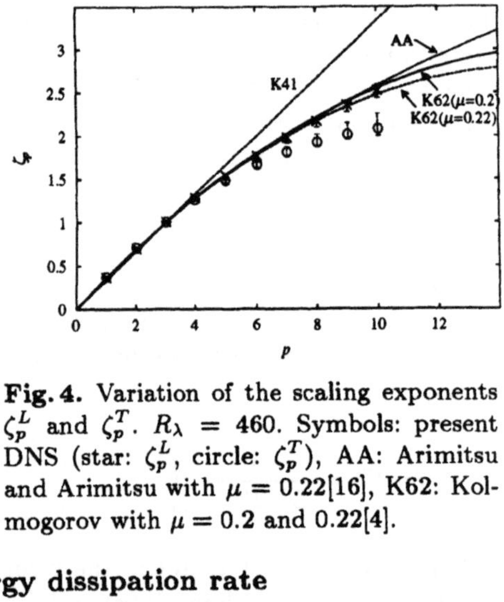

Fig. 4. Variation of the scaling exponents ζ_p^L and ζ_p^T. $R_\lambda = 460$. Symbols: present DNS (star: ζ_p^L, circle: ζ_p^T), AA: Arimitsu and Arimitsu with $\mu = 0.22$[16], K62: Kolmogorov with $\mu = 0.2$ and 0.22[4].

3.2 Locally volume-averaged energy dissipation rate

When the locally volume-averaged energy dissipation rate is computed in the same way as the structure functions, a considerable amount of time is required, especially for large r on the order of the integral scale. To compute the local volume average efficiently, the following technique was used:

$$\begin{aligned}\epsilon_r(\boldsymbol{x}) &= \langle\epsilon(\boldsymbol{x})\rangle + \frac{1}{V_r}\int_{V_r}\widetilde{\epsilon}(\boldsymbol{x}+\boldsymbol{y})\mathrm{d}\boldsymbol{y} \\ &= \langle\epsilon(\boldsymbol{x})\rangle + 3\int_{V_r}\widetilde{\epsilon}(\boldsymbol{k})\left(\frac{\sin(kr)}{k^3r^3} - \frac{\cos(kr)}{k^2r^2}\right)e^{i\boldsymbol{k}\cdot\boldsymbol{x}}\mathrm{d}\boldsymbol{k} \qquad (11)\end{aligned}$$

where $V_r = 4\pi r^3/3$ and $\tilde{\epsilon}(\boldsymbol{x})$ is the fluctuation of the energy dissipation rate from its average.

Figure 5 shows the normalized PDF for $(\epsilon_r - \langle\epsilon_r\rangle)/\sigma(\epsilon_r)$ for various r at $R_\lambda = 381$, where r is the unit grid size $dx = 2\pi/1024$. The right tail of the PDF becomes longer with decreasing r, while the left tail becomes shorter. The normalized PDF for $(\ln\epsilon_r - \langle\ln\epsilon_r\rangle)/\sigma(\ln\epsilon_r)$ is shown in Fig.6 for the dissipation range scale, in Fig.7 for the inertial range scale, and in Fig.8 for scales larger than the integral scale. The general trend is that the body of the distribution for $\ln\epsilon_r$ is very close to Gaussian for scales in the inertial and dissipation ranges, but the tails decay faster than Gaussian and this deviation becomes larger as r increases. The degree to which the PDF of $\ln\epsilon_r$ is Gaussian can be observed quantitatively using the flatness factor

$$F = \left\langle(\ln\epsilon_r - \langle\ln\epsilon_r\rangle)^4\right\rangle / \left\langle(\ln\epsilon_r - \langle\ln\epsilon_r\rangle)^2\right\rangle^2. \tag{12}$$

The flatness factor is plotted against r/η in Fig.9. F is close to three in the inertial and dissipation ranges, and becomes smaller as r increases past the size of the integral scale. The degree to which the body of $P(\log\epsilon_r)$ is Gaussian is consistent with the observations by Wang *et al.* [12], but the growth of the deviation from Gaussian with increasing r is opposite. This is presumably due to the insufficient length of the inertial range in Wang's study. The same trend is also observed when $R_\lambda = 460$. Using experiments, Praskovski and Oncley found that the $\ln\epsilon_r'$ PDF was close to Gaussian in the inertial range when $R_\lambda = 3200$, but the distribution became narrower at $R_\lambda = 12700$ [9]. It is difficult to infer the asymptotic PDF for $\ln\epsilon_r$ at large Reynolds numbers using the present DNS data.

The faster decay of the $\ln\epsilon_r$ PDF for larger values of r can be explained as follows. Suppose that the velocity field is multivariate Gaussian with a given energy spectrum. Then the spectrum support of the dissipation field is η, *i.e.*, the dissipation field in physical space consists of an ensemble of volumes of size η over which the coherence of the dissipation field prevails. Taking a spatial average over the volume of $r^3(\gg\eta^3)$ results in a Gaussian PDF for ϵ_r^G by the central limit theorem because each volume of η^3 is statistically independent. Note that the central limit theorem is used in the sense of the spatial average, not of the scale of space. Now consider the velocity field in the actual turbulent flow. There are structures in the dissipation field as well as the vorticity field whose largest length scale is on order of L. Taking the spatial average of $\epsilon(\boldsymbol{x})$ over the volume $r^3(\gg L^3)$, we obtain a Gaussian PDF for ϵ_r again by the central limit theorem, because a volume of r^3 contains many volumes of L^3 that are statistically independent. The bodies of the ϵ_r PDFs with $r > L$ are close to Gaussian, although the tails decay somewhat faster (figure not shown). When the Gaussian ϵ_r PDF at this large scale r is written in terms of the logarithm of ϵ_r, the tail of the $\ln\epsilon_r$ PDF falls off much more rapidly. On the other hand, when r is on order of $\eta(\ll L)$, the number of cascade steps in scale space is large and the central limit theorem for the breakdown coefficient works well, provided that an argument similar to that of Gurvich and Yaglom is valid [17]. We can

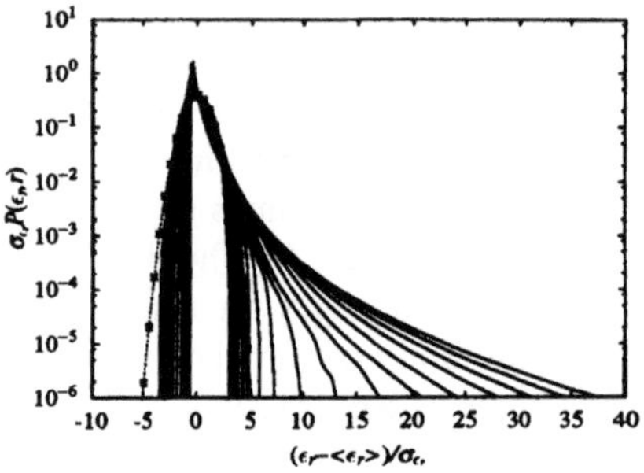

Fig. 5. Normalized ϵ_r PDF for the dissipation range scale at $R_\lambda = 381$. From the outermost curve, $r_n/\eta = 2^{n-1}dx/\eta = 2.38 \times 2^{n-1}, n = 1, \cdots, 5$, and $r_n/\eta = 2.38 \times 32n, n = 1, \cdots, 10$.

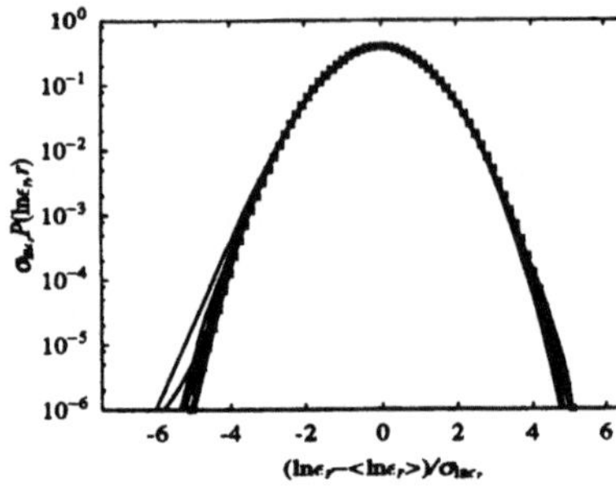

Fig. 6. Normalized ϵ_r PDF for the dissipation range scale at $R_\lambda = 381$. From the outermost curve, $r_n/\eta = 2^{n-1}dx/\eta = 2.38 \times 2^{n-1}, n = 1, \cdots, 5$.

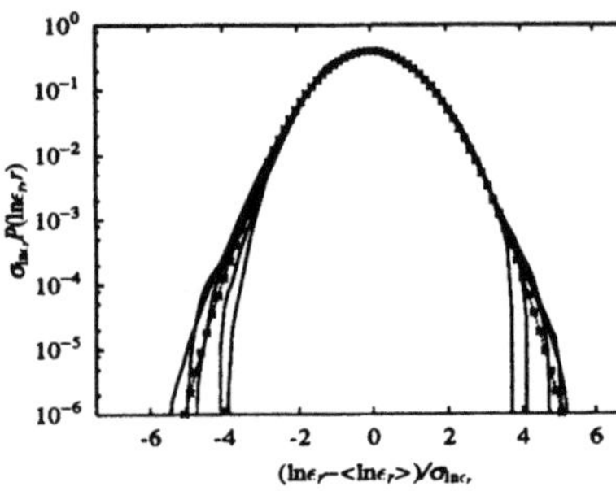

Fig. 7. Normalized ϵ_r PDF for the inertial range scale at $R_\lambda = 381$. From the outermost curve, $r_n/\eta = 32ndx/\eta = 2.38 \times 32n, n = 1, \cdots, 5$.

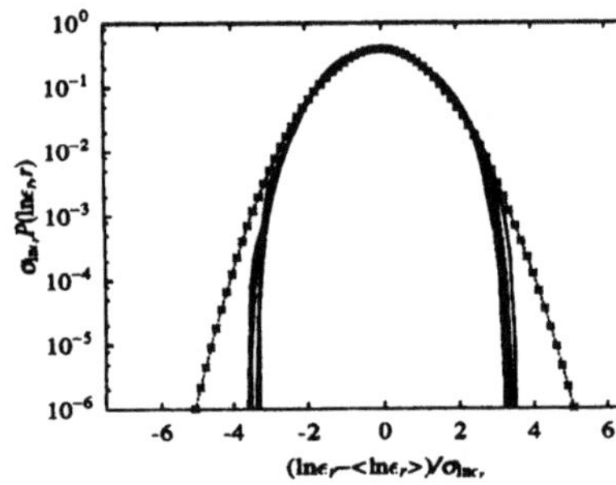

Fig. 8. Normalized ϵ_r PDF for scales larger than the upper inertial range at $R_\lambda = 381$. From the outermost curve, $r_n/\eta = 32ndx/\eta = 2.38 \times 32n, n = 6, \cdots, 10$.

estimate the number of steps required to reach η by assuming that the size of the scale is decreased by a factor of two for every step. For example, $L/\eta \approx 440 = 2^n$ yields $n \approx 8.8$ for $R_\lambda = 380$, and $L/\eta \approx 577 = 2^n$ gives $n \approx 12.5$ for $R_\lambda = 460$. Although these values are not large enough for the central limit theorem to apply, the actual $\ln \epsilon_r$ PDF is very close to Gaussian.

The log-normal distribution for ϵ_r yields the moment

$$\langle \epsilon_r^q \rangle = A_q \bar{\epsilon}^q \left(\frac{L}{r} \right)^{\mu q(q-1)/2} \tag{13}$$

for the inertial range scale r. The variance of ϵ_r is plotted against r in Fig.10 using compensation r^μ for three values of $\mu = 0.2, 0.22, 0.25$. For the inertial range r ($\lambda < r < L$), the value $\mu = 0.2$ yields the best horizontal curve, although the difference between $\mu = 0.22$ and 0.2 is very small. The present DNS value $\mu = 0.2$ is consistent with experimentally observed values, reported as between 0.2 and 0.25. It is interesting to compare μ with the scaling exponent of $\langle \Pi(r)^2 \rangle \approx S_6^L(r)/r^2$, which is the second order moment of the energy transfer rate due

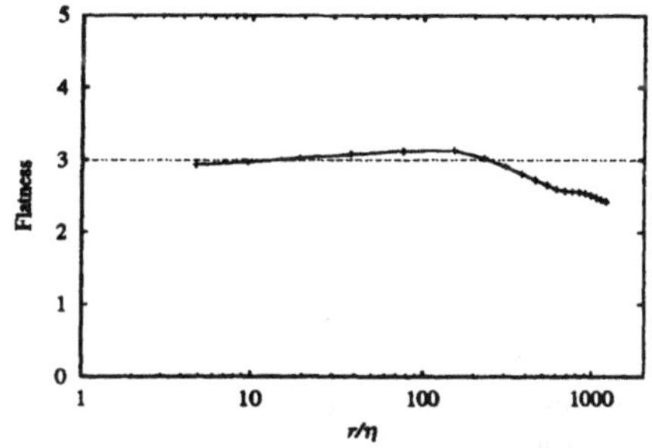

Fig. 9. Variation of the flatness factor $\langle(\ln\epsilon_r - \langle\ln\epsilon_r\rangle)^4\rangle/\langle(\ln\epsilon_r - \langle\ln\epsilon_r\rangle)^2\rangle^2$ with r/η at $R_\lambda = 381$.

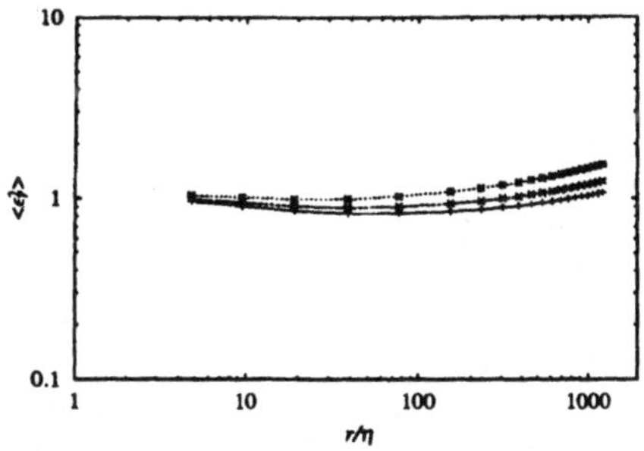

Fig. 10. Variation of the compensated variance $\langle\epsilon_r^2\rangle(r/\eta)^\mu$ with r/η at $R_\lambda = 381$. $\mu = 0.2, 0.22, 0.25$ from the lowermost curve.

to the nonlinear term of the Navier-Stokes equation. The present DNS gives $S_6^L(r) \propto r^{1.79}$ for $R_\lambda = 381$ ($r^{1.77}$ for $R_\lambda = 460$), *i.e.*, $\langle\Pi(r)^2\rangle \approx S_6^L(r)/r^2 \propto r^{-0.21}$. The exponent 0.21 is very close to $\mu = 0.2$ for $R_\lambda = 381$.

The scaling exponent for the velocity structure function with $\mu = 0.2$, obtained from the log-normal theory, is plotted in Fig.4 for $R_\lambda = 460$. The agreement is satisfactory when $p \leq 6$, but the difference becomes larger for higher orders. Arimitsu and Arimitsu derived the scaling exponents analytically using statistics based on the Havrda-Charvat-Tsallis (generalized) entropy [16]. They contain one parameter: μ. The curve with $\mu = 0.22$ is close to the present DNS results.

4 Concluding Remarks

It is well known that there are difficulties with the K62 theory. For example, it breaks the requirement that the scaling exponent is non-decreasing function of p. The log-normal theory, however, is a good approximation for moderate amplitudes of $\ln\epsilon_r$ for scales between the dissipation and inertial ranges at the values of R_λ studied here. The log-normal theory is derived by physical intuition of the energy cascade. More precise description of the ϵ_r PDF requires detailed analyses of the PDF using Navier-Stokes dynamics, which is a significant challenge for the turbulence community.

The author thanks the Nagoya University Computation Center and the Advanced Computing Center at RIKEN for providing the computational resources. This work was supported partly by the "Research for the Future" Program of the Japan Society for the Promotion of Science, under the project JSPS-RFTF97P01101, and by a Grant-in-Aid for Scientific Research (C-2 12640118) from the Japan Society for the Promotion of Science.

References

1. A. N. Kolmogorov, "The local structure of turbulence in incompressible viscous fluid for very large Reynolds numbers," Dokl. Akad. Nauk SSSR **30**, 9 (1941).

"Dissipation of energy in locally isotropic turbulence," Dokl. Akad. Nauk SSSR, **32**, 16-18 (1941).
2. A. C. Monin and A. M. Yaglom *Statistical Fluid Mechanics*, Vol. II, MIT Press, Cambridge (1975).
3. U. Frisch, *Turbulence: The Legacy of A. N. Kolmogorov*, Cambridge University Press, (1995).
4. A. N. Kolmogorov, "A refinement of previous hypotheses concerning the local structure of turbulence in a viscous incompressible fluid at high Reynolds number," J. Fluid Mech. **13**, 82 (1962).
5. G. Stolovitzky, P. Kailasnath, and K. R. Sreenivasan, "Kolmogorov's refined similarity hypotheses," Phys. Rev. Lett. **69**, 1178 (1992).
6. L. Stolovitzky and K. R. Sreenivasan, "Scaling of structure functions," Phys. Rev. E **48**, R33 (1993).
7. A. Praskovsky, "Experimental verification of the Kolmogorov refined similarity hypothesis," Phys. Fluids A **4**, 2589 (1992).
8. S. G. Saddoughi and S. V. Veeravalli, "Local isotropy in turbulent boundary layers at high Reynolds number," J. Fluid Mech. **268**, 333 (1994).
9. A. Praskovsky and S. Oncley, "Comprehensive measurements of the intermittency exponent in high Reynolds number turbulent flows," Fluid Dyn. Res. **21**, 331 (1997).
10. W. Van de Water and J. A. Herweijer, "High-order structure functions of turbulence," J. Fluid Mech. **387**, 3 (1999).
11. T. Zhou and R. A. Antonia, "Reynolds number dependence of the small-scale structure of grid turbulence," J. Fluid Mech. **406**, 81 (2000).
12. L. P. Wang, S. Chen, J. Brasseur and J. C. Wyngaard, "Examination of fundamental hypotheses in the Kolmogorov refined turbulence theory through high-resolution simulations, Part I. Velocity field," J. Fluid Mech. **309**, 113 (1996).
13. T. Gotoh and D. Fukayama, "Pressure spectrum in homogeneous turbulence," Phys. Rev. Lett. **86**, 3775 (2001).
14. T. Gotoh, D. Fukayama, and T. Nakano, "Velocity field statistics in homogeneous steady turbulence obtained using a high resolution DNS," Phys. Fluids, to appear, (2002).
15. K. R. Sreenivasan, "On the universality of the Kolmogorov constant," Phys. Fluids **7**, 2778 (1995).
16. T. Arimitsu and N. Arimitsu, "Analysis of turbulence by statistics based on generalized entropies," Physica A **295**, 177 (2001).
17. A. S. Gurvich and A. M. Yaglom, "Breakdown of eddies and probability distributions for small-scale turbulence," Phys. Fluids Suppl. **10**, S59 (1967).

Lyapunov Exponent of the System Described by Kuramoto-Sivashinsky Equation

Hiroshi Shibata

Department of General Education, Faculty of Engineering, Sojo University, Kumamoto 860-0082, Japan

Abstract. The dynamics of the system described by Kuramoto-Sivashinsky equation (SKSE) is studied in this paper. The Lyapunov exponent of the SKSE fluctuates around at the value of 0. Then the two time correlation function (TTCF) for the SKSE is calculated and it is shown that the TTCF decays algebraically. Those results strongly suggest that the large deviation statistics do not hold on the SKSE.

1 Introduction

The statistical characteristics in the turbulence described by partial differential equations have been studied extensively[1-8]. One of the systems is described by the Kuramoto-Sivashinsky (KS) equation[9-15]. The KS equation describes the phase turbulence such as reaction diffusion process, flame front, liquid on the tilted surface, etc. [1,3]. The KS equation is written explicitly as

$$\partial_t \psi + \eta\psi + \partial_{xx}\psi + \partial_{xxxx}\psi + 2\psi\partial_x\psi = 0, \tag{1}$$

where ∂_t and ∂_x mean the partial differentiation with respect to time t and space x, respectively. The solution of the KS equation, Eq.(1), is in the form of

$$\psi(x,t) = |\epsilon|^{\delta} f(|\epsilon|^{\mu+\frac{1}{2}}x, |\epsilon|^{\nu+2}t), \tag{2}$$

where ϵ is a small parameter and exponent δ, μ, and ν take the values of 1, $\frac{1}{2}$, and 2, respectively[9]. η is the damping constant and an unstable mode exists when $\eta < \frac{1}{4}$. The 3rd term $\partial_{xx}\psi$ gives rise to the instability in the long wave length regime. The 4th term $\partial_{xxxx}\psi$ suppresses the fluctuation in the short wave length regime. The interaction between the long and short wave length regimes is expressed by the nonlinear term $2\psi\partial_x\psi$. This nonlinear term $2\psi\partial_x\psi$ contributes to the global stability of the system. It should be noticed that the magnitude of every term is same.

In this paper we study the system described by the KS equation (SKSE) from the viewpoint of the dynamics. Especially the finite dimensional space is introduced and the state of the SKSE is defined in this space.

This paper is organized as follows. In Section 2, the Lyapunov exponent of the SKSE is studied numerically. In this section it is shown that the Lyapunov exponent of the SKSE fluctuates around at the value of 0. Following Section 2, the two time correlation function (TTCF) for the SKSE is studied in Section 3. It becomes clear that the TTCF for the SKSE decays algebraically. Conclusions are given in Section 4.

2 Lyapunov exponent of Kuramoto-Sivashinsky equation

In this section we introduce the finite dimensional space where the state of the SKSE is defined. Then we calculate numerically the time series of Lyapunov exponent[16-18] from the expansion rate.

First we set the space where we define the state of the SKSE. When we solve Eq.(1) numerically with the finite difference method, we prepare the discrete space sites. We write explicitly the solution at a discrete time step n as

$$\boldsymbol{\psi}^n = (\psi_0^n, \psi_1^n, \psi_2^n, \cdots, \psi_{N-1}^n)^T, \tag{3}$$

where N is the system size in order for us to solve Eq.(1) numerically. Here we introduce N dimensional space spanned by the sites for the solution. We can assume that the state of the SKSE is described in this N dimensional space.

Then we calculate the Lyapunov exponent for the SKSE. At a discrete time step n, the state is described by the state vector (3). We put a short length vector $\delta\boldsymbol{\psi}^n$ to the state vector $\boldsymbol{\psi}^n$. It should be noticed that the length of the vector $\delta\boldsymbol{\psi}^n$ is small enough not to change its stretching rate. Here we adopt 0.001 as the length of the vector. We set the periodic boundary condition

$$\psi_{i+N}^k = \psi_i^k \quad (i = 0, 1, 2, \cdots, N-1) \tag{4}$$

and the initial conditions

$$\psi_i^0 = 1.0 \sin(2\pi \frac{i}{N}). \quad (i = 0, 1, 2, \cdots, N-1) \tag{5}$$

The 4th order Adams-Bashforth and the 2nd order central difference of space are used in order to solve Eq.(1). The system size N is 320 and its size length is 16π. The space site width is $\frac{16\pi}{320}$ and the time step width is 10^{-5}. The end point of the short length vector is set as

$$\psi_i^{\prime k} = \psi_i^k + (\frac{DL}{N})^{\frac{1}{2}}, \quad (i = 0, 1, 2, \cdots, N-1) \tag{6}$$

where DL is the length of the short length vector ,i.e., 0.001 here and N is the system size. Then we calculate the expansion rate DL' of the short length vector. It is noticed that we reset the end point of the short length vector as

$$\psi_i^{k+1} + (\psi_i^{\prime k+1} - \psi_i^{k+1})/DL' \tag{7}$$

at every time step. The Lyapunov exponent is calculated as

$$\lambda_k = \frac{1}{DT} \ln(DL'), \tag{8}$$

where DT is the time step width and 10^{-5} here. The results are shown in Fig.1.

Fig.1 shows that the Lyapunov exponent of the SKSE fluctuates around at 0. This means that the chaoticity for the SKSE is weak[19-21] and it predicts that the dynamics for the state vector for the SKSE show weaker decay than exponential one. In addition, it is easily realized that the magnitude of the fluctuation of the Lyapunov exponent for the SKSE becomes small as the damping constant η becomes large.

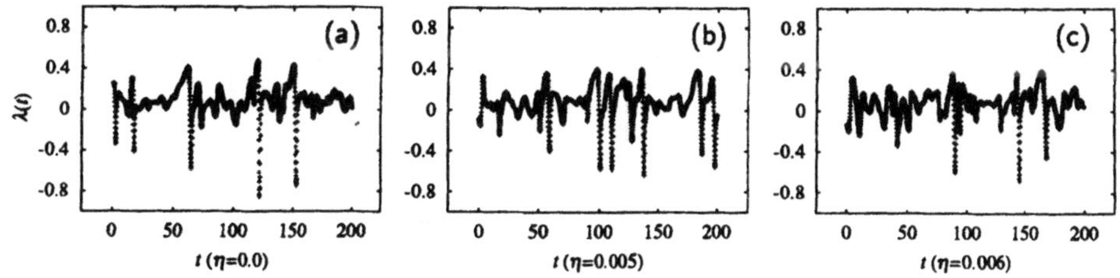

Fig. 1. Time series of Lyapunov exponent obtained by Eq.(8). (**a**) is for the damping constant $\eta = 0.0$. (**b**) and (**c**) are for $\eta = 0.005$ and 0.006, respectively. They show that the fluctuation of the time series decreases as η becomes large.

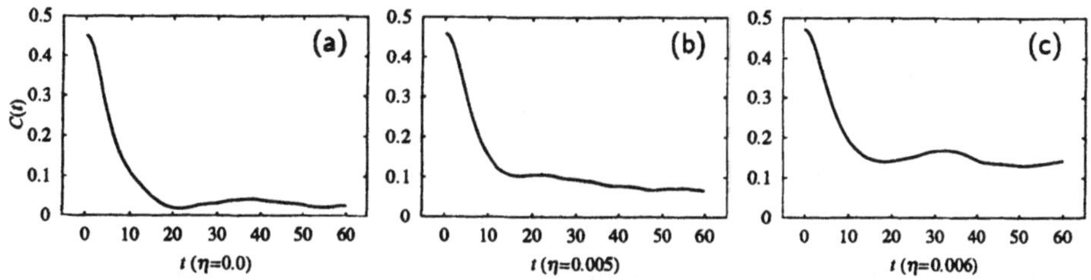

Fig. 2. Two time correlation function obtained by Eq.(10). (**a**), (**b**), and (**c**) are for $\eta = 0.0$, 0.005, and 0.006, respectively. The decay becomes weak as we increase η.

3 Two time correlation function of Kuramoto-Sivashinsky equation

In this section we calculate numerically the two time correlation function (TTCF) for the state vector of the SKSE. The dependency of the TTCF for the SKSE on the damping constant η is also studied.

The two time correlation function (TTCF) of the SKSE[22] is written as

$$C(t) = \frac{1}{L}\int_0^L dx(<\psi(x,t)\psi(x,0)> - <\psi(x,t)>^2), \tag{9}$$

where $< \cdots >$ means the long time average along the trajectory of a solution for Eq.(1). Rewriting Eq.(8) to the discrete form concerning space and time, we obtain

$$C(t) = \frac{1}{N}\sum_{i=0}^{N-1}(<\psi_i^n\psi_i^0> - <\psi_i^n>^2), \tag{10}$$

where n is $\frac{t}{\Delta t}$. Δt is the time step width and 10^{-5}. N is the system size and 320. Eq.(10) describes the TTCF of the state vector between time t and 0. The boundary condition and the initial conditions are same as the ones when we calculate the Lyapunov exponent. Of course, the method in order to solve Eq.(1) is same as the one in Section 2. The results are shown in Fig.2.

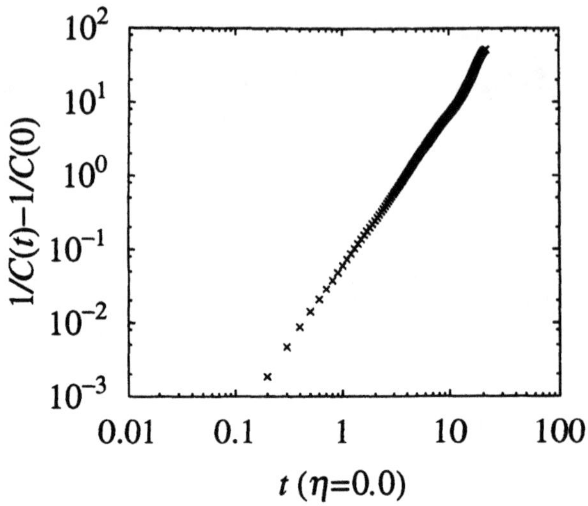

Fig. 3. $1/C(t) - 1/C(0)$ vs t for the SKSE at $\eta = 0.0$ with log-log plot. The slope of the line in this figure gives the exponent α in Eq.(11).

The TTCF of the state vector for the SKSE decays until time $t = 20$. After $t = 20$, the TTCF saturates its value. As we increase the value of η, the damping constant, the decay becomes weak as shown in Fig.2. In order to investigate the decay of the TTCF in detail, we study $1/C(t) - 1/C(0)$ vs t with the log-log coordinates, we obtain the clear line as shown in Fig.3.

The slope of this line is 2.18 in the regime from $t = 0$ to $t = 10$ and 3.45 in the regime from $t = 10$ to $t = 20$. From the results in Fig.3, we obtain the functional form of the TTCF of the state vector for the SKSE as

$$C(t) = \frac{C(0)}{At^{\alpha} + 1}, \tag{11}$$

where α equals 2.18 until $t = 10$ and 3.45 from $t = 10$ to $t = 20$.

4 Conclusions

We have studied the dynamics of the system described by the Kuramoto-Sivashinsky equation (SKSE). First we have realized that the Lyapunov exponent of the state vector for the SKSE fluctuates around at 0. This means that the SKSE is in the state of weak chaos. Second we have obtained the functional form of the two time correlation function of the state vector for the SKSE. The functional form is written by Eq.(11). In addition, the above results predict that the statics of the state vector for the SKSE deviate from the large deviation statistics[23-34]. This deviation is going to be confirmed.

Acknowledgements

The author would like to thank Research Institute of Applied Mechanics of Kyushu University for financial support No.13 ME-3.

References

1. T.Bohr, M.H.Jensen, G.Paladin, A.Vulpiani, Dynamical Systems Approach to Turbulence, Cambridge University Press, Cambridge, 1998.
2. U.Frisch, Turbulence, Cambridge University Press, Cambridge, 1995.
3. M.C.Cross, P.C.Hohenberg, Rev. Mod. Phys. 65(1993)851.
4. X.-L.Qui and P.Tong, Phys. Rev. Lett. 87(2001)094501.
5. A.La Porta, G.A.Voth, A.M.Crawford, J.Alexander, E.Bodenschatz, Nature 409(2001)1017.
6. B.I.Shraiman, E.D.Siggia, Nature 405(2000)639.
7. D.A.Egolf, I.V.Melnikov, W.Pesch, R.E.Ecke, Nature 404(2000)733.
8. H.Fujisaka, H.Suetani, T.Watanabe, Prog. Theor. Phys. 139(Suppl.) (2000)70.
9. H.Mori, Y.Kuramoto, Dissipative Structures and Chaos, Springer, Berlin, 1998.
10. P.Maneville, Dissipative Structures and Weak Turbulence, Academic Press, Boston, 1990.
11. Y.Kuramoto, Chemical Oscillations, Waves, and Turbulence, Springer, Berlin, 1984.
12. H.Chaté, P.Manneville, Phys. Rev. Lett. 58(1987)112.
13. P.Manneville, in: O.Pironneau (Ed.), Macroscopic Modeling of Turbulent Flows, Lecture Notes in Physics, Vol.230, Springer, Berlin, 1985, p.1.
14. G.I.Sivashinsky, Acta Astronaut 4(1977)1177.
15. Y.Kuramoto, T.Tsuzuki, Prog. Theor. Phys. 55(1976)356.
16. A.Katok, B.Hasselblatt, Introduction to the Modern Theory of Dynamical Systems, Cambridge University Press, Cambridge, 1995.
17. E.Ott, Chaos in Dynamical Systems, Cambridge University Press, Cambridge, 1993.
18. H.G.Schuster, Deterministic Chaos, VCH, Weinheim, 1988.
19. G.M.Zaslavsky, Physics of Chaos in Hamiltonian Systems, Imperial College Press, London, 1998.
20. C.Anteneodo, C.Tsallis, Phys. Rev. Lett. 80(1998)5313.
21. R.Cafiero, A.Valleriani, J.L.Vega, Eur. Phys. J. B 4(1998)405.
22. K.Sneppen, J.Krug, M.H.Jensen, C.Jayaprakasch, T.Bohr, Phys. Rev. A 46(1992)R7351.
23. P.Gaspard, Chaos, Scattering and Statistical Mechanics, Cambridge University Press, Cambridge, 1998.
24. C.Beck, F.Schlögl, Thermodynamics of Chaotic systems, Introduction, Cambridge University Press, Cambridge, 1998.
25. R.S.Ellis, Entropy, Large Deviations, and Statistical Mechanics, Springer-Verlag, New York, 1985.
26. D.Ruelle, Thermodynamic Formalism, Addison-Wesley, Mass., 1978.
27. O.E.Lanford, in: A. Lenard (Ed.), Statistical Mechanics and Mathematical Problems, Lecture Notes in Physics Vol.20, Springer, Berlin, 1973, p.1.
28. H.Shibata, Green-Kubo formula derived from large deviation statistics, Preprint.
29. H.Fujisaka, M.Inoue, Phys. Rev. A 41(1990)5302.
30. H.Mori, H.Hata, T.Horita, T.Kobayashi, Prog. Theor. Phys. 99(Suppl.) (1989)1, and references cited therein.
31. H.Fujisaka, M.Inoue, Prog. Theor. Phys. 77(1987)1334.
32. M.Sano, S.Sato, Y.Sawada, Prog. Theor. Phys. 76(1986)945.
33. R.Ishizaki, T.Horita, H.Mori, Prog. Theor. Phys. 89(1993)947.
34. R.Ishizaki, T.Horita, T.Kobayashi, H.Mori, Prog. Theor. Phys. 85(1991)1013.

Part IV

Geophysical Turbulence

Toward a Statistical Ocean Dynamics

Greg Holloway[1]

Institute of Ocean Sciences, Sidney BC, Canada

Abstract. Ocean variability occurs across such a wide range of scales that no computer, existing or presently foreseen, can directly represent the huge number of degrees of freedom. Traditionally the problem has been approached in terms of ad hoc subgridscale mixing hypotheses. In this article we explore first steps toward reformulating the question from a perspective of nonequilibrium statistical dynamics. The degrees of freedom explicitly represented by ocean models are seen as moments of probabilities of possible realizations of oceans. Influence of the many unrepresented degrees of freedom appear as "generalized thermodynamic forces" due to gradients of entropy with respect to explictly represented moments. Examples of practical implementation of these ideas are described for (a) effects of eddy-topography interaction in mean ocean circulation and (b) effects of different conduction coefficients for heat and for salt.

1 Oceans are too big

Advances in computing capacity along with advances in numerical methods provide ever greater power to solve for flows in oceans, lakes and duck ponds. In principle we suppose we know a good approximation to equations of motion on some scale, e.g. the Navier-Stokes equations coupled with heat and salt balances under gravity and rotation. In practice we cannot solve for oceans, lakes or most duck ponds on the scales for which these equations apply. For example, the length scales over which ocean salinity varies are often shorter than 1 mm. We try to solve for fields represented on grids (or other bases) that are far larger than the scales to which 'known' equations apply. Then we are compelled to guess equations of motion.

Guessing equations is uncomfortable, often causing us to assume without question equations used by some previous author. When we are brave, we realize this too is uncomfortable. It is natural to wish that, as computers grow ever more powerful, we guess less and less. What would be needed from the computer? In the oceans there are about 1.36×10^{18} m^3 of water. If we felt that variability was unimportant within volumes of $O(10^{-9}$ m$^3)$ then the computer should track $O(10^{27})$ volumes each described by several degrees of freedom. Clearly one can fiddle these numbers. Today's 'big' computer models, e.g. for weather forecasting or turbulence research, may advance $10^7, 10^8$ or 10^9 variables. Over time we are assured the computers will become bigger yet. Even if we imagine computer models advancing 10^{12} variables (not on my desk in my lifetime!), we still face the situation that, for each one variable we track, we must guess how that variable interacts with 10^{15} variables about which we are uninformed. Limiting ourselves to coastal oceans or lakes, the mismatch in degrees of freedom might reduce to 10^8 or less. A suitably modest duck pond could – someday – prove computable? Maybe.

This enterprise is like seeking to reinvent the steam engine from molecular dynamics simulation of water vapour. What a brave but bizarre thing to attempt. For oceans, lakes and ponds the circumstance is even worse than the dismal numbers above suggest. As computing power opens more explicit degrees of freedom, solutions exhibit chaos in the sense that arbitrarily nearby solutions at some time become entirely uncorrelated at later time. This invites ensemble forecasting to allow 'meaningful' results such as means, variances, most probable results, etc. We are led to ask what is an appropriate size of ensemble. 3? No. Perhaps 10^3? Maybe.

The point of scary big numbers is: for the numerical ocean, lake or duck pond to be workable, huge numbers of degrees of freedom must be somehow 'slaved' to a far fewer number. Indeed we suppose this is true. But the keyword is 'somehow', and that is the motivation for this paper.

2 The arrow of time

Our goal is to represent subgridscale variability, often seen in eddy viscosity or mixing parameterizations. A wealth of experience in such matters has shown, arguably, greater and lesser success. Numerical ocean models 'run', and outputs are not conspicuously absurd (sometimes). Successes have depended in parts on choices of eddy viscosities and mixing schemes, drawn from history of trial-and-error tuning. Our purpose here will not be to re-tune traditional eddy viscosities but rather to focus upon underlying fundamentals which can lead to results strikingly different from traditional schemes.

Ocean modeling rests upon geophysical fluid dynamics (GFD), a branch of classical mechanics applied to continua under gravity and rotation. Apart from the role of viscosity or mixing (which are 'added on' to GFD without derivation), GFD inherits from classical mechanics a marvelous geometric quality about space and time. 'Past' and 'future' in t are not fundamentally different from 'left' or 'right' in x. An evolution from past to future is no more usual than from future to past (bearing in mind that Earth's rotation reverses for a time reversal experiment). But real flows do not evolve 'backwards' from future to past. This is true in global oceans, lakes, duck ponds and tea cups. (Milk stirred into tea turns brown. Reversing the stirring does not re-separate the milk from the tea.)

How should GFD learn about time's arrow from past to future? The question is at the heart of moving from classical to statistical mechanics, e.g. as Coveney and Highfield, [1]. Two key ideas come to bear: probability and entropy.

Symbolically, let $\mathbf{y}$ denote a vector in phase space (of dimension 10^{27} or whatever), governed by $dy/dt = \mathbf{F}(\mathbf{y}) + \mathbf{G}$. $\mathbf{y}$ is simply the collection of all variables needed to fully describe the ocean (or whatever) while $\mathbf{F}$ and $\mathbf{G}$ express terms in the forced equations of motion. Depending upon one's interest, $\mathbf{y}$ may include also biological and chemical components. Computation of the highly erratic (presumably chaotic) $\mathbf{y}(t)$ in 10^{27} is not feasible or even desirable. We consider instead probability $p(\mathbf{y})dy$ for actual "$\mathbf{y}$" within phase volume dy of any $\mathbf{y}$. Then we would seek an equation for $dp(\mathbf{y})/dt = ...$ But this is even less feasible than attempting to solve $d\mathbf{y}/dt$.

Aside: one sometimes solves 'toy' stochastic models, such as a Fokker-Planck equation $\partial_t p = -\partial_y J = -\partial_y A(y,t)p + 1/2\partial_y\partial_y B(y,t)p$ where J is a flux of probability and A and B are modeled after some properties estimated from F and G. For realistic ocean modeling, actual derivation of A and B from F and G are not feasible (to my knowledge), and Fokker-Planck equations are explored as intuition-building models.

Practical progress results from reducing the problem of $p(\mathbf{y})$ to that of seeking a chosen set of moments $\mathbf{Y} = \int \mathbf{y}dp$ or, generally, $\mathbf{K} = \int \mathbf{k}(\mathbf{y})dp$ for functions $\mathbf{k}(\mathbf{y})$. An advantage is that one may choose as few $\mathbf{Y}$ as one likes, perhaps 10^8 or perhaps only 10, 'integrating out' as much of $\mathbf{y}$ as one wishes. The outstanding problem is that we do not have equations for $d\mathbf{Y}/dt = ...$ Progress is made by carrying out probability averaging (expectation) over the components of $\mathbf{y}$ retained in $\mathbf{Y}$, yielding

$$d\mathbf{Y}/dt = \mathbf{F}'(\mathbf{Y}) + \mathbf{G}' + ... \tag{1}$$

where $\mathbf{F}'(\mathbf{Y})$ includes linear dynamics and couples explicitly the retained components of $\mathbf{Y}$ while discarding the myriad couplings among $\mathbf{Y}$ and $10^{27}\mathbf{y}$. $\mathbf{G}'$ are the expectations of $\mathbf{G}$ for the retained $\mathbf{Y}$. Eq. (1) expresses traditional GFD symbolically with '...' supplied from ad hoc mixing. The manifest successes of GFD are due largely to circumstances near forced linear dynamics that are fully captured by $\mathbf{F}'$ and $\mathbf{G}'$.

Comment: Suggesting that 'textbook' equations of GFD are wrong often meets resistance. The books have been in use for so long, and the work of GFD turns to applications and properties of these equations, not to questioning the equations. How could such well-used equations be 'wrong'? The problem, I suggest, has been too little attention to the dependent variables we mean to describe. Viewing output of numerical models, it is easy to see maps of 'flows' or 'temperatures' or whatever. Pausing for a moment, we retreat to something like 'grid-cell and time-step averaged flow', etc. Yet if we try to define space-time filtering operators to pass over underlying equations, it is deeply problematic to obtain traditional GFD. The enterprise is far more troubled if we advance to modern computing with eddy-active, presumably chaotic, output. Then what meaning do we attach to 'temperature', say, in some cell over some step in some specific (chance!) realization? Perhaps what we should mean is 'temperature moment over probabilities of possible states'. That's a lot to say. However, if we then asked 'what is the equation of motion of the temperature moment over probabilities of possible states?', we would be far more ready to suppose that a textbook equation for 'temperature' might not be right.

Apart from '...', eq. (1) expresses the marvelous symmetry between past and future inherited from classical mechanics. Time is without its arrow, bringing us to our second key idea: entropy. Entropy is most familiar in thermodynamic context, often $dH = dQ/T$ where dH is the increment of entropy resulting from supply of an increment of heat dQ to a system with constant volume at temperature T. Entropy has also entered common language as an allusion to disorder. We should be careful to be precise. For a processe defined by probability function p, entropy is

$$H = -\int dp \log(p) \tag{2}$$

to within an arbitrary multiplicative constant and within an arbitrary additive constant. Describing probabilistically the field of molecular chaos, H from eq. (2) is the thermodynamic entropy $dH = dQ/T$. In another context, describing probabilistically the likelihoods of transmitted messages, H from eq. (2) is the entropy of information theory.

Aside: while 'log' in eq. (2) is considered natural $\log_e$, this is sometimes $\log_2$ for information application, rendering units of H in bits. Equivalence of thermodynamic H and information theoretical H is discussed in texts such as Katz [2]. However, the duality of entropy under statistical physics and information theory has not been exploited in the context of oceans, e.g. relating dynamics and state estimation.

Finally, the arrow of time arrives with the 2nd Law of Thermodynamics. For 'internal' or 'free' interactions (interactions not subject to external influence), p evolves such that for $t_2 > t_1$, $H_2 \geq H_1$. We are left to instruct GFD that '...' in eq. (1) should provide this property – appropriately!

3 Dynamics of moments of probable flows

The alert reader already asks: do not traditional eddy viscosity or mixing as '...' already break time reversal symmetry? They do. The challenge is that traditional schemes, while irreversible, may not satisfy $H_2 \geq H_1$ for $t_2 > t_1$. In many cases, traditional schemes, guided by intuition and experience, are ok. Milk stirred into tea turns brown and eddy diffusion accomplishes such irreversible browning. H for the milk-tea mix is increased after browning. If intuition and experience worked this reliably, the present paper would be unnecessary and we could instead invest effort refining estimates of eddy diffusion coefficients. But intuition, and GFD, go way wrong as two illustrations (eddy-topography interaction and stably stratified mixing) will show. For now we continue developing the equations of motion.

There have two approaches which we consider briefly. In these approaches the aim is not to complete eq. (1) itself but rather to characterize solutions of eq. (1).

An approach by Paltridge [3,4] hypothesized that the mean state $(\mathbf{Y})$ of the atmosphere is such as to maximize the production of (thermodynamic) entropy. This has been further advanced by Ozawa and Ohmura [5] in context of Earth's atmosphere and shown to be a plausible descriptor for other planetary atmospheres by Lorenz et al. [6]. Of particular note for ourselves are (1) the role of thermodynamic entropy, absent in other work below, and (2) a principle of maximum entropy production (MEP) to which we will return.

The second approach to solutions of eq. (1) resulting from '...' is seen in Salmon et al. [7] and in a host of papers since, referenced in parts by Holloway [8] and Salmon [9]. In these cases the mean states of idealized oceans are described by the maximizing entropy due to uncertain macroscale eddies. These studies have focused upon expectations for potential vorticity $q = (\boldsymbol{\Omega} + \nabla \times \mathbf{u}) \cdot \nabla\rho$, where $\boldsymbol{\Omega}$ is Earth's rotation, $\nabla \times \mathbf{u}$ is vorticity, and $\nabla\rho$ is density gradient. Calculations assume no external forcing and no internal dissipation. Thus the macroscale flow field does not communicate with the field of molecular chaos and thermodynamic entropy is not included. For such unforced, non-dissipative, ideal dynamics, H is maximized subject to conserved integrals of the motion. In early studies, conservation constraints included domain-integrated energy, potential vorticity (q) and enstrophy (q^2). Subsequent studies by Miller [10], Robert and Sommeria [11] or Robert and Rosier [12] have considered the roles of further invariants derived from advection of potential vorticity. Despite such idealizations, many aspects of these solutions seemed to capture realistic features. However, construction of such solutions does not yet provide the missing '...' in eq. (1). For later reference, we denote these maximum entropy (ME) solutions as $\mathbf{Y}^*$.

In many cases, our concern is not only to characterize stationary or equilibrium solutions to eq. (1), such as addressed by MEP or ME, but rather to ask how solutions to eq. (1) evolve from assigned initial conditions and under assigned external forcing. For this we need to complete the equations of motion, i.e. to represent '...' in eq. (1).

Two paths to '...' have been opened. One path is seen in recent research by Chavanis and Sommeria [13], Kazantsev et al. [14] and Polyakov [15] based upon MEP for subgridscale potential vorticity. The promise and challenges along this path are not yet clear. For the present I would refer the interested reader to the cited sources while we here turn to another path for which there is simply more experience to date. It will be seen that the two approaches have conceptually much in common and indeed recent work of Polyakov [15] finds also quantitative results with much in common.

The other path, sketched below, follows 'generalized thermodynamic force' (GTF) after Onsager [16,17] or Onsager and Machlup [18].

If entropy depends upon some set of macroscopic parameters, $\mathbf{X}$ (e.g., the expectations $\mathbf{Y}$), then a forcing acts upon $\mathbf{X}$ due to gradient of H with respect to $\mathbf{X}$, i.e., $\mathbf{C} \cdot \nabla_{\mathbf{X}} H$ where $\mathbf{C}$ is a coefficient tensor projecting $\nabla_{\mathbf{X}} H$ onto $d\mathbf{X}/dt$. As an illustration consider a gas in two concentrations in two chambers separated by a membrane with X, say, the fraction of total gas in one chamber. Derivatives of H with respect to available volume (per molecule) give thermodynamic pressures in the two chamber, yielding nonzero dH/dX when pressures are not equal. dH/dX can force dX/dt depending upon the extent of perforations in the membrane, etc., described by C. Importantly, there are two parts to GTF: $\nabla_{\mathbf{X}} H$

and **C**, which, together provide '...'

$$d\mathbf{Y}/dt = \mathbf{F}'(\mathbf{Y}) + \mathbf{G}' + \mathbf{C} \cdot \nabla_{\mathbf{Y}} H \tag{3}$$

Clearly GTF and MEP have very much in common. Each seeks to drive **Y** as rapidly as possible toward higher H, the physics challenge focussed upon 'as rapidly as possible', considering availability and skill of different methods of estimation.

We return to eq. (3) in section 6. Meanwhile there is an intermission.

4 Eddies and topography, the wrong way

Consider interaction among eddies and mean flows in basins with complex topography. First let's get a wrong answer, coincidentally the answer provided by nearly every major ocean model at ever major institution on Earth. Because it can be difficult to assess 'wrong' in realistic circumstances of many complicated, poorly known inputs while outputs are compared with limited data, let us instead pose a thought experiment. Suppose we only 'see' the ocean on larger scales amenable to numerical modeling, perhaps some 10s to 100 km. Suppose on these scales we find our ocean to be at rest, motionless with flat density surfaces. We cannot see but are aware that on smaller scales the ocean is filled with ubiquitous eddies. Maybe the eddies are from previous episodes of forcing, or maybe they are driven by smaller scale forcing we cannot see, perhaps by enthusiastic goldish. Suppose on the large scale we observe there is no imposed forcing. We wish to predict the future ocean on the scales we can see. We poll the major ocean models worldwide, asking: if the ocean is stably at rest and no forcing acts, what is the future? With extraordinary unanimity across different models, the answer is: nothing. That unanimous answer is dead wrong. Let's see.

We can test the answer within the same models. For a given model, we get a bigger computer allowing resolution on the eddy-active scales not previously seen. From our awareness that small scales eddies exist, we randomly excite the newly realized small scales. The test remains the same as before: on larger scales the ocean is at rest and no large scale forcing acts. We run our newer, higher resolution model to compare the previous prediction that no large scale flow will occur. We find instead (as the reader with computer access – or a friend – can check) that large scale flows emerge and these flows have a definite sense (shallow to right in northern hemisphere).

How did (nearly) all the models at all the major institutions (including those advising governments about climate, etc.) fail this test? It was all in '...' What the models did, based on intuition, experience and simply getting the models to run, was to replace '...' with some manner of eddy viscosity, perhaps even of fancier iterated-Laplacian sorts. When those eddy viscosities saw fields of no motion, they took no action. (If there was some slight motion, it would be damped anyway.) That was wrong. While eddy viscosity did serve to break time symmetry in eq. (1), it broke time symmetry the wrong way, driving H the wrong way (to be shown below).

Traditional GFD with traditional eddy viscosities violates 2nd Law, assuring wrong answers.

Aside: Apparently a 'solution' is to use enough computing power to resolve eddies. The challenge is: how much is enough? While modern models are often termed 'eddy-resolving' the more apt term is 'eddy-admitting'. That is, the resolution is sufficiently fine, allowing explicit damping terms sufficiently weak, that model dynamics support internal instabilities, admitting eddies. But ... eddies are only dynamically 'resolved' when further increase in resolution does not lead to systematic changes to statistics of the eddies. Importantly it is the feedback of eddies upon larger scale mean flow which (I speculate) is most difficult to achieve from refined resolution. Ultimately – in principle – we may suppose computers approaching the molecular dynamics simulation of steam engines or even duck ponds. Our aim in this paper is to seek another way.

5 Eddies and topography, the hard way

That ocean models would fail the eddy-topography test (above) was known theoretically for nearly three decades after a comprehensive theory set out by Herring [19] following spectral-based statistical closure after Kraichnan [20]. A simpler spectral closure theory by Holloway [21], after Kraichnan [22], was consistent with the results from Herring [19]. Despite effort to simplify, these closure theory calculations are hugely difficult and a brave reader is referred to the cited references.

Briefly we recall only relevant aspects from the simpler calculations by Holloway [21]. To render the problem tractable, barotropic quasigeostrophic (QG) dynamics were considered. Potential vorticity $q = (2\boldsymbol{\Omega} + \nabla \times \mathbf{u}) \cdot \nabla \rho$ is approximated $q = \zeta + h$ where ζ is the vertical component of QG vorticity $\nabla \times \mathbf{u}$ and $h = f\delta h/h_o$ represents variation δh of total depth, h_o is a constant reference depth, and f is a constant vertical component of $2\boldsymbol{\Omega}$. The fluid is considered of uniform ρ with $\nabla \rho$ replaced by the inverse of total depth.

Supposing fluctuations of ζ and h are spatially statistically homogenous, evolution of $\langle\zeta\zeta\rangle$ and $\langle\zeta h\rangle$ in spectral domain are predicted for assumed statistics of $\langle hh\rangle$ where $\langle\rangle$ denote probability expectation. Compared with direct numerical simulations, such closure theories showed reasonable skill. Moreover, considering an ensemble of realizations of ζ for given realization of h, theory easily showed that $\langle\zeta\zeta\rangle$ confined initially to small scales would readily force nonzero $\langle\zeta\rangle = \langle\zeta h\rangle/h$ on all scales for which $h \neq 0$. The thought experiment posed in sect 4 was sure to fail.

Eddy-topography closure was taken a little farther in Holloway [23] to include nonzero spatially uniform flow U and admit simple $f = f_o + \beta y$. This allowed calculation of pressure-topography 'form drag' with dynamically responding dU/dt. If, in addition to external forcing applied to U, there were assumed sources of eddy energy (e.g., stochastic wind forcing), then the pressure-topography forces could systematically propel U. Use of the term 'form drag' has been largely replaced by 'form stress' or 'topographic stress' to recognize that this force may not simply retard mean flow but may force mean flow. Numerical experiments confirmed that, with mean wind forcing applied to U of sign $U > 0$, the response tended toward $U < 0$. This counter-intuitive result gave rise to the label 'neptune effect' (whimsically suggesting the phenomenon was otherwise inexplicable).

Although some of the above-mentioned results are satisfying theoretically, they are very limited from practical perspective. A large amount of tedious calculation is needed to obtain results restricted to statistically homogenous, barotropic QG flow. So what? Clues to a way forward were buried in Holloway [21] then made powerfully clear in a key paper by Carnevale et al. [24] who showed that for entire classes of closure theory after Kraichnan [20,22] the theories strictly assured nonlinear interaction terms yielding $dH/dt \geq 0$, driving the system monotonically toward the ME solution $\mathbf{Y}^*$. (Other terms due to external forcing or dissipation could decrease H, preventing $\mathbf{Y}^*$.) Thus the outcome of closure theories were to show in detail how dynamics drove systems from any $\mathbf{Y}$ toward $\mathbf{Y}^*$. Can we use this property to motivate highly simplified approximations to closure theory?

6 Eddies and topography, the easy way

Apparently the roles of external forces and internal dissipation are to drive realistic $\mathbf{Y}$ away from $\mathbf{Y}^*$, inducing entropy gradient $\nabla_{\mathbf{Y}} H$ which, if unchecked, would force $\mathbf{Y}$ toward $\mathbf{Y}^*$. In terms of GTF this can be viewed as a Taylor expansion $\mathbf{C} \cdot \nabla_{\mathbf{Y}} H \approx \mathbf{C} \cdot \nabla_{\mathbf{Y}} \nabla_{\mathbf{Y}} H \cdot (\mathbf{Y} - \mathbf{Y}^*)$ about $\mathbf{Y} = \mathbf{Y}^*$ where $\nabla_{\mathbf{Y}} H = 0$. More carefully we would admit we do not know if typical $\mathbf{Y}$ are sufficiently close to $\mathbf{Y}^*$ and we would be daunted seeking to estimate $C \cdot \nabla_{\mathbf{Y}} \nabla_{\mathbf{Y}} H$. But a simple scheme emerges. Denoting $\mathbf{K} \equiv \mathbf{C} \cdot \nabla_{\mathbf{Y}} \nabla_{\mathbf{Y}} H$, eq. (3) becomes

$$d\mathbf{Y}/dt = \mathbf{F}'(\mathbf{Y}) + \mathbf{G}' + \mathbf{K} \cdot (\mathbf{Y} - \mathbf{Y}^*) \tag{4}$$

For application we need $\mathbf{K}$ and $\mathbf{Y}^*$. $\mathbf{Y}^*$ can be inferred after Salmon et al. [7] who find at ME a relation between QG streamfunction, Ψ, defined by $\nabla^2 \Psi = \zeta$ and topography, h, viz. $\Psi = L^2(\nabla^2 \Psi + h)$ where L^2, occurring as a ratio of Lagrange multipliers in maximization of H, has units of length^2 related to coherence scales in the eddy vorticity field. If our scales of interest, perhaps as model resolved, are significantly larger than L then the relation for Ψ simplifies further to $\Psi \approx L^2 h$. However this is still based upon QG for which $h = f\delta h/h_o$ requires $|\delta h/h_o| \ll 1$ contrary to the actual ocean whose depth varies by the full depth itself.

For implementation into realistic ocean models, ambiguities involve both h_o and the interpretation of Ψ as velocity or transport (integral of $\mathbf{u}$ over ocean depth D). These questions were considered in Holloway [25] then implemented in Alvarez et al. [26] by taking a ME transport streamfunction to be $\Psi^* = -fL^2 D$, from which the barotropic component of ME velocity $\mathbf{u}^*$ is $\mathbf{u}^* D = z \times \nabla \Psi^*$. The remaining question is how to represent $\mathbf{K}$, presumed to be a scale-dependent operator governing the rate at which eddy interactions can force $\mathbf{Y}$ toward $\mathbf{Y}^*$. Simple choices (with view to practicality) include $-K$ (a damping constant) or

$A\nabla^2$ (Laplacian diffusion). In practice the choice has been to assign momentum tendency as $A\nabla^2(\mathbf{u}-\mathbf{u}^*)$ with constant coefficient A.

It must be clear that choices in the previous paragraph are not strictly derived. They were available, practical choices at a time (circa 1992) offering advances over the common practice of eddy viscosity, c.f. $A\nabla^2\mathbf{u}$. Substituting $A\nabla^2(\mathbf{u}-\mathbf{u}^*)$ with $\mathbf{u}^*$ from $\mathbf{u}^*D = z \times \nabla\Psi^*$ where $\Psi^* = -fL^2D$ became known as the 'neptune parameterization', with L a length scale presumed to take values from some few to several km.

A number of papers have explored applications of neptune for cases ranging from estuarine to global ocean (Alvarez et al. [26]; Eby and Holloway [27]; Fyfe and Marinone [28]; Holloway et al. [29]; Pal and Holloway [30]; Sou et al. [31]; England and Holloway [32]; Marinone [33]; Nazarenko et al. [34]). Two items in particular are (1) a study of neptune impact upon global skill (Holloway and Sou [35]) measured against current meter records, and (2) the Arctic Ocean study by Nazarenko et al. [34] compared with MEP calculation by Polyakov [15].

The point of simple (easy!) schemes like nepune is to allow present-day ocean models to produce more skillful results from physics closer to statistical dynamics. Such simple schemes are not 'right'; they are only 'less wrong' than traditional models. This should spur further effort. Among such efforts, Holloway [36] considered baroclinic extension from neptune, including 'thickness' transports in layer models. Merryfield [37] considered ME QG with continuous stratification. Frederiksen [38] carefully re-examined to closure theory to see how to evaluate terms such as $\mathbf{K}$ that are only guessed in neptune applications. Merryfield et al. [39] extended ME without assuming QG, helping overcome ambiguity stemming from QG forms of h and Ψ, proposing $\Psi^* = fL^2D_o^2/D$ with reference depth D_o. Such efforts, both to establish fundamentals and to devise practical parameterizations will continue, presumably conjoined by newer work such as from MEP.

7 Mixing heat and salt in bi-stably stratified flow

Before closing we turn to a very different phenomenon on an entirely different scale. In part we seek to test how robust are statistical mechanical approaches. As well we could ask: if the statistical mechanical apparatus were brought to bear only to aid the eddy topography problem, maybe that's not so worthwhile. For other circumstances perhaps simpler intuitions suffice? Let's see.

Stratification of sea water is due to heat and salt scaled by corresponding density coefficients. At the level of 'molecular' conductivity, sea water is about 100 times as diffusive for heat as for salt. This leads to interesting effects. If the water column is stably stratified with respect to temperature, T, say, but unstably with salinity, S, while the overall density stratification remains stable, one may encounter spontaneous instabilities called 'salt fingers'. Contrariwise, if stable with respect to S but unstable with respect to T, there are instabilities called 'layering'. Much of the ocean interior is however stably stratified with respect both to T and S. Then it is assumed that heat and salt behave similarly and that ambient turbulence, perhaps on account of internal wave breaking, mix the two with a single apparent diffusivity, κ. It has also been known for some time that this is not true.

In early experiments, Turner [40] mechanically agitated a fluid stratified with respect to T and, separately, a fluid stratified with respect to S, taking care that the two stratifications were initially the same. Under the same mechanical agitation, it was found that T was mixed more efficiently than S by an amount greater than could attributed to molecular conduction. Later, Altman and Gargett [41] performed similar experiments in a tank was stably stratified with respect to both T and S, arranged so that both made the same initial contribution to stratification. Again it was seen that T mixed more readily than S by amounts exceeding molecular conduction. This is termed 'differential diffusion'. Observation of differential diffusion in the ocean is technically difficult, with results reported by Nash and Moum [42]. While the phenomenon is now observed, why does it happen?

Let's try intuition and experience. Omitting at first gravity, turbulent stirring of a (passive) tracer is like the milk-into-tea example. Heat and salt would be stirred down their respective background gradients, with the fluxes in vertical ($w'T'$ and $w'S'$) dominated by scales of turbulent energy with lesser contributions at smaller scales (subject to viscous and diffusive cutoffs). Because 'molecular' diffusion of T is faster than S, the short scales of $w'T'$ are suppressed more strongly than for $w'S'$. Hence we might expect the total turbulent salt transport to be greater than heat transport, the opposite of what is observed.

Including gravity, what changes? Mainly the turbulence is suppressed, exhausting its energy by working against gravity (in traditional thinking). But that is only to say we expect a weaker version of stirring-milk-into-tea and the previous (wrong) result that salt transport should exceed heat transport is still expected. Where did we go wrong?

Numerical simulations at first in 2D (motion only in a vertical plane) by Merryfield et al. [43] then fully in 3D by Gargett et al. [44] reveal what happens. Contributions to $w'T'$ and $w'S'$ reverse sign to counter-gradient in the shorter scales. When larger 'molecular' diffusion of T preferentially cuts off $w'T'$, the surviving $w'S'$ fluxes are of counter-gradient sense and hence subtract from the overall down-gradient S flux. The result is S flux weaker than T flux, as observed.

While numerical simulations have showed what happens, we are left asking: why? Especially why the prevalence of counter-gradient (backwards!) fluxes at shorter scales? I think such counter-gradient transports are generic to stably stratified turbulence. Closure theory (Holloway [45]) in 2D (vertical plane) anticipated very well (quantitatively) the 2D numerical simulations and (qualitatively) the 3D simulations. Importantly, that closure theory is of the broad class which, per Carnevale et al. [24], strictly satisfies $dH/dt \geq 0$ (apart from external forcing and dissipation). Then entropy gradient forcing, or GTF, explains each aspect of what happens.

At larger scales (relative to 'molecular' diffusive cutoff scales) a source of turbulent kinetic energy (KE) is assumed. Under gravity, two things happen.

In part KE is redistributed to smaller scales due to entropy gain from such redistribution (a GTF view of 'turbulent cascade'!) In part KE is converted to potential energy (PE) stored in T'^2 and S'^2 (as density variances times gravity), driving the ratio KE:PE toward higher entropy (per GTF). Large scale conversion of KE to PE is by down-gradient $w'T'$ and $w'S'$. Large scale T'^2 and S'^2 are also redistributed to shorter scales by another GTF 'cascade'. Things get a little complicated though. Recall GTF works as $C \cdot \nabla_{\mathbf{Y}} H$ involving $\mathbf{C}$ as well as $\nabla_{\mathbf{Y}} H$. The C for redistribution of KE is weaker than for redistribution of PE due to the role of pressure forces maintaining incompressibility $\nabla \cdot \mathbf{u} = 0$ for vector field $\mathbf{u}$ whereas scalar fields T and S are unrestricted. A consequence is that PE is more rapidly transferred from large to small scales causing the ratio KE:PE on small scales to favor entropy production by converting PE to KE, forced by $\partial H/\partial$(KE:PE). Conversions PE $\Rightarrow$ KE are by counter-gradient $w'T'$ and $w'S'$ whence stronger diffusion of T' leads to overall stronger $w'T'$.

8 Outlook

Broadly, entropy calculus helps clarify and organize the work that must be done so that traditional ocean models, based in classical mechanics plus ad hoc mixing, may acquire a consistent arrow of time. In some cases, traditional mixings and eddy viscosities happen (willy nilly?) to point the right way. Too often such guessed-at schemes point oppositely to time's arrow and then go quite wrong. No surprise! Illustrations above are drawn from two extremes: (1) the characteristics of ocean currents of scales of 10s and 100s of km, and (2) the nature of mixing on scales of cm to mm and smaller. In each case approaches from entropy calculus are sketched. It must be admitted though these are barely sketches. Very little intellectual resource has yet been invested on the statistical dynamical side compared with investment on the classical (traditional) ocean dynamics side. Among tasks ahead are (1) ongoing effort to build confident fundamentals and (2) brave efforts to bring statistical dynamics into practical ocean modeling even while further efforts build and refine fundamentals. Practical steps will be understood as steps along the way, in place only until they can be superceded. But the sweep of what may be done, with prospects for improvement both in understanding and in practical skill, are powerful motivations for the tasks ahead.

Acknowledgement.

I am grateful to Yukio Kaneda and Toshiyuki Gotoh for organizing this fascinating meeting and for inviting the present lecture. The thoughts expressed above were inspired from interactions with students and faculty during the meeting. Research upon which this lecture was based has been supported by the Office of Naval Research.

References

1. P. Coveney, R. Highfield: *The Arrow of Time*, (Fawcett-Columbine, New York 1992, 378 pp.)
2. A. Katz: *Principles of Statistical Mechanics* (W. H. Freeman & Co., San Francisco 1967, 188 pp)
3. G. W. Paltridge: Quart. J. R. Met. Soc. **101**, 475 (1975)
4. G. W. Paltridge: Quart. J. R. Met. Soc. **104**, 927 (1978)
5. H. Ozawa, A. Ohmura: J. Climate **10**, 441 (1997)
6. R. Lorenz, J. I. Lunine, P. G. Withers, C. P. McKay: Geophys. Res. Lttrs. **98**, 415 (2001)
7. R. Salmon, G. Holloway, M. C. Hendershott: J. Fluid Mech. **75**, 691 (1976)
8. G. Holloway: Ann. Rev. Fluid Mech. **18**, 91 (1986)
9. R. Salmon: *Lectures on Geophysical Fluid Dynamics* (Oxford U. Press 1998, 378 pp.)
10. J. Miller: Phys. Rev. Lett. **65**, 2137 (1990)
11. R. Robert, J. Sommeria: J. Fluid Mech. **229**, 291 (1991)
12. R. Robert, C. Rosier: J. Stat. Phys. **86**, 481 (1997)
13. P. H. Chavanis, J. Sommeria: Phys. Rev. Lett. **78**, 3302 (1997)
14. E. Kazantsev, J. Sommeria, J. Verron: J. Phys. Oceanogr.**28**, 1017 (1998)
15. I. Polyakov: J. Phys. Oceanogr. **31**, 2255 (2001)
16. L. Onsager: Phys. Rev. **37**, 405 (1931)
17. L. Onsager: Phys. Rev. **38**, 2265 (1931)
18. L. Onsager, S. Machlup: Phys. Rev. **91**, 1505 (1953)
19. J. R. Herring: J. Atmos. Sci. **34**, 1731 (1977)
20. R. H. Kraichnan: J. Fluid Mech. **5**, 497 (1959)
21. G. Holloway: J. Phys. Oceanogr. **8**, 414 (1978)
22. R. H. Kraichnan: J. Fluid Mech. **47**, 512 (1971)
23. G. Holloway: J. Fluid Mech. **184**, 463 (1987)
24. G. F. Carnevale, U. Frisch, R. Salmon: J. Phys. A. **14**, 1701 (1981)
25. G. Holloway: J. Phys. Oceanogr **22**, 1033 (1992)
26. A. Alvarez, J. Tintore, G. Holloway, M. Eby, J. M. Beckers: J. Geophys. Res. **99**, 16053 (1994)
27. M. Eby, G. Holloway: J Phys Oceanogr **24**, 2577 (1994)
28. J. Fyfe, G. Marinone: B.C. Atmos-Ocean **33**, 613 (1995)
29. G. Holloway, T. Sou, M. Eby: J. Mar. Res. 53, 539 (1995)
30. B. K. Pal, G. Holloway: Cont. Shelf Res. **16**, 1591 (1996)
31. T. Sou, G. Holloway, M. Eby: J. Geophys. Res. **101**, 16449 (1996)
32. M. H. England, G. Holloway: J. Geophys. Res. **103**, 15885 (1998)
33. S. G. Marinone: J. Geophys. Res. **103**, 18437 (1998)
34. L. Nazarenko, G. Holloway, N. Tausnev: J. Geophys. Res. **103**, 31003 (1998)
35. G. Holloway, T. Sou: J. Phys. Oceanogr. **26**, 1088 (1996)
36. G. Holloway: J. Phys. Oceanogr. **27**, 1153 (1997)
37. W. J. Merryfield: J. Fluid Mech. **354**, 345 (1998)
38. J. S. Frederiksen: J. Atmos. Sci. **56**, 1481 (1999)
39. W. J. Merryfield, P. F. Cummins, G. Holloway: J. Phys. Oceanogr. **31**, 1880 (2001)
40. J. S. Turner: J. Fluid Mech. **33**, 639 (1968)
41. D. B. Altman, A. E. Gargett: Amer. Soc. Civil Engr. (E. List and G. Jirka, eds.,1990)
42. J. D. Nash, J. N. Moum: J. Phys. Oceanogr., subm. (2002)
43. W. J. Merryfield, G. Holloway, A. E. Gargett: Geophys. Res. Lttrs **25**, 2773 (1998)
44. A. E. Gargett, W. J. Merryfield, G. Holloway: J. Geophys. Res., subm. (2002)
45. G. Holloway: Dyn. Atmos. Oceans. **12**, 107 (1988)

Internal-Wave-Packet Propagation and Breaking

George F. Carnevale[1], Marco Briscolini[2], Paolo Orlandi[3], and Rudolf C. Kloosterziel[4]

[1] Scripps Institution of Oceanography, University of California San Diego, 9500 Gilman Drive, La Jolla, California 92093, USA
[2] Dipartimento di Meccanica e Aeronautica, University of Rome, "La Sapienza," via Eudossiana 18, 00184 Roma, Italy
[3] IBM Italia S.p.A., Via Shangai 53, 00144 Roma, Italy
[4] School of Ocean and Earth Science and Technology, University of Hawaii, Honolulu, 1000 Pope Rd, Hawaii 96822

Abstract. Packets of internal waves propagating vertically through the ocean can locally overturn the fluid producing turbulence and mixing. In order to explore this phenomena, two kinds of numerical simulations are performed. In the first, internal-wave packets are followed as they propagate. It is found that the breaking of wave crests within the packet can lead to overturning events on the scale observed in the ocean, and the subsequent turbulence can form a continuous wake. In the second kind of simulation, an attempt is made to capture the transition from breaking internal waves to the small-scale turbulence they generate. Evidence is presented for a transition in the energy spectra from the anisotropic k^{-3} buoyancy range to the small-scale $k^{-5/3}$ isotropic inertial range. Density structures that form during wave breaking are analyzed and regions of mixing associated with the breaking events are visualized.

1 Introduction

Recent observations [1] of fluctuations of density and velocity in the upper few hundred meters of the ocean suggest that the propagation of internal-wave packets through this region may play an important role in creating overturns, that is patches with heavy fluid above light. These overturns are dynamically unstable and their breakdown produces small-scale turbulence and mixing. Alford and Pinkel [1] observed coherent regions of strong oscillatory vertical strain rate that travel vertically through 100 m or more of the thermocline. These propagating structures had an internal wave structure with vertical wavelength of about 10 m, and the coherent region could be described by an envelope of about 50 m in vertical extent. These coherent structures were strongly correlated with 2 m overturns. Given the complicated nature of the flow, with strong components of large-scale advection, it is difficult to verify whether these propagating structures are indeed internal wave packets. Here we validate that identification to a certain extent by showing how an idealized packet would propagate through a simplified model of the thermocline, and by showing that the observed overturning scales could be consistent with internal-wave packet propagation. In addition, we will study the transition from waves to turbulence initiated by the formation of overturns by internal waves.

Two-dimensional simulations can yield a wealth of information about internal wave-packet propagation, dispersion and to some extent their stability [24] [25]. Here we will use such simulations to follow the evolution of packets constructed to the specifications suggested by the observations [1]. However, to understand how the breaking waves evolve with the production of small-scale turbulence, we will need to use three-dimensional simulations. Alford and Pinkel's [1] data suggests that overturns may be produced by energetic waves with vertical wavelengths on the order of 10 m. Large scale fluctuations, say 10 m and above in vertical scale can be described reasonably well as internal waves. These are the waves described by the Garret-Munk [9] spectrum. For much smaller scales, say 1 m and below, the flow is probably better described in terms of nearly isotropic turbulence. Intermediate between the large-scale wave dynamics and the small-scale turbulence is a transition regime in which there is a competition between waves and turbulence. It is this intermediate range, often called the buoyancy range, that contains the overturning activity. Since the observations are essentially one-dimensional in space, a direct numerical simulation which could faithfully describe events in this range would help toward understanding the full three dimensional flow structures behind the observations.

Given current computer resources, it would not be practical to simulate all of the scales that are relevant to the breaking of waves in the buoyancy range. The forcing of the buoyancy range is thought to result from a complicated interaction of many internal waves with scales ranging in the vertical from tens of meters to hundreds of meters or more, and typical horizontal internal wave scales can be even much greater than these. On the small-scale end of the spectrum, the observed breaking is occurring on vertical scales of the order of 1 m, and these breaking events produce turbulence that extends down to a viscous cutoff on the order of 1 cm. Thus direct numerical simulation, of the entire range of scales is still impractical. As mentioned above, we will break the problem into separate studies of the propagation of internal-wave packets and the transition of waves to turbulence. For the wave-packet propagation, we will use a two-dimensional spectral code with hyperviscosity. This will show the propagation of the packet and the formation of the overturns. For the transition from waves to turbulence, we will use a combination of three-dimensional LES modeling and an artificial model of the large-scale forcing in order to reduce the spectral range that we will need to cover. To attack the problem of how waves at the short-scale end of the Garret-Munk [9] spectrum go unstable and break in the buoyancy range, we have used an artificial forcing with length-scales fixed at 20 m in the vertical and 20 m in the horizontal to represent the effect of all larger scales. At the small-scale end of the simulation, we have introduced an eddy viscosity with a cutoff at the 16 cm level in both horizontal and vertical directions. Thus our model does some violence to the true physics at the large and small-scale ends of the simulated range. However, the hope is that it will do justice to the evolution in the buoyancy range. This model does prove capable of capturing the transition from the buoyancy to the inertial range.

2 Internal-wave packets

The observations of Alford and Pinkel [1] show vertically propagating structures at depths from 150 to 350 m which they suggest may be internal wave packets. These structures have vertical extent of about 50 m with internal vertical wavelengths of about 12 m. They were observed to propagate vertically with a velocity on the order of 10 m/hr. Within the structures, there were overturning events with vertical scales of about 2 m. These overturns are found primarily in regions of high vertical strain rate. Theoretical analysis [27] suggests the possibility that the small-scale turbulence generated by the overturns in a packet may be left behind in a continuous 'scar' much longer than the size of the packet. Here we describe a numerical investigation undertaken to examine the validity of the identification of the structures observed in the data with internal-wave packets.

Assuming a constant background Brunt-Vaisala frequency N and ignoring the effects of the earth's rotation, the intrinsic dimensional frequency for internal waves is

$$\sigma = N\frac{k_h}{k}, \tag{1}$$

where $k_h = \sqrt{k_x^2 + k_y^2}$ is the horizontal wavenumber, and $k = \sqrt{k_x^2 + k_y^2 + k_z^2}$ is the total wavenumber. The observed frequency for one of the wavepackets in the Alford and Pinkel [1] data is 4 cph (cycles per hour). This is higher than the ambient $N \approx 3$ cph. Since $\sigma_{max} = N$, it is assumed that the observed frequency for this packet is the sum of the intrinsic frequency plus a Doppler shift. To predict this shift, it is necessary to know the wavelength of the packet, the magnitude of the ambient current and its direction relative to the packet propagation direction. Alford and Pinkel [1] suggest that the intrinsic frequency for their packet with observed frequency of 4 cph is near 0.14 cph which leads one to a wavelength of 180 m. This suggests that the horizontal wavelengths in both directions are much larger than the vertical wavelength. For our numerical modeling, this represents a difficulty. We are reluctant to introduce anisotropic grids for fear of the distortions that might result, especially when applying simple sub-grid scale models. Thus, in this preliminary work, we decided to consider only the case in which horizontal and vertical wavelengths were equal. The corresponding intrinsic frequency would then be about 2 cph which would still be consistent with the observed packet, just requiring less of a Doppler shift to match the observed frequency.

As for the amplitude of the observed packets, this can be given in terms of the peak magnitude of the observed strain rate $\partial w/\partial z$. The maximum value of vertical strain rate in the Alford and Pinkel [1] observations is approximately N, and in the case of the particular packet discussed above, it seems that the maximum is about $0.38N$.

In what follows, we will examine the evolution of a particular wave packet with two dimensional simulations. In an attempt to reproduce the kind of behavior evident in the observations, we used simulations in a domain of 200 m in both width and depth. We used a packet with non-dimensional wavenumbers of 12 in both directions, corresponding to vertical and horizontal wavelengths of (200 m)/12 ≈ 17m. Our 2D simulations had an effective resolution corresponding to a cutoff wavelength of ≈ 0.8 m. To follow this phenomenon in DNS with all relevant scales well resolved would require resolution from 200 m down to a few cm, which is somewhat impractical. Since the subgrid scale model used in the 3D simulations is not appropriate in 2D, we had recourse to hyperviscosity (with the Laplacian taken to the eighth power). The simulations illustrated here are from a spectral code dealiased with the 2/3 rule [21]. Although there are 768 wavevectors used in each direction, after application of the 3/2 rule this leaves only 512 active modes in each direction.

3 Linear dispersion of packets

The linearized version of the Boussinesq evolution equations can be used to obtain a model of the internal wave packet. The vorticity and density of a plane internal wave can be written dimensionally as

$$(\omega_x, \omega_y, \omega_z, \rho') = A\mathbf{e_k} \exp i(\mathbf{k}\cdot\mathbf{r} - \sigma t), \tag{2}$$

where A is an arbitrary amplitude and $\mathbf{e}$ is the eigenvector

$$\mathbf{e_k} = (gkk_y/Nk_h, -gkk_x/Nk_h, 0, \rho_0). \tag{3}$$

Taking a linear superposition of such waves distributed continuously in wavevector space and centered on a particular wavevector, say $\mathbf{k}_0$, would produce an internal wave packet. For example,

$$(\omega, \rho') = \mathcal{R}e \int G(\mathbf{k} - \mathbf{k}_0)\mathbf{e_k} e^{i(\mathbf{k}\cdot\mathbf{r} - \sigma t)} d^3k, \tag{4}$$

with

$$G(\mathbf{p}) \equiv A \exp\left(-a^2p_x^2 - b^2p_y^2 - c^2p_z^2\right), \tag{5}$$

where a, b, and c are length scales, represents a propagating ellipsoidal packet. A slight generalization based on simple coordinate rotations will also permit an arbitrary choice for the orientation of the ellipsoidal envelope relative to the crests internal to the packet. Within the envelope, the vorticity and density fields will have a phase velocity in the direction of $\mathbf{k}_0$ and group velocity

$$\mathbf{c}_g = \nabla_\mathbf{k}\sigma_\mathbf{k}, \tag{6}$$

which is perpendicular to the phase velocity.

By varying the dimensions a, b and c, we can change the shape of the packet as needed. A likely candidate for the packets whose effects are observed in Alford and Pinkel's data [1] would suggest that at least one of these lengthscales is very large. For the present calculations we take a to be infinite. Then we chose b and c and the orientation of the system to be such that the envelope is an ellipse with

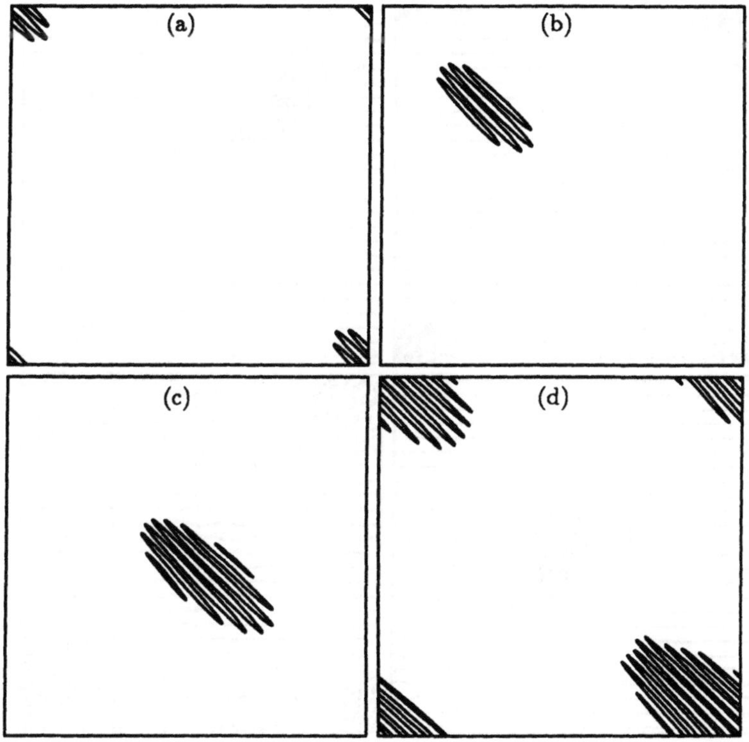

Fig. 1. Contours of the magnitude of the perturbation density $|\rho'/\rho_0|$ from a simulation of the linear propagation of a wavepacket. The domain size is 200 m on each side. The vertical axis is depth. The only contour level drawn is that at 0.5 of the maximum field value. The time sequence of the panels is a) t=0, b) t=3 hr, c) t=6 hr, and d) t=11 hr.

major axis aligned along the direction of propagation. Other choices may also be of interest, but that will be explored in future work. With the ellipse as chosen, the phase velocity is directed along the short axis and the group velocity along the long axis. In a numerical simulation, the packet can only be approximated, with the integral replaced by a discrete sum of wavevectors. By using (4) and (5) with $t = 0$, we are able to construct the initial condition for a packet that is both reasonably confined in space and well resolved internally.

The first issue that we need to address is the dispersive spreading of the wave packet. Simple arguments suggest that the physical extent of the wave packet will grow as $\Delta c_g t$ in the direction of the group velocity, where Δc_g represents the spread in group velocities calculated for the individual wavevectors that contribute significantly to the wave packet. We can make some crude dimensional estimates for the rate of dispersion by setting $c_g \sim N/k_0$ and $\Delta c_g \sim (N/k_0^2)\Delta k_0$, where Δk_0 measures the spread of wavenumbers in the packet. If we call Δx_0, the initial length of the wavepacket, then the change in the size of the packet

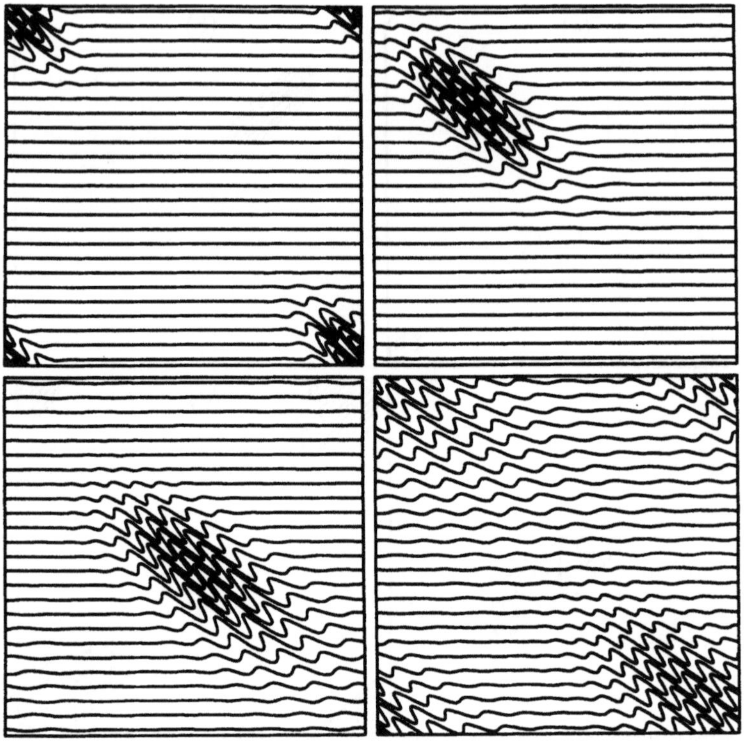

Fig. 2. Contours of ρ/ρ_0 from a simulation of the linear propagation of a wavepacket. The domain size is 200 m on each side. The vertical axis is depth. The time sequence of the panels is the same as in figure 1. The contour increment is such that the vertical separation between unperturbed isopycnals is 8 m.

can be crudely taken as

$$\Delta x - \Delta x_0 = \Delta c_g t. \tag{7}$$

The packet would then double in size by a time $t_d \sim \Delta x_0/\Delta c_g$, and the distance that the packet can travel before doubling in size is

$$x \sim c_g\, t_d \sim \frac{k_0}{\Delta k_0}\, \Delta x_0.$$

For the packet used in the simulations this predicts a doubling after propagation of about 200 m.

In figure 1, we show the evolution of the density perturbation field during the propagation of our packet following purely linear dynamics. In each panel, only the contour level corresponding to 0.5 $max|\rho'/\rho_0|$ is drawn. Positive and negative values have not been indicated, but clearly the sign of ρ' will alternate from one wave crest to the next. We see the packet propagates along the diagonal. This is in agreement with the fact that the wavevector is $\mathbf{k} = (12, 12)$ and that the group velocity is perpendicular to this. It is less obvious from the few panels

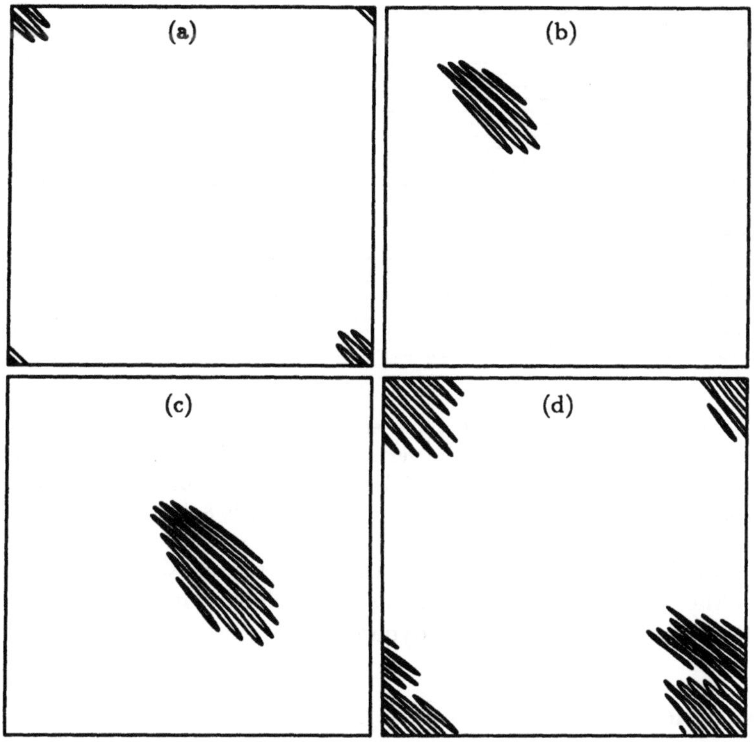

Fig. 3. Contours of the magnitude of the perturbation density $|\rho'/\rho_0|$ from a simulation of the nonlinear propagation of a wavepacket with max $\partial w/\partial z \approx 0.38N$. The domain size is 200 m on each side. The vertical axis is depth. The only contour level drawn is that at 0.5 of the maximum field value. The time sequence of the panels is the same as in figure 1.

that we can include here that the phase of the waves within the packet advances in the direction of $\mathbf{k}$. The average speed of the packet in propagating from one corner of the domain to the opposite corner is correctly given by $|c_g|$. This speed is consistent with that of the structures observed in the data [1]. Furthermore, we see that the width and length of the packet grow to a little more than double their original values in the time it takes to cross from one corner of the domain to the other, and this is correctly predicted by the formula (8). During the period of evolution illustrated, the peak amplitude of the packet decays to 25% of its initial value.

Although the amplitude of the packet can be changed arbitrarily in this purely linear simulation, we may simply assign an amplitude to see the effect of such a packet on the full density field. This is done in figure 2. The amplitude used represents fluctuations in $\partial w/\partial z$ about five times the maximum actually observed in the Alford and Pinkel data [1]. Nevertheless, we have used this packet with exaggerated amplitude to more clearly illustrate the nature of the linear

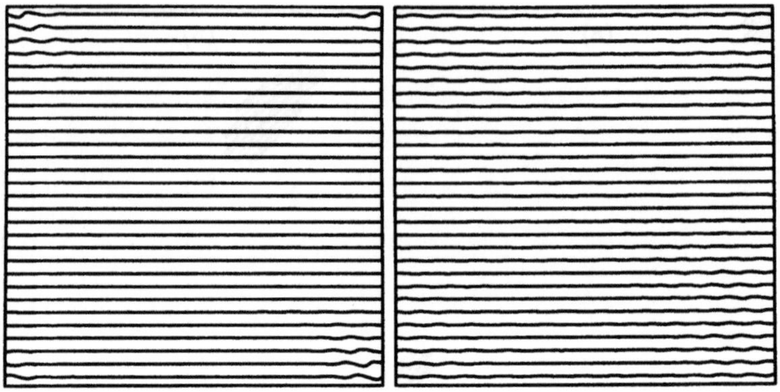

Fig. 4. Contours of ρ/ρ_0 from the same simulation as represented in figure 3. The domain size is 200 m on each side. The vertical axis is depth. The two times illustrated correspond to the first and last times of figure 3. The contour increment is such that the vertical separation between unperturbed isopycnals is 8 m.

propagation. In such a strong packet, there are regions of strong overturning, which, if the packet is not propagating too rapidly, would develop convective instability under the full nonlinear dynamics.

4 Nonlinear propagation of packets

Having determined that our packet propagates correctly under linear dynamics, we then investigated its evolution with the complete Boussinesq equations. The amplitude of the observed packet discussed in the introduction is such that the maximum value of the strain rate $\partial w/\partial z$ is about $0.38N$. With the packet amplitude set to match this value as its maximum $\partial w/\partial z$, we performed the simulation illustrated by contour plots of ρ'/ρ in figure 3. This figure should be compared to the corresponding figure for linear evolution, figure 1. The times represented are the same in each figure. By the time of panel (b) a clear asymmetry in the form of the packet has developed in the nonlinear case and there is some clear distortion of the packet in the final panel. Nevertheless, the overall evolution of this nonlinear packet is not very different from the linear case. This packet is so weak that the initial condition is not overturning anywhere and the Richardson number is above 1 everywhere. Thus, the classical criteria for convective instability and shear instability are not satisfied in this packet. This continues to be the case throughout the simulation in spite of small-scale generation by nonlinear wave-wave interactions. An idea of how weak this packet is can be obtained graphically from the plots of the density contours as illustrated in figure 4.

The next case that we will treat is one for which the amplitude of the packet is just above the threshold for overturning. The amplitude of this packet in terms of its maximum strain rate is $\partial w/\partial z = 0.76N$. In figure 5, we display the contour

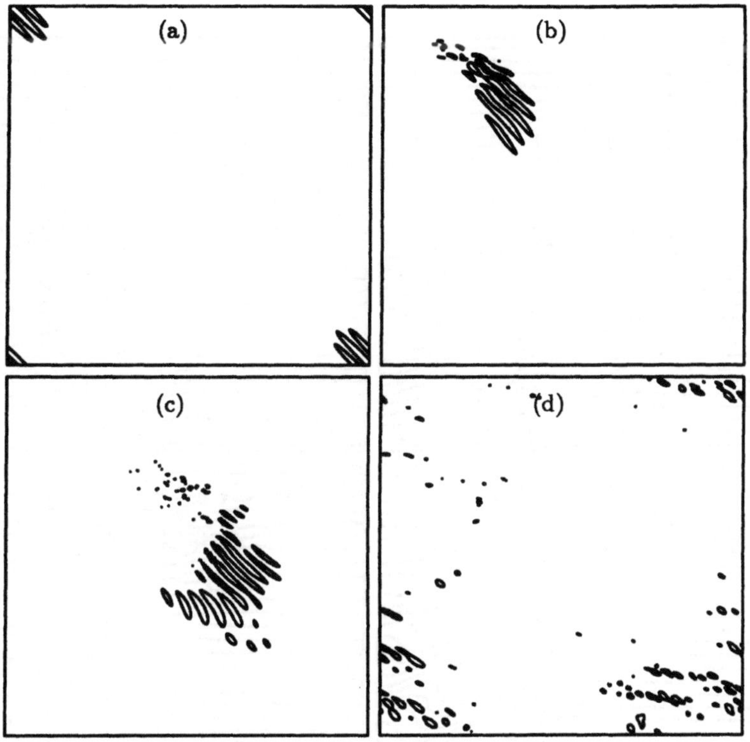

Fig. 5. Contours of the magnitude of the perturbation density $|\rho'/\rho_0|$ from a simulation of the nonlinear propagation of a wavepacket with max $\partial w/\partial z \approx 0.76N$. The domain size is 200 m on each side. The vertical axis is depth. The only contour level drawn is that at 0.5 of the maximum field value. The time sequence of the panels is the same as in figure 1.

plots for the perturbation density at the same times as in the previous figures. We see that there is some early production of small scales that are evident in the wake of the packet. By $t = 6$ hr the packet itself has become badly distorted, and by $t = 11$ hr, it has degenerated into small-scale structures, although these still retain to some extent an organization and alignment related to the original structure of the packet. To better illustrate the decay of this packet, we display contour plots of the full density field from $t = 2.5$ hr to $t = 4.8$ hr in figure 6. Each frame is an enlarged image centered on the wave packet, showing only a portion of the domain (a square of size 200/3 m on a side). In panel (a) we see an early stage in which the wave is overturning at points, but there has not yet been any strong production of energy in scales smaller than 2 m (note that the spacing between the unperturbed isopycnals is 2 m). There are four relatively strong crests evident in panel (a). These crests are advancing from bottom-left to top-right in these figures. The weakest crest (bottom-left) is just entering the packet in panel (a). In the linear evolution as each crest passes through the packet

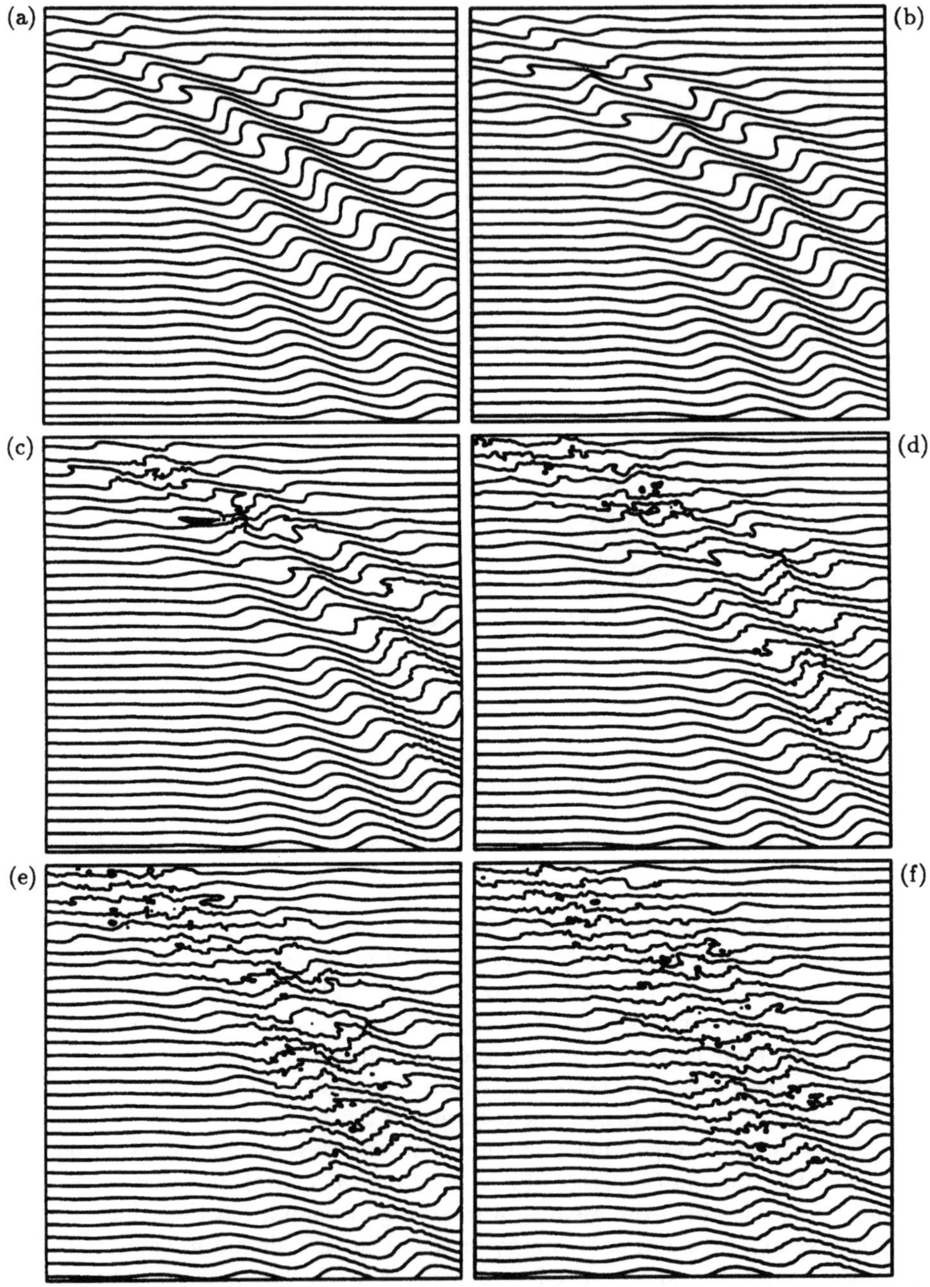

Fig. 6. Contours of ρ/ρ_0 from a simulation of the nonlinear propagation of a wavepacket with initially max $\partial w/\partial z \approx 0.76N$. Only a portion of the computational frame is show, and this corresponds to a square 200/3 m on each side. The contour increment is such that the vertical separation between unperturbed isopycnals is 2 m. The times corresponding to the panels are (a) 2.5, (b) 2.9 (c) 3.5, (d) 3.9, (e) 4.4 and (f) 4.8 hr.

from bottom-left to top-right, its amplitude first increases and then decreases. As envisioned by Thorpe [27], the crests amplify as they move toward the center of the packet and break leaving small-scale perturbations behind that link up with the 'debris' produced by the passage of previous crests. The period of the sequence of panels shown here is long enough for the weak crest on the lower-left side of the packet in panel (a) to move completely through the packet, finally becoming the weak crest on the upper-right side. In the case we have simulated here, the crests do produce overlapping zones of small-scale perturbations that form a somewhat continuous scar, a possibility suggested by Thorpe [27]. One should note, however, that during the period when a particular crest is actually breaking, the overturning and small-scale production is not uniform along the length of the crest, as assumed in Thorpe's idealized model, but rather appears in spots along the crest (see panels (c) and (d)). Also the breaking and subsequent scar formation does not continue indefinitely. The strength of the packet is both dispersed and dissipated, so that by $t = 210N^{-1}$ the process of scar formation has ceased.

We have also performed 3D simulations of the propagation of these wave packets. The general evolution exhibited in the 2D simulations is also found in 3D, although in 3D we did not have sufficient resolution adequately capture the 2 m overturns. Further details can be found in Carnevale and Orlandi [5].

5 Overturns from wave-wave interactions

Our goal here is to determine to what extent our simulations can capture the transition from the buoyancy range to the inertial range in the energy spectra. Constructing a theory of this transition is complicated because of the anisotropy of the buoyancy range. To make progress, some theoretical formulations have represented the entire spectrum as depending only on wavenumber k. The model for the kinetic energy spectrum in the buoyancy range is then

$$E(k) = \alpha N^2 k^{-3} \tag{9}$$

where α is an empirical constant and N is the Brunt-Vaisala frequency, which measures the strength of the stratification. The Brunt-Vaisala frequency is defined by

$$N^2 = -\frac{g}{\rho_0}\frac{\partial\overline{\rho}}{\partial z}. \tag{10}$$

where g is the acceleration of gravity, $\overline{\rho}$ is the background density profile, assumed stable (i.e. $\partial\overline{\rho}/\partial z < 0$), and ρ_0 is the volume average of $\overline{\rho}$. From the observed spectra of vertical shear, the constant α is determined to be about 0.47, but it will be more convenient for us to consider the two components of the horizontal velocity (u, v) separately, and, assuming horizontal isotropy in the observations, this would suggest $\alpha \approx 0.2$ for the spectrum of either component [10], [7]. The inertial range kinetic energy spectrum is given by

$$E(k) = C_K \epsilon^{2/3} k^{-5/3} \tag{11}$$

where ϵ is the turbulent dissipation rate of total kinetic energy and C_K is the empirical Kolmogorov constant. A reasonable value to assume for C_K is 1.5 [18]. For the energy of one component of the velocity field, there would simply be a pre-factor of 1/3 multiplying this isotropic spectrum. The Ozmidov (or buoyancy) wavenumber is then estimated by simply matching these two spectra at wavenumber k_b. The result is, up to an order one multiplicative constant [14], [10]

$$k_b \equiv \sqrt{N^3/\epsilon}. \tag{12}$$

The model for the potential energy spectrum in the buoyancy range is similar to that for the kinetic energy spectrum. The empirical constant α for the temperature spectrum is found to have value of about 0.2 [10], [11]. The spectral model for the inertial range of density fluctuations is the Corrsin-Obukhov spectrum, which involves the decay rate of density fluctuations as well as ϵ. For our purposes, we prefer to write the spectrum directly in terms of the turbulent decay rate of potential energy, which we shall write as ϵ_{pe}. Then the Corrsin-Obukhov spectrum for the potential energy takes the following form:

$$PE(k) = C_o \epsilon_{pe} \epsilon^{-1/3} k^{-5/3}, \tag{13}$$

where C_o is the Corrsin constant.

For all of the simulations presented here, we have used the Boussinesq approximation. The evolution equations may be written as

$$\frac{\partial \mathbf{u}}{\partial t} + \mathbf{u} \cdot \nabla \mathbf{u} + \frac{1}{\rho_0} \nabla p' - \frac{\rho'}{\rho_0} \mathbf{g} = \nu(\nabla^2) \nabla^2 \mathbf{u}, \tag{14}$$

$$\nabla \cdot \mathbf{u} = 0. \tag{15}$$

$$\frac{\partial \rho'}{\partial t} + \mathbf{u} \cdot \nabla \rho' + w \frac{\partial \overline{\rho}}{\partial z} = \kappa(\nabla^2) \nabla^2 \rho', \tag{16}$$

where $\nu(\cdot)$ and $\kappa(\cdot)$ are considered functions of the Laplacian operator and are used to represent eddy parameterizations in general [13] and $\mathbf{g} = -g\hat{\mathbf{z}}$. We have neglected the effect of rotation, which should not play a major role at the small scales with which we are concerned. The total density is given by

$$\rho = \overline{\rho}(z) + \rho'(x, y, z, t), \tag{17}$$

where $\rho'(x, y, z, t)$ is the deviation from the horizontally averaged density $\overline{\rho}(z)$. ρ_0 is the average of $\overline{\rho}(z)$ over z. The pressure p' is the deviation from the background mean pressure. p' can be determined in terms of $\mathbf{u}$ by taking the divergence of (14) under the assumption that the velocity field is divergenceless.

We simulate these dynamical equations with a spectral code with triply periodic boundary conditions. As a sub-grid scale parameterization, we have used the large-eddy simulation model of Lesieur & Rogallo [17]. This eddy viscosity $\nu_t(k)$ is approximately constant throughout the buoyancy range and the large-scale end of the inertial range, but increases rapidly with k in the vicinity of the spectral cutoff k_c. Due to the spectral shape of the eddy viscosity, this model is sometimes called the cusp model. It seems reasonable in modeling the buoyancy range to use such a model since it does not completely neglect the effects of unresolved eddies on the buoyancy range, but, at the same time, it puts the strongest eddy viscosity in the inertial range near the cutoff.

We should emphasize the point that the size of the eddy viscosity depends on the amount of energy at the cutoff scale. If the resolution of the simulation of a given physical flow is increased, that is if k_c is increased, then the eddy viscosity will be correspondingly smaller. The total viscosity used in the simulations is the sum of the eddy viscosity and the constant molecular viscosity ν_{mol}. Thus the $\nu(\nabla^2)$ in equation (14) in the spectral simulation is taken as the total viscosity:

$$\nu(k) = \nu_{mol} + \nu_t(k) \tag{18}$$

The choice of turbulent diffusion depends on the choice of values for various parameters that enter into the closure model for stratified turbulence. For simplicity, we have just taken the turbulent Prandtl number $Pr_t(k)$ to be a fixed constant independent of k in our simulations. We determined this constant by examining the evolution of the potential energy spectrum for decaying stratified turbulence that is initially highly excited at all scales. More specifically, we started with an initial spectrum in which the GM spectrum was continued to scales below 10 m as in the decay simulations of Siegel and Domaradzki [23]. With $Pr_t = 0.55$ our simulations of decaying turbulence produced spectra with the high wavenumbers obeying the $k^{-5/3}$ law for both velocity and density fluctuations.

Next we turn to the question of the forcing. The large-scale flows that actually drive the buoyancy range are predominantly the waves of the Garret-Munk range. The full range where internal wave dynamics dominates includes scales of kilometers in the horizontal and hundreds of meters in the vertical. Because of lack of resolution, we cannot provide a full representation of the effects of all large-scale internal wave forcing on the buoyancy range. In our model, of necessity, we perform a drastic reduction in modeling the forcing; we replace the driving of all of the GM waves by a linear standing wave at one wavelength. Bouruet-Aubertot, Sommeria and Staquet [2], [3] in two dimensional simulations of a stratified turbulence excited a standing wave of the type we use, but they allowed this wave to decay, whereas we maintain its amplitude at the same level throughout the simulation.

To give the form of the forcing used, let us first introduce non-dimensional units. We will take all lengths to be scaled by $2\pi/L$, where L is the length of one side of our computational domain. Time will be scaled by $1/N$. Then the frequency of linear internal waves is given by

$$\sigma = \pm\frac{k_h}{k}. \tag{19}$$

One particular linear standing wave is

$$\mathbf{u} = (u, v, w) = A\frac{g^*}{\sqrt{2}}(0, \sin y \sin z, \cos y \cos z) \sin\frac{t}{\sqrt{2}}, \tag{20}$$

$$\frac{\rho'}{\rho_0} = A \cos y \cos z \cos \frac{t}{\sqrt{2}}, \tag{21}$$

where A is an arbitrary amplitude and g^* is the non-dimensional gravity. Note that the dimensional period of this wave, which is the forcing period, is given by

$$T_F = \sqrt{2}\frac{2\pi}{N}. \tag{22}$$

To give some idea of the structure of this standing wave, we show in figure 7 a contour plot of the density field in a vertical $y - z$ cross section. Note that the density field in this standing wave has no variation along the x direction. In this figure, we see an instantaneous representation of the iso-density surfaces. When $t/\sqrt{2}$ is an odd multiple of $\pi/2$, these isopycnals will all be flat. The degree to which they deviate from that at other times depends on the value of A as well as t. The instant of time represented here is such that $t/\sqrt{2}$ is an integer multiple of π and, hence, one of maximum distortion of the density contours. Note that the density field in (21) has two nodal planes, represented by two thick contour lines in the figure, at $z = \pm\pi/2$ (non-dimensional). During the forcing cycle, these planes remain flat and fixed in position. The fluid above and below these planes vertically approaches and retreats from them depending on the phase in time and the y position considered. Thus the points on the nodal planes at $y = 0$ and $y = \pm\pi$ are the centers of regions of oscillating high strain rate. On the other hand the points where the isopycnals are steepest, that is where $y = \pm\pi/2, \pm 3\pi/2$ and $z = 0, \pm\pi$ the magnitude of the shear $\partial v/\partial z$ is highest. Thus one advantage of the standing wave forcing is that the points of highest shear and highest strain rate remain fixed in space making it easier to differentiate the kinds of overturning events associated with these extremes. This will be convenient for making comparisons with Alford and Pinkel's [1] analysis. In particular, they noted that there were regions of high shear where classical shear instabilities often, but not always, resulted in overturning. Even more interestingly, there were regions of high Ri in which overturns were also observed. In more than half of these cases Ri was even greater than 2 suggesting that the typical shear instability ($Ri < 1/4$) is unlikely. Many of these overturns were in regions of high vertical strain rate. With our standing wave forcing, the regions of high shear and high strain rate are separate and each occur in the same location during each forcing oscillation. This helps simplify the analysis.

In two-dimensional numerical studies [3], the standing wave becomes unstable and generates turbulence. This would also happen in our three-dimensional simulation, but the turbulence would be highly constrained since there is yet no source of x-variation in our flow. To break the two-dimensional symmetry of the flow, while maintaining the basic structure of the large scale, we add a weak component of forcing with x-variation. We have tried this in various ways: adding a random initial perturbation at all scales, randomly forcing the modes with $k = 1$ at each time step, adding another large-scale standing wave, adding a propagating wave, and so on. The results are similar to each other if the perturbations are sufficiently weak. For the simulations discussed below, we have

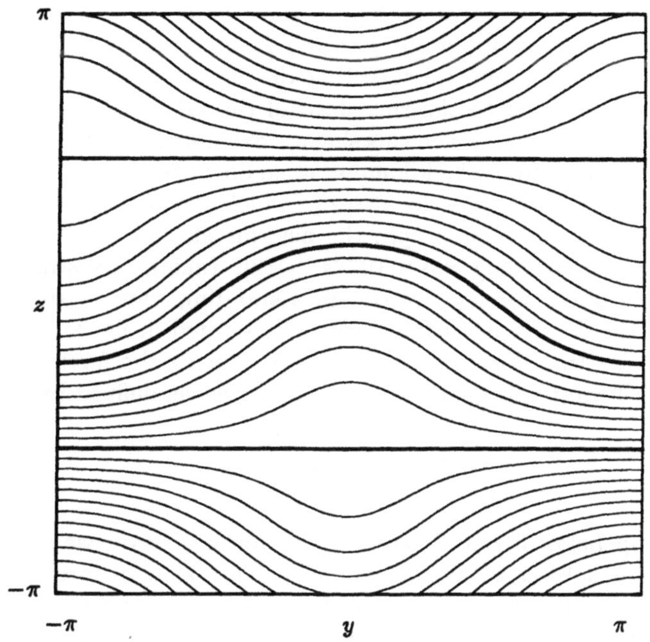

Fig. 7. Contour plot of the density field in a vertical $y - z$ cross section through the center of the domain. The width and height of the cross section are each 20 m, which corresponds to 2π in non-dimensional units. The instant shown corresponds to maximum displacement of the isopycnals for the forced standing wave.

added to the primary forcing wave only a small amplitude standing wave of the same spatial scale. Specifically, we added the following perturbation:

$$\mathbf{u} = A' \frac{g^*}{\sqrt{2}} (\cos(x+z), 0, -\cos(x+z)) \sin \frac{t}{\sqrt{2}}, \tag{23}$$

$$\frac{\rho'}{\rho_0} = A' \cos(x+z) \cos \frac{t}{\sqrt{2}}. \tag{24}$$

Thus in the simulations discussed below the forcing occurs only at $k = \sqrt{2}$. The coefficient A' was taken to be $A/20$, and, hence, the energy in the perturbation is only $1/400$ that of the primary forcing wave.

We performed a series of experiments in which the size of the computational domain and the amplitude of the forcing were varied. The initial studies were at resolution 64^3 and showed that for sufficiently large amplitudes A for which the forcing wave itself was overturning, a $k^{-5/3}$ spectrum extending over most of the spectral range could be established. For weaker forcing, a steeper spectrum approximating k^{-3} was found [4]. For intermediate amplitude forcings, we were able to observe, at least intermittently, cases which do appear to exhibit the transition from the buoyancy range to the smaller scale inertial range. Weak and strong forcings are measured relative to shear amplitudes typical in the thermocline. The best results were obtained with a forcing amplitude that could actually be considered representative of wave amplitudes in the thermocline. Specifically, the forcing amplitude that we refer to as intermediate, is for a value of A in equations (20) and (21) such that the maximum shear during a cycle of the forcing is equivalent to the rms shear of the GM spectrum at the scale of our computational domain. The rms shear is calculated by integrating the shear of the GM spectrum from the kilometer scale down to the scale of interest [12]. Our best results tended to be for cases in which the vertical wavelength of the forcing was 20 m. For $N = 3$ cph, the net rms shear from the GM spectrum for this scale is $S_{GM}(20\text{m}) \approx 3 \times 10^{-3}\text{s}^{-1}$ [12]. Taking this value to determine the amplitude of our forcing, we obtain a standing wave in which the largest deviation of the density isosurfaces are as illustrated in figure 7. Thus we have a standing wave that does not itself overturn during the forcing cycle, and, in addition, the Richardson number of the forcing wave, defined by

$$Ri \equiv -\frac{\frac{g}{\rho_0}\frac{\partial \rho}{\partial z}}{\left(\frac{\partial u}{\partial z}\right)^2 + \left(\frac{\partial v}{\partial z}\right)^2} \tag{25}$$

does not drop below 3.125. Therefore, the forcing wave itself is convectively stable and not subject to shear instability. This kind of forcing is consistent with the picture that the GM waves themselves are not convectively or shear unstable, but through wave-wave interactions will produce smaller scale waves that are unstable by these criteria. Choosing a stronger forcing wave that is itself convectively or shear unstable would miss the important cascade process that produces the unstable waves of the buoyancy range, but would rather produce turbulence directly resulting in an inertial range [4].

For all of the simulations discussed below, we used a resolution 128^3 and a computational cube of 20 m on a side. Our isotropic spectral cutoff is at wavenumber 60, and the smallest resolved wavelength is about 33 cm (with grid spacing $20\text{m}/128 \approx 16$ cm). The forcing amplitude was fixed so that the max shear in the forced wave is $S_{GM}(20\text{m})$, and the Brunt-Vaisala frequency was taken to be 3 cycles per hour, which is a typical oceanic value.

A long simulation was performed with realistic values for the molecular viscosity and diffusivity. The kinematic viscosity was set to $\nu_{mol} = 0.01\text{cm}^2/\text{s}$ and the molecular Prandtl number at $Pr_{mol} = 7$ [8]. We can calculate a Reynolds number for the oceanic flow for vertical motions on the 20 m scale by using the rms shear. Thus we can write

$$Re = S_{GM}(L)L^2/\nu_{mol} \tag{26}$$

For $L = 20$ m, this Reynolds number would be approximately 10^5. By including the molecular viscosity, the simulation is an attempt to represent flow with this Reynolds number. We will see that there is not much difference with results obtained by neglecting the molecular components of viscosity and diffusivity. That is to say that over the range of scales simulated (20 m to 33 cm) the difference between infinite Reynolds number flow and that for $Re = 10^5$ is small.

We can think of our standing wave forcing as the linear superposition of a set of propagating internal waves. To be precise, the combination of the two

(a) (b)

(c) (d)

(e) (f)

(g) (h)

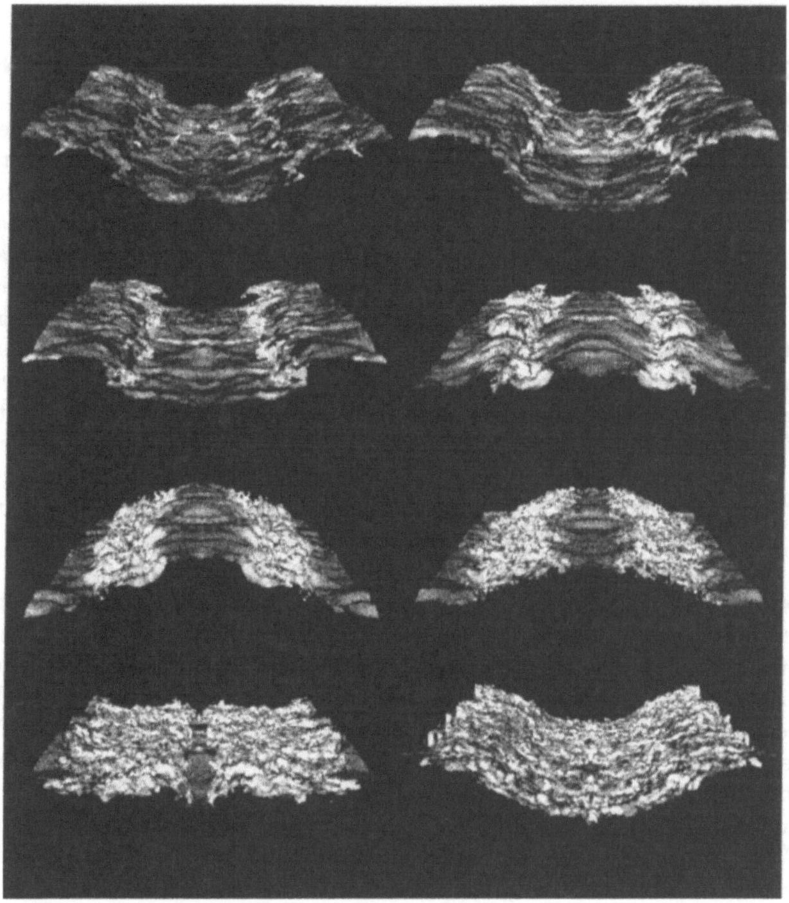

Fig. 8. The evolution of the $\rho = \rho_0$ isopycnal during one cycle of the forcing. The frames are ordered by time from left to right and top to bottom. The first corresponds to $t = 11.39T_F$, and the interval between frames is $\Delta t = T_F/7$.

standing waves given in (20) and (23) consists of 12 propagating plane waves. These wave interact nonlinearly producing smaller-scales that eventually fill out the entire spectrum. The early evolution is essentially just that of the nearly two-dimensional standing wave. During this time there are only sinusoidal waves on the most disturbed isosurface, but these waves then fold over forming elongated overturns. These regions are convectively unstable and break. At this point the three dimensionality of the flow becomes evident.

After about five cycles of the forcing, the large-scale wave breaks repeatedly, however, not necessarily during each forcing period. The wave breaking on the most disturbed isopycnal occurs roughly symmetrically with large-scale overturning occurring nearly at the same values of y and z each time and along lines of constant x, respecting in the large scales the symmetry of the main part of the forcing. However, no two breaking events with the subsequent evolution during the forcing cycle are the same. In figure 8, for one such cycle, we show eight instantaneous images of this isosurface using a perspective three-dimensional plot. The frames are ordered temporally from left to right and top to bottom. The first frame in the upper left hand corner corresponds to $t = 11.39T_F$, and the interval between frames is $\Delta t = T_F/7$. Thus the first and last frames correspond to the same phase of the forcing. The first frame captures the moment when breaking is just beginning. Let us say that the first four frames represent the breaking event, and the last four the aftermath. We see that during the breaking event, heavy fluid spills over lighter fluid, crashing down with the creation of small-scale structures all along the lines of the two breaking regions. Similar behavior is in laboratory experiments with standing-wave forcing [26] [20]. Afterwards, the region of the small-scale turbulent structures spreads, eventually 'contaminating' the entire isosurface. If we compare the final frame with the first frame, we see that the final surface is much rougher, filled with small-scale structures everywhere, and that there is no larger scale folding-over of the surface as there was in the first frame. In the later evolution, the wave will break again, but only after a refractory period, in this case of about two forcing cycles.

Next we will consider the energy spectra for the flow at the same times as those illustrated in figure 8. Since the energy is highly anisotropic at scales larger than those in the inertial range, plotting the total energy as a function of the isotropic wavenumber tends to obscure the transition between small and large scales. To most clearly display the transition, we have found it useful to consider the spectrum $E_v(k)$ of v, the y component of velocity, which is the horizontal component that is directly affected by the forcing. Along with the spectra, we have also drawn lines corresponding to the inertial range spectrum $(1/3)C_K\epsilon^{2/3}k^{-5/3}$ and buoyancy range spectrum $0.2N^2k^{-3}$. For each frame, ϵ is taken as the total kinetic energy dissipation rate at that time. We have included a factor of $(1/3)$ which is appropriate for a single component in the isotropic inertial range. For the Kolmogorov constant, a value of 1.5 was used in each case. We should emphasize that no attempt is made here to fit the data, but the coefficient is just taken as this standard value *a priori*. For the buoyancy range spectrum, we have used the coefficient $\alpha = 0.2$ in all cases. In each frame shown in figure 9, we see a fairly good match at wavenumbers greater than about 20 (that is for scales below about 1 m) to the Kolmogorov inertial range spectrum. The main deviation is at wavenumbers near $k = 60$, the cutoff wavenumber, and this is to be expected from previous experience with the cusp model [17]. The spectrum below wavenumber 20 is naturally far more irregular than that above due to the much smaller number of modes in the lower spectral bands. If we neglect the first few wavenumbers, then there is some evidence here for a steeper spectral range for wavenumbers below about $k = 20$, that is for scales larger than about 1 m, at least in the frames that correspond to times during the breaking of the wave (first four panels). In the aftermath of breaking, the spectra tend to be somewhat flatter (the last four panels). The best representative of the transition between buoyancy and inertial range is found in panel (c), which

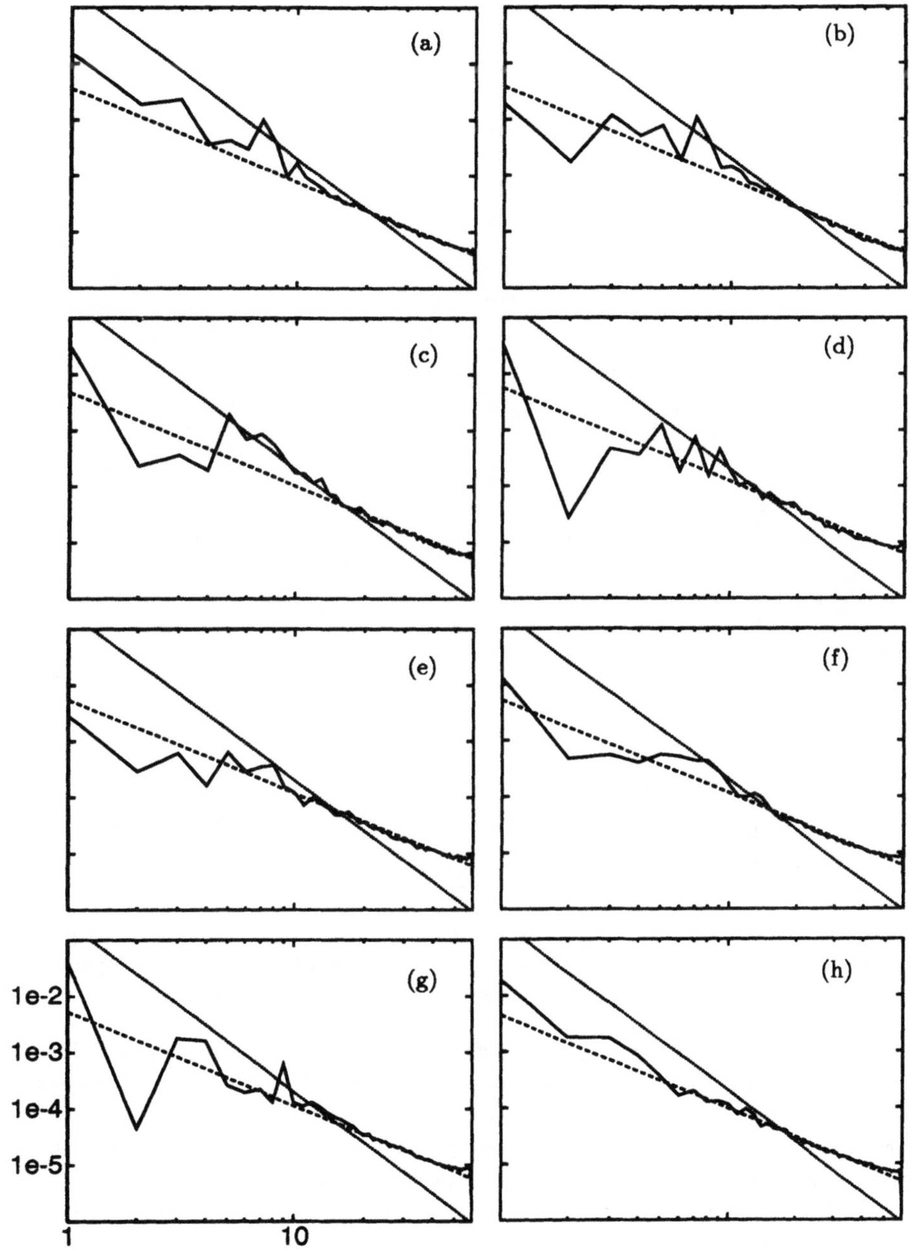

Fig. 9. Kinetic energy spectra for the v component of the velocity during one cycle of the forcing. The dashed lines correspond to the Kolmogorov spectrum $(1/3)C_K\epsilon^{2/3}k^{-5/3}$ with $C_K = 1.5$ and the saturation spectrum $0.2N^2k^{-3}$. The time interval between frames is $\Delta T_F/7$ and the first frame corresponds to time $t = 11.39T_F$. These are log-log plots of $E_v(k)$ in units of $N^2(L/2\pi)^3$ vs. k in units of $2\pi/L$. All plots have the scales as indicated in panel (g).

(a)

(b)

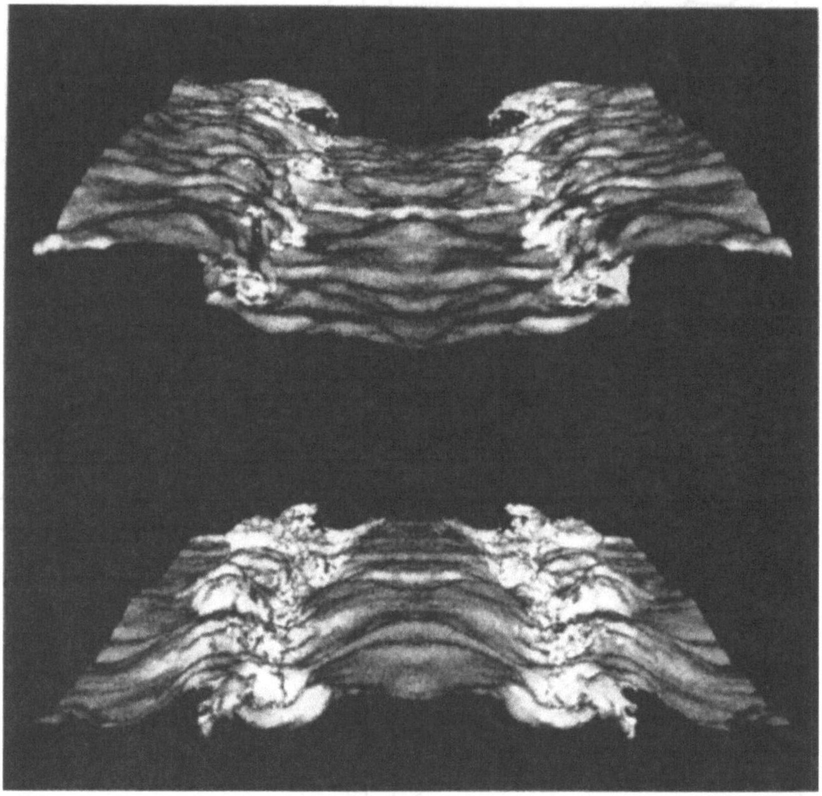

Fig. 10. A breaking event visualized on the $\rho = \rho_0$ isopycnal. These are enlargements of the images shown in the composite figure 8 in panels (c) and (d), corresponding to times (a) 11.68 and (b) 11.82 T_F (one seventh of a forcing period apart).

corresponds to a time when the enstrophy is near a local maximum. Here the buoyancy range spectrum makes a reasonably good fit in the range of scales from about 4 m down to about 1 m. From the forcing scale (20 m vertical) to about the 5 m scale, there is a dip in the energy that has also been seen in the spectra from similar two-dimensional simulations of the decay of a standing wave [3].

In this experiment it appears that the expected spectral signature of a transition between a buoyancy range at large-scale and the inertial range at small-scale occurs only for periods during which there is active breaking. Indeed, it appears that wave-wave interactions repeatedly build up energy in the buoyancy range until a k^{-3} spectrum is achieved. At that point significant breaking occurs and energy drains from the buoyancy range. Let us focus on the breaking event. In figure 10a, is the image of the $\rho = \rho_0$ isosurface at the time identified as the best for illustrating the spectral transition from the buoyancy to the inertial range.

It shows the curling over and spilling down or plunging of the heavier fluid over lighter, while figure 10b suggests mixing by the appearance of many small scale structures along the two parallel lines of the breaking wave. The corresponding spectra for all three components of kinetic and for the potential energy are shown in figure 11. First we notice that although the spectra are highly anisotropic from the forcing scale (20 m) down to about the 1 m scale, there is an approximate 'return' to isotropy for the smaller scales. This is particularly evident in the kinetic energy spectra for t=11.82 T_F (panel c). In panels (a) and (c), we have made an attempt to draw the best fit inertial range spectra to determine the appropriate Kolmogorov constants (C_K) that fit these data. We did this for the $E_v(k)$ spectra, obtaining the best fit 'by eye' from enlarged portions of the small scale spectra. The result that was used to draw the inertial range model spectra in panels (a) and (c) is $(C_K) = 1.4$. In panels (b) and (d), the potential energy spectra are drawn. In these panels the small scales were fit to the Corrsin-Obukhov spectrum to determine the appropriate Corrsin constant. In panels (b) and (d) the Corrsin constants used to draw the model Corrsin-Obukhov spectrum were $C_o = 0.83$ and 0.8 respectively. In all panels the model buoyancy range spectrum drawn is $0.2N^2k^{-3}$. Thus the Kolmogorov constant found here is somewhat smaller than the empirical values of 1.5 and the Corrsin constant is somewhat larger than the empirical value of 0.67. Nevertheless, the values are remarkably close to the empirical values, given that the spectral width of the inertial range here only covers wavelengths from about 1 meter to about 33 cm. Also the near collapse of the three kinetic energy spectra for small scales is encouraging. Thus it seems that the subgrid scale model is working well at small scales and that the dynamics of the transition from anisotropic buoyancy to the isotropic inertial range is acting as imagined in theoretical models. Finally, we should note that the value of ϵ from the simulations is about one third of the value observed [1] for $N = 3$ cph. This appears quite reasonable given the level of modeling we have had to employ for the forcing and subgrid scale vortices.

Besides the kinetic and potential energy spectra, we can also find predictions for the buoyancy flux spectrum in both the theory of Lumley-Shur [19], [22], [28] and the theory of Holloway [15], [16]. The modal spectrum of the buoyancy flux can be written as

$$-g\mathcal{R} < w^*_{\mathbf{k}}\rho'_{\mathbf{k}} > /\rho_0. \tag{27}$$

If this quantity is positive, then for wavevector $\mathbf{k}$ there is conversion of potential energy to kinetic energy, and vice versa if it is negative.

The prediction of the Lumley-Shur theory for the buoyancy flux spectrum in the buoyancy and inertial ranges is

$$BF(k) = -2D\frac{\epsilon_0}{k_b}\left(1 + D(k_b/k)^{4/3}\right)^{1/2}(k_b/k)^{7/3} \tag{28}$$

where k_b is as defined in (12) and D is a constant. Lumley [19] assumed the buoyancy flux to be negative and, hence, D to be positive. In displaying his final result, Lumley incorporated D into his definition for k_b, but we will leave it explicit. Lumley's prediction of negative buoyancy flux through the buoyancy

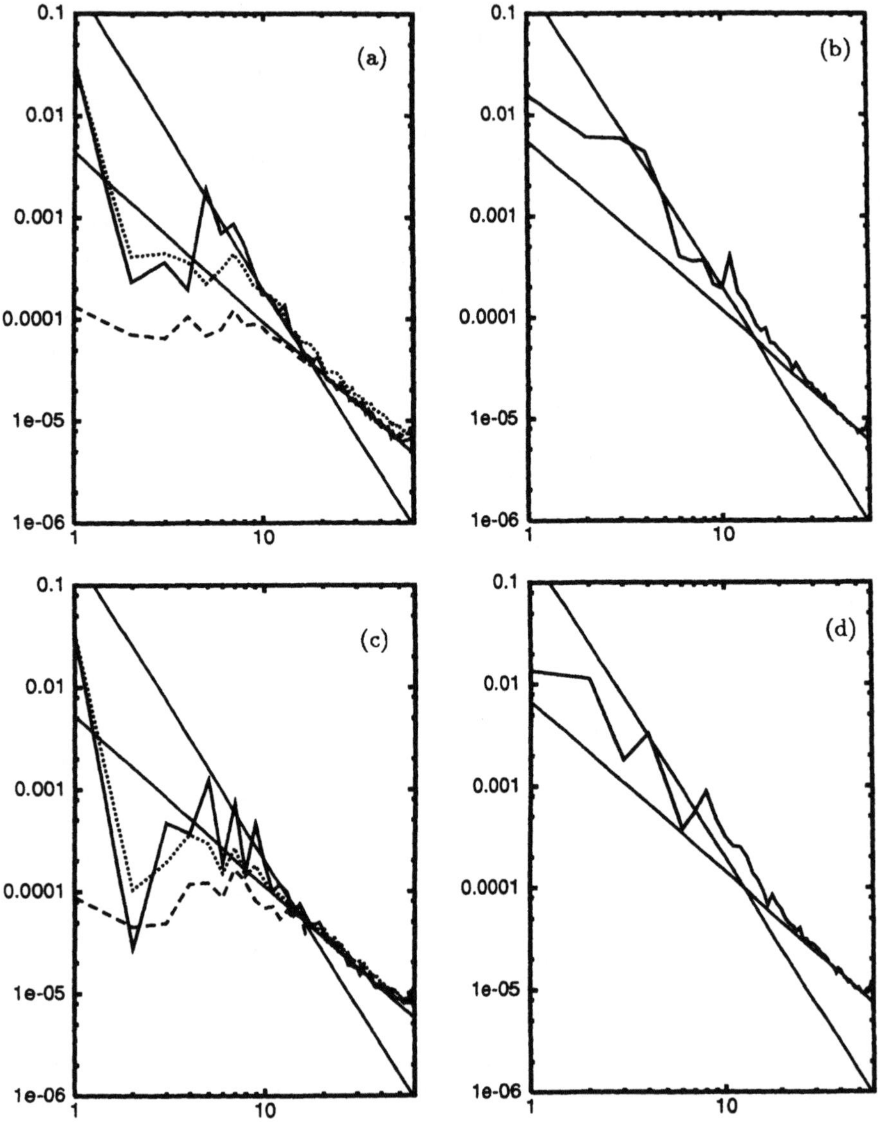

Fig. 11. a) Kinetic energy spectra for all three components of the velocity at time $t = 11.68T_F$. The thick long dashed, solid and short dashed lines correspond to the energy spectra for the u,v, and w components respectively. The thin solid lines correspond to the Kolmogorov spectrum $(1/3)C_K\epsilon^{2/3}k^{-5/3}$ with $C_K = 1.4$ and the saturation spectrum $0.2N^2k^{-3}$. b) Potential energy spectrum at time $t = 11.68T_F$. The thick solid line corresponds to the potential energy spectrum. The thin solid lines correspond to the Corrsin-Obukhov spectrum $C_o\epsilon_{pe}\epsilon^{-1/3}k^{-5/3}$ with $C_o = 0.83$ and the saturation spectrum $0.2N^2k^{-3}$. c) As in (a) but for t=11.82 T_F and $C_K = 1.4$. d) As in (b) but for t=11.82 T_F and $C_o = 0.80$.

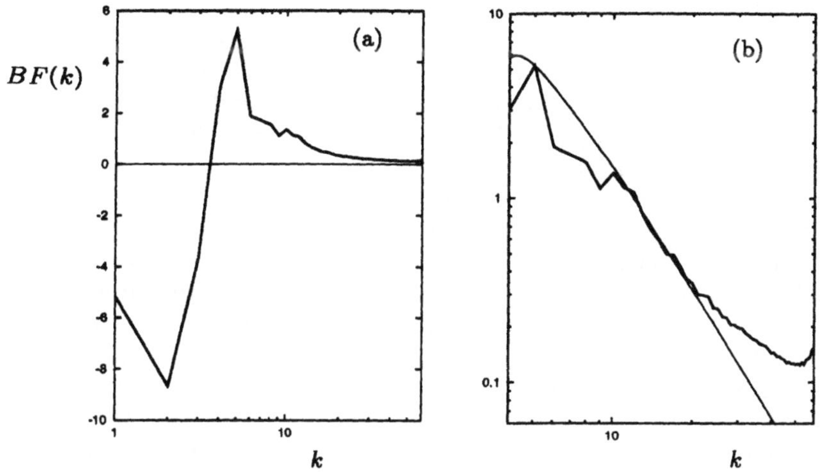

Fig. 12. (a) Graph of the buoyancy flux spectrum averaged over 6 periods of the forcing, with 10 samples per period. (b)as in (a) but only the positive portion of the spectrum plotted in log-log format to compare with the theoretical spectrum of equation (28) with negative coefficient D. The result from the simulation is represented by the thick line, while the theoretical spectrum, based on $\epsilon_0 = \bar{\epsilon}$ where the overbar represents time averaging, is drawn as a thin line. All graphs in (a) and (b) are normalized by $\bar{\epsilon}/k_b$.

and inertial ranges is just the opposite of what we have found numerically for our wave-forced problem. All of the ingredients for an alternative prediction of the buoyancy flux are given in [15], and based on this we have derived the same prediction as given in (28, but with the sign of D clearly arbitrary [6].

In figure 12(a), we plot the buoyancy flux spectrum from our simulation as a function of k. This is a time averaged spectrum, where we have averaged over a period of $6T_F$, with time increment of $0.1T_F$. The time averaging is necessary to remove temporal fluctuations in the large-scales. Note that the buoyancy flux spectrum is negative for large scales ($1 < k < 3$), and positive for smaller scales. This implies a transformation of kinetic to potential energy at large scales (closest to the forcing scale $k = \sqrt{2}$) and a transfer of potential to kinetic energy at all smaller scales. Since our observed buoyancy flux spectrum is positive through both the buoyancy and inertial ranges, it can be compared to the theoretical prediction given by (28) only by choosing a negative value for D. To define the constant D, we note that the wavenumber where the buoyancy flux vanishes is determined by D. Here we shall choose D so that the zero value occurs at $k = 3.5$ (corresponding in our simulation to a wavelength of 5.7 m) since our buoyancy flux was found to vanish between $k = 3$ and $k = 4$. The theory will apply only above this wavenumber, and we can think of this as the lower limit on the buoyancy range, or the upper wavenumber of the Garret Munk spectrum in the schematic shown in our introduction. To compute k_b, given by (12), we use the time averaged dissipation rate $\bar{\epsilon}$. Thus all parameters in the theory are determined by the wavenumber where the buoyancy flux vanishes and the values of N and $\overline{\epsilon}$ (in this case $k_b \equiv (N^3/\overline{\epsilon})^{1/2} \approx 34.1$). The resulting theoretical buoyancy flux spectrum is compared in figure 12(b) to the results from our simulation. For wavenumbers in the buoyancy range, the match between theory and simulations is reasonably good. For the theoretical curve, the decay with k is approximately $k^{-7/3}$ for all k above about 10. The simulation data follows the theoretical curve fairly well up to about wavenumber 20, where the simulation spectrum begins to deviate from $k^{-7/3}$, and is clearly much shallower than this for $k \geq 30$. This shallowness of the simulation spectrum for k greater than about 30 is probably an indication that the buoyancy flux is not captured properly by the SGS model near the high wavenumber cutoff. The cusp model viscosity grows rapidly with k for wavenumbers above about $k = 30$ and is largest at k_{max}. This is just the range where our buoyancy flux spectrum becomes very shallow. It is very possible that the artificial damping of the high k modes that the model performs to mimic transfer of energy beyond k_{max} does not allow for the proper treatment of the buoyancy flux in that region. But this is not unexpected for such a subgrid-scale model.

6 Structures in regions of high strain rate

The main structure of interest in the buoyancy range evident in the density isosurfaces presented in the last section is the overturn produced by the curling over of the isosurface in a manner familiar from surface wave breaking. The overturning region shown in the breaking wave illustrated in figure 8(c) has a vertical scale of about 2 m. This is similar in size to overturns found in oceanographic measurements in the buoyancy range. Alford and Pinkel [1] made an inventory of more than 2200 overturns. They found a median Thorpe scale, a measure of the vertical extent of the overturn, of 1.88 m. Note that this is not greatly different from the scale suggested by the transition point in the spectra shown in figure 11, where the associated length scale is about 1.2 m. Since the observational data is primarily one-dimensional in space, it is difficult to form a three dimensional image of those overturns. The ability to perform three-dimensional analysis of such structures is one of the benefits of numerical simulation.

Examining the full density field more thoroughly, we also find interesting structures of a rather different nature than those associated with strong vertical shear. These can be represented well by the deformations of the density surfaces that are the flat nodal surfaces of the forcing wave. We shall just refer to these surfaces as the nodal surfaces even when perturbed and deformed by eddies. The most basic motion of the fluid in the nodal surfaces is alternately toward and away from the centers of high strain; however, the combination of the large-scale background straining motion and small-scale eddies produces localized deformations of the nodal surface that can result in overturning and mixing in a manner different from the overturns discussed in the last section. By plotting simultaneously three density isosurfaces (one 'nodal surface' and the most strongly perturbed isosurfaces above and below it) for a sequence of times during the

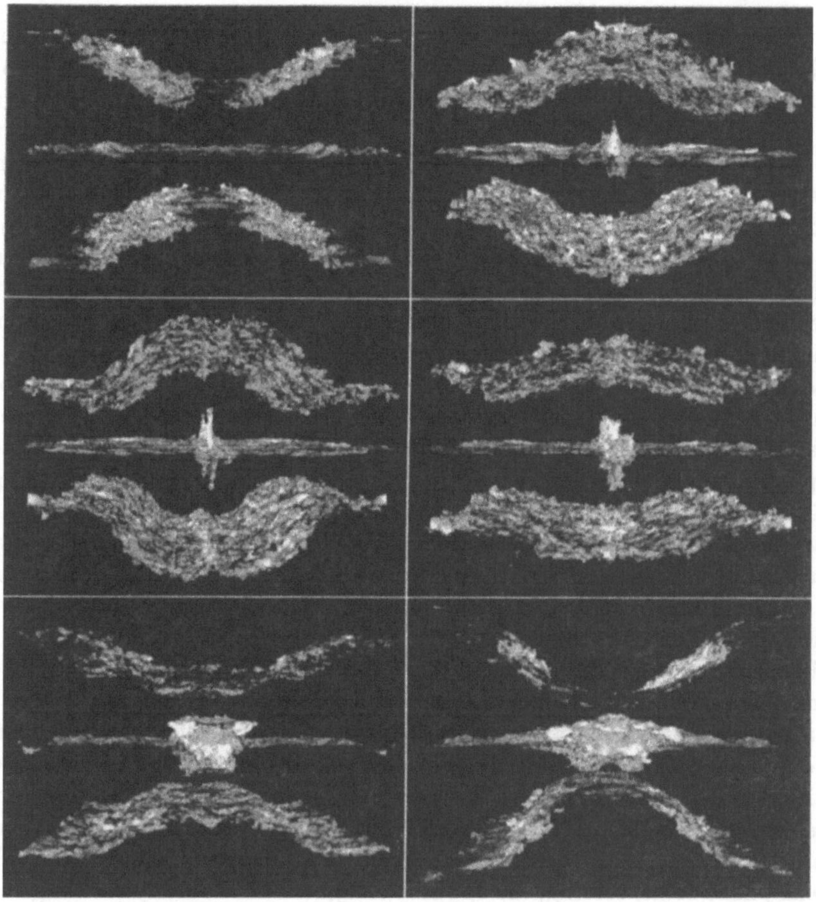

Fig. 13. The evolution of three density isosurfaces ($g^*(\rho - \rho_0)/\rho_0 = 0, \pi/2, \pi$) showing the evolution of 'spouts' from a 'nodal surface' and their subsequent collapse with considerable broadening and mixing. Times represented are $t = 12.1$, 12.4, 12.5, 12.7, 12.8, and 13.1 T_F.

forcing cycle, we can get some understanding of the nature, formation, evolution and fate of these structures. In figure 13, the sequence proceeds from left to right, top to bottom. We have shifted our view of the computational domain by an amount in the vertical sufficient to center the upper 'nodal surface' in the image. Above and below the 'nodal surface,' the most strongly displaced isosurfaces are shown. The upper and lower isosurfaces move vertically but always in opposite directions at any horizontal location. The combined effect of the motion of these surfaces above and below the 'nodal surface' produces vertical 'dilation' and 'compression' centered on the 'nodal surface' without producing large-scale sinusoidal displacement of that surface. In panels (a), (b) and (c), the upper/lower surface is moving upward/downward in the middle of the domain (i.e. at $y = 0$, where y is the horizontal coordinate), and oppositely at the left and right ends of the domain. This is associated with the vertical straining of the nodal surface in the middle and at the left and right ends of the domain. In panel (b) the isosurface 'erupts' with elements moving up and down along a midline pointing into the plane. The eruption reaches its maximum extension when the upper and lower surfaces stop their motion, and reverses direction around the time of panel (c). At $y = 0$ on the 'nodal surface' this is a time of maximum vertical dilational strain but zero strain rate (where $\partial w/\partial z$ is the vertical strain rate). The structures formed by these eruptions represent localized intrusions of heavy fluid into light fluid and vice versa. We shall refer to them as 'spouts.' As the upper and lower isosurfaces move back toward the 'nodal surface,' the sense of straining motion is reversed and the spouts that were formed are flattened. This causes a spreading out of these structures, which in some cases results in tossing elements of the spouts to the right and left of the midline. This leads to the kind of pattern seen in panel (e) which is in part an elongated horizontal structure as opposed to the elongated vertical structures originally produced during the vertically dilational phase of the large-scale straining motion. The final panel (f) shows the isosurface a short time after the upper and lower surfaces have again reversed their direction of vertical motion. This is a phase of the motion near to that of the initial panel (a), but now there is a mixed patch of fluid at the mid section ($y = 0$) of the 'nodal surface.'

To summarize, we can say that the 'spouts' originate from small-scale deviations of the nodal surface created by turbulent flow at the nodal surface. Once perturbations pull structures from the nodal planes vertically, these elements are subject to advection due to the large-scale straining motion of the forcing wave. At times and positions where the straining is highly dilational in the vertical, these deviations from the flat plane elongate vertically and narrow horizontally, forming 'spouts.' Then, during the vertically compressional and horizontally dilational phase of the forcing, the spout is elongated horizontally creating regions of convectively unstable overturned fluid. Note that if the large scale forcing were the only field acting on the spout, then the growth of the spout would simply have been reversed when the sense of the straining motion was reversed. Thus the presence of the eddy field must play an important role in this irreversible process. The distortions of the spout by the eddy field are enhanced during the horizontally dilational phase of the evolution.

7 Conclusions

With some very idealized numerical modelling, we were able to show that an internal-wave packet with approximately the correct vertical structure as observed by Alford and Pinkel [1] can propagate vertically over a hundred meters without dispersing radically and with the production of overturns on the scale of about 2 m, which is entirely consistent with Alford and Pinkel's suggestions. Further, we were able to simulate during a breaking event the transition from buoyancy to inertial range that reproduces features of the observed spectra. We were able to give three-dimensional structure to a classical shear type instability much as found in the observations and at about the correct vertical scale. In addition, we described an unexpected kind of overturning and mixing that occurs in regions of high strain rate. The correlation for overturns and high strain rate was suggested by Alford and Pinkel [1], and yet it is not clear at this point whether the structures we were able to simulate are directly related to any of the overturns reported in that paper.

8 Acknowledgments

This research has been supported by Office of Naval Research grants N00014-97-1-0095 and N00014-96-0762. Numerical simulations were performed at the San Diego Super Computer Center and the Center for Turbulence Research, Stanford. We thank the staff of the San Diego Supercomputer Center for help in preparing the three-dimensional images for this paper and the animation used in this research. Additional computing support was provided by the Office of Naval Research and the University of Rome. We thank Matthew Alford, Lou Goodman, Myrl Hendershott, Greg Holloway, James Lerczac, Robert Pinkel and David Siegel for very helpful discussions and suggestions.

References

1. M. Alford, R. Pinkel: J. Phys. Ocean **30**, 805-832 (2000)
2. P. Bouruet-Aubertot, J. Sommeria, C. Staquet: J. Fluid Mech. **285**, 265-301 (1995)
3. P. Bouruet-Aubertot, J. Sommeria, C. Staquet: Dyn. Atmos. Oceans **23**, 357-69 (1996)
4. G.F. Carnevale, M. Briscolini, 'Large Eddy Simulation of Oceanic Fine Structure,' In: *'Aha Huliko'a, Internal Wave Modeling, Proceedings, Hawaiian Winter Workshop, University of Hawaii, January (19-22, 1999* eds. P. Muller and D. Henderson (1999)
5. G.F. Carnevale, Orlandi, P., 'Propagation of internal wave packets in the thermocline,' In: *Proceedings of the (2000 Summer Program, Center for Turbulence Research, Stanford University pp. 1(19-130* (2000)
6. G.F. Carnevale, M. Briscolini, P. Orlandi: J. Fluid Mech. **427** (205-239 (2001)
7. A.E. Gargett, P.J. Hendricks, T.B. Sanford, T.R. Osborn, A.J. Williams: J. Phys. Ocean. **11**, 1258-1271 (1981)
8. A.E. Gargett: J. Fluid Mech. **159**, 397-407 (1985)
9. C. Garret, W. Munk: J. Geophys. Res., **80**, 291-297 (1975)
10. C.H. Gibson: J. Fluid Mech. **168**, 89-117 (1986)
11. M.C. Gregg: J. Phys. Ocean. **7**, 436-454 (1977)
12. M.C. Gregg: J. Geophys. Res., **94**, 9686-9698 (1989)
13. J.R. Herring, O. Metais: J. Fluid Mech., **235**, 103-121 (1992)
14. Holloway, G. 'Theoretical approaches to interactions among internal waves, turbulence and finestructure' In: *Nonlinear Properties of Internal Waves, AIP conference proceedings No. 76,* (ed. B.J. West. AIP, New York.) (1981)
15. G. Holloway: Dyn. Atmos. Ocean **21** 107-122 (1983)
16. G. Holloway: J. Physical Oceanography **16** 2179-2183 (1986)
17. M. Lesieur, R. Rogallo: Phys. Fluids A **1**, 718-722 (1989)

18. M. Lesieur: *Turbulence in fluids* (Dordrecht; Boston : Kluwer Academic Publishers (1997)
19. J.L. Lumley: J. Atmos. Sci. **21**, 99-102 (1964)
20. A.D. McEwan: J. Fluid Mech. **128**, 47-57 (1983)
21. S.A. Orszag: J. Atmos. Sci., **28**, 1074 (1971)
22. O.M. Phillips, 'On the Bolgiano and Lumley-Shur theories of the buoyancy subrange' In: *Atmospheric Turbulence and Radio Wave Propagation* Proceedings of the International Colloquium, Moscow (1965 (A.M. Yaglom and V.I. Tatarsky, Eds, Nauka, Moscow, (1967) pp. 121-128.
23. D.A. Siegel, J.A. Domaradzki: Journal of Physical Oceanography, **24**, 2353-2386 (1994)
24. B.R. Sutherland: Phys. Fluids **11**, 1081-1090 (1999)
25. B.R. Sutherland: J. Fluid Mech. **429**, 343-380 (2001)
26. J.R. Taylor: J. Fluid Mech. **239**, 309-340 (1992)
27. S.A. Thorpe: J. Phys. Ocean., **29**, 1085-1095 (1999)
28. J. Weinstock: J. Phys. Ocean. **15**, 475-477 (1985)

Pattern Formation in Two-Dimensional Turbulence on a Rotating Sphere

Shigeo Yoden[1], Keiichi Ishioka[2], Michio Yamada[2], and Yoshi-Yuki Hayashi[3]

[1] Department of Geophysics, Kyoto University, Kyoto, 606-8502, Japan
[2] Graduate School of Mathematical Sciences, University of Tokyo, Tokyo, 153-8914, Japan
[3] Division of Earth and Planetary Sciences, Hokkaido University, Sapporo 060-0810, Japan

Abstract. Series of numerical experiments on two-dimensional decaying turbulence are performed for a non-divergent barotropic fluid on a rotating sphere to survey the nature of pattern formation from random initial fields. Time evolution of the flow field is highly dependent on the rotation rate of the system. In non-rotational cases, isolated coherent vortices emerge in the course of time development as in the planar two-dimensional turbulence. As the rotation rate increases, however, the evolution of the flow field changes drastically, and a westward circumpolar vortex appears in high-latitudes as well as zonal band structures in mid- and low-latitudes. Dependence of these features on the initial energy spectrum is investigated and the dynamics of such pattern formations is discussed with a weakly nonlinear Rossby wave-zonal flow interaction theory.

1 Introduction

An interesting feature of two-dimensional turbulence is the emergence of isolated coherent vortices found by McWilliams [1,2], which is due to a self-organizing mechanism involved in the disorder two-dimensional motion. As an application to geophysical fluid, Rhines [3] firstly studied the two-dimensional turbulence on a β-plane [4] to understand the effect of differential rotation of planet. His numerical result showed the anisotropic growth of flow field and the emergence of alternating bands of mean zonal current. These features of the β effect have been reconfirmed in recent numerical studies with much higher resolution models [5,6]. Williams [7] took account of the exact spherical geometry in his numerical experiment on forced two-dimensional turbulence, aiming at a reproduction of the zonal band structure of Jovian atmosphere due to the β-effect. He obtained a clear band structure of zonal flow for a stochastic vorticity forcing. However, both longitudinal periodicity and equatorial symmetry were assumed in his experiment to reduce the computational domain to 1/16 of the entire sphere.

Advancement of computing facilities in 1990's enabled us to do these computations with higher resolutions in full spherical geometry [8–13]. In the decaying turbulence experiments [8], the emergence of a westward circumpolar vortex was found in high-latitudes for high-rotation cases, in addition to the spontaneous formation of the zonal band structures in mid- and low-latitudes. Such formation of a circumpolar vortex was also observed in the forced turbulence experiment [10] and a laboratory experiment with rotating tank [14].

In this article, recent studies on the two-dimensional decaying turbulence on a rotating sphere by ourselves [15,16] are summarized with emphasis on the pattern formation due to full nonlinear upward cascade of turbulent energy in small scales and the role of Rossby waves in angular momentum redistribution. Possible application to some phenomena in real geophysical fluids is also discussed.

2 Numerical Model and Experiments

Freely-evolving two-dimensional non-divergent flow on a rotating sphere is governed by the following vorticity equation:

$$\frac{\partial \omega}{\partial t} + \frac{1}{a^2} J(\psi, \omega) + \frac{2\Omega}{a^2} \frac{\partial \psi}{\partial \lambda} = (-1)^{p+1} \nu_{2p} \left(\Delta + \frac{2}{a^2} \right)^p \omega, \tag{1}$$

where $\psi(\lambda, \mu, t)$ is a streamfunction field, $\omega(\lambda, \mu, t)$ the vertical component of vorticity ($\equiv \Delta\psi$), λ the longitude, $\mu \equiv \sin\phi$ with the latitude ϕ, t the time, $J(A, B)$ the horizontal Jacobian, Δ the horizontal Laplacian, a the radius of the sphere, Ω the rotation rate of the sphere. On the right hand side hyper-viscosity of the order of p is assumed with a viscosity coefficient ν_{2p}. In the case of $\nu_{2p} = 0$, (1) means the conservation of potential vorticity (hereafter denoted as PV; in the present case it is equivalent to absolute vorticity) $q \equiv \omega + 2\Omega\mu$ following Lagrangian motion.

To construct a spectral model of (1), we expand a dependent variable $\psi(\lambda, \mu, t)$ into a set of spherical harmonics,

$$\psi(\lambda, \mu, t) = \sum_{n=2}^{N} \sum_{m=-n}^{n} \psi_n^m(t) P_n^m(\mu) e^{im\lambda}, \tag{2}$$

and truncate it at the total wavenumber $N = 341$ (i.e., triangular truncation T341). A spectral transform method [17] is used to compute the nonlinear Jacobian term; grids for the spectral transformation is 1024(longitude)×512(latitude).

Time evolution of $\psi(\lambda, \mu, t)$ is computed from an initial random flow field, of which energy spectrum is given by

$$E(n, t{=}0) = \frac{An^5}{e^{-n/2}}, \qquad \text{or,} \qquad E(n, t{=}0) = \frac{An^{\gamma/2}}{(n + n_0)^\gamma}. \tag{3}$$

Here the energy density is given by $E(n, t) = \frac{1}{2} n(n+1) \sum_{m=-n}^{n} |\psi_n^m(t)|^2$. Each complex amplitude $\psi_n^m(t{=}0) = (-1)^m (\psi_n^{-m}(t{=}0))^*$ (where * denotes complex conjugate) is determined so that it has random amplitude and random phase under the above constraint. The former expression in (3) was assumed in [8] while the latter was used in [9]. A constant A is set so that the total kinetic energy is unity without dimension: $\mathcal{E}(t{=}0) = \sum_{n=2}^{N} E(n, t{=}0) = 1$.

Table 1. Description of the initial parameters for each experimental set and decay rates of total energy, $\mathcal{E}(t=5)/\mathcal{E}(t=0)$, and enstrophy, $\mathcal{Q}(t=5)/\mathcal{Q}(t=0)$. These ratios depending on the rotation rate Ω are listed in %, which are ensemble averages of 10 runs from the initial conditions made with different random numbers

set name	yy1993 [8]		cpn010		cpn050 <control>		cpn100		cpg100	
n_0	—		10		50		100		50	
γ	—		1000		1000		1000		100	
$\mathcal{Q}(t=0)$	180.0		111.7		2590.7		10262.1		2992.0	
Ω	$\mathcal{E}$[%]	$\mathcal{Q}$[%]	$\mathcal{E}$[%]	$\mathcal{Q}$[%]	$\mathcal{E}$[%]	$\mathcal{Q}$[%]	$\mathcal{E}$[%]	$\mathcal{Q}$[%]	$\mathcal{E}$[%]	$\mathcal{Q}$[%]
0	99.8	25.9	99.9	35.4	97.5	10.3	89.7	5.3	97.1	10.5
25	99.8	24.8	99.9	40.6	97.4	7.6	89.6	4.3	97.0	7.8
50	99.8	29.9	99.9	51.3	97.3	5.5	89.5	3.1	96.9	5.6
100	99.9	40.8	100.0	67.9	97.3	4.2	89.3	1.8	96.9	4.0
200	99.9	53.3	100.0	78.4	97.3	5.1	89.3	1.4	96.9	4.6
400	99.9	65.3	100.0	87.7	97.4	7.6	89.3	1.9	96.9	6.5

Computations are done in a non-dimensional form. The radius of the sphere is taken as a length scale (L); $a = 1$ in the non-dimensional form. Advective time scale is adopted; the time scale (T) is equal to the length scale divided by square root of mean kinetic energy (or velocity scale U). If we take a rough scaling for the atmosphere as $U \sim 10$ m/s and $L = 6 \times 10^6$ m, the time scale is $T \sim L/U = 6 \times 10^5$ s ~ 7 days. Six values of Ω are taken as an experimental parameter: 0, 25, 50, 100, 200 and 400. The sphere has $\Omega/(2\pi)$ rotations per unit time interval. If we take the above scaling, $\Omega = 50$ is close to the present rotation of the earth.

The parameter values of the initial energy spectrum used in the present study are listed in Table 1 and the spectra are shown in Fig.1. Five vertical dash-lines in the figure indicate a characteristic wavenumber $n_\beta \equiv \{\pi\Omega/(4U)\}^{1/2} = \{\pi\Omega/(4\sqrt{2\mathcal{E}})\}^{1/2}$ for each Ω; at the spatial scale with n_β, the linear rotation term is comparable to the nonlinear Jacobian term in (1). Note that the initial energy spectrum used in [8], denoted by a solid line, has much energy in large scales of $n < n_\beta|_{\Omega=400}$; even in the highest rotation case of $\Omega = 400$ the linear term dominates from the initial time. On the other hand, the spectral form used in [9], the latter in (3), can limit the initial energy distribution near n_0 for a large value of γ. Here we use five sets of initial conditions including the reexamination of [8] with higher resolution(denoted as yy1993); the control experiment(cpn050) is given by $(n_0, \gamma) = (50, 1000)$. The spectral peak of initial energy is located in much larger wavenumber than n_β in the three experimental sets with $n_0 = 50$ or 100(cpn050, cpn100 and cpg100).

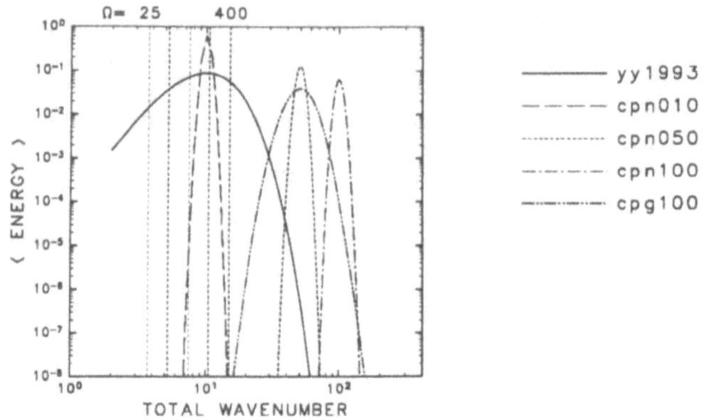

Fig. 1. Five initial energy spectra given by (3) with parameter values listed in Table 1. Vertical dash-lines indicate the characteristic wavenumber $n_\beta = \{\pi\Omega/(4\sqrt{2})\}^{1/2}$ for five values of Ω; 25, 50, 100, 200 and 400

The hyper-viscosity is set with $p = 8$ and $\nu_{2p} = 1 \times 10^{-38}$. Ten initial conditions are made with different random numbers for each experimental set, and time evolutions from these are computed until $t = 5$ by the fourth-order Runge-Kutta method with the time increment of $\Delta t = 1 \times 10^{-3}$. All of the computations are done in double precision.

3 Results

Time variations of the total energy and the total enstrophy are not very sensitive to the value of Ω. The total energy dissipates about 2% by $t = 1$ and it is nearly conservative afterward in the control run [15]. On the other hand, the total enstrophy has a maximum dissipation rate around $t \sim 0.2$ and decays as $Q(t) \propto t^{-1}$ after that time. Decay rates of the total energy and the total enstrophy for the integration period are also listed in Table 1 for the five experimental sets. The energy is almost conserved for the cases in which the spectral peak of the initial energy is located in a low wavenumber (yy1993 and cpn010), while the energy decay is about 10 % when the spectral peak is at $n = 100$ (cpn100). On the other hand, the enstrophy decay is dependent on the initial energy spectra. The decay rate is smaller if the spectral peak is at a lower wavenumber, particularly for the cases of large Ω.

Time evolution of the energy spectrum $E(n, t)$ is dependent on the initial spectrum and the rotation rate as reported in [15]. For the high-rotation case of $\Omega = 400$, however, a power law close to n^{-4} is obtained for all cases in a wide range, extending to low wavenumbers less than n_β.

Time evolution of the flow field can be viewed by movies of the PV and the vorticity fields [18]. Figure 2 shows some snapshots of the vorticity field for a non-rotational case in the control experiment. The initial vorticity field is homogeneous and isotropic due to random initial condition. At $t = 1$, high-gradient,

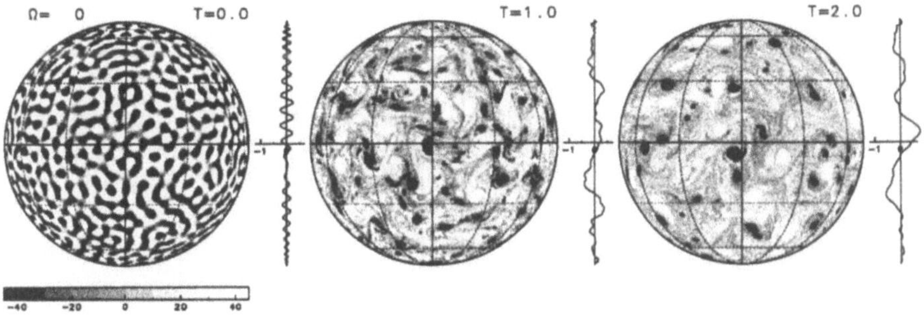

Fig. 2. Time evolution of the vorticity field for $\Omega = 0$ in the control experiment: (left) $t = 0$, (center) $t = 1$, and (right) $t = 2$. Orthographic projection from $\lambda = 0°$, $\phi = 0°$ is used and lines of meridians and parallels are drawn for every 30°. Solid line on the right side of each panel shows the zonal mean zonal angular momentum as a function of sine latitude

filamentary structures are created, which are indicative of the dissipation of enstrophy in small scales, and eventually isolated coherent vortices emerge as in the planar two-dimensional turbulence [1,2] through mergers of vortices with the same sign of vorticity.

The time evolution is largely dependent on the rotation rate Ω. Figure 3 is an example to show the dependence of vorticity field at $t = 5$ evolved from the same initial condition shown in Fig.2(left) on the rotation rate. In the non-rotational case (left), number of the coherent vortices decreases from $t = 2$ shown in Fig.2(right) owing to their mergers, and filamentary structures dim out after large enstrophy dissipation. In the high-rotation case of $\Omega = 400$(right), on the other hand, the vorticity field is not isotropic but shows zonally elongated structures. The zonal mean zonal angular momentum shows the emergence of westward circumpolar vortex, particularly in the southern hemisphere for this run, as well as zonal band structures in mid- and low-latitudes. Movies of the PV field for rotational cases clearly show westward propagation of the disturbance as Rossby waves generated through the upward energy cascade [18]. Note that the energy spectrum with n^{-4} power law should be related to the vorticity distribution with sharp contrast shown in Fig.3(right).

Figure 4 shows time evolutions of the zonal mean zonal angular momentum for six values of Ω starting from the same initial condition in the control experiment. The formation of westward circumpolar vortex in high latitudes and zonal band structure of alternating mean zonal jets in mid- and low-latitudes can be clearly seen as the increase of Ω. The band structure develops from an early stage of the time evolution. At the beginning of integration, the mean zonal angular momentum profile has a rather high characteristic latitudinal wavenumber, and the characteristic wavenumber evolves rapidly toward lower value. Around $t \sim 0.5$, the band structure comes to a steady or very slowly evolving state, and their latitudinal positions are hardly changed with time afterwards. Generally the number of the jets increases while the width of them decreases, as

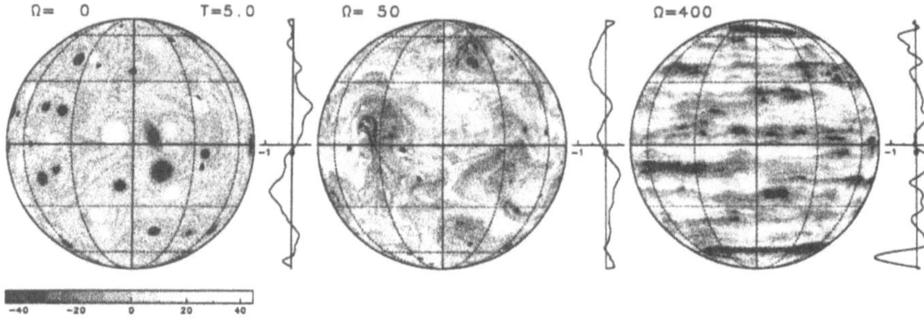

Fig. 3. Vorticity field at $t = 5$ depending on the rotation rate in the control experiment: (left) $\Omega = 0$, (center) $\Omega = 50$, and (right) $\Omega = 400$. The same projection as in Fig.2 is used. The zonal mean zonal angular momentum is also shown

the rotation rate Ω increases. The characteristic wavenumber of the zonal band structure seems to be roughly proportional to n_β as shown in [10] for the forced two-dimensional turbulence on a rotating sphere. It should be noted that the latitudinal positions of mean zonal jets are crucially dependent on the initial condition made with 10 different random numbers.

Dependence of the zonal mean zonal angular momentum at $t = 5$ on Ω is shown in Fig.5 for the five experimental sets with different spectral profiles. The formation of westward circumpolar vortex for large Ω is a robust result independent of the choice of the initial energy spectrum. As Ω increases, the intensity of the westward vortex becomes large and its maximum position shifts toward higher latitudes. Note that the westward vortex for $\Omega = 400$ is stronger for the larger value of n_0 from the set cpn010 to cpn100. As a result of the ensemble average, eastward flow appears in mid- and low-latitudes for large Ω cases owing to the conservation of total angular momentum. However, each realization shows a zonal band structure in these latitudes. The ensemble average masks the band structure because the latitudinal positions of the jets depend rather randomly on each realization.

4 Discussion

It was confirmed that the westward circumpolar vortex is a robust feature of two-dimensional decaying turbulence on a rapidly rotating sphere. Hayashi et al. [16] investigated the dynamics of the formation of the circumpolar vortex with the framework of a weakly nonlinear Rossby wave-zonal flow interaction theory, and the essence is summarized here.

The important quantity that characterizes two dimensional non-divergent flow on a rotating sphere is the transition wavenumber defined as a local wavenumber at which the magnitude of the nonlinear Jacobian term in (1) is comparable to that of the linear advection term of the planetary vorticity(the third term in (1), so called β term) [3,6,19]. The transition wavenumber may be written as

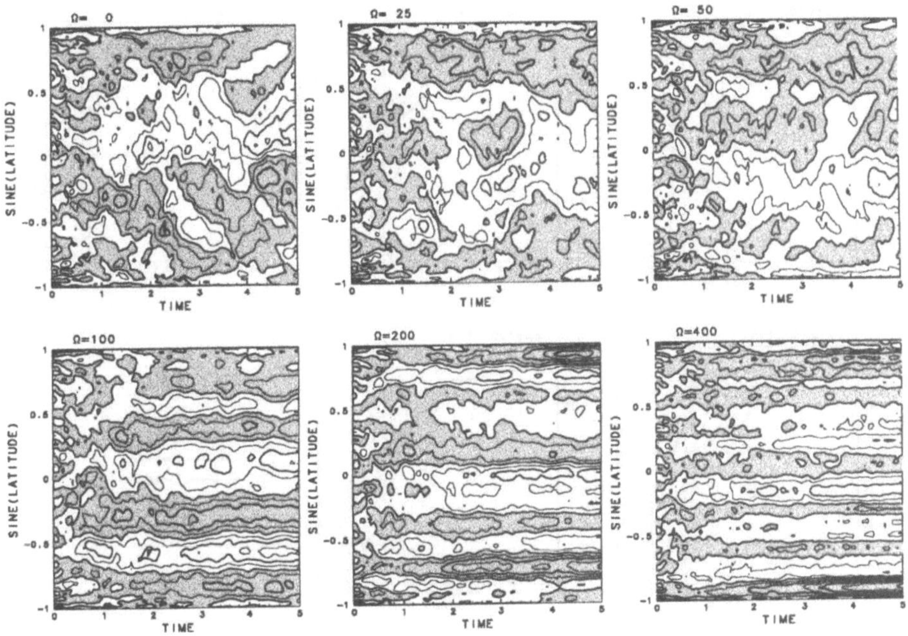

Fig. 4. Time evolutions of the zonal mean zonal angular momentum from the same initial condition of cpn050 for $\Omega = 0$, 25 and 50(top), while 100, 200 and 400(bottom). Contour interval is 0.25 and negative westward flow is shaded

$k_\beta(\phi) = \sqrt{2\Omega \cos\phi}$. When a local characteristic wavenumber of the flow is larger than $k_\beta(\phi)$, the nonlinear term dominates and turbulent motion appears, while Rossby waves dominate when it is smaller than $k_\beta(\phi)$. On a rotating sphere the transition wavenumber vanishes at the poles, while it is maximum at the equator since the β effect is largest at the equator. This implies that, for a disturbance with a given characteristic scale, flow becomes turbulent more easily in the polar region than in the lower latitudes. Provided that the characteristic wavenumber of the flow is $\tilde{n}$, a Rossby-wave dominant region appears around the equator between latitudes $\phi_t = \pm\cos^{-1}[\tilde{n}^2/(2\Omega)]$ when the rotation is sufficiently large.

In all the runs for high rotation cases, the characteristic wavenumber $\tilde{n}$ is about 10 or less, and the transition latitude from wavy to turbulent regime becomes $\phi_t \geq \pm 83°$ for $\Omega = 400$. The formation of the westward circumpolar vortex may be considered as a result of turbulent mixing of PV in the polar region where the flow regime is regarded as turbulent. On a rotating sphere, the homogenization of PV due to mixing of the higher PV around the pole and the lower one in the lower latitudes causes westward flow in the polar region [14,19], provided that there is no net positive angular momentum supply from the outside of the polar mixed region.

When the rotation rate is high and the outside of the polar mixed region is regarded as in the Rossby wave regime, it can be demonstrated that the

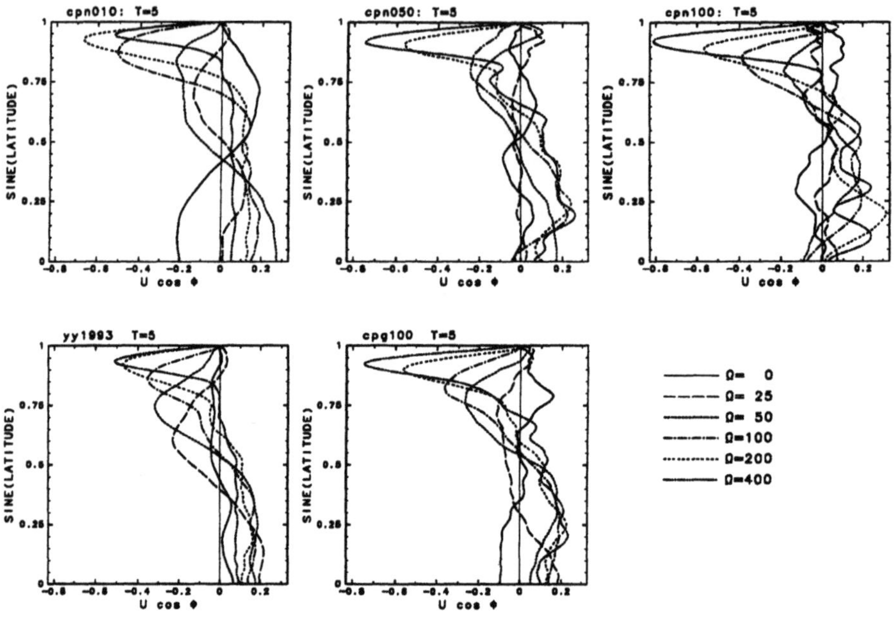

Fig. 5. Dependence of the zonal mean zonal angular momentum at $t = 5$ on the rotation rate Ω for the five initial energy spectra: (top) cpn010, cpn050 and cpn100, (bottom) yy1993 and cpg100. Ensemble average of 2×10 runs is shown under the assumption of equatorial symmetry

net angular momentum supply from the outside of the polar mixed region is negative(westward). As is expected from the knowledge that the pseudo angular momentum associated with Rossby waves is negative, the key issue is whether the polar mixed region absorbs Rossby waves from the wavy region to gain westward momentum or radiates Rossby waves toward the wavy region to gain eastward momentum.

Weak nonlinear theory on the wave mean-flow interaction of Rossby waves on a sphere applied for the present experimental framework gives the result of westward momentum accumulation in high latitudes [16]. WKBJ theory of Rossby wave propagation on a rotating sphere [20,21] shows that a Rossby wave packet propagates along a great circle as seen from the frame of rotation with the speed of eastward phase velocity of the wave packet. A Rossby wave packet whose pitch angle at the equator is sufficiently large can propagate into the polar region where β is small. The propagation of such wave packets induces the second order westward flow, particularly in higher latitudes due to the spherical effect, since the angular momentum associated with the wave packet is conserved during its propagation. If the wave mean-flow interaction takes place, we can expect that large westward fluctuation of mean zonal flow in the higher latitudes enhances the critical latitude absorption there, which contributes the westward momentum accumulation in the polar region.

The above weak nonlinear scenario was examined by a numerical comparison between the nonlinear and weakly nonlinear evolutions starting from the same initial condition [16]. The weakly nonlinear system gives very similar results as the full nonlinear ones for the cases of $\Omega = 400$; the accumulation of the westward angular momentum occurs in the polar region very similarly, and the eastward flow acceleration continues steadily. The resemblance between the nonlinear and weakly nonlinear evolutions is fairly good in the cases of $\Omega = 200$, while the evolution of the mean zonal flow is occasionally different in mid- and low-latitudes in the cases of $\Omega = 100$.

Pattern formation in two-dimensional turbulence on a rotating sphere may be applied for understanding some phenomena in real geophysical fluids. The formation of zonal band structure from a random flow field has been considered as a mechanism to explain the bands in the atmospheres on the giant outer planets such as Jupiter, Saturn, Uranus and Neptune [7,22]. Westward zonal flow in high latitudes in the sun [23] has superficial similarity to the circumpolar vortex obtained in the present study. Recently numerical study of the solar tachocline was also done by investigating freely evolving stratified turbulence in a thin rotating spherical shell [24].

Annular variability observed in both hemispheres of the atmosphere [25,26] is a modulation in the strength of the circumpolar vortex in the troposphere and lower stratosphere with intraseasonal and interannual time scales. Variations of the mean zonal jets with significant barotropic components have much connection with time variations of meridional and vertical PV flux [27], and key dynamics of the annular variability may be identical with that of the formation of the present circumpolar vortex [14].

5 Conclusions

Series of numerical experiments on the two-dimensional decaying turbulence on a rotating sphere were performed with a high-resolution spectral model. Pattern formation due to a self-organizing mechanism in turbulent motion is dependent on the rotation rate of the sphere. The main results are summarized as follows:

1. The flow field becomes anisotropic owing to the effect of rotation. A westward circumpolar vortex appears in high latitudes and zonal band structures of alternating mean zonal jets appear in mid- and low-latitudes for high-rotation cases.
2. The formation of the westward circumpolar vortex is a robust feature of the two-dimensional decaying turbulence on a rapidly rotating sphere and insensitive to the choice of the initial energy spectrum.
3. The zonal band structures are discernible from an early stage of the time evolution, and once established their latitudinal positions are hardly changed with time. As the rotation rate increases, the number of the jets increases while the width of them decreases.

References

1. J.C. McWilliams: J. Fluid Mech. **146**, 21 (1984)
2. J.C. McWilliams: J. Fluid Mech. **219**, 361 (1990)
3. P.B. Rhines: J. Fluid Mech. **69**, 417 (1975)
4. J. Pedlosky: *Geophysical Fluid Dynamics*, 2nd edn. (Springer, New York 1987)
5. M.E. Maltrud, G.K. Vallis: J. Fluid Mech. **228**, 321 (1991)
6. G.K. Vallis, M.E. Multrad: J. Phys. Oceanogr. **23**, 1346 (1993)
7. G.P. Williams: J. Atmos. Sci. **35**, 1399 (1978)
8. S. Yoden, M. Yamada: J. Atmos. Sci. **50**, 631 (1993)
9. J.Y-K. Cho, L.M. Polvani: Phys. Fluids **8**, 1531 (1996)
10. T. Nozawa, S. Yoden: Phys. Fluids **9**, 2081 (1997)
11. T. Nozawa, S. Yoden: Phys. Fluids **9**, 3834 (1997)
12. H.P. Huang, W.A. Robinson: J. Atmos. Sci. **55**, 611 (1998)
13. K. Ishioka, M. Yamada, Y.-Y. Hayashi, S. Yoden: Nonlinear Processes in Geophys. **7**, 105 (2000)
14. P.B. Rhines: 'Rossby Waves and the Polar Vortex'. In: *http://www.ocean.washington.edu/research/gfd/rossby.html* (1998)
15. S. Yoden, K. Ishioka, Y.-Y. Hayashi, M. Yamada: Il Nuovo Cimento C **22**, 803 (1999)
16. Y-Y. Hayashi, K. Ishioka, M. Yamada, S. Yoden: 'Emergence of circumpolar vortex in two dimensional turbulence on a rotating sphere'. In: *IUTAM Symposium on Developments in Geophysical Turbulence*. ed. by R.M. Kerr and Y. Kimura (Kluwer Academic Pub., Dordrecht, 2000) pp. 179–192.
17. W.M. Washington, C.L. Parkinson: *An introduction to three-dimensional climate modeling* (University Science Books, Mill Valley 1986)
18. K. Ishioka, M. Yamada, Y.-Y. Hayashi, S. Yoden: Nagare Multimedia(Nagare) **18**, (1999). Also in: *http://www.nagare.or.jp/mm/99/index.htm*
19. P.B. Rhines: Chaos **4**, 313 (1994)
20. B.J. Hoskins, D.J. Karoly: J. Atmos. Sci. **38**, 1179 (1981)
21. Y-Y. Hayashi, T. Matsuno: J. Met. Soc. Japan **62**, 377 (1984)
22. J.Y-K. Cho, L.M. Polvani: Science **273**, 335 (1996)
23. T.E. Dowling, E.A. Spiegel: *NonLinear Astrophysical Fluid Dynamics, Ann. NY Acad. Sci.* 190 (1990)
24. M.S. Miesch: Astrophys. J. **562**, 1058 (2001)
25. D.W.J. Thompson, J.M. Wallace: Geophys. Res. Lett. **25**, 1297 (1998)
26. D.W.J. Thompson, J.M. Wallace: J. Climate **13**, 1000 (2000)
27. D.L. Hartmann, J.M. Wallace, V. Limpasuvan, D.W.J. Thompson, J.R. Holton: Proc. Nat. Acad. Sci. **97**, 1412 (2000)

Quasi-Geostrophic Turbulence in a One-Layer Ocean affected by Horizontal Divergence

Akira Masuda[1] and Akira Okuno[1]

Dynamics Simulation Research Center, Research Institute for Applied Mechanics, Kyushu University, Kasuga 816-8580, Japan

Abstract. Two topics of quasi-geotstrophic (QG) turbulence are presented as regards the effect of *horizontal divergence* denoted by $F = \lambda^2$, where λ^{-1} is the radius of deformation. Part I shows that strong F suppresses the so-called Rhines effect on a β-plane, which effect refers to a characteristic pattern of zonal currents and the suppression of red cascade by Rossby waves. That is, when F is large and the tendency toward a barotropic flow is prohibited, QG turbulence on a β-plane behaves as if on an f-plane; energy cascades up isotropically though in a slow pace due to F. The conclusion is not only confirmed by numerical experiments, but also proved by a limiting form of governing equation obtained by a transform of variables. Part II deals with three subjects of self-similar spectral evolution of QG turbulence on an f-plane affected by horizontal divergence. First, a similarity form is proposed for freely-decaying turbulence, which is reduced to that of Batchelor (1969) when $F = 0$. Next, inertial subranges are argued. The power of the stream function or kinetic energy is preferred to that of total energy. The kinetic energy spectrum in the inertial subrange is the same irrespective of F: $k^{-5/3}$ for energy cascade and k^{-3} for enstrophy cascade, where k denotes a wavenumber. Finally self-similar spectral evolution is discussed in terms of the ordinary differential equations (ODEs) for the peak wavenumber, total energy and total enstrophy. The dynamics based on the ODEs gives a comprehensive explanation to the self-similar evolution of QG turbulence on an f-plane with and without horizontal divergence in various situations.

1 Introduction

Quasi-geostrophic (QG) turbulence in the ocean or atmosphere is discriminated from two-dimensional (2D) turbulence by two factors : (1) β-effect due to the meridional variation of the Coriolis parameter and (2) horizontal divergence due to density stratification. These two factors have been and will be the main subjects of geophysical fluid dynamics. However, they have not been fully explored yet in particular as regards QG turbulence.

This paper presents a sketch of two recent studies of ours, which focus on the effect of horizontal divergence on QG turbulence. The first is about the suppression of the β-effect or the Rhines effect by horizontal divergence when the tendency toward a barotropic flow is inhibited *a priori*. This effect of horizontal divergence is demonstrated by numerical experiment. Also the physical and mathematical reasons are argued based on a limiting form of governing equation obtained by a transform of variables.

The second topic is concerned with the self-similar spectral evolution of QG turbulence on an f-plane with and without horizontal divergence, where β-effect is discarded. In case of no horizontal divergence $F = 0$, QG turbulence on an f-plane is reduced to the ordinary 2D turbulence. Finite F, however, makes such a dimensional argument as in 2D turbulence difficult. Three subjects are discussed: (1) self-similar freely-decaying QG turbulence, (2) spectral forms in the inertial subrange and (3) dynamics in terms of ordinary differential equations for the spectral peak wavenumber, energy and enstrophy.

The basic equation common to the two topics is the Charney–Hasegawa–Mima (CHM) equation with β-effect (see e.g. Horton and Hasegawa 1994)

$$\frac{\partial}{\partial t}(\nabla^2 - F)\psi + J(\psi, (\nabla^2 - F)\psi) + \beta\frac{\partial \psi}{\partial x} = -\nu_h \nabla^6 \psi, \tag{1}$$

where (x, y) denote the (eastward, northward) Cartesian coordinates, t time, ψ the quasi-geostrophic stream function, β a constant y-derivative of the Coriolis parameter, The parameter $F = \lambda^2$ expresses the effect of horizontal divergence, and λ^{-1} is the radius of deformation. Other notations are as usual. We consider a periodic domain of a model ocean on a β-plane. The surface layer is assumed to be active above the quiet deep layer, so that the barotropic flow was prohibited *a priori* to purify the effect of horizontal divergence. See Okuno and Masuda (2002) for details. It is worthwhile to mention that the CHM equation does not represent exactly the quasi-geostrophic turbulence in a stratified ocean; Eq. (1) describes the evolution of a single vertical mode and no conversion is allowed among different vertical modes.

2 Part I — Suppression of the Rhines effect by horizontal divergence

Among others, the paper of Rhines (1975) is one of the best known with respect to the β-effect on 2D turbulence. It revealed that for a barotropic flow β-effect prohibits energy from cascading up to lowest wavenumbers. Moreover zonal currents dominate the final stage of evolution. This effect of β on 2D turbulence has been called the Rhines effect. The transition occurs around a wavenumber called the Rhines wavenumber, where the westward phase speed of Rossby waves is comparable with the characteristic velocity scale U of turbulence; below this wavenumber, the flow field behaves as waves rather than as turbulence. Rhines (1977, 1979) also argued the effect of stratification or horizontal divergence on QG turbulence over a flat bottom on a β-plane. According to his results the red cascade enlarges the scale to the internal deformation radius, at which the flow becomes *barotropic*. Once this transition occurs, the current behaves as in a barotropic ocean described by Rhines (1975), and finally the Rhines effect appears.

In the experiments of Rhines, however, pure effect of horizontal divergence was hidden behind the barotropic flow at the final stage. Our previous experiment suggested that irregular bottom topography induces surface intensification

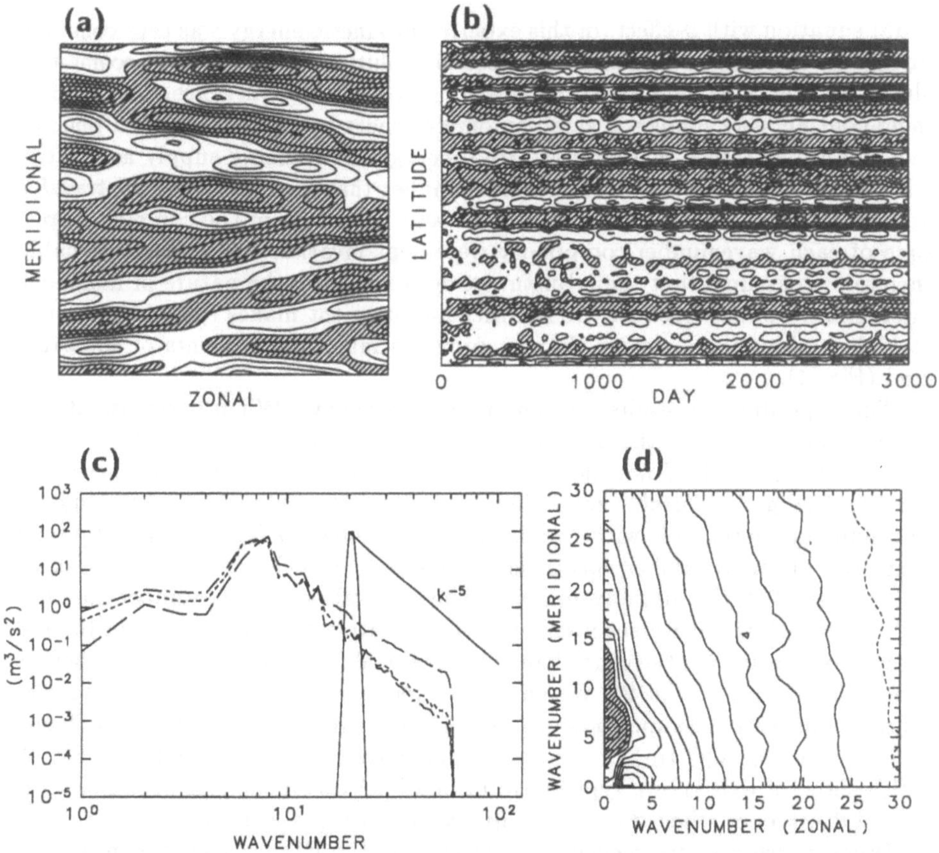

Fig. 1. Results for small F or weak divergence ($\alpha = FU/\beta \sim 3.0 \times 10^{-8}$): (a) Instantaneous stream function on the 3000th day. The contour interval is $1.07 \times 10^3 m^2 s^{-1}$. The hatched areas represent negative contours. (b) Contours of the zonal mean of the zonal velocity as a function of time and latitude. The hatched areas represent negative values. The contour interval is $1.21 \times 10^{-2} ms^{-1}$; (c) Evolution of the one-dimensional wavenumber spectrum of kinetic energy. The solid, long dashed, dashed and dash-dotted lines denote the spectrum on the 0th, 600th, 1800th and 3000th day, respectively. Except for the initial one, each spectrum is the average over 200 days before the nominal date; and (d) Contours of the two-dimensional wavenumber spectrum of kinetic energy on a logarithmic scale, which are the average from the 2800th to the 3000th day corresponding to the one-dimensional spectrum denoted by the dash-dotted line in Panel (c). The contour interval is 0.5 and the hatched area indicates the region where the spectral density is higher than $10^5\ m^4\, s^{-2}$.

of kinetic energy (Takase and Masuda 1996, Masuda et al. 1997), which works against the tendency toward a barotropic flow. It has not been explored well what kind of influence horizontal divergence has in case when the flow remains baroclinic as in an ocean over irregular bottom topography (see also Cox 1987). Therefore we carried out numerical experiment in which the tendency toward a barotropic flow is prohibited *a priori* by using a 1.5-layer model based on the

CHM equation with β-effect. In this experiment kinetic energy was kept constant artificially in order to promote the spectral evolution, which would become too sluggish to be practical on account of strong horizontal divergence. This adjustment continuously supplies a small amount of energy to all wavenumber bins; in fact this process even increases the total energy. The energy supply associated with this adjustment, however, is much smaller than nonlinear energy transfer at each step. Moreover, the spectral form and the phase of the complex amplitude of each wavenumber component were kept invariant. It seems reasonable therefore to assume that the adjustment does not alter the nature of nonlinear interaction. As a by-product, this artificial adjustment makes the k^{-3} spectrum at high wavenumbers clearer than in previous many experiments of the same kind (Fig. 4).

The experimental results are summarized as follows. When horizontal divergence is weak or when the deformation radius is larger than the Rhines scale ($\alpha \equiv UF/\beta \ll 1$), the flow is nearly nondivergent and the ordinary Rhines effect is evident (Fig. 1). Durable zonal flows prevail with a characteristic meridional wavenumber comparable with the Rhines wavenumber. After a sufficiently long time, the 1D wavenumber spectrum of kinetic energy has a spectral peak at the vicinity of the Rhines wavenumber. The spectral peak wavenumber does not shift down below the Rhines wavenumber, indicating the suppression of the red cascade of energy. On the other hand, the spectral form at high wavenumbers is proportional to k^{-5}. The 2D wavenumber spectrum of kinetic energy has a significant anisotropy; components of zero zonal wavenumbers are dominant. These Rhines effects on QG turbulence on a β-plane are to be contrasted with Fig. 2, which shows a well-known typical evolution of 2D turbulence for negligible horizontal divergence when $\beta = 0$: isotropic field and rapid red cascade.

When horizontal divergence is moderate and the Rhines scale is comparable with the radius of deformation, there appears again a band of zonal flows (Fig. 3). Also the 1D wavenumber spectrum of kinetic energy is proportional to k^{-5} at high wavenumbers. However, the zonal flows are a little wider than those in the previous case of weak horizontal divergence. In addition, eddies are less elongated zonally. As a whole, the β-effect or the Rhines effect is still dominant, though it is slightly weakened by horizontal divergence.

For the case of strong divergence or deformation radius smaller than the Rhines scale, the β-effect is suppressed: even after the 3000th day, there appeared no significant zonal flows and the spectral peak was still moving toward low wavenumbers (Fig. 4). Moreover, the spectral peak wavenumber then is already smaller than the Rhines wavenumber. At high wavenumbers, the 1D spectrum is proportional to k^{-3} instead of k^{-5}. Also the 2D spectrum is isotropic. Thus we observe the same behavior as in 2D turbulence shown in Fig. 2 except for the pace of evolution which is slowed down greatly by large horizontal divergence. In other words, β-effect has little influence on QG turbulence when horizontal divergence is strong enough.

Then let us investigate why large F suppresses β. We first introduce new coordinates moving westward at the phase speed of long nondispersive Rossby

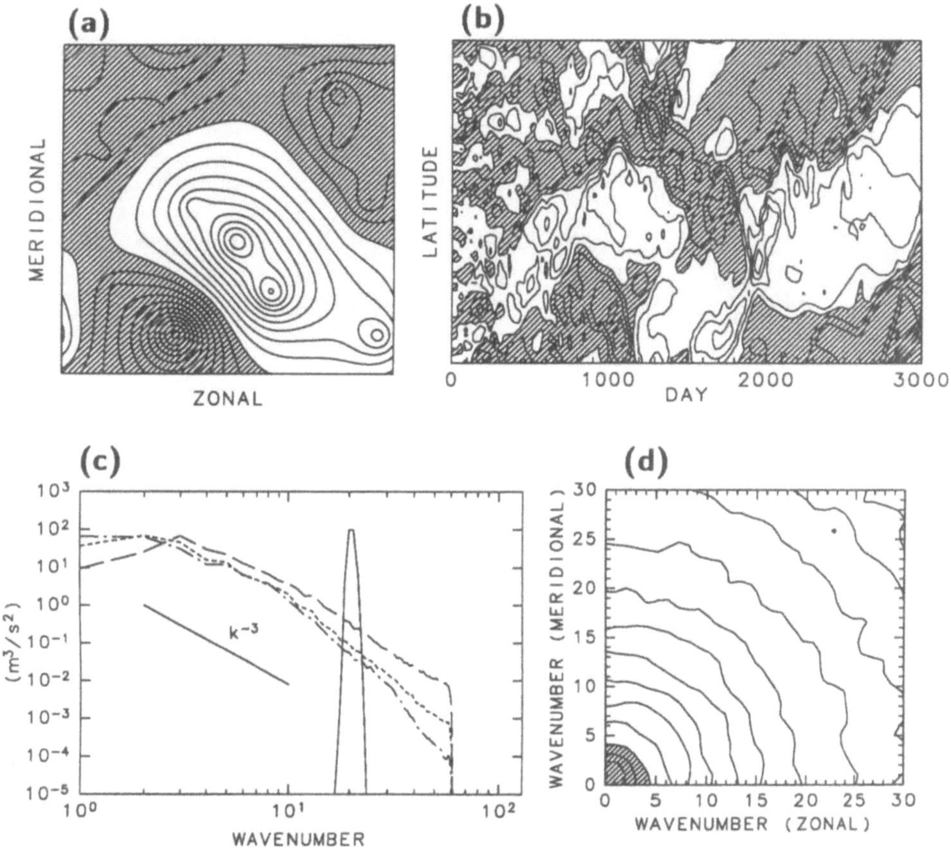

Fig. 2. The same as Fig.1 except that $\beta = 0$. The contour intervals are (a) $2.12 \times 10^3 \mathrm{m^2 s^{-1}}$, (b) $1.21 \times 10^{-2} \mathrm{ms^{-1}}$, and (d) 0.5, respectively.

waves β/F as

$$\begin{cases} \xi = x + \frac{\beta}{F} t \\ \tau = t \end{cases} . \tag{2}$$

This transformation of variables does not alter the wavenumber spectrum of kinetic energy. The CHM equation becomes

$$\frac{\partial}{\partial \tau}(\nabla_\xi^2 - F)\psi + J_\xi(\psi, \nabla_\xi^2 \psi) + \frac{\beta}{F}\nabla_\xi^2 \frac{\partial \psi}{\partial \xi} = -\nu_h \nabla_\xi^6 \psi, \tag{3}$$

where ∇_ξ denotes the gradient operator in the new coordinates. As F increases to infinity, the equation approaches

$$\frac{\partial}{\partial \tau}(\nabla_\xi^2 - F)\psi + J_\xi(\psi, \nabla_\xi^2 \psi) = -\nu_h \nabla_\xi^6 \psi, \tag{4}$$

proving that large F suppresses β.

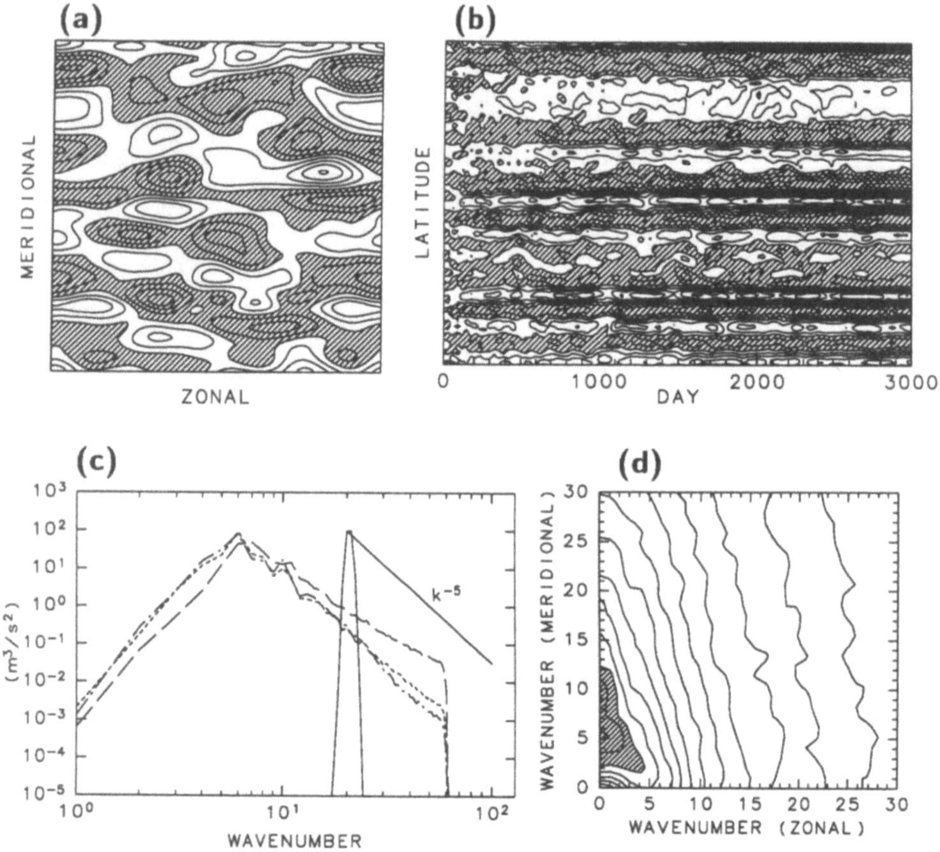

Fig. 3. The same as Fig.1 except that horizontal divergence or F is moderate ($\alpha \sim 0.2$). The contour intervals are (a) $1.36 \times 10^3 \mathrm{m}^2\mathrm{s}^{-1}$, (b) $1.06 \times 10^{-2} \mathrm{ms}^{-1}$, and (d) 0.5 , respectively.

Since $q \equiv (\nabla^2 - F)\psi \sim -F\psi \gg \nabla^2\psi$, we may rewrite the limiting form of equation as any of Eq. (1), Eq. (3) or

$$\frac{\partial q}{\partial \tau} + \frac{1}{F^2} J_\xi(q, \nabla_\xi^2 q) = \frac{\nu_h}{F} \nabla_\xi^6 q \qquad \left(\frac{\beta}{F} \to 0\right). \tag{5}$$

We carried out three kinds of simulation based on Eqs. (1), (3) and (5). The results confirmed effectively the same evolution of turbulence, though not shown here.

The most important result here is that horizontal divergence F totally suppresses the β-effect or the Rhines effect when the tendency toward a barotropic flow is inhibited *a prior* as in the CHM equation. This conclusion is supported by a fluid-dynamical argument as follows; the physics is related with the mathematics mentioned above. When horizontal divergence is absent as in barotropic flows, the phase speed of Rossby waves increases with wavelengths. As the scale of eddies increases due to red cascade of energy, the current speed of eddies even-

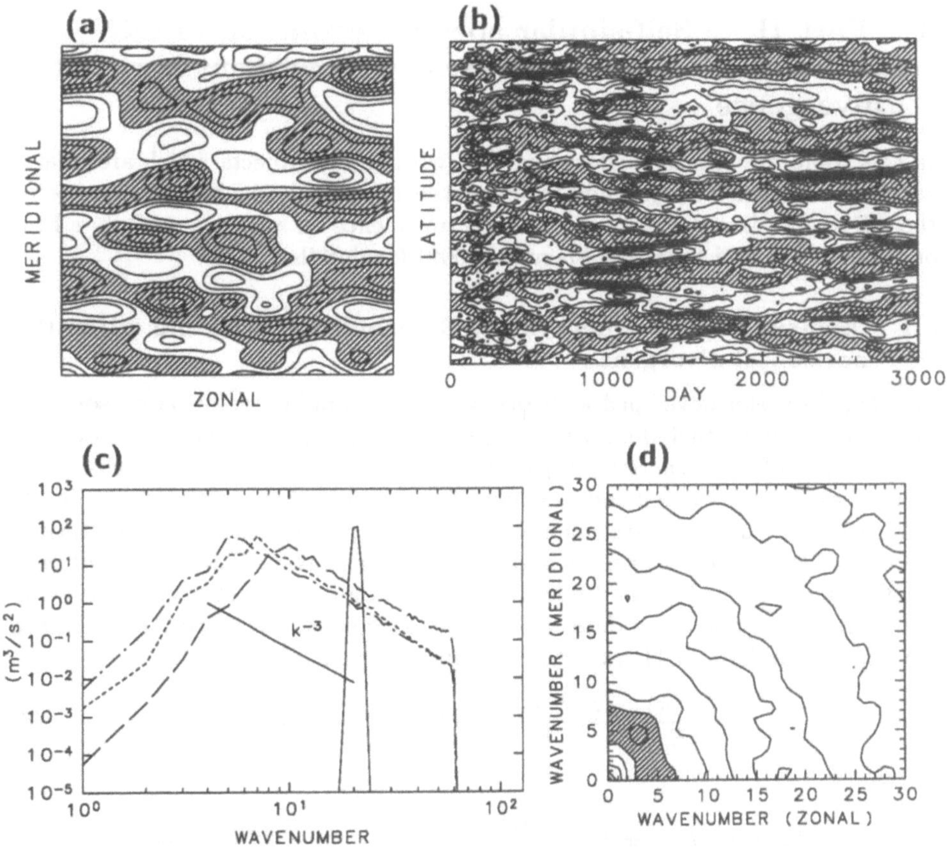

Fig. 4. The same as Fig.1 except that horizontal divergence or F is large ($\alpha \sim 2.8$). The contour intervals are (a) $1.40 \times 10^3 m^2 s^{-1}$, (b) $0.59 \times 10^{-2} ms^{-1}$, and (d) 0.5 , respectively.

tually becomes smaller than the phase speed of long Rossby waves. Hence the behavior as Rossby waves becomes dominant; it hinders the red cascade of energy and produces a band of zonal flows. On the other hand, for baroclinic Rossby waves with a finite deformation radius, there is an upper limit of the phase speed at β/F, the phase speed of long baroclinic Rossby waves. When $\alpha = UF/\beta$ is large, the characteristic speed of turbulence U is larger than the phase speed of long baroclinic Rossby waves. Then the flow tends to behave as turbulence rather than as waves. The turbulent nature becomes more conspicuous in the coordinates moving westward with long (baroclinic) Rossby waves.

3 Part II — Self-similar spectral evolution of QG turbulence on an f-plane with and without horizontal divergence

In this topic we discard β-effect and deal with three subjects which are related with one another closely. They focus on the effect of horizontal divergence on the self-similar isotropic evolution of QG turbulence on an f-plane. Only a brief outline is described here. See Masuda (2002) for details.

3.1 Self-similar freely-decaying QG turbulence with and without horizontal divergence

In 1969, Betchelor developed a theory on the self-similar evolution of isotropic, freely-decaying 2D turbulence. He considered the asymptotic behavior for large t in the limit of small energy dissipation. He assumed the presence of self-similar spectral evolution which is independent of the details of the initial condition. In that case, there are two external and one internal parameters: the characteristic velocity of turbulence U, time t and wavenumber k. Here U is equivalent with kinetic energy $E = U^2/2$ and remains constant on account of energy conservation. Supposing that they determine the evolution of turbulence, he obtained the self-similar form of 1D (kinetic) energy spectrum $E(k)$ as

$$\frac{E(k)}{E_p} = \frac{E(k)k_p}{U^2} =^{\exists} e\left(\frac{k}{k_p}\right) = e\,(Ukt)\,, \tag{6}$$

where

$$E_p = E_p(U,t) = U^3 t = \frac{U^2}{k_p} \qquad \text{and} \quad k_p = k_p(U,t) = \frac{1}{Ut} \tag{7}$$

are the peak spectral density and the spectral peak wavenumber (or the characteristic wavenumber), respectively.

We want to generalize this kind of theory to turbulence with non-zero F on a β-plane for geophysical application. As a first step let us include F, but discard β. When there is horizontal divergence ($\lambda \neq 0$), total energy is composed of kinetic and potential parts. We therefore have to discriminate the spectrum of kinetic energy $E^{(\text{k.e.})}(k) \equiv k^2\phi(k)$ and that of total energy $E(k) \equiv (\lambda^2 + k^2)\phi(k)$, where $\phi(k)$ denotes the power spectrum of the stream function ψ. Also we have the enstrophy spectrum $W(k) \equiv k^2 E(k)$. When $\lambda \neq 0$, there are four independent parameters that characterize the QG turbulence affected by horizontal divergence. It is impossible therefore to extend the purely dimensional arguments as in Batchelor (1969). Nevertheless we obtain a plausible formula of spectral evolution as follows.

Self-similarity alone yields

$$\frac{\lambda^2 + k_p^2}{\lambda^2 + k^2}\frac{E(k,t)k_p}{U^2} = \frac{\phi(k,t)k_p}{\frac{U^2}{\lambda^2+k_p^2}} =^{\exists} e\left(\frac{k}{k_p}\,;\frac{\lambda}{k_p}\right), \tag{8}$$

$$k_pUt \qquad = {}^{\exists}f(\lambda U t), \tag{9}$$

beyond which purely dimensional argument is helpless. To proceed further, we require another assumption, as is suggested by the governing equation. That is, we suppose that k_p is determined by a combination of three external parameters U, t and λ, but that they appear in the form of $U^2/(\lambda^2+k_p^2)$ and $t/(\lambda^2+k_p^2)$. Then the functional form of f is determined as

$$\frac{\lambda}{k_p} \sim \left[\frac{\lambda \cdot Ut}{(1+\frac{k_p^2}{\lambda^2})^{\frac{3}{2}}}\right]^{\frac{1}{4}} \sim \begin{cases} \left[\frac{\lambda^3}{k_p^3}(\lambda \cdot Ut)\right]^{\frac{1}{4}} = \lambda \cdot Ut & (k_p \gg \lambda) \\ (\lambda \cdot Ut)^{\frac{1}{4}} = \lambda \cdot \left(\frac{Ut}{\lambda^3}\right)^{\frac{1}{4}} & (k_p \ll \lambda) \end{cases}. \tag{10}$$

In general the spectral form e depends on λ/k_p as

$$\frac{\phi(k,t)k_p}{\frac{U^2}{\lambda^2+k_p^2}} \sim e\left(\frac{k}{k_p};\frac{\lambda}{k_p}\right) = \begin{cases} e\left(Ukt; \frac{\lambda}{k_p} \to 0\right) & (k \gg \lambda) \\ e\left(\left[\frac{Ut}{\lambda^3}\right]^{\frac{1}{4}} k; \frac{\lambda}{k_p} \to \infty\right) & (k \ll \lambda) \end{cases}. \tag{11}$$

As before we again assume that the power spectrum of stream function ϕ is determined by the combination of U, t and λ in the form of $U^2/(\lambda^2+k_p^2)$ and $t/(\lambda^2+k_p^2)$. Then the universal form e is determined irrespective of λ/k_p. This form is not inconsistent with the spectral form in the inertial range discussed later. Even though this assumption may be doubtful at small wavenumbers (Iwayama et al. 2001), we can consider that Eq. (11) refers to either limit of $\lambda/k_p \longrightarrow 0$ or $\lambda/k_p \longrightarrow \infty$.

The wavenumber range dominated by enstrophy dissipation is proved to be connected with the inertial subrange even when $F \gg 1$, though details are omitted here. The energy or enstrophy flux and others are derived as in Batchelor (1969) in terms of the universal function e.

Thus the present formulation turns out to include Batchelor's theory when $F = 0$. In the opposite case of large horizontal divergence, this formula yields the characteristic length scale $\sim 1/k_p \propto t^{1/4}$ and the peak spectral density $E_p \propto t^{1/4}$. This conclusion agrees with the numerical experiment and argument of Watanabe et al. (1997, 1998). Later we will give the dynamical foundation of the above dimensional arguments.

3.2 inertial subranges of 2D turbulence with F

Watanabe et al. (1998) studied the forced problem, in which they argued that the (total) energy spectrum E in the energy cascade region should be

$$E(k) \sim \frac{\lambda^2 \epsilon^{\frac{2}{3}}}{k^{\frac{11}{3}}} {}^{\exists}g\left(\frac{k}{\lambda}\right) \tag{12}$$

$$g(x) \sim \begin{cases} x^2 & \text{for } x \gg 1 \\ \text{const} & \text{for } x \ll 1. \end{cases} \tag{13}$$

Let us begin with the balance of terms in the CHM equation as in Watanabe et al. (1998). With respect to the nonlinear interaction term, however, we make a somewhat different estimation. That difference turns out to provide a simpler expression for the spectrum in the inertial range.

The balance of nonlinear term and time derivative in Eq.(1) yields

$$\frac{(\lambda^2+k^2)\psi}{t} \sim k^4\psi^2 \Rightarrow \psi \sim \frac{(\lambda^2+k^2)}{k^4 t} \Rightarrow \psi \approx \frac{(\lambda^2+k^2)}{k^4 t} \exists f\left(\frac{k}{\lambda}\right). \tag{14}$$

The spectrum determined by the energy flux ϵ in the wavenumber space is obtained as follows. From Eq. (1) energy flux is estimated to be

$$\epsilon \sim \frac{(\lambda^2+k^2)\psi^2}{t} \sim k^4\psi^3. \tag{15}$$

Then we have

$$\frac{\lambda^2+k^2}{t} \sim \epsilon^{\frac{1}{3}} k^{\frac{8}{3}} f^{-1}\left(\frac{k}{\lambda}\right). \tag{16}$$

We therefore have

$$\phi(k) \sim \frac{\psi^2(k)}{k} \sim \frac{(\lambda^2+k^2)^2}{k^9 t_\epsilon^2} f^2\left(\frac{k}{\lambda}\right) = \epsilon^{\frac{2}{3}} k^{-\frac{11}{3}}. \tag{17}$$

The function f vanishes in this way of estimation. The argument is the same for enstrophy flux η.

The results assert that the spectrum of kinetic energy (or power of the stream function) is independent of λ in the intertial subrange. This conclusion is valid for both forced and freely-decaying turbulence.

The spectrum determined by the energy flux (dissipation) ϵ becomes

$$\phi(k,t) \sim \epsilon^{\frac{2}{3}} k^{-\frac{11}{3}}, \quad E^{(\mathrm{k.e.})}(k,t) \sim \epsilon^{\frac{2}{3}} k^{-\frac{5}{3}}, \quad E(k,t) \sim \epsilon^{\frac{2}{3}}(\lambda^2+k^2)k^{-\frac{11}{3}}. \tag{18}$$

Likewise the spectrum determined by the enstrophy flux η is

$$\phi(k,t) \sim \eta^{\frac{2}{3}} k^{-5}, \quad E^{(\mathrm{k.e.})}(k,t) \sim \eta^{\frac{2}{3}} k^{-3}, \quad E(k,t) \sim \eta^{\frac{2}{3}}(\lambda^2+k^2)k^{-5}. \tag{19}$$

It is simpler to express the spectrum of kinetic energy or the power of the stream function than that of total energy. In reality, numerical experiment shows that there is no kink or such around $k \sim \lambda$ in the spectrum of kinetic energy (Fig. 4).

3.3 Direct investigation of self-similar spectral evolution –dynamics

Attempt of formulating finite dissipation of energy led to the concept of the "dynamics" of self-similar spectral evolution. There are two basic assumptions of this idea. One is that the spectrum evolves keeping a self-similar spectral form. The other is that the spectrum at high wavenumbers is determined by the enstrophy dissipation.

We will sketch the idea by taking an example of the spectral evolution for the freely-decaying turbulence when λ is large and ϵ is small enough. The enstrophy spectrum at high wavenumbers is expressed two ways as

$$W(k) \sim \eta^{\frac{2}{3}} \lambda^2 k^{-3} \sim W_p \left(\frac{k}{k_p}\right)^{-3}. \tag{20}$$

Also self-similarity gives an enstimate of total enstrophy

$$W \equiv \int W(k)\, dk \sim W_p k_p. \tag{21}$$

As a result the dissipation of total enstorphy W is expressed as

$$\eta \sim \frac{k_p^3}{\lambda^3} W^{\frac{3}{2}}. \tag{22}$$

The ODE governing the evolution of total enstrophy is

$$\frac{dW}{dt} = -\eta \sim -\frac{k_p^3}{\lambda^3} W^{\frac{3}{2}}. \tag{23}$$

Since $\epsilon \longrightarrow 0$ we have

$$\frac{W}{k_p^2} \sim U^2 = U_0^2 \sim \frac{W_0}{k_{p0}^2}, \tag{24}$$

where the suffix 0 indicates the initial values. Eliminating k_p from this equation we rewrite the ODE as

$$\frac{dW}{dt} \sim -\frac{W^3}{U^3 \lambda^3}. \tag{25}$$

The solution becomes

$$\frac{1}{W^2} \sim \frac{2t}{U^3 \lambda^3}. \tag{26}$$

Then it follows that

$$k_p \sim \frac{W^{\frac{1}{2}}}{U} \sim \left(\frac{Ut}{\lambda^3}\right)^{-\frac{1}{4}} \tag{27}$$

$$E_p \sim \frac{U^2}{k_p} \sim U^2 \left(\frac{Ut}{\lambda^3}\right)^{\frac{1}{4}}. \tag{28}$$

These results agree with the argument before and numerical experiment and arguments of Watanabe et al. (1998).

Another illustrating example is the opposite limit of $\lambda \rightarrow 0$. Using the same argument as above, we have

$$\frac{dW}{dt} = -\eta \sim -W^{\frac{3}{2}}, \tag{29}$$

the solution of which is

$$W \sim \frac{1}{t^2}. \tag{30}$$

Then we obtain

$$k_p \sim \frac{W^{\frac{1}{2}}}{U} \sim \frac{1}{Ut}, \qquad E_p \sim \frac{U^2}{k_p} \sim U^3 t \tag{31}$$

The results are none other than those in Batchelor (1969).

In these arguments the point is to derive ordinary differential equations describing the evolution of total energy E, total enstrophy W and k_p, on the basis of self-similarity. This method can deal with forced problems and finite dissipation of energy. We applied this method to various situations of QG turbulence with and without horizontal divergence such as freely-decaying and forced turbulence. The dynamics turns out to explain systematically all situations of evolution examined, which have been discussed in various ways so far. For instance, dimensional argument alone may not be helpful in understanding the similarity form of Batchelor for freely-decaying 2D turbulence. On the other hand the dynamics proposed here provides a comprehensible explanation.

There is an interesting situation that has not been examined so far. That is, the adjustment process where total kinetic energy is conserved for large λ, as was made already in the experiment of Part I. Though details are omitted, the dynamics here predicts

$$k_p \propto t^{-\frac{1}{3}}, \tag{32}$$

which was confirmed by numerical experiment. Even when finite dissipation of energy is taken into account, it is possible to calculate the evolution numerically if we use the ODEs as are presented here.

One may notice here that this method is analogous to an argument on freely-decaying 3D isotropic turbulence. The basic equation is for the dissipation of energy

$$\frac{dE}{dt} = -\epsilon \sim -k_p E^{\frac{3}{2}}. \tag{33}$$

In contrast to QG-turbulence, however, threre is no simple relation between E and k_p in 3D turbulence, so that we have to use either Loitsiansky's or Birkhoff's assumption for the global structure of turbulence. In the former case, we have

$$\frac{E}{k_p^5} = \frac{E_0}{k_{p0}^5}. \tag{34}$$

Then, the equation becomes

$$\frac{dE}{dt} \sim -\frac{k_0}{E_0^{\frac{1}{5}}} E^{\frac{17}{10}}. \tag{35}$$

The solution leads to a well-known result as

$$E \sim \left(\frac{k_0}{E_0^{\frac{1}{3}}} t \right)^{-\frac{10}{7}} . \tag{36}$$

We see that in this method QG turbulence is easier to deal with than 3D turbulence because energy conservation gives an explicit relation between W (or E) and k_p without appealing to the debatable assumption of Loitsiansky or Brikhoff on the global structure of turbulence.

4 Summary and discussion

Two topics were presented of the effect of horizontal divergence on the QG turbulence governed by the CHM equation. In part I, numerical experiments were carried out to examine the effect of horizontal divergence on almost freely decaying geostrophic turbulence on a β-plane. For moderate horizontal divergence or a moderate radius of deformation, the β-effect was confirmed to induce a highly anisotropic field characterized by a band of zonal currents with a wavenumber of the order of the Rhines wavenumber $\sqrt{\beta/2U}$, where U is the characteristic velocity of turbulence. Besides, for wavenumbers k higher than the spectral peak wavenumber, kinetic energy has a spectrum approximately proportional to k^{-5}. On the other hand, when horizontal divergence is large enough, turbulence was found to behave just as the geostrophic turbulence on an f-plane in the absence of the β-effect: (1) the field becomes isotropic with no significant zonal currents; (2) the inverse cascade of energy is not hindered by the β-effect though it takes a longer time for turbulence to transfer energy to longer scales; and (3) the spectrum of kinetic energy (not total energy) is proportional to k^{-3} at high wavenumbers. The third result of k^{-3} spectrum of kinetic energy suggests that nonlinear interaction occurs as in the ordinary two-dimensional turbulence, except that energy is shared mostly by potential energy. The experiments and arguments in Part I thus revealed that large horizontal divergence suppresses the β-effect, making the geostrophic turbulence on a β-plane almost the same as two-dimensional turbulence.

Part II describes the self-similar spectral evolution of QG turbulence on an f-plane. First, an attempt was made to generalize Batchelor's theory for freely-decaying non-divergent 2D turbulence to the case affected by horizontal divergence. Second, order estimation and dimensional analysis showed that the power spectrum of stream function has a universal smooth form irrespective of λ: $\phi(k) \sim k^{-\frac{11}{3}}$ in the subrange of energy cascade and $\phi(k) \sim k^{-5}$ in the subrange of enstrophy cascade. It is straightforward to derive the spectrum of either total energy $E(k) = (k^2 + \lambda^2)\phi(k)$ or kinetic energy $E^{(k.e.)}(k) = k^2\phi(k)$. According to the present analysis, both $\phi(k)$ and $E^{(k.e.)}(k)$ have no kinks around $k \approx \lambda$, which was supported by numerical experiments. Third, ODEs are presented to describe E, W and k_p, allowing finite dissipation of energy. The dynamics based on the ODEs explains systematically well the so-far results of either freely-decaying and forced 2D turbulence with and without F. In the case when kinetic energy is conserved, as in the experiment in Part I, the ODE yields that $k_p \sim t^{-\frac{1}{3}}$. This prediction was confirmed by numerical experiment. Thus Part II sheds light on self-similar isotropic spectral evolution of QG turbulence affected by horizontal divergence.

It seems quite difficult to develop an adequate theory that may connect QG turbulence on a β-plane with the ordinary 2D turbulence. Probably the most serious is that we have to discuss the two-dimensional wavenumber spectrum, because Rossby waves and the Rhines effect give rise to strong anisotropy and add wavy nature to turbulence.

Acknowledgements This work has been partially supported by Grant-in-Aid for Science Research on Priority Areas (B) from the Ministry of Education, Culture, Sports, Science and Technology of Japan.

References

1. G. K. Batchelor: Phys. Fluids, **12**, Supple. II, 233 (1969)
2. M. D. Cox: J. Phys. Oceanogr., **17**, 1044 (1987)
3. W. Horton, A. Hasegawa: Chaos, **4**, 227 (1994).
4. T. Iwayama, T. Watanabe, H. Fujisaka: J. Phys. Soc. Japan, **70**, No.2, 376 (2001)
5. A. Masuda: in preparation (2002)
6. A. Masuda, H. Takase, A. Okuno: 'Turbulent mesoscale eddies over random bottom topography driven by random stress curl'. In: *Proc. CREAMS'97 International Symposium, Fukuoka, 1997*, ed. by J.-H. Yoon, pp. 229–232
7. A. Okuno, A. Masuda: submitted to Phys. Fluids (2002)
8. P. B. Rhines: J. Fluid Mech., **69**, 417 (1975)
9. P. B. Rhines: 'The dynamics of unsteady currents', In: *The Sea*, Vol. 6 (Wiley, New York 1977), pp. 189–318.
10. P. B. Rhines: Ann. Rev. Fluid Mech., **11**, 401 (1979).
11. Y. Takase, A. Masuda: Rep. Res. Inst. Appl. Mech. (in Japanese), **79**, 23 (1996)
12. T. Watanabe, H. Fujisaka, T. Iwayama: Physical Review E, **55**, No.5, 5575 (1997)
13. T. Watanabe, T. Iwayama, H. Fujisaka: Physical Review E, **57**, No.2, 1636 (1998)

Self-Similarity of Decaying Two-Dimensional Turbulence governed by the Charney–Hasegawa–Mima Equation

Takahiro Iwayama[1], Theodore G. Shepherd[2], and Takeshi Watanabe[1,3]

[1] Graduate School of Science and Technology, Kobe University, Kobe 657-8501, Japan
[2] Department of Physics, University of Toronto, Toronto M5S 1A7, Canada
[3] Department of Systems Engineering, Nagoya Institute of Technology, Nagoya 466-8555, Japan

Abstract. In decaying two-dimensional Navier–Stokes turbulence, Batchelor's similarity hypothesis fails due to the existence of coherent vortices. However, it has recently been shown that in decaying two-dimensional turbulence governed by the Charney–Hasegawa–Mima (CHM) equation

$$\frac{\partial}{\partial t}\left(\nabla^2\varphi - \lambda^2\varphi\right) + J(\varphi, \nabla^2\varphi) = D,$$

where D is a damping, the one-point probability density functions of various physical fields are well described by Batchelor's similarity hypothesis for wave numbers $k \ll \lambda$ (the so-called AM regime) [1]. In this report, we extend this analysis to the dynamics of spectral energy transfers. It is shown that the energy flux exhibits self-similar scaling, and the relation between the energy spectrum and the energy flux predicted by the similarity theory holds well for scales larger than that of the energy maximum. However, this relation breaks down for scales smaller than that of the energy maximum, where the observed downscale energy transfers would, according to the similarity theory, require negative energy spectra.

1 Introduction

The motion of geophysical fluids, such as the atmosphere and oceans, is strongly affected by both planetary rotation and density stratification. Planetary rotation acts to homogenize the fluid motion along the axis of rotation, while density stratification suppresses vertical overturning. Thus, both planetary rotation and density stratification tend to make the fluid motion horizontally two-dimensional, at least in a layerwise sense. Two-dimensional (2-D) turbulence has therefore been of considerable interest because it is believed to be an apt paradigm for geophysical flows [2]. It is also more accessible to numerical simulation than is three-dimensional (3-D) turbulence.

It is well known that the 2-D Navier–Stokes (NS) equation has peculiar characteristics that distinguish it from the 3-D NS equation. In particular, it has an infinite number of inviscid invariants: the kinetic energy K and the spatial integral of any function of vorticity ω. Of the latter the enstrophy Z (half the mean-square vorticity) is thought to be the most important quantity. In the limit of high Reynolds number, the dissipation of kinetic energy vanishes while that of enstrophy remains finite. Therefore, Batchelor [3] hypothesized that the kinetic energy K is the unique invariant of decaying 2-D NS turbulence and that the temporal evolution of the system depends only on K and time t. Based on dimensional arguments, he then proposed the self-similar energy spectrum $K(k)$, where k is the horizontal wavenumber, and the decay of enstrophy according to $\mathsf{Z} \sim t^{-2}$. His work, together with that of Kraichnan [4] and Leith [5], has been frequently referred to as the classical theory of 2-D NS turbulence, or the KLB theory, in the literature.

However, direct numerical simulation of decaying 2-D NS turbulence has not supported Batchelor's prediction [6–11]. For example, the decay of enstrophy is well described by the vortex scaling theory of Carnevale et al. [7], in which the decay is much slower than Batchelor's prediction [7,8,12–18].

The extent of validity of Batchelor's similarity hypothesis for 2-D NS turbulence was comprehensively discussed by Bartello and Warn [9]. They directed their attention to the one-point probability density function (PDF) of the vorticity ω and its n-th order moments. Using Batchelor's [3] assumption that the evolution of the system depends only on K and t, they derived the self-similar form of the vorticity PDF and the time evolution of its n-th order moments. Direct numerical simulations of decaying 2-D NS turbulence performed by Bartello and Warn [9] showed that the derived formula for the n-th order moments is valid only for the low-order moments ($n \leq 0.4$), and breaks down for the higher-order moments, including the enstrophy: the decay rate of the higher-order moments is a constant independent of n, consistent with the vortex scaling theory. They showed that the existence of a second asymptotic invariant ω_{ext}, which is characteristic of the amplitude of the most intense vortices, is responsible for the failure of Batchelor's similarity hypothesis. This invariant is inseparably related to the existence of coherent vortices [6]. Therefore, the existence of coherent vortices is the origin of the failure of Batchelor's similarity hypothesis for decaying 2-D NS turbulence.

This raises the question of whether there are any 2-D turbulent systems for which Batchelor's similarity hypothesis is applicable. Watanabe et al. [19] suggested that 2-D turbulence governed by the Charney–Hasegawa–Mima (CHM) equation [20] in the AM regime $k \ll \lambda$ — equivalently the quasi-geostrophic (QG) potential-vorticity equation for an equivalent-barotropic fluid in the limit of small deformation radius [2] — is a candidate system for which Batchelor's similarity hypothesis may be applicable.

In a recent paper [1], we comprehensively examined the validity of Batchelor's similarity hypothesis for CHM turbulence in the AM regime in order to confirm the above conjecture. For this purpose, we followed the study of Bartello and Warn [9]. That is, the self-similar forms of the one-point PDFs of various physical fields — the stream function φ, the potential vorticity q, the two components of the velocity u and v, and the vorticity ω — and the time evolution of their n-th order moments, were derived. We also performed direct numerical simulations of the CHM equation to check the validity of the theoretical estimates, as well as the self-similarity of the energy spectrum. The results of the simulations coincided with the theoretical results within numerical errors. In particular, the decay rates of the n-th order moments of the various PDFs were found to agree well with the theoretical estimates from $n = -1/3$ to $n = 8$.

The success of Batchelor's similarity hypothesis for CHM turbulence in the AM regime found in our previous study [1] motivates us to further analysis of CHM turbulence using Batchelor's similarity hypothesis. In this report, we derive the self-similar form of the energy spectrum evolution equation, and assess the validity of this form using the numerical solutions of the CHM equation performed by Iwayama et al. [1].

2 Application of Batchelor's similarity hypothesis to the energy spectrum evolution equation of CHM turbulence

We consider the CHM equation in the strong turbulent state neglecting the effects of waves, equivalently the QG potential-vorticity equation on the f-plane, which can be written as follows:

$$\frac{\mathrm{D}q}{\mathrm{D}t} \equiv \frac{\partial q}{\partial t} + J(\varphi, q) = D, \tag{1a}$$

$$q = \omega - \lambda^2 \varphi, \quad \omega = \nabla^2 \varphi \tag{1b, c}$$

where all quantities are made nondimensional, $J(\cdot,\cdot)$ is the 2-D Jacobian operator, ∇^2 the 2-D Laplacian, $\varphi(x, y)$ the electrostatic potential for the plasma case or the variable part of the free surface of the fluid for the geophysical case, and the right-hand side of (1a) is a damping term. The constant λ is either the ratio of the horizontal length scale of interest L to the ion Larmor radius in the plasma case, or the ratio of L to the Rossby deformation radius in the geophysical case.

In the limit of $\lambda \to \infty$, (1a) is reduced to

$$\lambda^2 \frac{\partial \varphi}{\partial t} + J(\omega, \varphi) = -D. \tag{2}$$

Since (2) has been called the asymptotic model (AM) [21], we shall call this regime the AM regime. There are similar characteristics between the vorticity equation derived from the 2-D NS equation and from (1a). Eq. (1a) likewise has two quadratic inviscid invariants: the total energy $\mathsf{E} = -\langle \varphi q \rangle /2$ and the potential enstrophy $\mathsf{Q} = \langle \omega q \rangle /2$, where the angle brackets denote a spatial average. The energy spectrum evolution equation of the CHM equation can be written as

$$\frac{\partial}{\partial t} E(k) = T(k) + D(k), \tag{3}$$

where $E(k)$ is the total energy spectrum defined as $\mathsf{E} = \int_0^\infty E(k)\,\mathrm{d}k$, $T(k)$ is the energy transfer function, and $D(k)$ is the dissipation spectrum.

The scale transformations $(x, y) \to a(x, y)$ and $t \to bt$ leave the equation (2) with $D = 0$ unchanged provided that $\varphi \to \lambda^2 a^4 b^{-1}\varphi$. Therefore the stream function φ in the AM regime is scaled as

$$\varphi \sim \lambda^2 L^4 T^{-1}, \tag{4}$$

where L and T are characteristic scales of length and time, respectively. Since in the AM regime the energy is approximately expressed in terms of φ according to

$$\mathsf{E} \approx \frac{\lambda^2}{2}\langle\varphi^2\rangle, \tag{5}$$

the length scale L is expressed in terms of E, λ and T as

$$L \sim \lambda^{-3/4}\mathsf{E}^{1/8}T^{1/4}. \tag{6}$$

Using the relations (4)–(6) and $E(k) \sim L\mathsf{E}$, one obtains the similarity form of the energy spectrum

$$E(k) = \lambda^{-3/4}\mathsf{E}^{9/8}t^{1/4}G\,(x)\,, \tag{7}$$

$$x \equiv k\,\lambda^{-3/4}\,\mathsf{E}^{1/8}\,t^{1/4}, \tag{8}$$

where G is a positive definite function of universal form. Similarly, the self-similar forms of the energy transfer function $T(k)$ and the energy flux $\Pi(k)$, which is defined by

$$\Pi(k) = \int_k^\infty T(k')\,\mathrm{d}k', \tag{9}$$

are obtained as

$$T(k) = \lambda^{-3/4}\mathsf{E}^{9/8}t^{-3/4}G_1(x) \tag{10}$$

and

$$\Pi(k) = \mathsf{E}t^{-1}G_2(x), \tag{11}$$

respectively. Here, G_1 and G_2 are also functions of universal form. The relation between them can be obtained from (9) as

$$G_2(x) = \int_x^\infty G_1(x')\,\mathrm{d}x'. \tag{12}$$

Now consider the energy spectrum evolution equation (3), neglecting the dissipation term $D(k)$. Changing the independent variables in (3) from k and t into x and t, and using (7) and (10), (3) reduces to

$$G_1(x) = \frac{1}{4}\frac{\mathrm{d}}{\mathrm{d}x}\left\{xG(x)\right\}. \tag{13}$$

Using (12), (13) is rewritten (assuming $xG(x) \to 0$ as $x \to \infty$) as

$$G_2(x) = -\frac{1}{4}xG(x). \tag{14}$$

Eq. (13) gives the relation between the energy spectrum and the energy transfer function, while (14) gives the relation between the energy spectrum and the energy flux. Since $G(x)$ is a positive definite function, (14) implies that the energy flux has negative values, corresponding to an upscale cascade of energy. Moreover, if one assumes that $E(k) \propto k^\delta$, then (13) gives $T(k) \propto k^\delta$. This relation between $E(k)$ and $T(k)$ with $\delta = 5$ is verified in the infrared range by Iwayama et al. [22].

In order to examine the validity of the above discussion, we analyze the numerical results of a high-resolution simulation from our earlier study [1] and compare them with the theoretical results. The number of grid points used in the simulation was 512^2. The damping term adopted in the simulation was $D = -\nu_2\nabla^4 q$. In the previous study, two initial forms of the energy spectrum, specified by

$$E(k) \propto \frac{k^6}{(k+60)^{18}} \tag{15a}$$

or by

$$E(k) \propto \frac{k^{30}}{(k+30)^{60}}, \tag{15b}$$

were considered. Both initial spectra have a sharp peak at $k = 30$. The former is frequently used in decaying 2D NS turbulence as a narrow-band spectrum [6,8,11]. The latter is similar to the initial spectra used in [19], [21], and [22]. We note that so far as the narrow spectrum is concerned, this is a very traditional set-up for decaying turbulence studies, because one can then study the dual cascade. We have checked that the following results are independent of the dissipation term within reasonable limits on ν_2, and of the form of the initial energy spectrum. In what follows we present the results of the simulation evolved from the spectrum (15a).

We show the energy spectra, the energy transfer functions, and the energy fluxes at times $t = 16$, 32, 60, 110 and 196 in terms of similarity variables using the instantaneous total energy in Figs. 1, 2, and 3, respectively. Since the value of λ used in the simulation was $\lambda = 100$, the wave number range $k > \lambda$ is beyond the scope of the above theoretical consideration. Therefore, the similarity variables corresponding to the wave number range $k < \lambda$ are drawn in all figures.

All three quantities appear to collapse onto universal functional forms with time, although there is considerable noise in the upscale energy transfer leading to variations in the maximum upscale energy flux.

Figure 4 shows the self-similar energy spectra predicted by the energy fluxes using the relation (14). For comparison, the energy spectrum at $t = 196$ in terms of similarity variables is also shown. In the low wavenumber range below the peak of the energy spectrum, x . 1.5, the relation (14) is well satisfied. However, in the

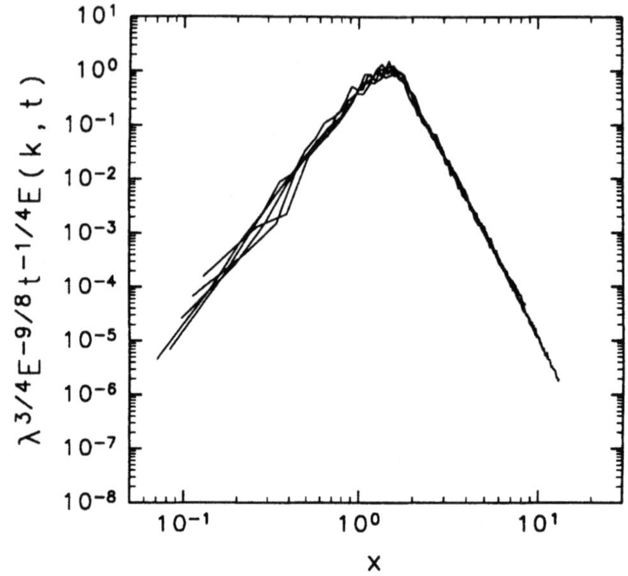

Fig. 1. Energy spectra at $t = 16$, 32, 60, 110 and 196 in terms of similarity variables.

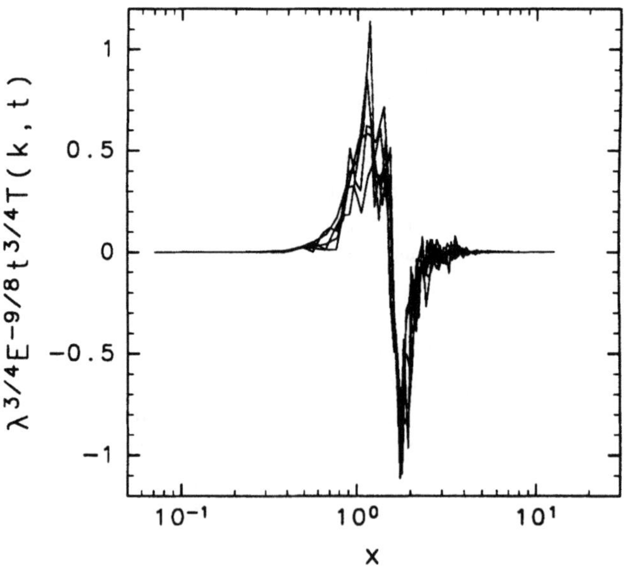

Fig. 2. Energy transfer functions at $t = 16$, 32, 60, 110 and 196 in terms of similarity variables.

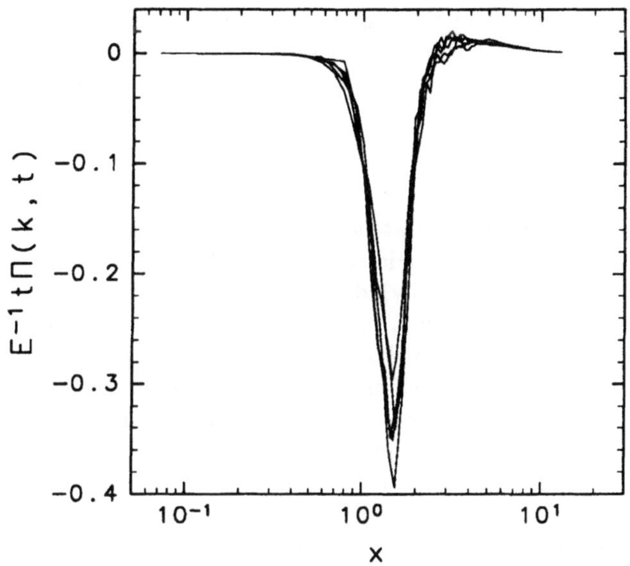

Fig. 3. Energy fluxes at $t = 16$, 32, 60, 110 and 196 in terms of similarity variables.

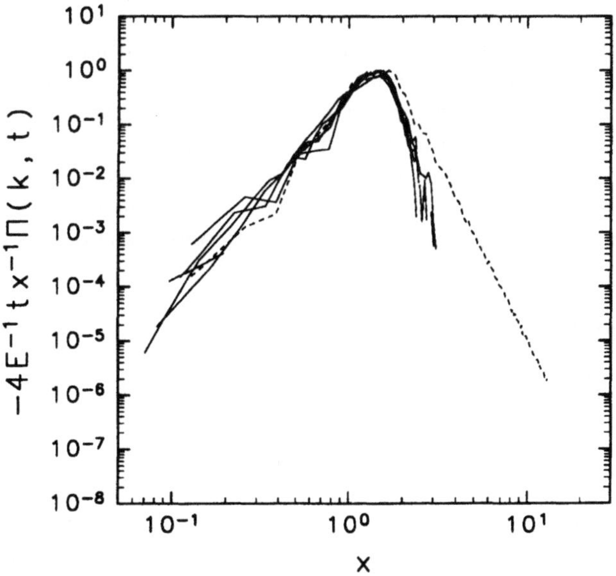

Fig. 4. The self-similar energy spectra calculated from (14) at $t = 16$, 32, 60, 110 and 196 (solid). The energy spectrum at $t = 196$ in terms of similarity variables is also shown (dashed).

high wavenumber range beyond the peak of the energy spectrum, the relation breaks down. Indeed, since in the high wavenumber range $x \gtrsim 2$ the simulated energy fluxes have positive values, (14) predicts a negative energy spectrum. This is a major failure of (14).

3 Summary

In decaying 2-D NS turbulence, Batchelor's similarity hypothesis describes only the low-order statistics of the vorticity field, and breaks down for higher-order statistics (including the enstrophy) due to the existence of coherent vortices [9]. However, we have shown in a previous study [1] and in this report that Batchelor's similarity hypothesis works well for decaying CHM turbulence in the AM regime. We derived the self-similar forms of the energy transfer function and the energy flux in this report, and checked the validity of the theoretical predictions using the numerical solutions obtained in our previous study.

Furthermore, we applied Batchelor's similarity hypothesis to the inviscid energy spectrum evolution equation (3), and obtained the relation (14) between the self-similar forms of the energy spectrum and the energy flux. The numerical results showed that in the low wavenumber range below the peak of the energy spectrum, this theoretical relation was very successful. However, in the high wavenumber range the relation was violated. Since Batchelor's similarity hypothesis is valid only in the inviscid limit, the failure of the present analysis may well stem from the effects of numerical dissipation. However, the potential enstrophy dissipation in the numerical solution was primarily limited to the highest wavenumbers, while the relation (14) was violated even in the vicinity of the peak of the energy spectrum. It would be good to resolve this point in a future study. A more serious problem with the similarity theory would seem to be the fact that decaying 2-D turbulence must involve a spread of energy both to larger and to smaller scales (this is required by the arguments of Fjørtoft [23]), but the relation (14) cannot possibly represent downscale transfers of energy.

Acknowledgements

The GFD-DENNOU Library was used for drawing figures.

References

1. T. Iwayama, T. G. Shepherd, T. Watanabe: J. Fluid Mech. **456**, 183 (2002)
2. J. Pedlosky: *Geophysical Fluid Dynamics*, 2nd edn. (Springer, New York 1987)
3. G. K. Batchelor: Phys. Fluids Suppl. **12**, II-233 (1969)
4. R. H. Kraichnan: Phys. Fluids **10**, 1417 (1967)
5. C. E. Leith: Phys. Fluid **11**, 671 (1968)
6. J. C. McWilliams: J. Fluid Mech. **219**, 361 (1990)
7. G. F. Carnevale, J. C. McWilliams, Y. Pomeau, J. B. Weiss, W. R. Young: Phys. Rev. Lett. **66**, 2735 (1991)

8. J. B. Weiss, J. C. McWilliams: Phys. Fluids A **5**, 608 (1993)
9. P. Bartello, T. Warn: J. Fluid Mech. **326**, 357 (1996)
10. J. R. Chasnov: Phys. Fluids **9**, 171 (1997)
11. A. Bracco, J. C. McWilliams, G. Murante, A. Provenzale, J. B. Weiss: Phys. Fluids **12**, 2931 (2000)
12. G. F. Carnevale, J. C. McWilliams, Y. Pomeau, J. B. Weiss, W. R. Young: Phys. Fluids A **4**, 1314 (1992)
13. R. Benzi, M. Colella, M. Briscolini, P. Santangelo: Phys. Fluids A **4**, 1036 (1992)
14. O. Cardoso, D. Marteau, P. Tabeling: Phys. Rev. E **49**, 454 (1994)
15. A. Siegel, J. B. Weiss: Phys. Fluids **9**, 1988 (1997)
16. T. Iwayama, H. Fujisaka, H. Okamoto: Prog. Theor. Phys. **98**, 1219 (1997)
17. A. E. Hansen, D. Marteau, P. Tabeling: Phys. Rev E **58**, 7261 (1998)
18. C. Sire, P.-H. Chavanis: Phys. Rev. E **61**, 6644 (2000)
19. T. Watanabe, T. Iwayama, H. Fujisaka: Phys. Rev. E **57**, 1636 (1998)
20. A. Hasegawa, K. Mima: Phys. Fluids **21**, 87 (1978)
21. V. D. Larichev, J. C. McWilliams: Phys. Fluids A **3**, 938 (1991)
22. T. Iwayama, T. Watanabe, T. G. Shepherd: J. Phys. Soc. Jpn. **70**, 376 (2001)
23. R. Fjørtoft: Tellus **5**, 225 (1953)

A Fast Method for the Calculation of the Fluid Flow on a Sphere using a Combined Compact Difference Scheme

Tomonori Nihei and Katsuya Ishii

Nagoya University, Furo-cho, Chikusa-ku, Nagoya 464-8603, JAPAN

Abstract. A spectral-like three-point combined compact difference (sp-CCD) scheme, which is originally proposed by Chu and Fan (1998, J. Comput. Phys. 140, 370), is utilized for the shallow water models on a sphere. A standard test set proposed by Williamson et al. (1992, J. Comput. Phys. 102, 211) is used to evaluate the accuracy of the new method. It is shown that the accuracy of the method is comparable to that of the spherical harmonics model as a whole. The application of the method to an multi-layer model shows that the method does work well. These results suggest that the method could be extended to the numerical atmospheric models.

1 Introduction

Recently, highly precise and effective computational methods are required for the global climate models [5]. The spherical transform method is widely used for its accuracy, but its computational cost rapidly increases as grid resolution is heightened. For example, the cost for two-dimensional problem on a sphere is proportional to N^3, where N is the number of grid points in the latitude direction.

Compact difference (CD) schemes, which are introduced in the 1930s, have shown its accuracy for the small scales significantly higher than traditional schemes with the same stencil size. Lele [4] suggested that the resolution, which means the accuracy in Fourier space, should be considered in addition to the formal accuracy, and proposed spectral-like compact difference schemes and compact schemes for interpolation and filtering. In 1998, Chu and Fan [1] proposed a 3-point combined compact difference (CCD) scheme that uses a combined derivative formulation, and showed that it has an improved spectral resolution with the cost which is comparable to that of the standard compact scheme.

In this paper, we describe a combined compact difference scheme with spectral-like resolution, and evaluate its performance using the test cases for the shallow water equations in spherical geometry proposed by Williamson et al. Then we apply the scheme to a multi-layer model and check its behavior.

2 Governing Equations

The governing equations that are used in the Williamson's test cases and multi-layer model are described in this section.

2.1 Shallow Water Equations

The shallow water equations are used in the Williamson's test cases. The shallow water equations on a sphere whose radius is a can be written in advective form as

$$\frac{\partial u}{\partial t} + \boldsymbol{v} \cdot \nabla u - \left(f + \frac{u}{a}\tan\theta\right) v + \frac{g}{a\cos\theta}\frac{\partial h}{\partial \lambda} = 0, \tag{1}$$

$$\frac{\partial v}{\partial t} + \boldsymbol{v} \cdot \nabla v + \left(f + \frac{u}{a}\tan\theta\right) u + \frac{g}{a}\frac{\partial h}{\partial \theta} = 0, \tag{2}$$

and

$$\frac{\partial h^*}{\partial t} + \boldsymbol{v} \cdot \nabla h^* + \frac{h^*}{a\cos\theta}\left(\frac{\partial u}{\partial \lambda} + \frac{\partial v\cos\theta}{\partial \theta}\right) = 0. \tag{3}$$

where h^* is the depth of the fluid and h is the height of the free surface above a reference sphere. $h = h^* + h_s$ where h_s is the height of the underlying topography. Here $\boldsymbol{v}$ is the horizontal vector velocity of the fluid with components u and v in the longitudinal (λ) and latitudinal (θ) directions, respectively, g the gravitational constant, f the Coriolis parameter given by $2\Omega\sin\theta$, where Ω is the rotation rate of the sphere, and the longitudinal, latitudinal, and outward radial unit vectors are $\widehat{\boldsymbol{i}}$, $\widehat{\boldsymbol{j}}$, and $\widehat{\boldsymbol{k}}$, respectively. The ∇ operator is the spherical horizontal gradient operator given by

$$\nabla(\bullet) \equiv \frac{\widehat{\boldsymbol{i}}}{a\cos\theta}\frac{\partial}{\partial\lambda}(\bullet) + \frac{\widehat{\boldsymbol{j}}}{a}\frac{\partial}{\partial\theta}(\bullet). \tag{4}$$

The relative vorticity ζ and horizontal divergence δ are defined by

$$\zeta \equiv \widehat{\boldsymbol{k}} \cdot (\nabla \times \boldsymbol{v}) = \frac{1}{a\cos\theta}\left(\frac{\partial v}{\partial\lambda} - \frac{\partial u\cos\theta}{\partial\theta}\right), \tag{5}$$

and

$$\delta \equiv \nabla \cdot \boldsymbol{v} = \frac{1}{a\cos\theta}\left(\frac{\partial u}{\partial\lambda} + \frac{\partial v\cos\theta}{\partial\theta}\right). \tag{6}$$

respectively.

2.2 Multi-layer Model

In this model, the hydrostatic approximation is applied in each layer. The density of fluid is constant within each layer and monotonically decreases from the bottom layer to the top layer. The sheer stresses between the layers and between the bottom layer and the underlying topography are ignored, and the pressure at the top surface of the fluid is assumed to be constant. Then the governing equations of the multi-layer model on a sphere can be written in advective form as

$$\frac{\partial u_n}{\partial t} + \boldsymbol{v}_n \cdot \nabla u_n - \left(f + \frac{u_n}{a}\tan\theta\right) v_n + \frac{g}{a\cos\theta}\left(\sum_{m\le n}\frac{\partial h_m}{\partial\lambda} + \sum_{m>n}\frac{\rho_m}{\rho_n}\frac{\partial h_m}{\partial\lambda} + \frac{\partial h_s}{\partial\lambda}\right) = 0, \tag{7}$$

$$\frac{\partial v}{\partial t} + \boldsymbol{v}_n \cdot \nabla v_n + \left(f + \frac{u}{a}\tan\theta\right) u_n + \frac{g}{a}\left(\sum_{m\le n} \frac{\partial h_m}{\partial \theta} + \sum_{m>n} \frac{\rho_m}{\rho_n}\frac{\partial h_m}{\partial \theta} + \frac{\partial h_s}{\partial \theta}\right) = 0, \tag{8}$$

and

$$\frac{\partial h_n}{\partial t} + \boldsymbol{v}_n \cdot \nabla h_n + \frac{h_n}{a\cos\theta}\left(\frac{\partial u_n}{\partial \lambda} + \frac{\partial v_n \cos\theta}{\partial \theta}\right) = 0. \tag{9}$$

where ρ_n is the density, h_n the depth, and $\boldsymbol{v}_n$ the horizontal vector velocity of the n-th layer with components u_n and v_n in the longitudinal and latitudinal directions, of the n-th layer, respectively.

3 Numerical Method

3.1 A CCD Scheme with Spectral-like Resolution

Consider a uniform grid with spacing Δx and the value of function at i-th grid point x_i is given by $f_i \equiv f(x_i)$. In CCD schemes, the derivatives of the function at a grid point are approximated by using them and the value of function and its derivatives at the neighboring points. Chu and Fan proposed a six-order and eighth-order scheme [1]. The eighth-order scheme can be written as

$$f_i' + a_1\left(f_{i+1}' + f_{i-1}'\right) + b_1 \Delta x \left(f_{i+1}'' - f_{i-1}''\right) + c_1 (\Delta x)^2 \left(f_{i+1}''' + f_{i-1}'''\right) = \frac{d_1}{\Delta x}\left(f_{i+1} - f_{i-1}\right), \tag{10}$$

$$f_i'' + \frac{a_2}{\Delta x}\left(f_{i+1}' - f_{i-1}'\right) + b_2\left(f_{i+1}'' + f_{i-1}''\right) + c_2 \Delta x \left(f_{i+1}''' - f_{i-1}'''\right) = \frac{d_2}{(\Delta x)^2}\left(f_{i+1} - 2f_i + f_{i-1}\right), \tag{11}$$

$$f_i''' + \frac{a_3}{(\Delta x)^2}\left(f_{i+1}' + f_{i-1}'\right) + \frac{b_3}{\Delta x}\left(f_{i+1}'' - f_{i-1}''\right) + c_3\left(f_{i+1}''' + f_{i-1}'''\right) = \frac{d_3}{(\Delta x)^3}\left(f_{i+1} - f_{i-1}\right). \tag{12}$$

The maximum formal accuracies of (10) – (12) are eighth-order, eighth-order, sixth-order, respectively. As pointed by Lele, considering about the resolution of the scheme is also important. The modified wavenumber shows the resolution characteristics of the scheme. It is related the phase speed error in the linear advection equation. The modified wavenumber of the first derivative $w'(w)$ for this CCD scheme is given by

$$w'(w) = \frac{2\begin{bmatrix}\{(1+b_2c_3+3c_2b_3)d_1+(2b_1-b_1c_3+c_1b_3)d_2-(3b_1c_2+c_1b_2)d_3\}\sin w\\ +\{(b_2+c_3)d_1-(b_1-2b_1c_3+2c_1b_3)d_2-c_1d_3\}\sin 2w\\ +\{(b_2c_3-c_2b_3)d_1-(b_1c_3-c_1b_3)d_2+(b_1c_2-c_1b_2)d_3\}\sin 3w\end{bmatrix}}{\begin{bmatrix}1+2(a_1b_2+b_1a_2)+2(a_1c_3-c_1a_3)+2(b_2c_3+c_2b_3)\\ +2\{a_1+b_2+c_3+(a_2c_3-c_2a_3)b_1+3(a_1c_3-c_1a_3)b_2+(a_1c_2-c_1a_2)b_3\}\cos w\\ +2\{(a_1b_2-b_1a_2)+(a_1c_3-c_1a_3)+(b_2c_3-c_2b_3)\}\cos 2w\\ +2\{(b_2c_3-c_2b_3)a_1-(b_1c_3-c_1b_3)a_2+(b_1c_2-c_1b_2)a_3\}\cos 3w\end{bmatrix}}, \tag{13}$$

where w is the scaled wavenumber. The modified wavenumber for various schemes are plotted into Fig.1. The exact differentiation is shown as the straight line. CD and CCD schemes have better resolution than the ordinary central difference schemes. The spectral-like CD scheme has much better resolution than the other schemes. To research the higher resolution relationship, we relaxed the requirement for formal accuracy of (11) and (12) to set two parameters of d_2 and d_3 free. In this case, the other ten parameters are given as

$$\begin{aligned} a_1 &= \frac{19}{32}, & b_1 &= -\frac{1}{8}, & c_1 &= \frac{1}{96}, & d_1 &= \frac{35}{32}, \\ a_2 &= \frac{11d_2-15}{16}, & b_2 &= -\frac{3d_2-7}{16}, & c_2 &= \frac{d_2-3}{48}, & & \\ a_3 &= d_3, & b_3 &= -\frac{8d_3+15}{20}, & c_3 &= \frac{4d_3+15}{60}. & & \end{aligned} \tag{14}$$

respectively. We have determined values of two parameters as $d_2 = 19.4444$ and $d_3 = -4.8131$. As shown in Fig.1, the modified wavenumber for our CCD scheme almost coincides with that for the spectral-like compact difference scheme proposed by Lele [4]. We call these new class of CCD schemes spectral-like CCD (sp-CCD).

Equations (10) – (12) can be solved by inverting following the periodic block-tridiagonal system:

$$\begin{pmatrix} \mathbf{B} & \mathbf{C} & \mathbf{O} & \cdots & \cdots & \mathbf{O} & \mathbf{A} \\ \mathbf{A} & \mathbf{B} & \mathbf{C} & \mathbf{O} & & & \mathbf{O} \\ \mathbf{O} & \mathbf{A} & \mathbf{B} & \mathbf{C} & \mathbf{O} & & \vdots \\ \vdots & \ddots & \ddots & \ddots & \ddots & \ddots & \vdots \\ \vdots & & \mathbf{O} & \mathbf{A} & \mathbf{B} & \mathbf{C} & \mathbf{O} \\ \mathbf{O} & & & \mathbf{O} & \mathbf{A} & \mathbf{B} & \mathbf{C} \\ \mathbf{C} & \mathbf{O} & \cdots & \cdots & \mathbf{O} & \mathbf{A} & \mathbf{B} \end{pmatrix} \begin{pmatrix} \mathbf{x}_1 \\ \mathbf{x}_2 \\ \mathbf{x}_3 \\ \vdots \\ \mathbf{x}_{N-2} \\ \mathbf{x}_{N-1} \\ \mathbf{x}_N \end{pmatrix} = \begin{pmatrix} \mathbf{d}_1 \\ \mathbf{d}_2 \\ \mathbf{d}_3 \\ \vdots \\ \mathbf{d}_{N-2} \\ \mathbf{d}_{N-1} \\ \mathbf{d}_N \end{pmatrix}, \tag{15}$$

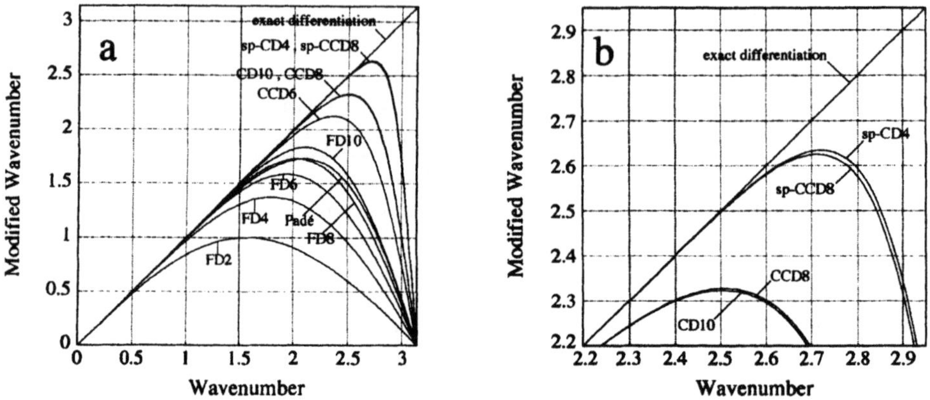

Fig. 1. (a) The modified wavenumber for the first derivative approximations: (FD2) second-order central differences; (FD4) fourth-order central differences; (FD6) sixth-order central differences; (FD8) eighth-order central differences; (FD10) tenth-order central differences;(Padé)standard Padé scheme; (CCD6) sixth-order combined compact difference scheme; (CCD8) eighth-order combined compact difference scheme; (CD10) tenth-order compact difference scheme; (sp-CD4) spectral-like fourth-order compact difference scheme; (sp-CCD8) new eighth-order combined compact difference scheme. (b) Magnification of a part of (a).

where

$$\mathbf{A} = \begin{pmatrix} a_1 & -b_1 \Delta x & c_1 (\Delta x)^2 \\ -\dfrac{a_2}{\Delta x} & b_2 & -c_2 \Delta x \\ \dfrac{a_3}{(\Delta x)^2} & -\dfrac{b_3}{\Delta x} & c_3 \end{pmatrix}, \mathbf{B} = \begin{pmatrix} 1 & 0 & 0 \\ 0 & 1 & 0 \\ 0 & 0 & 1 \end{pmatrix},$$

$$\mathbf{C} = \begin{pmatrix} a_1 & b_1 \Delta x & c_1 (\Delta x)^2 \\ \dfrac{a_2}{\Delta x} & b_2 & c_2 \Delta x \\ \dfrac{a_3}{(\Delta x)^2} & \dfrac{b_3}{\Delta x} & c_3 \end{pmatrix}, \mathbf{O} = \begin{pmatrix} 0 & 0 & 0 \\ 0 & 0 & 0 \\ 0 & 0 & 0 \end{pmatrix}, \tag{16}$$

$$\mathbf{x}_i = \begin{pmatrix} f_i' \\ f_i'' \\ f_i''' \end{pmatrix}, \text{ and } \mathbf{d}_i = \begin{pmatrix} \dfrac{d_1}{\Delta x}(f_{i+1} - f_{i-1}) \\ \dfrac{d_2}{(\Delta x)^2}(f_{i+1} - 2f_i + f_{i-1}) \\ \dfrac{d_3}{(\Delta x)^3}(f_{i+1} - f_{i-1}) \end{pmatrix}.$$

The operation count for this scheme is proportional to N, which is the same order as the ordinary difference scheme.

3.2 Grid System

The longitude-latitude grid system used by Fornberg [2] was utilized for computation. The grid points in this system are arranged so that no grid points are at the poles. This grid system has the advantage of avoiding singularity at the poles and enabling the periodic boundary conditions to be imposed in both longitudinal and latitudinal direction. It should be noted that the sign of the quantities must be adjusted properly when using the periodic boundary condition in latitudinal direction, according to whether the quantities are scalar or vector. The spatial derivatives of each physical quantity in the governing equations were calculated by sp-CCD scheme and the time integration was done by the fourth-order Runge-Kutta method.

3.3 Low-pass Spatial Filter

The concentration of the grid points near the poles causes numerical instability. A spatial filter is used to remove this instability. In the appendix of [4], Lele proposed a pentadiagonal implicit spatial filter in the form of

$$\begin{aligned} &\overline{f}_i + \alpha\left(\overline{f}_{i+1} + \overline{f}_{i-1}\right) + \beta\left(\overline{f}_{i+2} + \overline{f}_{i-2}\right) \\ &= af_i + \frac{b}{2}\left(f_{i+1} + f_{i-1}\right) + \frac{c}{2}\left(f_{i+2} + f_{i-2}\right) + \frac{d}{2}\left(f_{i+3} + f_{i-3}\right), \end{aligned} \tag{17}$$

where $\overline{f}_i$ is the filtered value at grid point x_i. The Fourier analysis of (17) gives the transfer function

$$T(w) = \frac{a + b\cos(w) + c\cos(2w) + d\cos(3w)}{1 + 2\alpha\cos(w) + 2\beta\cos(2w)}. \tag{18}$$

There are six parameters of a, b, c, d, α and β. These parameters are determined as

$$\begin{aligned} \beta &= \min\left(\frac{4 - 3\cos w_c + 2\cos^2 w_c + \cos^3 w_c}{16 - 2\cos w_c + 4\cos^2 w_c + 6\cos^3 w_c},\right. \\ &\qquad \left.\frac{3 + 3\cos w_c + 3\cos^2 w_c - \cos^3 w_c}{10 + 2\cos w_c + 10\cos^2 w_c - 6\cos^3 w_c}\right), \\ \alpha &= \frac{\cos w_c}{2\left(1 + \cos^2 w_c\right)}\left\{2\beta - 3 + (1 - 6\beta)\cos^2 c_w\right\}, \\ a &= \frac{3\alpha + 2}{4}, b = \frac{16\alpha + 10\beta + 9}{16}, c = \frac{\alpha + 4\beta}{4}, d = \frac{6\beta - 1}{16}. \end{aligned} \tag{19}$$

Fig.2 shows the transfer function for our filtering scheme with various w_c. In this study, w_c is chosen as

$$w_c = \begin{cases} w_{\max}\left\{1 - 0.75\left(1 - \cos\theta\right)^5\right\} & \text{(for longitudinal direction)} \\ w_{\max} & \text{(for latitudinal direction)} \end{cases}, \tag{20}$$

where $w_{\max}$ is set to $w_{\max} = 0.952\pi$. This filtering scheme is a pentadiagonal system and solved with the cost of proportional to N. The filtering is applied to the variable of u, v, and h in governing equations (1) – (3) and (7) – (9) for each time step and each step of the Runge-Kutta method.

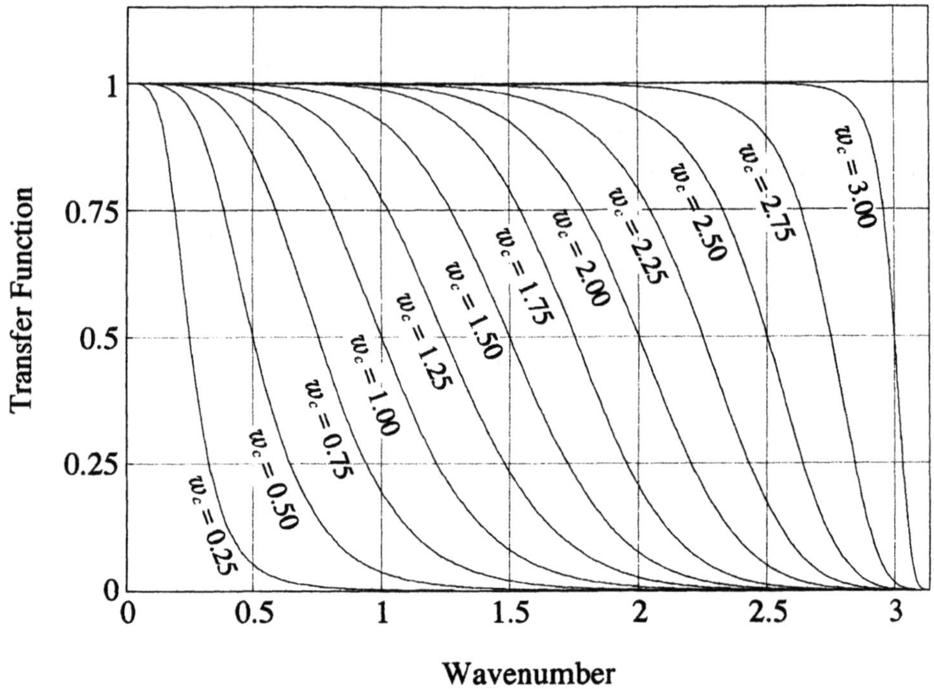

Fig. 2. Transfer functions for the compact filter scheme with various cutoff wavenumber.

4 Results

The standard test set proposed by Williamson et al. [5] consists of seven test cases. Detailed mathematical formulas of the test cases and requested performance measurements are intensively described in [5]. Figure 3 shows the height field for the Williamson's test case 7. The calculation was performed in double precision format and with the grid resolution of 128 × 64. The time step was set to 60 s. Test case 7 uses the observed height and wind field as initial conditions. Since test case 7 has no analytic solutions, numerical solutions of high-resolution spherical harmonics model are regarded as reference solutions. In this study, solutions of a T341 model are used. The spherical harmonics model was coded by using the subroutine library for scientific computing 'ispack' [3]. Figure 3 shows that the large scale structures are quite similar to the reference solution. The agreements with the reference solution are also good for the other test cases.

Figure 4 shows the vertical cross section of the multi-layer model at the latitude where the top of the mountain is. The initial height and wind fields are given as rigid body rotation with the geostrophic balance if the bell-shaped mountain located at (90° W,30° N) were ignored. The densities of the fluids are determined so that the buoyant frequency becomes constant. In this calculation, $la_m \sim 1$ and $lh_m \sim 0.2$, where a_m is the half-width of the mountain, h_m the

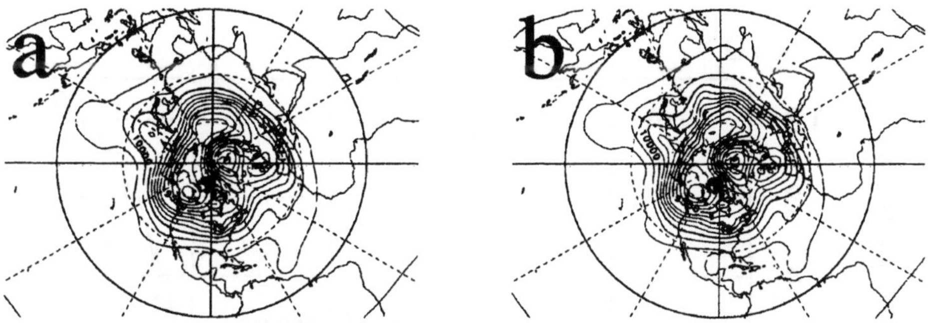

Fig. 3. Height fields of at day 5 for the Williamson's test case 7 of the 21 December 1978. The contour interval is 100 m: (a) our sp-CCD scheme.;(b) reference solution.

height of the mountain, and l the Scorer's parameter, respectively. The calculation was done with the grid resolution of 64×32 and with the 8 layer model. The time step was set to 10 s, and the time integration was done during 50 hours. The tilted straight line in Fig. 4 shows the crest of the internal wave obtained from the linear approximation, and its agreement with that calculated by the model is comparatively good. Figure 5 shows the time evolutions of the total mass and the total energy during the calculation. The conservation property is thought good enough in practical use.

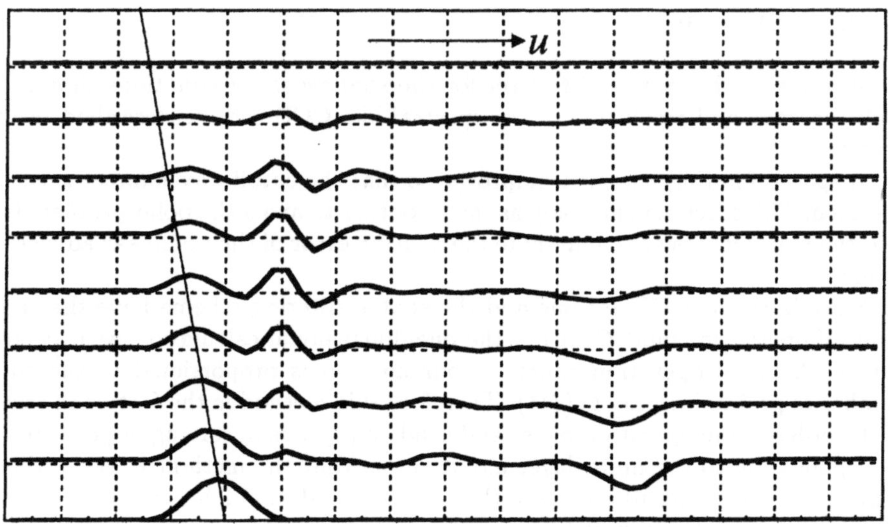

Fig. 4. The lee waves calculated by the multi-layer model. The tilted straight line shows the wave crest obtained from the linear approximation.

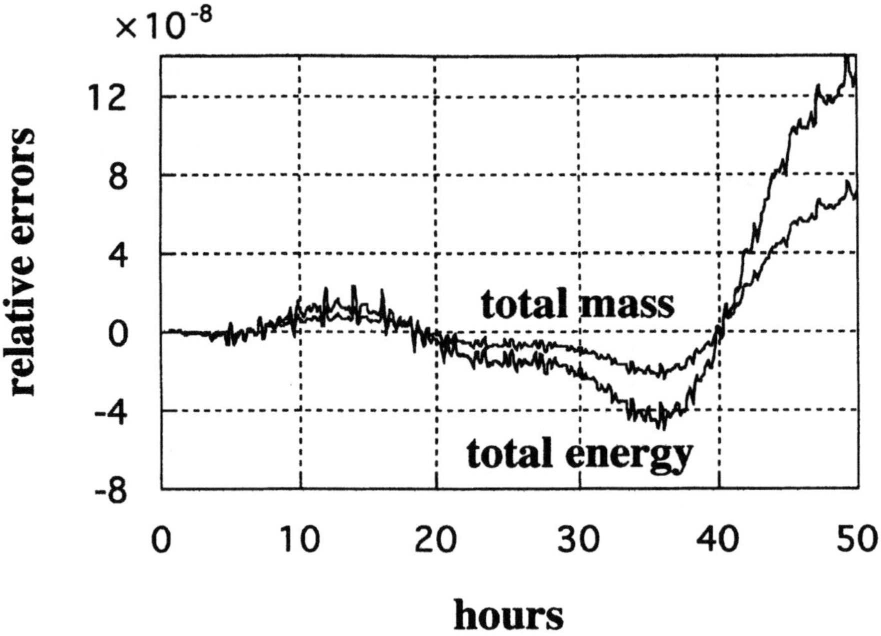

Fig. 5. The time evolutions of the total mass and the total energy.

5 Conclusions

Application to the standard test set for the shallow water equations shows the high accuracy and resolution of the spectral-like CCD scheme described in the Section 3.1.

A spatial filtering scheme is improved so that the cutoff wavenumber is easily specified. This filter acts not only as the smoother to avoid the polar problem but also the hyper viscosity to eliminate the aliasing error of the nonlinear advection terms.

Since both the CCD scheme and the spatial filtering scheme have the same order of operation count $O(N)$ in the one-dimensional calculation with N grid points, the total operation count for our method is proportional to the total number of grid points, i.e. $O\left(N^2\right)$. Therefore this method is the fastest in terms of the order of the operation count and is advantageous in solving large systems.

Application to the multi-layer model shows that the method works without a problem. These results suggest that the method could be extended to the numerical atmospheric models.

References

1. P. C. Chu and C. Fan: J. Comput. Phys. **140**, 370 (1998)

2. B. Fornberg: SIAM J. Sci. Comput. **16**(5), 1071 (1995)
3. K. Ishioka: http://www.gfd-dennou.org/arch/ispack/, GFD Dennou Club (2000)
4. S. K. Lele: J. Comput. Phys. **103**, 16 (1992)
5. D. L. Williamson, J. B. Drake, and J. J. Hack, R. Jakob-Chien, and P. N. Swarztrauber: J. Comput. Phys. **102**, 211 (1992)

Part V

Panel Sessions

Panel Session 1
Advanced Computational Approaches in Turbulence Research

Chairperson: Yukio Kaneda, Advocater: Javier Jimenez

Kaneda: Let's start Panel Session P1, "Advanced Computational Approach in Turbulence Research." The chairman of this session was supposed to be Dr. Rubenstein, but unfortunately, he cannot join to this workshop. Just after receiving his message that he cannot come, I was inclined to cancel this session for many reasons. But, being encouraged by my colleagues, I changed my mind and I decided to keep this session, because there are many questions we wish to discuss. For example, "How far must a man go to know turbulence by DNS?" or in more sophisticated English by Dr.Rubinstein, "How far must we go to understand turbulence by DNS?" Like this kind of question, there are a lot of questions of interest.

Here is a note on Session P1 by him. Let me read the first page: "The continual growth of computer power will have profound implications for scientific research in general, and for turbulence research in particular. Research tools inevitably influence the direction of research. Computers might seem to replace the traditional goal of scientific research to subsume particular phenomenon under a comprehensive theory, with the ability to simulate particular cases and to extract the relevant information from a database.

Let me add some comments. It is said that computer is an extension of the brain, whereas telescope is just an extension of the eyes, and microscope also. Even such a low-level tool, in today's standard, as telescope had made significant influence on science. According to the book "Science in History" by J. D. Bernal, the invention of telescope changed the basic ideas on celestial bodies; for example, the long-standing view à la Aristotle that the shape of the moon must be spherical, because celestial perfect shape must be "sphere". Just the very primitive way of observation, i.e., just looking through a telescope dramatically changed that scholastic view. Another example is microscope; just looking through a microscope shows that some diseases are not due to the devil or any supernatural power, but are caused simply by very small viruses or bacterium. The very primitive way of observation can convince us that nothing is necessary but very simple and objective observation. If we look back on this year in the future, say after 100 years from now, then this age may be thought to be at some revolutionary stage. Maybe, computational approaches will have significant effects on turbulence research.

Dr. Rubenstein kindly summarized the questions as the followings,

Question 1: Will turbulence theory, closures, and other analytical approaches have a role in the future?

Question 2: What developments in computer technology will be most important, and how can they be exploited?

Question 3: How close can DNS come to completely solving the problem of turbulence?

By the way, on the last page of the program are shown possible topics of Panel Session 2, and they will be discussed tomorrow. Today, I wish to have Session P1 on computational approaches. I asked Professor Jimenez to provide some materials on the role of computational approach, and he kindly accepted.

Jimenez: Thank you. Let me first tell you that this was supposed to be a panel with, I think, five people, and that there is no way that I can be a substitute for all of them. Bob Rubenstein posed a list of questions to the panel, and I wrote to myself a shorter list of those of his questions which I found personally most interesting. Even in that reduced list, many questions have parts about which I can say nothing useful.

I will therefore concentrate in a few of the questions, mostly having to do with things like "What can computational fluid dynamics say about the turbulence problem?" I remember one conference that I organized in Spain a long time ago in which the question of the turbulence problem came up, and nobody could agree on what the turbulence problem was, or on what we were talking about. Everybody had his own idea. Everybody has his own turbulence problem. So, first of all, I will try to define what the turbulence problem is for me.

Then I will talk about computers, and about what I think computers can tell us about turbulence. When will they be able to tell it to us? How far do we have to go? How big a computation do we have to make before we can say that "the turbulence problem is solved?"

First, what do I think is the turbulence problem? To me, turbulence is defined by a figure which is found in one of the first pages of the book by Townsend (1956, The structure of turbulent shear flow, Cambridge U. Press, page 37) where he shows how the energy and the dissipation are distributed in isotropic turbulence. I reproduce my version of the figure here, including the spectra of both. Each one has a peak at the wavenumbers where the respective property resides. The energy is at large scales, and the dissipation is at small scales. The fact that the energy is in one place and the gradients are in a different one, is to me what defines turbulence. That is the essence of turbulence. In between the two peaks there is something that takes energy from one to the other, from the scales where the energy is, to where it has to be dissipated by viscosity. And the fact that there is a gap in between in which there is neither energy

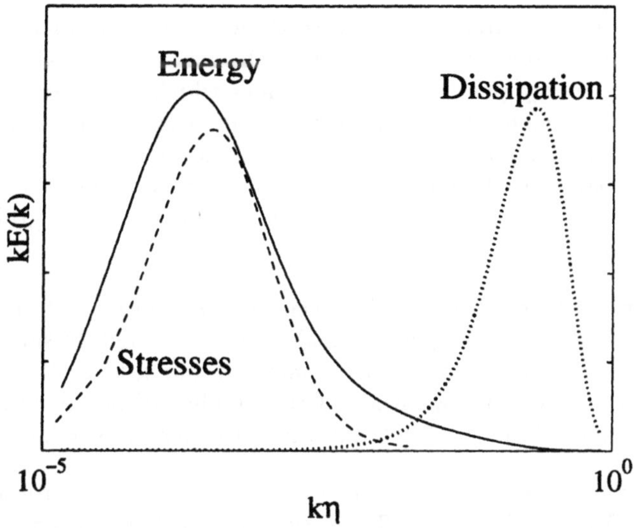

Fig. 1. Spectral distributions of dissipation and Reynolds stresses in shear turbulence.

nor dissipation, but where one is still transported into the other, is what makes turbulence special.

This is what defines a problem as turbulent. If we can compute something in which these two peaks are really separated from each other, then we have computed asymptotic turbulence. I mean, there is really nothing else left to compute, or almost nothing. When we have that, we can start making questions about that system.

By the way, that is also true of the Reynolds' stresses. The cospectrum of the Reynolds stresses in a shear flow is also included in the figure above. It is also concentrated in the large scales, and actually decays faster towards the smaller scales than the energy spectrum. In most cases, if we have been able to compute correctly the spectral energy peak, we have also computed correctly the distribution of the shear stresses.

Now, what is the size of the system that we would have to compute to get that? We can estimate the scale separation that we need from the spectrum, and it is roughly the ratio L/η between the integral scale L, which is the length scale of the peak of the energy spectrum, and the Kolmogorov length η, which is near the peak of the dissipation spectrum. We know from Kolmogorov's theory that this ratio scales with Reynolds number as $Re_\lambda^{3/2}$ and, in the figure above, where $Re_\lambda \approx 1500$, it is $L/\eta \approx 1000$. This is approximately the minimum Reynolds number for which the two spectral peaks are really separated. So to compute a real turbulent flow we need at least a grid which is 1000^3.

Actually, if we want to do things correctly we need to compute a bit more. It is not enough to compute from one peak to the other. There is energy in scales larger than the energy peak, typically in scales which are a factor of 10 larger; and similarly we have to include scales which are a factor of 5 smaller than the peak of the dissipation spectrum. So, if we think in the example above, we would need a scale separation of about 50,000, which, since we have to put at least two grid points in our smallest relevant scale, means that we need at grid of the order of $100,000^3$ points.

We can try to save a little, and reduce the resolution a bit at both ends of the spectrum. We could also perhaps make do with a slightly smaller Reynolds number, and in that way reduce the diameter of our grid to 10,000, instead of 100,000, but the order of magnitude of the number of points that we need to compute turbulence is $10^{12} - 10^{15}$, and these are clearly very large numbers.

So, does that make sense or not? Can we hope to compute a problem of that size in a foreseeable future? To answer that we have to consider computer requirements, and what can we expect of the evolution of computers.

Let me call N the number of points along one edge of our three-dimensional grid. We need something like 10 variables per grid point, and at least $\log_2(N)$ operations per variable in each time step (this is true as much in spectral codes as in most good finite-differences schemes, such as multi-grids). That means that we are going need a memory $M \approx 40N^3$ (bytes), and a number of operations per step $Q \approx 10N^3 \log_2(N)$. Both constrain our computer; the first in terms of central storage, and the second in terms of CPU speed.

Let me assume that we run the largest problem that would fit in our computer, so that M is the central storage size. We can estimate the computer time needed for one time step as

$$Q/S = 4\log_2(N)[M(\text{bytes})/S(\text{flops})] \approx [M/S] \quad (\text{min}),$$

where S is the computer speed in flops per second and the parameter $[M/S]$ is a figure of merit of the computer. Historically, this parameter has always been approximately equal to one; large computers have tended to have one Gbyte for each Gflop of computational speed. That means that the computer time needed run a single step of the largest possible turbulence DNS does not depend too much of the problem size, and has always been approximately 1 minute. That is probably true. When I think of the codes that I have been running since I was a student, the time per step has always been 1 minute, or maybe 30 seconds, independently of which machine I was using and of what was the problem that I was trying to solve. My personal memory has not been long enough to see a big effect of the logarithmic factor in the formula above.

Now, we have to be a little careful with this estimate, because it depends on the parameter $[M/S]$, which is something that computer manufactures decide

N	Re_λ	Memory	Speed	Total Time	Year
10^3	70	50 Gbytes	50 Gflops	2 weeks	1993
10^4	300	50 Tbytes	50 Tflops	6 months	2002
10^5	1500	50 Pbytes	50 Pflops	5 years	2015?

for us. I say this now because when I listen lately to computer engineers and to information scientists, they tell me that they want to bring this number down by a lot. They feel sure that they can give us a Tflop tabletop computer in a reasonable future, but when you ask them about the accompanying Tbyte, they tell you that maybe not, because most people do not need a Tbyte. We need a Tbyte. The market is now driven by people doing graphics and computer games. They need speed and not too much memory, but we need both speed and memory. Otherwise we will be computing in the future the same things that we are computing now, although much faster, but we will not be able to compute the higher Reynolds numbers that we want, because we will not have the memory to store those large flow fields.

So, whenever you talk to people who build computers, please insist that there is a market for large memory, and that fluid mechanics, which is engineering, is limited as much by Tbytes as by Tflops. And, as long as you have their ear, you might want to remind them that disk storage also has to grow at a pace.

But, if we are able to keep convincing the computer manufactures to give us our $[M/S] = 1$, we can estimate how much time it will take to compute a given Reynolds number, and when we are going to be able to do it.

For that we need to how many time steps we need to run, which has to do with how long do we have to run to compile good statistics. A good rule of thumb is that we need N time steps for a fluid particle to cross our computational box, and at least $20N$ steps to start believing the statistics. Using these estimates we get the following table.

We had a 50 Gflops supercomputer around 1993, and I am told that Japan will have one running at 50 Tflops next year. Historically a growth rate of 100 to 1000 times in computer speed every ten years has held for some time and, when I speak to computer hardware people, they seem confident that they can keep doing it. They don't see any big problem in the future. Nobody is saying that this is what is going to happen in 40 years, but for the moment this formula works, and the proper technologies seem to keep appearing at the right rate.

If we then believe that the same rate of growth will continue in the future, we will have a 50 Pflops computer by the year 2010-2020, and we will be able by then to start computing our first incontrovertible example of turbulence, with a real separation between the energy and the dissipation peaks. And for the first time we would be able to see without doubt whether our theories agree with it or not. In practice we may have to wait a little longer, because the total times given in the table are dedicated times using the full machine, and it is going to be difficult to convince the funding agencies to give us 6 months of the largest machine on Earth for a single scientific experiment. But it is clear that should be well into our way to computing asymptotic turbulence by 2020.

What happens after that? We can keep relying indefinitely on computer power to increase our Reynolds number, or we can decide that we don't want to do anything else. You see, once you have computed the whole spectral energy peak, and the whole Reynolds stress peak, turbulence theory tells us that you shouldn't need to do much more. In a sense, you have done the ultimate large-eddy simulation. All we need is to provide some dissipation, and the scales that you have computed should be asymptotically independent of the Reynolds number. In fact, if you look at the figure above, the width of the energy peak is only about 10^4, and you can save a factor of 10 because you don't need to resolve your dissipation peak very well. So, with a grid which is only $10,000^3$ you can do asymptotic LES, and it should be exactly the same as if you had used an infinite number of points. Looking at the table, we should be able to do that easily in 10 years, even if we don't get the whole dedicated computer.

Something else that we might want to do is an asymptotic Small-eddy simulation, by which I mean a simulation which resolves the whole dissipation peak, even if the energy injection from the large scales is done artificially. This is something that is not very practical from the engineering point of view, since engineers are usually more interested in energy. But from the physical point of view, you might want to pose questions like: Are the small scales independent of the character of the forcing? Is the wall independent of what happens outside? Is molecular mixing independent of the larger scales?

This should be easier to do than asymptotic LES because, if you look at the spectra, the dissipation peak is narrower than the energy peak, and a grid of 1000^3 should be enough. In fact this is what we have been doing for some time, since we tend to resolve our simulations well, and worry less about how we force them. There is some real physics that still needs to be done in this area, and we have to learn how to do the optimal forcing, and how important it is, but the grids needed are already possible, and we may be near doing our first "final" turbulence computation.

Now, once we can do all these things, what are we going to do with them? First we may want to do well-characterized experiments; this is obvious. There is a myth that experiments are always right, while computations have to be checked against them. That is nonsense. The experiments are as wrong as the computations. Experiments have walls, tunnels have inlets and outlets, and all experiments have noise and a lot of observational difficulties. You can make an argument that direct numerical simulations are actually a better way of characterizing what you are doing, even if what you are doing might not be exactly the same thing as the experimentalists. But you can run a well-defined computation, develop a theory for that case, and test your theory with less ambiguity than using an experiment.

You can also create a database with this data, which would be useful for people doing models and theory, but remember that each of these fields is 50 Tbytes. These are huge things that we cannot dream of sharing internationally right now, but probably in 10 or 20 years communications and storage technology will have solved the problem.

Finally, I think that the most useful application for such realistic simulations is to make them unrealistic. One of the best features of numerical analysis is that it allows you to manipulate the equations and the boundary conditions in ways which are not easily implemented experimentally. Physics has always made use of "thought experiments"?, which cannot really be done but which teach you important things about the system in question. Traditionally, those have been limited to modifications of the system whose consequence are easy to predict but, as phenomena become more complex, we want to make thought experiments whose answers are not clear, and we cannot draw conclusions until we know the results. Consider for example the questions about clocks that underlie relativity. It is reasonably easy to decide what would happen if one observer wants to synchronize her clock with another one, and has to make do with the limitations of the speed of light. From that, if you are very smart, you can develop relativity. But it is much harder to know what would be the consequences to the energy cascade if the quadratic nonlinearity in the Navier-Stokes equations were substituted by a cubic one. And, until you know the answer, you can not draw conclusions even if you are very smart. The big advantage of numerical against laboratory experiments is that you can actually try such experiments, and that you get answers.

Those are just a few suggestions of what you could do with the realistic turbulence simulations that will become possible in the next decades. I am sure that other people here have other ideas, and that they will come in the discussion. But before I sit down, I would like to make a minor comment about the far future. This is something that is not really important for me. It might become important for some of the young people here, and it will very probably be important for the students of some of those young people. So it is something that we have to think about as educators. We are getting close enough to the far future to start thinking about what is going to happen.

And the far future is the $100,000^3$ computation. This, remember, is the ultimate DNS. And before that we have the $10,000^3$ simulation, which is the ultimate LES. We should be able to do the first one by 2020, and it will take about 5 years. We could do the second one earlier, in ten years, and it will take only 6 months. Both are very long times, and those will be one-of-a-kind research computations. But think of what happens later. Computers keep getting faster, and we don't really want to compute anything larger. Even if we take the conservative estimate of an increase of computer speed by a factor of 100 every decade, we could do our ultimate LES in two weeks by 2020, in 2 days by 2030, in 4 hours by 2040, and in 30 minutes by 2050. After that, engineering turbulence becomes free. We will be able to answer all the questions about turbulent energy and Reynolds stresses by pushing a button and waiting for one minute. It will still take some computer time to do research the DNS, although not much, but will we really want to do it? We will know the answer to all practical questions without it.

So the question is, in 40 years, what is going to be the role of turbulence theory? I will present you now with a list of possibilities, and you will see that they start with the reasonable, and then degenerate into things which are less and less serious. But we should think about all of them, because they may happen, and each one among us may want to choose one or the other.

First is turbulence engineering, which is an obvious one. Turbulence is a really practical phenomenon that factors in many, many cases. It's not a question of turbulence per se. If you are doing geophysics, you are not necessarily interested in turbulence. But, the flows are turbulent, and you need to know the answer. In engineering is the same thing. If you want to compute the wind around a building, you are not doing turbulence theory, but you want to know how the wind flows. Most industrial flows happen to be turbulent, and you have to compute them.

There are many sciences that are already at this stage, and their users typically have formulas that they can apply. Think of the future traders that play the stock market, and bring it down every now and then. The basis was they do gave rise to two Nobel prizes. One was given to the man who developed the theory, which is a diffusion approximation to a set of stochastic differential equations. The other one was granted to the man who reduced that theory to a simple formula which could be programmed in a calculator. Future traders now use the calculator. They punch in their information about a certain stock, and out comes the answer that they have to buy a certain option at 25.625. Structural engineering is also like that, although a little more complicated. That doesn't mean that those people are useless, but they are certainly not doing research in the subject to which they are applying their formulas. So, this is one possible role for turbulence scientists in the future.

I offer you another option: turbulence humanities. These are fields which are intellectually satisfying, but basically useless. At the risk of offending somebody, let me give the example of pre-Socratic philosophy. It is nice to know what Anaximander said, and I have a lot of respect for the historians of philosophy but, from the practical point of view, it is essentially a luxury. That is not to say that humanities are not important, or that we should feel depressed about becoming humanists. In fact, humanities tend to be more popular than engineering. Consider number theory, which is also in this class. Fermat's last theorem is probably irrelevant for most people, including I am told for most mathematicians, but when somebody proved it, everybody got very excited, because it was a humanities thing.

Carnevale: Number theory treats the factorization of very large digits and this is very important for security applications.

Jimenez: Sure. The fact that there are poets who do poems doesn't preclude that there are also poetry engineers who do commercials, but you know that that is not why poems are done.

The next option that you might want to consider is turbulence sports. I have the theory that whenever machines become better than humans at doing something, that activity becomes a sport. Fighting, for example. A long time ago, hand to hand fighting used to be very important. But now, we don't need to do that anymore; we throw the javelin as a sport, and shoot things for real. We only throw the javelin because there is a personal satisfaction in throwing the javelin faster and longer than the opponent. When we really need to fight, we use other means. Martial arts in this part of the world are another beautiful example. It's not that you really expect to kill people by martial arts. You do it for personal development. So you can perhaps think of doing turbulence theory as mental gymnastics, just to keep in shape. It is certainly hard enough. You may even consider turbulence games, with opponents sitting across a table to see who is able to compute triple correlations faster with pencil and paper.

So in 40 and 50 years we may all be out of our jobs, or doing things which are completely different from what we are doing now, or doing the same things for completely different reasons. For the moment, however, there are many shorter-term questions that need to be addressed in turbulence, and I will stop talking here and let the younger people worry about the longer term.

Holloway: I guess I'm really going to try and ask the chairman, perhaps, for a bit of clarification. I think a great many of us here come from the turbulence problem, and we've just heard the turbulence problem addressed. We also understood we would be concerned about large-scale planetary flows. That is to recognize that we, and all of our children's children, will live on this planet. We undertake, in some way or another, to understand this planet, whether we can steward it or not. While immediate topic is turbulence, the larger overall challenge posed for this meeting is how advances in computation and in turbulence theory can help toward the urgent practical question of understanding this planet.

I'm a little concerned that we may slightly confuse that larger problem. One question is "what do we need to do, given advances in computation and in theory, to help comprehend the planet?" Another question is "What do we need to do to correct the turbulence problem?" I fear that we're acting as though those two questions are the same. I think that they are not at all the same.

If I can kind of indulge a little from the oceanographic perspective, if we went into the interior of the ocean, and you asked me to show you an example of turbulence, I would probably look on scales not much bigger than about a meter. If you then asked to find the dissipation scales of the turbulence in the ocean, I would look on scales not very much smaller than a meter, and you probably would not call it turbulence hardly at all. Yet I think this will show what is the bigger problem.

If only the ocean were a lot more turbulent, I think you'd have a good chance of cracking it with the kind of an exercise we've just seen. But it isn't. At a scale slightly larger than the meter of scale that we might call turbulence, we find something "else". There's some sort of non-linear waves, with some linear tendencies and that kind of compete with the turbulence. Somehow energy which has been supplied into the ocean on much larger scales, such as excited waves that have interacted with waves, has somehow delivered both density variance (heat and salt variance) and velocity variants to "turbulence" scales.

Somehow, those very large scales, – tens of meters and hundreds of meters, themselves interacting with scales on the order of kilometers and tens of kilometers – are ultimately providing energy down to scales that finally appear by most definitions to be turbulence. Then they are dissipated. But because those dissipation scales are so close to the scales that you might imagine as turbulence, a lot of the pleasantly asymptotic theories which might apply at high Reynolds number don't apply.

Before we say too eagerly, "Gosh, computers in 20 years may be big enough to solve the turbulence problem for us", we should be really cautioned that is not the planetary flow problem. Turbulence is a wonderful problem; I don't want to knock it. After all, I'd make a lot of enemies here in this room. But it's not the bigger problem. If we ask "What would the computer have to do to solve the planetary flow problem," we would be back to the kind of a lecture from the very beginning when I said "There's actually something like 10^{24} excited degrees of freedom," There are actually a few more than that.

If we use the calculations that were just put on the board for us and say, "Well, in that case, when will we solve that problem," We can kind of skip ahead to some of your advanced computers and ask how long they'd have to run to solve the problem of the planet and the answer is somewhat longer than the age of the universe.

[Several people talking at once]

I'm sorry, I'm sorry. I took the computer in 2010 and asked how long to solve the planetary flow problem then. That required the age of the universe, but of course computers advance. OK, so I'm not going to worry about that calculation except –

Jimenez: I am personally not worried.

Holloway: It's a nice reason for us to all go to the beach and let the computer in 2060 or something like that do the work.

Jimenez: Or not. Supposing you have to ..... in 40 years?

Holloway: Oh, no, sorry. We need some technicians to develop the equipment. OK. The only point I'm trying to raise, and maybe this invites other comment, is that we shouldn't leap to the idea that if only we can solve the turbulence problem, we're home free. In fact, I almost beg to suggest that the turbulence problem looks like the easy part.

[Response in background]

You stopped at 40, your comment that we must wait more than 20 years. Then, after 60 years, both you agree?

Jimenez: I am going to be able to retire before you do.

Kaneda: Do you think that after 60 years or so, we will be able to reach a point where we can treat that number of freedom?

Holloway: It is a wonderful hypothetical question to talk about where things will be in 20 or 30 or 50 years.

Questioner: But it's nothing to do with us. All most of us won't survive that long.

Holloway: No. What I'm really trying to do is to earnestly look back on the problem of the large-scale flow of the planet in a context which is in many ways beyond and different from the turbulence problem. So we don't lose track of that while we say, "Let's solve the turbulence problem." That's an important part of the planetary problem, but only a part.

Kaneda: Any other comment or question? Just a brief one.

Carnevale: Just a brief and very obvious comment. Yes, sure we can compute things, but that doesn't mean we understand them. For example, we can compute a chaotic system with three degrees of freedom all we want, and compute as long as we want; yet, people are still engaged in doing interesting theory about such systems. Yes, you can compute, you can make a hypothesis and change your experiment a little and do as many experiments as you want, but yet something is lacking.

As another analogy, consider that Faraday did many wonderful electromagnetic experiments, but it was the Maxwell Theory and Maxwell's equations that really gave us some satisfaction, really permitted us to think we understood what was going on. And I would hope that the same thing is true in turbulence. We simply can't say, "Ok, I can compute everything, therefore I understand it." You don't understand anything, you can compute it. In a sense we already compute the ocean, it's there. You can go out and look at it; it's a big giant computer, that's doing all the 10^{24} modes right now. It doesn't really help us, doesn't really give us any satisfaction. So in a sense, I think, turbulence theory will go on.

Staquet: To me, the only way to check that numerical simulations are correct is to have models whose results can be compared with those of the simulations. You spoke about the Maxwell equations, but in fluid mechanics as well, the deterministic hydrodynamic theories are extremely useful to check against your numerical results. Once you have performed a numerical simulation, you first check it against a simple and known analytical result, to validate your results. This approach holds also for full turbulence simulations: you will need theories of turbulence, or results of laboratory simulations, to check your results against those obtained in the latter cases. This will allow you to get confidence in your numerical work for cases where there are no theories or experiments (which are of course the cases where numerical simulations are useful). As you know it well, you can get anything but the true behavior with numerical simulations if stability conditions are satisfied but the accuracy of the code is too low.

Jimenez: Yes, you need simple theoretical results, as you say, to test your code. But there is no reason why you need a simple theoretical result for the full problem. Once your code is tested, there is a mathematical theory that tells you whether it is able to solve a particular partial differential equation under any conditions. You really don't need to know the answer. The mathematical theory of numerical analysis is well developed. There is nothing wrong with it. Once you certify a code, it can go to work, and you really know that the answer is going to be true to a given accuracy. That is a very important point that unfortunately has to be repeated again and again. It took mathematicians something like 20 years to be confident that the numerical analysis of partial differential equations was valid. Not only for linear partial differential equations, but for any kind of partial differential equations for which you can show existence.

Staquet: But numerical analysis of partial differential equations is only proved for linear systems.

Jimenez: No, sorry. Existence is of course not proved for all non-linear systems. What I mean is that there is a theory that says that, for any hyperbolic or even incompletely parabolic system of equations, such as the Navier-Stokes equations, if the solution exists and is unique, and if your numerical method is known to work for the linear system tangent to your set of equations at all the points in the orbit, which is usually easy to show, then your numerical solution is guaranteed to converge to the unique solution of the full nonlinear problem.

Kaneda: I have also asked Professor Kida to give some comments.

Kida: Before discussing, I have a question about the turbulence research. What is the purpose of it? I summarize the purpose here, but I am afraid we have no time to discuss about it. However, I want to hear your opinion later. In my view the main purpose is understanding of turbulence phenomena. Understanding itself is difficult to define, though. The understanding, prediction and control of turbulence are my purpose in turbulence research. If you agree with this, I'd like to proceed.

Kaneda: All right, I agree.

Kida: OK. To achieve it we have many approaches both in experiment and theory. Here, we focus on the theory.

[Voice in background]

Yes. But in any theory we usually make a model. The gap between reality and a model is small or big, as the case may be. The Navier-Stokes equation itself is a model. Smagorinsky model of turbulence has a bigger gap. In theory, there are at least two steps: First, we derive a model equation from the real system, and second, after getting this equation, we solve it analytically or numerically.

I think, in this meeting, we are focusing mostly on this part (i.e. numerical study and visualization). I have many things to say, but I here only stress that visualization should be important for the near future, at least 10 years or so. Another one is flow analysis. These are the studies I want young people to be involved in.

A turbulent flow is an unsteady 3-D vector field. It's hard to analyze and hard to understand. As we know, many people have already been doing it. For the visualization study, they use some common tools, modules or something else with isosurface, streamline, contour, vector, etc. on a high-speed computer. But I don't think this is sufficient to understand the turbulence. We must find some new quantity and new technique.

[Speaker away from microphone]

Kida: It was 40 years ago when I started to learn the turbulence theory. At that time many people were involved in the closure model as turbulence theory. But these days, young people may not be well educated about the turbulence theory. People in a turbulence research group now use computers and simulate or solve numerically the Navier-Stokes equation, and also analyze the data. But

I'm afraid that most people today cann't understand or even don't know the turbulence theory.

Turbulence theory, in my case, is based on the Navier-Stokes equation. But, as I said before, the Navier-Stokes equation itself is a model in a different level of approximation. Even for the Navier-Stokes equation, we must simplify it by some theory or something else to get useful information from it. In this meeting I heard many times such comments as "it is better than LRA (or DIA)". This is a good theory. But I am afraid that many students here may not understand the theory itself, and neither senior staffs, though there may be a few exceptions. I am suspicious that all the people here understand the true background and true derivation of it.

If you have time, I hope you to study a little more about the essence of the theory. If you don't care about it, I won't say anything. But if you are sincere about the theory, I suggest you at least to learn the derivation of this theory because there is a serious misunderstanding among experts. I dare say it because if you want to develop or you want to extend it further, for example, to shear flows or non-anisotropic case, or if you consider something new, you must at least understand the essence of this theory. Because people here do not seem very interested in the theoretical aspects of turbulence, I stop my comment just by reminding you of the problem.

Kaneda: I have many things to comment, but there is no time. Just one more comment please.

Herring: I just wanted to comment on the nature of theories and turbulence, and also generally in physics. For example, what should a turbulence theory be good for? I think the LES serves as a model. What we want to do is reduce the degrees of freedom–in some way by averaging out some other process–of the equations that you start with to something much simpler. And I would call that an advance–assuming the simpler equations have some accuracy–even though the simpler equations must be solved numerically, they should be solved quicker, and perhaps you can forecast what the solutions look like.

Kida: Yes. It is the second part of my purpose of understanding, prediction and control. I have stressed the first one, and you are talking the second.

Herring: That's right, yes. Sorry. Just one more point: The problem that most people see–you referred to the younger students–is that these equations (DIA, LRA, etc.), these closure equations, they look more complicated than those you started from the Navier-Stokes. They are integral differential equations, they look very tough. But, you have solutions, which, in fact, are averaged and therefore smooth. Numerical algorithms to obtain solutions can make use of smoothness which lets you cover may more scales of motion, and reach very high Reynolds numbers. Whereas with Navier-Stokes you can't do that.

Kaneda: The point is the smoothness.

Herring: That's right. And this was always a point in the early days, when Kraichnan was presenting his DIA in the sixties. People would say, "This is a horrible looking equation, why do you think that this easier to solve than Navier-Stokes?" And he would simply point out that it possessed a smoothness, lacking in Navier-Stokes, which had complicated fluctuations.

So, it's always a reduction, so the system that you wind up with (you call it "theory") must be simpler than the equations they try to explain, either by averaging or projecting them onto some other lower degrees of freedom and so forth. The philosopher, I think his name was Popper, said something like, "It's a bad theory that's more complicated than the thing it's trying to explain." I think, in terms of DNS, we have to remember that DNS is an experiment. It doesn't explain anything. You can test theories, but it's just another kind of experiment that's very useful.

Kida: The resulting equations of closure theories and the approximation methods should be discussed separately.

Herring: Yes.

Kida: You understand?

Kaneda: I'm sorry but we should go ahead. Next commentator is Dr. Ishihara.

Ishihara: Recently we have developed a turbulence visualization software which can render scalar data with 512^3 grid points at five frames per second on a conventional PC. I would like to give a practical demonstration of the software.

Here is a control panel of the software. I am going to show a vorticity field obtained by the DNS with 512^3 grid points. In this figure, green parts indicate the regions where the amplitude of vorticity is higher than some threshold. The threshold can be changed in this way. We can interactively change the point of view in this way and we can enter the vorticity field.

Kaneda: We are swimming in a turbulent sea.

Jimenez: From the data.

Kaneda: Actually, this visualization technique was first developed for assisting medical treatment, by experts in our school.

[Voices in background]

Ishihara: Now we have this kind of software. So the question is what kind of data do you want to visualize by using this software.

Holloway: Our question might be what would you do with the consequence of visualizing this kind of data?

Ishihara: Pardon?

Holloway: What will I do having succeeded in visualizing this kind of thing?

Jimenez: There is a point in that. The point is that the visualization is OK, but how do you convert it into numbers? This is a real question.

[Voices away from microphone]

Gotoh: I suppose that what we are looking at is the vorticity level.

Ishihara: Yes.

Gotoh: I'm not sure what the level of the present vorticity is. When we look at the second order statistics such as the energy in the inertial range we can find the values of the appropriate vorticity or velocity levels. Then we look at those figures visualized by the levels determined above. Then we can find what is the most responsible feature or shape at the second order level in the inertial range dynamics. Next we can increase the vorticity level. Then we may find shape or structure, which is most responsible for the higher order moment, right? This can apply to the third order, fourth order, and so on.

We can infer from this visualization that some of these particular structures in the natural range are responsible for the turbulence diffusion, etc. So, that is my understanding of how to use these kinds of graphics.

Jimenez: May I defend visualizations again? You need them to decide which questions do you want to ask. I think that the main purpose of visualization is to motivate you to ask questions. But the answer is not given by them, because the human brain is unable to keep track of so many things. If I look at the vorticity field in that screen, I really don't know what is going on; there are probably a million vortices in there, and I am not able to follow them individually. But I can see that there are huge holes in their distribution, and I find that very interesting; and that makes me want to know why.

Kaneda: Yes, and the structure may be very different from the individual sets.

Jimenez: The question is that once that you have seen that, you have to develop analytical measures.

Kaneda: Your suggestion is "Use visualization to find a good question"?

Kaneda: Kida-san? You encouraged the younger generation to go in the direction. Well, what is this direction, in your viewpoint? Is it just for entertainment?

[Response away from microphone]

Kida: No, it is for science.

Kaneda: So, how can it be useful, visualization?

Kida: Because we don't know the solution analytically, we cannot understand it only by the theories. By looking at many quantities in various ways, we can imagine what is going on and how the various quantities interact with each other. It is only after we look at the solution directly by our own eyes that we can really start to investigate it.

Kaneda: To get intuition and to get understanding?

Kida: Of course, you start ...

Kaneda: Start understanding.

Jimenez: I think that it is very dangerous to say that you get understanding from visualization. You start looking at things, and get ideas, and you can use those ideas to make questions to yourself, but what you really want to do is to eventually convert those into numbers.

Kaneda: OK. You say that unless it is transformed to numbers, the understanding is insufficient.

Jimenez: I think that is a fair thing to say. Otherwise, if you want to find anything in turbulence, you will. In a turbulent flow there is one of everything. If you are looking for a green dragon with wings, there is probably one in there somewhere. So, you've got to be careful that you don't see what you want to see, and say, OK, this is the answer. Once you think that you know what's happening, you have to check it. And you have to check it in a way that takes you out of the loop, to be as objective as possible.

Holloway: Visualization can also be a way to discover FORTRAN errors in hydrodynamic codes. Using intuition, if something seems to be too weird to be physical, it is probably FORTRAN. Or whatever language you program in.

Kaneda: Any other comment or question?

[Speaker away from microphone]

Claude Cambon: Going back to your historical example of the telescope: if you will look at the sky you can get new observations, to compare with existing models, and to ascertain new ones. Tycho-Brahé collected a lot of observations, then Kepler organized them in a 'kinematic model'. The first one only gave the 'numerical database'. I use modern terms to put in the context. As a result, a kinematic model, a splendid kinematic model, was obtained (the three laws of

Kepler for the planetary system). And later, Newton gave the dynamical explanation of these three laws. So in this case, in fact, the telescope stimulated a theory. But I cannot imagine how you can add all these data (from a numerical database) without organizing them. Perhaps the concept of theory, the concept of model can evolve in the future. But to completely drop the idea of organizing in a model or theory would be terrible for me. And, just about the last transparencies by Kida-san. I think that the theory has an educative role because you can understand a lot of things, events. About visualization, as Chantal said, visualization can guide intuition, but it could be a misleading guide because of the threshold you have to choose. you can have an impression, but if you have another threshold, you can have a different way. In fact, you have cut in the multi-parameter universe, and depending on the cut, you can have a particular view.

Gotoh: Well that's why you need a theory to decide what kind of quality must be visualized. We need a theory to look at. And also, by looking at those objects we can get some insight into new models. So, both can interact each other, and we can get one step ahead. So ....

Cambon: The theory can suggest that the objects are not completely autonomous, because in fact, the universe is a hierarchy of open systems, with a collection of 'holon', according to the philosopher Arthur Koestler. Any system (holon) contains inner sub-system and is connected with a wider system that includes it. For example, a geophysical flow is forced from the outside, and also includes internal effects to parameterize. But I will stop with cheap philosophy.

Kaneda: It's a pity that there remain so many questions of interest untouched. But, I'm sorry we have to stop now for the next session. I hope, tomorrow the session will be conducted in a more organized way by Professor Cambon. Thank you for your patience.

Panel Session 2
Turbulence Research for Geophysical Applications

Chairperson and Advocator: Claude Cambon

Cambon: Session two is about geophysical turbulence. I am not a specialist of geophysical turbulence, but in fact when you are interested in stratified, rotating, turbulence, and so on, it is easier to find application to geophysical flow than to tubomachinery, for example. As it was treated by in the last talk by Nagano-san, the rotating channel flow is not so complicated but rotating turbo-machinery is a very complicated problem in general.

I shall just follow the topics proposed by Yukio-san, since I didn't made a lot of work to prepare this panel session. There are five topics, and questions, in fact. First, about spectral modeling, linear versus non-linear effect, closure. The second is about LES subgrid-scale modeling, the third about parametric modeling, but particularly in the geophysical context with a Neptune effect. Four, applicability in climate models, weather prediction models. And five and last, what is lacking in the contemporary models?

I am not competent to discuss about all these topics. I will only discuss the first two. And even if I am the chairman, when I will start speaking I may be too long. It will be necessary to moderate the moderator, so I ask you to stop me in about five minutes. I will just discuss about the two first points, and after let the discussion continue by the distinguished specialists of geophysics we have here.

Unfortunately, in my talk, I just discussed rotating turbulence with the problem of angular drain of energy in rotating turbulence. But in stratified turbulence you can have different angular drain, and it will be interesting, first, to discuss in this general case rotating and stratified separately first, and after, a combined effect of rotation and stratification. And also to illustrate the topic 'linear versus non-linear', what is working and what is not working.

An important common point: either you use a linear approach, like RDT, or a non-linear approach like re-normalized or not perturbation theory, non-linear two-point closure. In both cases, it is interesting to look at eigenmodes, and this is related, if you have rotation and stratification, to a decomposition between wave and vortex modes.

1 Eigenmodes in Stably Stratified Turbulence

Cambon: Just to illustrate, in fact, the eigenmodes of a linear regime under system rotation and stratification are of two kinds. A non-oscillating mode, which include potential vorticity, but also as a particular case the vertically shared horizontal flow (VSHF, with a strong–only variability in the vertical direction). And you have oscillating modes, that include the wave part.

If you look at RDT in this case, which is a linear theory without prescribed complex mean flow (only space-uniform gradients), the first eigenmode is constant in the inviscid limit. For the oscillating ones, the wave part of the velocity field will exchange energy with the potential part. As soon as you have stratification, even weak, the potential part will exchange with that, leading to equipartition so that the final state, as I discussed yesterday, from any initial distribution of that modes, in terms of RDT, will be given by an equidistribution between wave and potential energies, the latter being linked to the temperature variance. So that, if you know the initial distribution in terms of the different modes, you will know what is the final ratio between kinetic energy and potential energy. The equipartition is mediated by phase mixing according to the dispersion relation for rotating stratified turbulence.

So, RDT is not very interesting in this case, but for a problem I will just comment later.

In a non-isotropic nonlinear theory like EDQNM-2 and 3, I will not go into detail, it isn't so important to have a non-linear interaction between all these modes. You will find, for example in stratified, or in dominantly stratified turbulence, that the energy will concentrate towards the VSHF mode, yielding layering because the VSHF mode is really the mode which corresponds to a horizontal flow which varies very rapidly in the vertical direction.

Holloway: You're saying that if you integrate your EDQNM equations, this tends to concentrate the energy? Is that right? Do we understand this is a transient tendency or long term?

Cambon: Yeah, OK, OK. And in pure rotation, you do not have this mode as an eigenmode; you have only wave modes, only the inertial wave modes from the complete basis. And it is completely different, you concentrate energy towards the two dimensional mode.

You can have an intermediate case if you add stratification and rotation, in which you can favor either a steady mode or other wavy mode. In fact, as another important particularity, not only these three kinds of mode, you have also an angular dependence in Fourier space. The vertically sheared horizontal flow corresponds to a pure vertical wave vector and thus, the energy transfer corresponds to an angular drain. The PV mode corresponds to a toroidal mode which involves horizontal velocity in physical space. Poloidal and toroidal components are related to the two sticks in the so-called Craya local frame of reference, which is nothing else than a polar spherical system of coordinates in spectral space. This is not a sphere in actual space, but in spectral space (see Figs.1 and Fig.11 of Cambon in this volume).

More physically, I shall comment a bit on the last transparency proposed by Jack. Perhaps you can also interpret the analysis of Billand and Chomaz[2]. Billand and Chomaz considered the stability of two columnar structures entering a strongly stratified field. Now it is rotating, too. But in the beginning it was only stratified. And in fact, the vertical coherence of the couple of vortices is very rapidly destroyed in terms of like a piling up of dishes, before a zigzag instability. And in fact, a good candidate for a zigzag mode is this vertically shared horizontal flow. Now consider the linear analysis of Billand and Chomaz[2]. Linear invariance, is like RDT invariance in which you have initially a strong mean PV mode. Then, you will add perturbation to three modes, like PV mode, VSHF mode and wave mode. And under a strong initial PV, it is this VSHF mode, which becomes unstable. And I think the instability of this mode will lead to layering, and the layering will break the vertical coherence of the initial vortex flow.

Thus, for a certain point of view, the linear analysis of Billand and Chomaz, but with prescribed mean flow containing PV, is consistent with the non-linear analysis we have using EDQNM-2 and 3. Perhaps, this gave a relevant comment to the last transparency presented by Jack's talk.

Now about RDT and a linear model. An important thing is to see what is the ability of RDT to predict complex anisotropy. When you have a shear, no matter–as shown also by our colleagues from Stanford Lee, Kim and Moin, it is enough to consider the linear effect of a shear on an isotropic realization of a turbulent flow to develop an elongated structure. The wall is not necessary, only the shear can develop elongated structure. There are not exactly streaks, of course. But you have an analogy with streaks.

Hence, as soon as you have a linear response to mean shear, you develop elongated structure. Now, if you have rotation, nothing appears. If you have an initial realization isotropic, it remains isotropic, rotation cannot break anisotropy. Hence, you cannot obtain this columnar structure expected in Proudman-Taylor two-dimensionalization. But, it's not finished because RDT can recover an interest if you look not at snapshots, or not at single time correlations but to two-time correlations, you can develop an anisotropy even if the underlying velocity field remains isotropic, it is capable of anisotropically dispersing particles.

And if you look at Lagrangian turbulent diffusion, with for example the length of excursion following a particle in the turbulent flow, you will recover a small anisotropy between the vertical and horizontal direction. So, no anisotropy appears for snapshots and at single time, anisotropy is found looking at two time correlations and using some simple assumptions (to relate Lagrangian to Eulerian correlations) as proposed by Yukio-san, and we have also collaboration with Christos Vassillicos, whose talk would have been devoted to that.

Even more important, stratification. Same thing as rotation, no structuring obtained by RDT. RDT cannot give pancake structures, but it can predict, looking at two-time correlations, that you do not have the same ability to disperse particles in the vertical and horizontal direction. In the horizontal direction, you can continue to increase the dispersion with time, and in the vertical direction you have a plateau and you block completely the vertical dispersion. But if you look at a simple argument like equating the Lagrangian and Eulerian two-time correlations (Kaneda), you have a ballistic regime for the dispersion in the vertical direction rather than a Brownian regime. But no matter, you obtain an anisotropy, which is significant and reproduce a large part of what is shown experimentally. And my interpretation is that the part of horizontal flow which includes PV as well as the VSHF, makes the work to disperse the particles in the horizontal direction. And, as shown first by Yukio, the dispersive phase mixing yield blocking the diffusion in the vertical direction.

These structuring effects are typically non-linear (see Figs. 2 and 3 of Cambon in this volume). But even without having elongated or pancake structure in the velocity field, you can understand something, using RDT, in Lagrangian diffusion. It's a bit paradoxical but now I think we understand this completely.

Cambon: This snapshot was presented by Chantal[9], closely related to the one by Kimura and Herring[11], of course, about pancake dynamics (see Fig. 13 of Cambon in this volume). And if you try to characterize the size (aspect ratio) of a pancake, you can use vorticity or velocity. For instance, if you look at a typical integral length scale in different directions, it is very important to note that only the non-linear anisotropic model like EDQNM-2, is in good agreement with 256^3 DNS to predict that the integral length scale which represents the horizontal length of a pancake increases, and that the vertical length scale is completely blocked. And if you use RDT, as I expected, you remain completely isotropic. It's another proof that looking at a single time, even two-point correlations, RDT cannot predict this structuring even if it can predict some element of Lagrangian diffusion.

2 Linear versus Nonlinear: Open problems

Cambon: Let me conclude about linear/non-linear. Finally, I will not take so long. I made progress, thank you. Pure linear, you can use inviscid RDT for having directly a statistical solution. It was illustrated by Hanazaki yesterday, but in fact, it is better in my opinion to gather the solution into oscillating and non-oscillating modes at the beginning and to use the eigenmode basis. But finally, same numerical results are obtained. In addition, viscous RDT can give new information if the Prandtl number is different from one. Just note that the decomposition in terms of oscillating and non-oscillating parts of the motion do not depend on the magnitude of the wavevector (only its orientation) in the inviscid case, but becomes k-dependent in the latter case.

Interest of two-point Eulerian statistic, I have illustrated. And in the talk of Vassillicos, you could have another illustration. To us non-linear theory, as I tried to illustrate, is needed, for example to predict pancake or columnar structuring. The base of eigenmode suggested by RDT can be relevant, however, for studying even the nonlinear flow. And I think it is better to try to interpret the result in terms of PV and wave mode decomposition than in terms of only the vertical and horizontal velocity component. Here you can understand more ....

Another thing, I spent one hour boring you with that (in my main talk two days ago). RDT can be used as a building block to improve non-linear theory, to introduce some structure, taking into account non-isotropic, no trivial zero-order response function using the Kraichnan terminology, for example. In fact, this is done by using the eigenmode decomposition and using an integrating factor technique in order to transfer the problem from the primitive variables to the set of slow amplitudes of PV, wave, and so on modes.

A problem which is not solved, I think the question by Herring, is the problem when G, the RDT Green's function, or the zero-order response function, is not only a combination of cosine, sine, but includes algebraic or exponential growth, algebraic growth when you have pure mean shear, exponential if you have elliptical flow or a hyperbolic flow (because RDT can completely give an answer to elliptical flow instability). Thus, in this case, you have the problem that the time-integral in the nonlinear transfer term is not convergent. It's the reason why we develop complex EDQNM theory (anisotropic) only with simple wave propagator G. I didn't succeed in developing a more sophisticated model, including shear or the exponential instability effect.

Just to finish, eigenmode decomposition suggested by the linear RDT is valid to analyze the results, and anisotropy is crucial, including angular dependence in wave-space.

A question I have no answer for, but I think at least three persons here will have an idea of that. Is it possible, not in the linear case but in connection with the non-linear model to connect the result to an entropy principle looking at the anisotropic result or so on, and looking at the result presented by Kimura and Herring, for example?

Well, I have only two transparencies about wave turbulence.

Cambon: OK. I will stop and you will continue, and if I have a short time, I will just move to my last two transparencies. Thank you.

You panelists can ask a question and make presentation as you want. I saw George with transparencies. But Jack proposed to make comment. Perhaps Jack will give a comment and after George will present his transparencies.

Herring: Thank you. I have only a short bit to add to what you said. First of all, it always seems to me in this problem of waves versus vortex motion that the waves would be fast and the vortex motion would be slow. That's a bit like low Mach number flows in which the acoustics, actually decouples from the incompressible flow, and you study the incompressible flow, just ignoring the acoustical flows. So that, perhaps in this way, this problem could be set in terms of the slow manifold, if you like, and you study only that, and throw away the acoustics, or the gravity waves.

The problem for pure stratified flows is that the frequency of the gravity waves extends down to zero because you have $\sin\vartheta$ equal to zero, in that case. So you can't really make a clean separation of slow versus fast. If you have any rotation at all then you can do that because the fast modes are limited, really, by the rotation rate.

Cambon: It's between twice Ω and N, and never zero. OK.

Herring: Right, never zero. There's that problem, and I think that makes the pure gravity wave problem more difficult.

The second point is, it's rather remarkable to me that if you look the movie of the vorticity field, you don't see any evidence of waviness at all. Why is this? I mean, it's rather strange in a way. If you looked at other quantities, perhaps strains or so on, you'd see evidence of waves.

And the third comment, that applies to the oceans and the atmosphere. Some years ago there was a discovery that the spectrum in a certain range of the atmospheric motion was $k^{-5/3}$. This was done by Balsley, Carter and people in Boulder. Ken Gauge and others proposed that this could be explained by inverse cascading two-dimensional turbulence (the forcing at smaller thunder storm scales)[16]. And the analysis of Reilly and his colleagues showed that there is a kind of two-dimensional layered structure that is natural to stratified turbulence.

This is what really motivated Olivier Métais and me to look at that case with DNS. But we didn't really produce an inverse cascade for purely stratified turbulence. Of course, quasi-geostrophic flow does inverse cascade. But the problem in the atmosphere is that, in order to get the quasi-geostrophic flows you have to have a small Rossby number. And if the spectrum is $k^{-5/3}$, then as you proceed down the spectrum to smaller and smaller scales, you run out of the conditions for validity of the quasi-geostrophic system before you reach the spectral region you want to explain in the atmosphere.

This was a disappointment, I think, to people, who wanted simple and clean answers to this issue. You can't provide rotations and solve the problem. Later, just last year perhaps, Lindborg, in a rather long but important paper in JFM[14], looked at third order structure functions (not energy spectra). The structure function is sensitive to the energy flux, and gets at the dynamics of the flow more directly. What he found from aircraft observations did not support the inverse cascade theory.

Holloway: It seems to me that what we are enjoying is a hugely stimulating science meeting in which scientists discuss their many results, as we are used to doing. Perhaps we can try again to ask, "Given the progress we're making in the science, what do we see as the path towards the practical applications?" We knew from the outset that would be a challenge, but I wonder if that's where we can try to move towards in this final panel session. Maybe .... we're careful to pay attention to the time, but try to do that?

Kaneda: Thank you. I have an afraid. Fundamental research is not another name of inapplicable, useless or peripheral research. If we wish to continue fundamental researches, we perhaps need to show some evidence of applicability or usefulness. For example, when we attend some meeting in an application-oriented community, in particular in the ones for global scale flows, turbulence often seems to play only quite a little role. It's disappointing. In some diagrams, turbulence is just at a small corner. Let me be provocative. What is turbulence research in the context of geophysical or global-scale flow phenomena ?

I wish to have some idea on the applicability or practical usefulness of turbulence research. Could you give me any answer?

Holloway: OK. Let me try a little bit if we can begin this discussion and if it doesn't work, it doesn't work.

Trying to motivate the question I thought about a different question which is, "Do we remember where we have come from?" Most of us don't think about it very much. So I made a little sketch. Hominids, our ancestors, have been on this planet for maybe something more than 3 million years. Here please understand this is only a sketch and I'm not an expert.

During the something-more-than 3 million years that we have been on this planet the temperature of the planet has varied erratically. I don't pretend the time scale on this cartoon is accurate. During most of this three million years, the earth has been in a series of glacial periods which have exhibited a lot of fluctuation. Briefly those glacial periods are interrupted with inter-glacials, which also show a high degree of variability. Then, something like 15,000 years ago, we climbed out of the last ice age and entered a highly unusual time in

Earth's history, at least according to the little I've hoped to understood from what people said.

Warming to 9,000 years ago, warming slightly more to 6,000 years ago and perhaps slightly cooling since then, with something funny happening within the last century (or less). Now the interesting thing to bear in mind is that these hominids have been around for 3 million years while Earth didn't do the kinds of things that it's done in the last few thousands of years. All of a sudden in the last few thousand years those hominids began to build cities, and pyramids; they began to develop industry, to dig up fossil fuels to power their industry, and they invented computers.

We don't know how this happened. What were the circumstances that have allowed us to become the civilized people we are today enjoying this meeting? And so we are a little bit motivated about application. Whatever has brought us to this point where we have the computers to perhaps address turbulence, yet we live on a planet and don't really understand how that planet has allowed the possibility that we are here today (with computers and turbulence theories).

This naturally leads to the question, "what's next?" Well, that's just a motivational sketch to say that not only we, the scientists in this room, but also the species we represent has endured, and just now benefits enormously from, the evolution of a planet we don't understand.

Admitting we don't understand it, I just quote what's imagined in climate research about the role of oceans in particular as part of this planetary evolution. Here I'm ever so glad that it was Chantal just this morning who introduced the thermohaline over-turning circulation as a significant part of this complex picture. Yet, that global scale thermohaline over-turning of the ocean is somehow crucially dependent upon (as observed in models) what you suppose about the small-scale mixing in the ocean. Now where are we? I push it just a little farther if I may, Chantal, and that is that the thermohaline over-turning depends upon a sensitive balance between what's happening in the heat and in the salinity fields in the ocean. Things that slightly disturb the relations between heat and salinity affect global thermohaline overturning.

Thus finally we're down to the scales at which we're trying to decide how mixing in the ocean may segregate heat and salt. Incredibly, such small scale mixing processes are part of how we have become the people on the planet we live on. And part of what may happen to us (and the planet) next! It's just a motivational comment, but I feel very strongly about this question. We gather the marvelous scientific intelligence that all of you bring to this meeting, and then somehow I want to go home with new ideas and do something better. Well, ok, onward then!

3 Turbulence for Geophysical Application

Holloway: The topic this afternoon is "Turbulence for Geophysical Application" and I underscore "Application". My own thoughts are that this problem may be much harder than the "mere" problems of turbulence at Reynolds number 10^5 that perhaps will be solved, if not in our careers then maybe in the next careers, or something like that. Then the geophysical problem is so much more difficult at another level.

Here are some of the things that could somewhat discourage us. In fact, the scales are not nearly as well separated between so-called turbulence scales and dissipation scales. Yet they are clearly multi-scale problems. There are many, many scales overlapping and competing, with linear and non-linear dynamics as part of all the scales we look at, all coupled. And then it's all chronically non-homogeneous, so that's a little discouraging for some of our ability with spectral theory.

Soon you can become so discouraged that you begin to ask, "Well, what should I do?" One possibility is, "Give up and go do something else." Only I don't think we're going to do that because we're not quitters and we're genuinely intellectually curious.

The next thing we could try is to say, "While we're engaged in our intellectual interests, lets just wait and a bigger computer will come and solve the problem for us." Indeed we wonder if the bigger computer will come and solve the turbulence problem; it may. But I think that what we're looking at in terms of understanding the planet we live on is way beyond the turbulence problem.

So "not in our lifetime" was my answer to that idea. What we can do is to continue to try. We not only try, but I think we're encouraged to be a bit daring. Not foolhardy, but daring. The point is to have courage. And, especially, this is fun. It's when you're having fun that you're probably doing some good science.

In terms of strategy, it seems clearly we work at what we can do well. Part of what we can do well are the sorts of science applications that have been discussed at this meeting. We can rightly celebrate those things. At the same time, we're motivated to try to reach beyond what we know how to do. That's where there is a problem. Most of us were taught in school to try to do things carefully, do them correctly and know that we've done them correctly. Only sometimes we can't do that. We are compelled to reach beyond what we understand and remember that we sometimes guess at things. By trying to make real-world application, sometimes we very soon come back and bring new insights to how we make our guesses about some of the work we're trying to do.

Finally, it is most important, as much as we can, to keep in mind fundamentals. It's possible, for example, to get very wrapped up in turbulence-like ideas within their own realm. But remember that turbulence sits on something more: fundamental physics. I guess that the fundamental physics that I try to focus upon are that internal degrees of freedom interacting among themselves should satisfy a kind of second-law principle. These may sound like some sorts of philosophical ideas, but of the sorts that could lead towards application.

4 New Approaches

Holloway: The question is, do we see some new approaches that might be helpful? While I don't want to get back into my would-be lecture of a couple of days ago, I think there is an argument to be made for the departure from higher-entropy acting as a forcing of systems.

This is an example where we make guesses, see how well they work, then come back and try to do something better. The important thing is to remember that what we're trying to do is to represent or approximate gradients of entropy as forces that we've not recognized in our usual force equations. Maybe, given the time this afternoon, this isn't something that especially bears elaborating. We'll see how time goes and decide how much we'd like to talk about some of the details of the approaches that we could make, intermediate between the careful science that we've been talking about for these three days and the "Golly, I'll have to guess and try it" where we don't know careful methods.

The only thing I want to show – it's something that Yukio asked me about – is an illustration of some of the concerns that you may like to see. It has happened in my lifetime that I've enjoyed two opportunities to collaborate on a paper with a Japanese co-author. Here is a picture from work with Kyohiko Mitsuzawa); my other opportunity was to collaborate with Yukio.

Now, this is just a schematic (Plate 1 from Mitsuzawa and Holloway (1998) showed surface and deep circulation for a region of the western North Pacific and the Japan and Okhotsk Seas[15]) that reminds you about what's going on off of the coast of the Western Pacific with a Kuroshio that comes, meeting an Oyashio. What we were trying to do was to relate the surface circulation to the sub-surface circulation. The reason to show this just now is from the perspective of why is Nagoya the way that it is. This has something to do with this warm saline current off the coast of Japan which, if it was different, would make Nagoya a different place and would make the global climate different.

Here we can see an example from among the various examples of what happens when you exercise the simplest neptune correction. Let me re-emphasize this is not "the answer" but only a step on the way. In the first case, if we have done things in a usual kind of modeling, you see some very usual model results. (Figure 7 from Eby and Holloway (1994) showed numerical output for a region off Japan from a portion of a global ocean circulation model[7]). Surface ocean currents and the net ocean-atmosphere heat exchange were obtained, first, from the model with conventional ocean physics then, second, from the same model but including the neptune representation of statistical mechanics forcing.

In this case, because the figure is old, the model was at quite coarse resolution. If you can recognize Honshu, then maybe I'll call this box Kyushu, is that correct? You wouldn't know that, would you? Hokkaido has gotten lumped in with Sakhalin and attached to the mainland because it "didn't fit". Okay. Importantly, and I think this is really characteristic of much of the effort to model this planet, a simulated Kuroshio carries warm and saline water too far north. The rate of annual mean heat loss from ocean to atmosphere, here occurring almost at the model representation of Tsugaru Strait, is hundreds of watts per square meter. That does not agree with observed climate.

If we make the simplest neptune modification, changing only a few lines of FORTRAN, we develop a Kamchatka-Kuril-Oyashio system which, in this crude representation, encounters a Kuroshio at a more reasonable latitude with more reasonable ocean heat loss. We could look at other examples, but this is an example very close to where we are here, and we can take a very simple (foolhardy perhaps?) little bit of a step. Then we see better where next we might be able to go, perhaps with more motivation to look at the closure theoretical treatments and a wealth of other ideas.

So that's my little effort to try to respond to the question of illustrating practical planetary-flow consequences from some of statistical fluids theory. We see that we can step forward even if don't know the way carefully. By making that effort, we can come back perhaps freshly motivated along our path. As we're all trying to see what we're trying to do this afternoon?

Cambon: Just a comment about the minus five-third law. You can consider that I represent Leslie Smith, I am not Claude Cambon, and Leslie Smith made a .... no, you don't accept. Leslie Smith made an interesting calculation with Fabian Waleffe forcing DNS from the small scale. Depending on the ratio f/N, there are three cases.

If f/N is close to 1, between 1/2 and two, the resonant inertia-gravity waves cannot occur, so you can suppose that PV and VSHF are completely dominant for the cascade. And Leslie and Fabian found, in fact, an inverse cascade for the PV mode, and concentration not too far from the two-dimensional part, or the barotropic part, of the PV mode. Thus, it was not too far from 2-D turbulence, with minus five-third spectrum. But it was only in this case or in the case of completely dominant rotation. In the case of completely dominant rotation, they find a minus three law, with two-dimensional large circulation structures appearing at large scale. We never had that in decaying (unforced) turbulence.

But as soon as N is dominant with respect to f, N larger like in a lot of geophysical applications, they have absolutely no inverse cascade of 2-D or PV and the concentration of energy towards the VSHF mode at large scale is built only by resonant interaction between gravity waves. I just communicate the information.

5 About Wave Turbulence

Cambon: And just before, perhaps, asking George to present the transparencies he was writing, may I just present one. And after we will discuss only the problem of LES. For example ..... It was about the close relationship, because Jack also was the first to mention the work by Caillol and Zeitlin[3], the close relationship between wave turbulence and turbulence theory.

A vanishing viscous term is also used in Wave-Turbulence theory to regularize the resonance operator. And what you obtain is very close to what I could call EDQNM-2 and 3. Why? Because the damping isn't important, and I can model the damping as I want. It is only non-zero, but as close to zero as I want.

But there is also a large community in wave turbulence in which you have a lot of different problems. It is larger than only recovering something close to EDQNM. You have first the possibility to consider resonance in the Lagrangian or Eulerian formulation. And I am not acquainted at all with a Lagrangian point of view. Zakharov et al. use an Hamiltonian formalism. The only example I have which, perhaps, is interesting in our domain of rotating stratified turbulence, is the work by Caillol and Zeitlin[3]: they don't use Lagrangian nor Hamiltonian formalism. I can also mention a recent model of weak interacting Alfven waves by Galtier, Nazarenko, Newell, Pouquet[8]; unfortunately, Annick Pouquet is not here. It is exactly EDQNM 2 or 3, with vanishing eddy damping, if you want.

But a more practical use of a Hamiltonian formalism, in fact, is to describe the transport of some slow amplitudes along the rays characteristics for the waves, as the wave packet presented by George and also by Chantal. The Hamiltonian operator is a generalized dispersion operator. Accordingly, you have these $\boldsymbol{X}$ and $\boldsymbol{K}$. $\boldsymbol{X}$ is a variable in physical space; $\boldsymbol{K}$ is a wave vector. You can have a mixed physical spectral space in a rational way. In other transport models with spectral information, developed in Los Alamos or in my own laboratory, people try to describe a weak inhomogeneity, with an energy spectrum slowly depending of $\boldsymbol{X}$.

If you look only at classic 'geometric optics', you have only a linear relation for the Hamiltonian system, but in some papers, George, with Fredrickson and Martin, proposed to include in the problem, some non-linearity consistent with Kraichnan's DIA. Additional assumptions are used, I don't know, but it is interesting to discuss such links with DIA, and with perhaps Lagrangian resonant interactions.

And using Poisson brackets for expressing nonlinearity (on the slide there is a fish pictogram), I'm sorry 'Poisson' is 'fish' in French ! But I didn't spend enough time to look at this, and George would be the best to discuss that and the other things about rotation you want to discuss. George, as you want.

Carnevale: At this point, I don't know what to talk about first. You mentioned rotation, and then there's entropy and anthropology. So I don't know where to begin. I've got transparencies on a few things, but let's go back to the original question that Claude raised at the end of his transparency during his presentation, the question of entropy. I think the picture that I have is fairly clear, and I'm a little puzzled by the idea that nonlinear interactions result in anisotropization while inviscid statistical theory would predict an isotropic statistically stationary state. I'm sure it's just a paradox. Claude presented an anisotropization, where the flow goes with rotation from being isotropic to being columnar, say.

A. Inviscid Rotating Dynamics

And that seems very sensible. However, let's think of it in terms of entropy and a maximization of entropy. Let me just schematically write the evolution equations. Here's our usual turbulence equation.

$$\dot{\boldsymbol{u}} = \boldsymbol{N} : \boldsymbol{u}\boldsymbol{u} + (-2\boldsymbol{\Omega} \times \boldsymbol{u}). \tag{1}$$

That is $\dot{\boldsymbol{u}}$ equals a non-linear interaction term. And I'm going to forget about viscosity, and just consider inviscid flow. And now I'm going to add to that a rotation term. Well the thing that's important in terms of statistical mechanics is simply the set of invariants of the flow, and you imagine that in long-term evolution, you equipartition the invariants amongst all possible modes. So, the invariants of the flow are, for non-rotating flow, just given by the non-linear term. You investigate that and you find that kinetic energy is invariant, OK? There's that. There's also the helicity invariant. But one can ignore that if you assume that your flow is going to be isotropic and homogeneous, or else you can just simply set the initial helicity to zero; it's not going to grow.

So, the main invariant to worry about from this non-linear term is the energy invariant. And from that one can say, "Well, if I maximize my entropy, and my only invariant that's constraining my non-linear interactions is kinetic energy, then what do you get?" You get the prediction the energy has to evolve to an equipartitioned spectrum, which means the modal energy,

$$U_{\boldsymbol{k}} = \left\langle u_{\boldsymbol{k}}^2 \right\rangle, \tag{2}$$

where $\boldsymbol{u_k}$ is the Fourier transform of the velocity field, for each wave vector $\boldsymbol{k}$, has to be a constant. You equipartition the energy amongst all your possible modes.

So, if you're doing a simulation, it's a truncated simulation. You have a finite number of modes. You're simply going to get equipartition, if you run long enough, and if you have no dissipation and your code is conservative; you've got to relax to modal energy is constant. In other words, your isotropic energy spectrum is proportional to k^2.

Fine, so that's the usual answer for isotropic turbulence. OK? And numerical codes do this all the time, it's no problem. Now what happens if we add a linear term?

$$\dot{\boldsymbol{u}} = \boldsymbol{N} : \boldsymbol{u}\boldsymbol{u} + \boldsymbol{L} \cdot \boldsymbol{u}. \tag{3}$$

Well, at most it can remove an invariant. You can't add an invariant by adding a linear term here. The problem is going to be more restricted in terms of our search for invariants. So, now when I look at this, what are my invariants? Now, consider $\boldsymbol{L}\cdot\boldsymbol{u}$ to correspond to the effect of rotation. Well, rotation doesn't change the energy invariant. So, we're back to the same statistical problem since we've got the same invariant. So, what's the prediction from statistical mechanics? The prediction seems to be the same result, isotropic energy distribution.

B. EDQNM

OK, so that's the beginning of the puzzle. And it's compounded by EDQNM, because there is a theory for EDQNM that shows you that if you have an eddy-damped quasi-normal Markovian equation, if you have any kind of Markovian closure, you must relax to equipartition. Let me just write out the structure a little bit. If I write the velocity filed as $\boldsymbol{u}_1$ this index '1' means the wave vector index and the component index that is the index distinguishing u, v, and w. Let me throw that all into one index. And so I can write the evolution equations of that field as a quadratic plus some linear term:

$$\dot{\boldsymbol{u}}_1 = \boldsymbol{N}_{123} : \boldsymbol{u}_2\boldsymbol{u}_3 + \boldsymbol{L}_{12} \cdot \boldsymbol{u}_2, \tag{4}$$

with summation over repeated indices. And here I'm not going to allow dissipation, just rotation type linear terms for L.

And then, I can write the modal energy as something like this, an average over an ensemble: $\boldsymbol{U}_{12} = \langle \boldsymbol{u}_1\boldsymbol{u}_2 \rangle$. And then I can write down the general EDQNM evolution equation in some sort of formal schematic sense like this:

$$\dot{\boldsymbol{U}} = \theta \boldsymbol{N}\boldsymbol{N}\boldsymbol{U}\boldsymbol{U} + \boldsymbol{L} \cdot \boldsymbol{U}, \tag{5}$$

where all of the indices are left implicit (for details see Carnevale et al.[4]). So, this is completely inhomogeneous, completely anisotropic; all of the indices are represented (implicitly) and there's no simplification here.

The point is that this θ, a six dimensional array, is the triad interaction time, and under most reasonable assumptions, it's a positive term or a positive definite matrix. And under those conditions alone simply by assuming this θ is a positive definite matrix, one can prove that $\boldsymbol{U}$ must evolve to the state of equipartition predicted by the quadratic invariants of the system. This is done by introducing the entropy, defined as $\int P \ln P$, but now restricted to quadratic moments. If all you know about your system is the correlation matrix $\boldsymbol{U}$, then the entropy can be written, up to an additive constant, as one half the logarithm of the determinant of this matrix:

$$S = \frac{1}{2} \ln \det U. \tag{6}$$

So, one can define the entropy S for this EDQNM, and then one can prove that the dS/dt or the entropy rate must always be positive or zero, and you can prove that that maximum entropy occurs only at the equipartition state that we originally predicted given the quadratic invariants of the problem. In other words, for turbulence with rotation in a truncated system, EDQNM must also predict that eventually, if you wait long enough, you go isotropic. This is assuming all interactions occur, OK? So if you assume all interactions occur, you can prove this theorem. If you've made some other specific restraints on the problem like the restriction to resonant interactions only, then it may not hold, since such restrictions could alter the invariants.

C. Experiment

Now finally, let me just end with a thought experiment one, or two. Let me suggest a numerical experiment. This should be fairly easy to do. And the experiment is simply this; you start off with an isotropic spectrum where this $\boldsymbol{U}_{\boldsymbol{k}}$ is constant for each mode. It's just a constant; every mode is equally populated in your numerical simulation. You've got no viscosity, you've got a truncation, your energy spectrum looks like that. And you let the simulation evolve in time. But our statistical argument says, "OK, it (the spectrum $\boldsymbol{U}_{\boldsymbol{k}}$) stays like that," nothing happens in terms of energy distributions. And my guess is that that's OK. I don't see how that can fail, with or without rotation.

Perhaps the root of the problem or the resolution of the paradox rests in the next experiment. If I start with some sort of isotropic energy distribution that is not equipartitioned, then, if I write $E_u(k)$, $E_v(k)$ and $E_w(k)$ as energy spectra of the three components of the velocity field, these are all equal initially by assumption, but not equal to the equipartition spectrum k^2. I can't tell you that as time proceeds these might not separate, that is become different. In other words, you could get a different distribution for the horizontal energy and a different distribution for the vertical energy. The entropy arguments only tell you that at time equal infinity you're going to wind up with an isotropic equipartition spectrum.

Holloway: But, George, I think you want to assert more. Can't you also say that as you move from t_0 to intermediate t that, too, corresponded to an increase in entropy?

Carnevale: That's right. The entropy would have to increase with time, but the theory does not tell us whether this is accompanied by anisotropization at intermediate times or not. I suspect there would be.

Holloway: OK, I'm just saying that the entropy tendency must be positive, even during early transient change, though it might be less intuitively clear.

Carnevale: Yeah, that's right. So this anisotropic intermediate state has to have higher entropy than this isotropic initial state. But for $t \to \infty$ the spectra would go to equipartition. You can break all of this by adding a little viscosity, or allowing energy to cascade to infinity because you could cascade one quantity faster than another and all these sorts of things would come in. But just in terms of a numerical simulation with the codes we've got right now, I think this would be what would happen in an inviscid run.

And so, maybe if you add in a little bit of theory about resonant interactions, maybe you would wind up with an anisotropic state as a nearly stable situation for quite some time. But for $t \to \infty$ this code should go to the isotropic equipartitioned state. So, that was my thought on this paradox.

Cambon: Yeah, yeah. Very interesting. And in fact, I expected this because I saw Bob Rubinstein in Lyon some time ago and he spoke about that kind of computation with truncated spectra without viscosity with Yamazaki and Kaneda. And he spoke about the fact that anisotropy was transient. At decreasing Rossby number, the transient zone extended, which would seem to indicate that this transient zone could be relevant if the Rossby number is significantly small. It contradicts, of course, what we obtain with EDQNM and what other people obtain with DNS or LES.

But I have no confidence in certain DNS; I prefer not to mention them. In Grenoble, for example, but not Chantal and her group. In forced turbulence, Smith and Waleffe found an inverse cascade and concentration of energy towards the two-dimensional plane, and so on. Hence, I am ready to accept your argument. When I saw some simulation by Morinishi and coworkers, they obtained transient, also, anisotropy and eventually the anisotropy decayed. But for me, it was because it was a decaying turbulence, and in decaying turbulence, the Reynolds number is smaller and smaller. So that's the non-linearity, which is the initial engine for anisotropization, is less and less. And your argument is very interesting. But you may consider that EDQNM cannot be completely wrong because it is consistent with a weak wave interaction theory and you can present this like Caillol and Zeitlin, perhaps it's more fashionable than to present it as a two-point closure. What is ....

Holloway: There is an absolutely essential point to make here, and that is to say if you were to insist upon resonant interaction theory, this may prevent achieving absolute equilibrium limit. So you have to remember off resonances, although they may have very small off-resonant width. That small but finite off-resonant width is ultimately the only path to the long term equilibrium.

Cambon: Yeah, yeah. OK, OK.

Carnevale: The conservation laws for those resonant interactions are different from the conservation laws for the full set of interactions.

Holloway: I wonder if perhaps we could invite Shigeo to comment a little on this. All of us today saw the marvelous simulations you're making of planetary flows, which are clearly becoming highly anisotropic. Now dissipation is present, but you do we really suppose that if your code would run conservatively with no viscosity we would collapse into these bands? But left to run, they would return to a completely statistically isotropic flow on the sphere? It's what we expect should happen?

Cambon: Yeah.

Holloway: So we're content with what happened. It is OK?

Cambon: Have you a comment?

Jimenez: I may want to comment something about this, because this is basic turbulence theory. And I agree with your argument, but it has nothing to do with turbulence. Turbulence is a viscous phenomenon.

Carnevale: Yes.

Jimenez: I mean that you really have to take in account that your energy is going out, and is being dissipated by viscosity. What you have to look at is the ratio of the rates at which the energy is being transferred along different wave directions. Equilibrium arguments give you the directions along which things want to go, but they tell you nothing about the rate of energy transfer, which is absolutely essential for turbulence. It has to be there. You have left it out of your argument completely, and this is not turbulence.

Carnevale: No, no, no. This was just a question about the way the code should work. The point was that Claude's results indicated that the nonlinear terms were necessary for the anisotropization, but this theory tells you that, even in turbulence, the nonlinear terms will tend to drive the system toward isotropy, and hence the paradox.

Cambon: But is the viscosity all?

Kaneda: I agree with you. We, nevertheless, spent some time in playing this game. We did the simulation. In that respect, I think the transient state may be more interesting. It shows the tendency of the non-linearity; "What does the non-linearity wish?", "In which direction does it incline to guide ? ". Eventually by viscosity, it decays out. If initially the isotropic energy spectrum is concentrated at low wave numbers, it of course spreads. But under the existence of rotation, the transient state is anisotropic.

And if we start with energy spectrum with a peak wave number at high wave number, then the scenario is very different. By playing this game, we can see some tendency of non-linearity. If there's dissipation, then it eventually decays. At least in the final stage, the simulation is, as far as we understand, consistent with the maximum entropy state. But the transient states are different. And the energy transfer is much delayed by Ω or $\Omega \sin\theta$

Carnevale: Yes, but the main point is that these kinds of arguments are useful to understanding the role of the non-linear terms, the role of the linear terms, and whether viscosity is at all important. I think the question that you come up with is, "Is this just done by non-linear terms, and the interaction of non-linear terms with the rotation terms?" And now we see that those alone (at least in a truncated system) don't do it, because the nonlinear terms don't know about the rotation and they will just try to drive the system to equipartition. So, this points out that it's necessary either to have some viscosity, or to have some cascade, or to realize that the anisotropization may be just a transient phase.

Jimenez: But that the effect of viscosity is very important is clear because the spectrum decreases with wavenumber, not the other way around. Turbulence is not an equilibrium phenomenon.

Carnevale: Yeah, but see you're contradicting the premises.

Jimenez: Yeah I know.

Carnevale: Because your saying that the invariants don't hold.

Jimenez: And I know it.

Jimenez: I am not saying that. I am saying that there is a viscous monster that is eating all the energy that is transferred towards your highest wavenumbers.

Carnevale: But certainly the nonlinear terms are still trying to drive the system toward equipartition even though you can't reach it.

Jimenez: The question of how fast it is going to get there, because viscosity is going to kill everything–whatever you get in there is going to be killed anyway. So, it's how fast you can get it there. If you are transferring energy slower in one direction of wave space than in others, you may end up with solutions that are anisotropic, as you say.

Carnevale: Yeah, exactly. But again, I think these kinds of arguments are useful in exploring these questions. And I would just add to what you were saying, that you could even have no viscosity if you had infinite resolution.

Kaneda: This system is very different from the reality.

Cambon: Just a comment. Imagine that viscosity is really pivotal. At the beginning, I was surprised that in wave-turbulence theory, there is absolutely no viscosity. For the consistency, we have not to take this kind of eddy damping , they say, it's for engineers. In order to clarify, Staquet and Sommeria have discussed that in an annual review about the dynamics of gravity waves[17]. And I realize, in fact, that in the mathematical approach (e.g. Caillol and Zeitlin[3]), apparently without viscosity, they need a damping parameter to regularize the resonant operator, so that even if the parameter is very small and insignificant, it is not exactly zero. It has to be, to exist.

And in fact, the pivotal problem of turbulence is why finite dissipation is obtained in the limit of vanishing viscosity. OK? And so, even a bit of viscosity as a mathematical artifact can break what you expected from the entropy principle, or ....

Very interesting. Now, have you some comment about Hamiltonian formalism or related ?

Carnevale: Yeah. I didn't write any slides about this, but there was some work done about 20 years ago about going from EDQNM for statistically homogeneous turbulence to the next step for inhomogeneous flows.

This is a simple enough procedure; you can imagine that you have some two-point closure that you're trying to write. And instead of just assuming that the two-point correlations are just functions of the difference in positions $\boldsymbol{x} - \boldsymbol{x}'$, which is the case with the homogeneous statistics, we can just keep all of the $\boldsymbol{x}$ and $\boldsymbol{x}'$ variation. You see, you write the correlations $U(\boldsymbol{x}, \boldsymbol{x}') = \langle u(\boldsymbol{x})u(\boldsymbol{x}')\rangle$ not as a function of just $\boldsymbol{x} - \boldsymbol{x}'$ but also of the mean position $\boldsymbol{X} = (\boldsymbol{x} + \boldsymbol{x}')/2$. Thus $U = U(\boldsymbol{x} - \boldsymbol{x}', \boldsymbol{X})$ without approximation. And then you can expand everything in gradients of $\boldsymbol{X}$. You take the Fourier transform of this form of U to obtain $U = U(\boldsymbol{k}, \boldsymbol{X})$, and then you can construct a theory based on this Fourier spectrum that also depends on position through $\boldsymbol{X}$. This is done without any multiple scale analysis; it is rather more straightforward. You can expand all terms in the derivation of the EDQNM about the homogeneous case. Then, you will generate new terms involving gradients in $\boldsymbol{X}$. You quickly generate terms that look very sensible, like you'd wind up with the equation for the rate of change of the energy spectrum, which now depends on space.

For example, following this course for the case of Rossby waves in two-dimensional flow yields an equation of the form

$$\frac{\partial U_k}{\partial t} + \frac{\partial \Omega_k}{\partial k} \cdot \frac{\partial U_k}{\partial X} - \frac{\partial \Omega_k}{\partial X} \cdot \frac{\partial U_k}{\partial k}$$
$$= 4 \iint \delta^2(k + p + q)\theta_{kpq}[a_{kpq} U_p U_q + b_{kpq} U_k U_q] \mathrm{d}^2 p \mathrm{d}^2 q, \quad (7)$$

where a_{kpq} and b_{kpq} are kinematic coefficients involving only the wavevectors and θ_{kpq} is the triad interaction time that depends on X as well time. The terms on the left look like the usual results from the Eikonal or ray tracing approximation, only Ω is now the usual linear wave frequency plus a nonlinear correction that depends on X. The nonlinear interaction term on the right is only the lowest order term in an expansion in gradients of X. From here you can include higher order terms as needed.

Cambon: Ω plays the role of a Hamiltonian operator in this case?

Holloway: It's a little more complicated.

Cambon: No, no. It's formal analogy not natural analogy. OK.

Herring: Bérengère Dubrulle and her colleagues–I think Nazarenko was one of them–developed a formalism like this. They were able to obtain the logarithmic layer. They have papers on this in *Physica. D*, **110**, (1997), 123.

Cambon: I know only the starting point. Sumimasen [Japanese]. I know only the starting point of the theory, which was a kind of WKB form, which in fact is consistent with a zonal form of rapid distortion theory, which was proposed by Lifschitz and Hameiri[13], and that we use also in Lyon. But I think it was a starting point, and I imagine they continue in the way? I don't know.

Herring: In fact this works rather well in two dimensions because the spectrum is rather steep ($E(k) \sim k^{-3}$). In three dimensions it's hard to argue that this approximation is reasonable because you're never guaranteed that this kind of WKB approximation is valid. If you look at actual flows, lengths derived from the mean shear rate are similar to the correlation length. The theory may gives plausible results, but be untrustworthy quantitatively.

Carnevale: Well this is just our first expansion beyond homogeneous turbulence theory for EDQNM.

Herring: I think this is something that David Leslie called semi-homogeneous.

Carnevale: Some references on this are Refs.[6] and [5].

If anyone's interested, I can send copies.

Cambon: It was an illustration of a weakly homogeneous continuation, in fact, of this story, which is also sometimes consistent with geometric optics because we can follow the characteristic curve along which the wave energy propagates using the group velocity and not only the advection velocity. OK. I will move with LES.

Staquet: I just wanted to make a comment on LES, that ....

Cambon: I will move to LES. And after I stop ....

6 LES and SGS

Cambon: Yeah. In fact, I saw a very interesting proposal from the Nagoya team about the non-isotropic subgrid scale model. And just a brief comeback about pure scalar subgrid viscosity model. In fact, it is well known after Kraichnan that it is possible to derive a subgrid scale model and subgrid scale viscosity from almost all isotropic versions of two-point closure, DIA, EDQNM, as you want. And Kraichnan proposed to scale into the dissipation rate and k_c, k_c is the cut off wave number. And in fact, Leslie and Quarini[12], in 1979, proposed to introduce the scaling in terms of the spectrum $E(k_c)$ and k_c. (Incidentally, before Lesieur and coworkers). If you assume this form of sub-grid viscosity, even the prefactor could be obtained very rapidly by a simple argument of equilibrium; in fact the flux of energy coming from the large scales has to be dissipated immediately. And with this simple argument, the prefactor can be immediately related to the Kolmogorov constant.

Hence, we've only dimensional analysis and simple argument of balance between what you have to dissipate and the chosen prefactor. You can immediately get rid of complicated theory, and justify such a spectral model.

Another version, in fact more used now, is the dynamical model in which you conserve a pure scalar subgrid viscosity. But, in this case, you try to have also a flux, which is exactly dissipated by the Sub-grid-scale model. But you can adjust the flux dynamically with two cut offs in the resolved part, it's a dynamical Germano model, and so on.

Just to say that, using a pure scalar subgrid scale viscosity, it is not possible to make a lot of new things. And I was happy to hear that, consistent with even weak but anisotropic form of inertial range, people here propose a new sub-grid-scale model, and even possibly associated with a random force. And, in fact, our experience with Chantal with high resolution DNS, EDQNM and all this stuff, for example in stratified turbulence, is that the anisotropy is very important in the inertial range and extend towards small scales. Of course, the Reynolds number is limited, but I am sure that the anisotropy, even angular anisotropy in the sense of spectral angular anisotropy, is very important in stratified turbulence.

Hence, I have no proposal but I wanted to mention this interesting way. And also, recently, David McComb, who is also a specialist of turbulence theory, two-point closure and renormalization groups. He has a lot of criticism about the Yakhot-Orszag method[18], and he considers that the re-normalization group may have only two interesting domains of applications, a very small wave number, in fact it's an infrared cut-off, or very high. But he is convinced that only the very high wave number limit is relevant for RNG theory. And he proposed to renormalize from the dissipative range until matching the inertial range, with direct application to the subgrid scale model. Yakhot's proposal for subgrid scale modeling looks strange since it was based on a RNG theory valid for k tending to zero. It is not inconsistent with Kraichnan's evaluation for the constant part of the subgrid viscosity, however. I think David McComb had an interesting idea, and he showed me a non-isotropic renormalized expression, which could take into account mean velocity gradient, possibly rotation and shear.

7 Memory and Computing time

Cambon: And to finish I will, just speak about the time of computation. Often, when we have a simple model, easy to compute like isotropic EDQNM or isotropic LRA, the computational cost has nothing to do with DNS. I don't specify the scale in my scheme, it's symbolic. In DNS, or even in LES, the computational cost increases dramatically with the Reynolds number. In closure, the computational cost increases very slowly with the Reynolds number. What is time (and memory) demanding in closure models: It's taking into account the anisotropy because, in fact, you have to add angular parameter, and because it is difficult to extend pseudo-spectral techniques for solving the velocity field in DNS to two-point closure. You can ask to Gotoh and Kaneda! They say that they know what it is there, it is complex, it is difficult. In fact, if you have no well-adapted scheme, if you don't change the numerical scheme, you increase the cost when you take into account more and more angular degrees of freedom. These degrees are angular harmonics, angular positions, as you want, their number increases when you increase anisotropy. Accordingly, particularly at very small Reynolds number, the cost of the computation can be higher than the cost of a DNS. On the other hand, DNS can only accommodate moderate Reynolds numbers, so that, even if you have high anisotropy, because the slope is very slow in my scheme (cost versus Reynolds number), eventually at a sufficiently high Reynolds number, the cost of a closure is much smaller than the cost of a DNS, or a DNS is no longer possible. For instance, Godeferd and Staquet are exploring a large range[9], with a possibility to have a cost higher in closure than in DNS at low Reynold's number (because of anisotropy), but with closure running for Reynolds numbers which remain unreachable by DNS. I will finish with that.

Jimenez: Can you please leave that in there? I just want to make a point, which I made yesterday, about the second Reynolds number in most of LES is a constant, you see dependent on Reynolds number. So please, could you maybe strike that LES up there? It is true because velocity is there.

Cambon: OK, no LES.

Jimenez: I'm sorry, ...

Cambon: I agree. I agree. But LES is matching between our statistical theory and DNS. Well–I agree.

[Response in background]

But, in fact, you have an older argument. Anisotropy is very time demanding, but increasing in the Reynolds number is not costly in closure. You have a smooth quantity; so that statistical spectra do not need such good resolution to be represented than the velocity field itself.

But nevertheless, even if the closure is less time demanding than the LES, it is very important to develop analytical laws, e.g. with asymptotic analysis, not only to treat EDQNM or LRA as a black box. It's very, very important. And analytical results would be very useful for parametrics today, as Chantal told me yesterday. You have a problem in the DNS. If you have even three parameters or four parameters, the exploration of the whole range of the parameters is very time consuming. Even if a single run is not too time consuming, if you have to make twenty, fifty runs, it is boring, I think. Now I let you make some comments about that.

Staquet: So first, I'm not a specialist of turbulence modeling because I'm mainly working on mechanisms. There are still two things I would like to say. First of all, the fact that conducting a parametric study with DNS is time consuming and not illuminating (and moreover boring). Such a study may be possible if a few (three say) parameters are involved; but remember that, in any case, what you get are only numbers. Inferring the dependence of the flow upon the three parameters is then very difficult without any theoretical model. But when the number of parameters is increased to any arbitrary large value, even 10 or 15, how can you manage the problem? Only a theoretical model, even a simple one that retains the most significant parameters, can provide a hint on the flow behavior and its dependence on these parameters.

Now a second thing, which you started to mention: when an external force is acting upon a turbulent flow, the features of turbulence will be modified. Let me consider the case of a buoyancy force, which I know better. After a few Brunt-Vaisala periods, when transfers of energy have been killed or strongly suppressed by the stable stratification, the smallest scales themselves become anisotropic. So, I think there is a need to develop a subgrid scale model for anisotropic turbulence. The non existence of such a model is actually one of the reasons for which I am so reluctant to use LES. All subgrid scale models proposed on the market rely indeed upon an isotropic assumption. I still wonder whether the model proposed by Germano, which involves the flux of energy at the cut-off scale, would not be more satisfactory in this sense. But still, it relies upon an isotropic modeling of the subgrid scales. I think this is one direction future research should go to.

[Voice in background]

Kaneda: Thank you for mentioning our recent model. There is one hidden purpose of our anisotropic modeling. It is related to the comments by George and Jack. Perhaps George's expansion is based on this assumption.

If there is wide enough inertial range, then we may choose a large enough wavenumber k_c such that $k_c \gg k_M \equiv 1/l_M, k_E \equiv 1/l_E$. Here l_M and l_E are the characteristic length scales of the mean flow and energy containing eddies of the fluctuating turbulent field, respectively. The anisotropic spectrum proposed in our presentation is obtained by retaining just the leading term for small $\delta \equiv k_M/k_c \sim k_l/k_c$. Here we assume $k_M \sim k_L$.

I think this assumption of the smallness of δ is a good one. In the limit $\delta \rightarrow 0$, the energy spectrum is given by just the isotropic Kolmogorov spectrum. δ may be in general small. However, in practice, it may be not very small. It may be therefore necessary to include the term reflecting the effect of finite δ. And the energy spectrum due to the finite δ is anisotropic as shown in the presentation by Dr. Yoshida.

Herring: Maybe I can say one thing, about Kraichnan, there. Of course what I say is just a subjective impression. I regret he is not here to give you directly his thoughts. What you say is quite correct, and for many years I think another tone of pessimism in his attitude was the fact that if you look at the renormalized perturbation theory, you can ask, "What's the convergence rate of this kind of procedure?" If you start with primitive perturbation theory it's expansion in Reynolds number. The renormalized perturbation theory converts this into a possibly asymptotic series. But there were no error estimates.

However, the equations that he proposed (the DIA) is an exact solution of a model system–the random coupling model. And they looked very plausible, as the Boltzmann equation also does. The DIA also had plausible ingredients. And so he was very attached to these things in some way.

Much later he invented the decimation scheme, which can bound errors. The DIA was in that case the leading order in such procedure. And, in fact, there was an in principle bound to the error of DIA. So I think, at that point, he became much more satisfied with what he had done. That's what a theory should be: it should propose equations and error bounds for things estimated. And actually, he didn't call it a theory, he called it an "approximation."

Gotoh: So, I have a few comments; I am a little bit discouraged after that pessimistic opinion. I have been involved in developing an efficient method for solving the two-point closure. Yukio-san asks me a lot, but my progress is very slow, and at present, my feeling is that "Okay, maybe we can expand EDQNM equations around the isotropic state for weak anisotropic case." Yes. But the labor is very huge.

Cambon: I drop spherical harmonics for the same reason. And also I ...

Gotoh: Yes, of course. We developed some method for using the FFT method. But that costs a lot, almost comparable to the DNS. But the analytical reduction is very low. So, in that sense, there is an opportunity to develop a new method to efficiently compute the spectral equation. But when we consider the high wavenumber range, there might be small anisotropy, but the decrease of anisotropy may be weak. In that case, the estimation of the eddy viscosity or anisotropic viscosity in terms of this theory may work. So that this is my present hope.

And for the larger scales, of course there might be some anisotropy but that can be exactly treated. And one thing, maybe we have 10^5 times powered computers, and then the smallest size come down to the scale of 5 kilometers or 10 kilometers. Anyway. I talked about that kind of thing. The model of the eddy viscosity accounting for the small scale of turbulence is very much important for long prediction or climate prediction. But at the present level of the computer powers, the eddy viscosity effect might be smaller than the other factors. But I believe that when computer power becomes larger and larger, then the role of the eddy viscosity or turbulence dynamical effect will be much more important. That's my thinking.

Holloway: I have a question about this. I think Jack has thought about this, and perhaps you have as well. That is, suppose we forget about the computer demand and say "I would like to evaluate an LRA or an EDQNM, but I wish to do so with no assumption about isotropy." Thus I imagine simply summing in wavevector space, summing all the triads. Maybe you say "Ah, but that's a terrible thing to imagine doing." Only, perhaps it's not such a terrible idea because the second-order moments in wave-space, after all, are imagined to vary smoothly. The question would be if you use some kind of Monte Carlo subsampling of space to make an efficient closure solver.

Kaneda: Yes, We have tried.

Holloway: You've tried? And it doesn't work?

Kaneda: It worked.

Holloway: Well, then, that's great!

Kaneda: We have done also recently some tests. Here are some examples of the achievements for 3D turbulence. The results depend on the number of realizations used in the Monte-Carlo method. The difference in these curves comes from that of the number of realizations, N. Here N equal $1, 2, 3, \ldots$. The convergence is quite fast and even very few number of realizations, even $N = 1$, may give satisfactory results.

There are two motivations in our trying the Monte-Carlo method. One is related to the complexity of the closure equations. Even for the isotropic and homogeneous incompressible turbulence, the closure equations seem to be sufficiently complicated. If the symmetries are broken, then they become quite complicated. With the increase of the number of dependent variables, the equations would be further complicated. For example, if one wishes to treat MHD turbulence, then the closure equations contain not only the velocity but also the magnetic field, and even writing down the closure equations, not the numerical codes, correctly is difficult, at least for me.

The other motivation is related to the gap between the progress in solving the original equations of motion and that for the closure equations. It seems that the latter has been so far quite separated from the former. There have been extensive studies and developments in algorithms to efficiently evaluate the terms in the original equations of motion, such as the Navier-Stokes equations. On the other hand, the progress in solving the closure equations has been painfully slow. It would be nice, if the progress in the former is immediately transferred to the latter.

In the Monte-Carlo method we are trying, we need not write down the closure equations, but the method itself generates the random field consistent with the closure equations. Moreover, since in the solution process of the method the principle structure of the original equations of motion is kept, any code or algorithm made for the evaluations of terms in the original equations of motion can be transferred, or available to the Monte-Carlo method. Thus, in my hope, the progress in closure solvers may be linked to that in CFD.

Holloway: Does anything limit you ....., like anisotropy ?

Kaneda: In principle, no.

Cambon: With no restriction, with a symmetrical–even when we use EDQNM2 or such sorts of stuff, we take into account this symmetry to reduce and you do not have a specific symmetry. Very interesting. You can try.

Kaneda: So far, we have tested mostly in isotropic cases. When the symmetry, such as isotropy is lost, I expect, larger number of realization would be required.

Holloway: But it's not huge.

Kaneda: Another problem we need solve is that of time integration.

Holloway: Well then what if we only solve Markovian problems but with the full anisotropy. Already that's an enormous step along the way, supposing for example you limit attention to Markovian closure so you could deal with full anisotropy.

Kaneda: Yes. The Monte-Carlo method is applicable for integrating the wave-vector integration at each time step. We should introduce another cheating trick for time integration. And for that purpose, we tried some techniques based on the Taylor expansion in powers of time separation of the two-time correlations.

Cambon: I think you published something.

Kaneda: Yes, but about ten years ago.

Cambon: Ten years ago. Why did you stop? Because you are no longer working on that?

[Several voices at once,]

Kaneda: No we didn't stop, but the progress has been very slow.

Cambon: I'm afraid I have no longer clever things to say. But if you want to conclude, to comment on other aspects? In fact, if I remember the five questions, there was a question about the Neptune effect, and parameterization, if LES is parameterization. Oh, yes. What is your final?parametric modeling it's connected with LES, *yes * , but parametric modeling in Neptune, it's a global modeling. And applicability in the climate model, weather prediction model and what is lacking in contemporary models? I have nothing to say about that, but what do you think.

Ah, yeah. Oh.

Holloway: Perhaps these questions are for Yukio's annual Nagoya workshop.

Cambon: May I mention something about that? In Lyon last year, we invited some participants, like Jack, you. And we had in mind a kind of international informal club of specialists. At the beginning it was two-point closure; two-point closure is too restrictive, we can extend to subgrid modeling, and all what is not only DNS, or what is not only for a person using silicon intelligence, but a club for carbon intelligence, and so on. But *you have here *, you are?for twenty or thirty years, you can be in the club. ***. Thus, everybody invited to this meeting, can be in the club.

And we will try to launch a kind of group that we called a special interest group in the ERCOFTAC European association (European Research Community on Flow Turbulence and Combustion). We are not specialists of combustion, but it was possible even in the European frame to extend towards a thematic network. And I heard that a thematic network could be open to scientists, overseas scientists. So that, ... I have no idea, no *present * organization to propose. But we could bear in mind to continue to form an informal club, and to have a meeting in Japan, in USA, in Europe ....

Gotoh: Please keep in mind to involve the younger scientists. Otherwise, it would be ....

Cambon: To involve?

Gotoh: To involve the younger scientists. The younger scientists, we are not new blood.

Cambon: Oh, yeah. Yeah.

[Conversation away from microphone]

Holloway: Young scientists so this not become a geriatric club!

Jimenez: This is not necessarily for Japanese scientists. All the scientists are very good but basically, we have to move younger scientists around; that was my point.

[Comment away from microphone]

Cambon: Thank you.

[Comment away from microphone]

Cambon: I think it's time up. Finally, we have discussed a lot of things. And yesterday, we had a very philosophical discussion about a long time for the future, and long time in the past. Thank you.

Kaneda: As you see, it's a pity that time is finite. We have only finite time. So we must now end this workshop. I have many things to say, but let me say here only "Thank you very much for your participation".

References

1. Balsley, B. B. and Carter, A. A., The Spectrum of atmospheric velocity fluctuations at 8km and 86km. *Geophys. Res. Lett.*, **9**, 465–468, (1982).

2. Billand, P. and Chomaz, J. M., Experimental evidence for a new instability of a vertical column vortex pair in a strongly stratified fluid, J. Fluid. Mech, **18**, 167–188, (2000).
3. Caillol, P. and Zeitlin, V., Kinetic equations and stationary energy spectra of weakly nonlinear internal gravity waves, Dyn. Atmos. Oceans, **32**, 81-112, (2000).
4. Carnevale G. F., Frisch U., and Salmon R. J. Phys. A: Math. Gen. **14**, 1701-1718, (1981).
5. Carnevale G. F. and Frederiksen J. S. 1983, Geophys. Astrophys. Fluid Dyn., **23**, 175-207, (1983).
6. Carnevale G. F. and Martin P. C. 1982, Geophys. Astrophys. Fluid Dyn., **20**, 131-164, (1982).
7. Eby, M., and G. Holloway, Sensitivity of a large scale ocean model to a parameterization of topographic stress. J Phys. Oceanogr, **24**, 2577-2588, (1994).
8. Galtier, S., Nazarenko, S., Newell, A. C., Pouquet, A., A weak turbulence theory for incompressible MHD, J. Plasma Phys., 63, (2000).
9. Godeferd, F. S. and Staquet, C. Statistical modelling and DNS of decaying stably-stratified turbulence, Part II: large and small scales anisotropy, J. Fluid Mech., submitted. (2000)
10. Herring, J. R. and Métais O. Numerical experiments in forced stably stratified turbulence*J. Fluid Mech.*, **202**, 97–115, (1989).
11. Kimura, Y. and Herring, J. R., Particle dispersion in rotating stratified turbulence, in proc. FEDSM99, FEDSM99-7753, (1999).
12. Leslie, D. C. and Quarini, G. L. The application of turbulence theory to the formulation of sub-grid modelling procedures. J. Fluid Mech., **91**, 65–91, (1979).
13. Lifschitz, A. and Hameiri, E. Local stability conditions in fluid dynamics. Phys. Fluids, **A**, 3, (1991).
14. Lindborg, Can the atmospheric kinetic energy spectrum be explained by two-dimensional turbulence? J. Fluid Mech., **388**, 259–288, (1999).
15. Mitsuzawa, K. and G. Holloway, Characteristics of deep currents along trenches in the northwest Pacific. J. Geophys. Res., **103**, 13,085-13,092, (1998).
16. Nastrom, G. D., Gage, K. S., and Jasperson, W. H. Atmospheric kinetic energy spectrum, $10^0 - -10^4$km *Nature*, **310**, 36–38, (1984).
17. Staquet, C. and Sommeria, J., Internal gravity waves: from instabilities to turbulence, Annu. Rev. Flyid Mech., **34**, 559–593. (2002).
18. Yakhot, V. and Orszag, S. Renormalization group analysis of turbulence. I Basic theory. J. Sci. Comp., **1** (1), 3–51, (1986).

FSC
www.fsc.org
MIX
Papier aus verantwortungsvollen Quellen
Paper from responsible sources
FSC® C105338

If you have any concerns about our products,
you can contact us on
ProductSafety@springernature.com

In case Publisher is established outside the EU,
the EU authorized representative is:
Springer Nature Customer Service Center GmbH
Europaplatz 3, 69115 Heidelberg, Germany

Printed by Libri Plureos GmbH
in Hamburg, Germany